The Best of bulk solids handling 1981—1985
Volume F/86

Bulk Handling in Open Pit Mines & Quarries

Reinhard H. Wöhlbier
Editor-in-Chief

1986

TRANS TECH PUBLICATIONS

This book contains articles published originally during 1981—1985 in

bulk solids handling

— The International Journal of Storing and Handling Bulk Materials —

Bulk Handling
in Open Pit Mines & Quarries

The Best of bulk solids handling
Volume F/86 (1986)

by
TRANS TECH PUBLICATIONS
P.O. Box 1254
D-3392 Clausthal-Zellerfeld
Federal Republic of Germany

ISBN 0-87849-065-5

Layout and phototypeset
by schriftsatz-studio kurt grohs · D-3401 Landolfshausen
Printed in the Federal Republic of Germany

Contents

ystems

Crawler-mounted spreader for overburden

Complete Opencast Mining Equipment with all these Systems:

- Bucketwheel-Excavators
- C-Miner
- Draglines
- Mobile Crushers
- Stationary and Shiftable Belt Conveyor Systems
- Drive Stations
- Train Loading Facilities
- Belt Wagons
- Belt Tripper Cars
- Spreaders
- Cross Pit Systems
- Open Pit Auxiliary Equipment:
- Walking Mechanisms
- Transport Crawlers
- Carrier Units for Special Tasks
- Cleaning Equipment
- Garland Changing Machines

Cross-pit:
a further alternative in opencast mining

OPEN PIT MINING EQUIPMENT

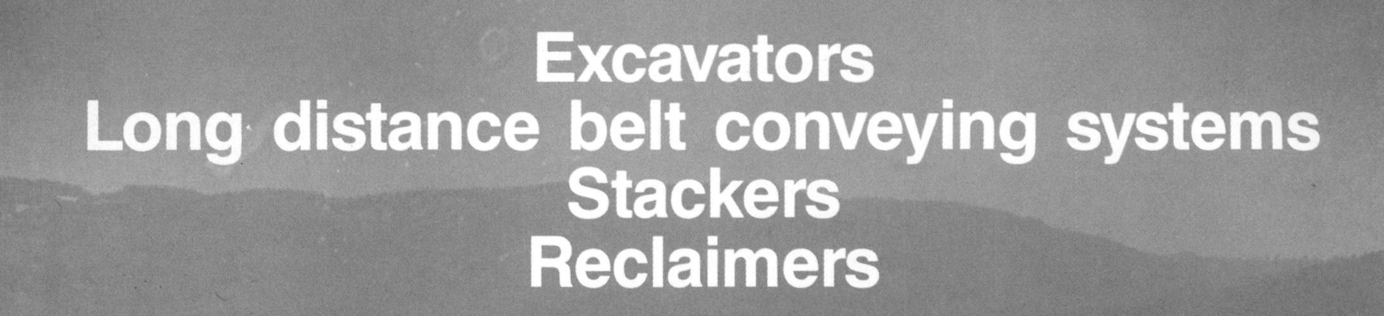

Type KU 800 bucket-wheel excavator

Belt wagon PVZ

Type RK 5000 bucket-chain excavator

Type ZP 6600 stacker

Belt conveying system

Exporter:

STROJEXPORT
PRAHA - CZECHOSLOVAKIA

113 26 Praha 1
Václavské nám. 56
Phone: 21 31, Telex 1 21 408
Czechoslovakia

Manufacturer:

Vítkovice k. p.
Ostrava

For large-scale earth moving – spanning great distances, working continuously:
Spreaders from O&K.
O&K BEA
Mountains are moved in the biggest open-cuts in the world.
Huge material flows have to be directed.
They flow continuously – up to 240,000 bm³ per day.
This work is done by giant conveying equipment – spreaders and tripper cars from O&K.
Trade mark
O&K
In Australia: O&K Australia Pty. Ltd., P.O.Box 232, Toowong. Qld. 4066
In Canada: O&K Orenstein & Koppel Canada Ltd., 4905-99 Street, Edmonton, Alberta T6E 4Y1
In the U.S.A.: O&K Orenstein & Koppel Inc., 1726 Cole Blvd. Golden, Colorado 80401
O&K Tagebau- und Schiffstechnik
Einsiedelstr. 6, D-2400 Luebeck 1
Telephone (451) 45011, Telex 26 823
Cables orenkop, luebeck
SR 93/4e Sachse

 Volume 2, Number 4, December 1982

The Development of Bucket Wheel Excavators During the Last Fifty Years

Walter Durst, Germany

Summary

This article reviews the important development steps in the manufacture of bucket wheel excavators during the past 50 years. New developments and technical improvements in the areas of crawler drives, slewing gear, the bucket wheel, and the conveyor belt of the wheel boom are presented in detail. A section is devoted to the safety equipment of these machines. Finally, the author describes special applications of bucket wheel excavators in Canadian open pit oil sand mines, in chalk mines, in various open pits in Germany and other countries, as well as the construction of the Chasma-Jehlum Canal in Pakistan.

1. Introduction

The development of open pit mining over the past 50 years made it possible to mine economically raw materials from ever increasing depths. A prerequisite for this is the capability of stripping and dumping larger and larger masses of overburden at reasonable costs.

Open pit mining technology gave impulses for the development of suitable equipment for digging, dumping and transporting large masses, overburden as well as the pay mineral itself. This development took place simultaneously for the so-called conventional (discontinuous) mining equipment such as front end loaders, shovels and draglines as winning machines with heavy trucks as transport units, and for the continuously operating equipment such as bucket wheel and bucket chain excavators as digging elements, conveyor lines for the transport of the mined material, and spreaders for dumping the overburden.

Conventional equipment is primarily used when mining pay zones which are not covered with high overburden depths and where the overburden can be overthrown directly across the stripped raw materials into the mined out zones of the pit.

In Germany, where the overburden has always been of considerable thickness and, therefore, direct overthrow of the overburden was not possible, the application of conventional equipment did not appear to promise any economic success. Therefore, the development of continuously working equipment, principally of bucket wheel excavators, was accelerated.

Fig. 1 shows the first bucket wheel excavator, built by O&K in 1933, in comparison with the biggest excavator in the world built by O&K with a daily output of 240,000 m^3 (bank) which was put into operation at the end of the seventies. This makes the development in size apparent of equipment built during the last fifty years. The development of a great number of constructional elements and their combination into a serviceable whole was the consequence of such considerable increases in size.

These individual developments were mainly carried out by the manufacturers of continuously working equipment, and often completely new methods had to be adopted for which no empirical knowledge existed. Therefore, one can say that genuine pioneer work was done in technical development. It must be emphasized that responsibility for the feasibility of these developments initially had to be borne by individual firms (maker and user). Substantiated by experience in operation, these developments became commonly accepted in equipment technology.

Since it was founded, the firm of O&K was concerned with the problem of mining masses and has played an important role in the technical development of bucket wheel excavators since 1933. At that time, the Lübeck works of this company operated under the name of Lübecker Maschinenbau Gesellschaft (LMG).

Here, the excavators were built which were so successfully used for the construction of the Kiel Canal, called the Kaiser-Wilhelm Canal at that time. LMG had a team of competent engineers for the development in all fields of technology. Many details for the construction of bucket wheel excavators which are taken for granted were invented in the technical office and used in the new machines. They often proved to be correct and useful. In the following, some of these developments are described which originated from the Lübeck works.

2. Crawler Travel Gears

When using mining equipment in open pits, it was soon found that the original rail travel gear which was taken over from older mining machines caused difficulties in operation. Therefore, the step was taken to equip the machines with crawler travel mechanisms.

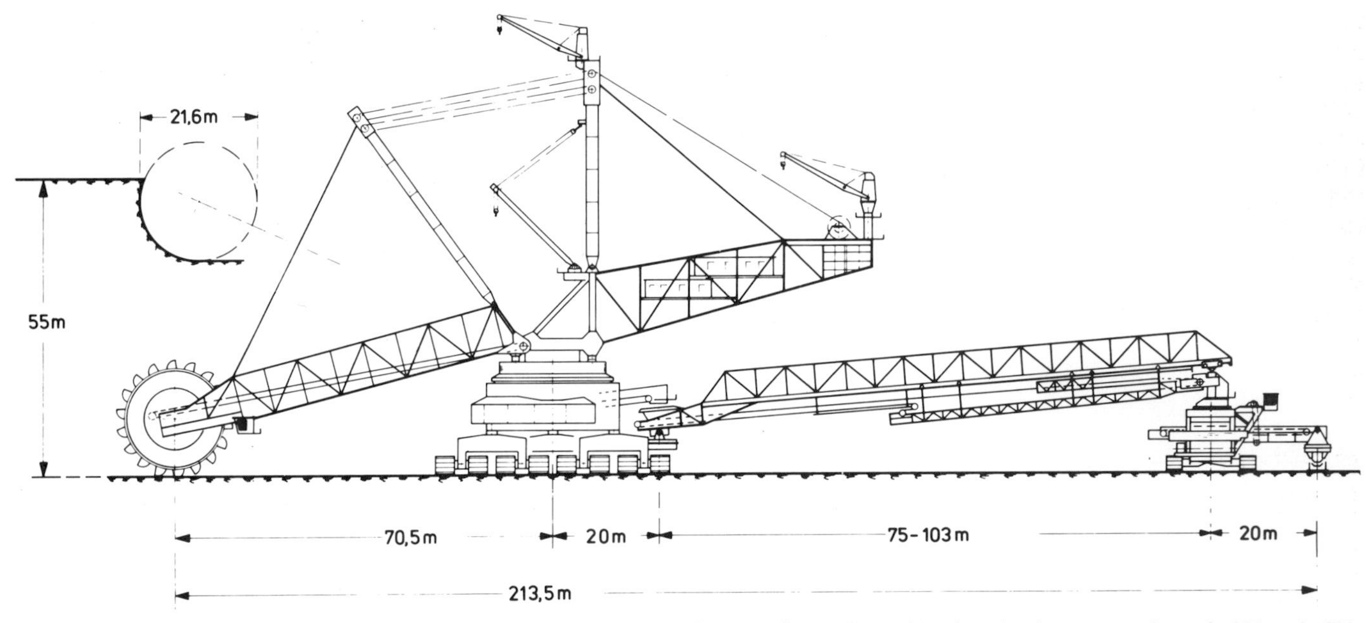

Fig. 1: Comparison of bucket wheel excavator sizes of 1933 and 1979:

Top:

Bucket wheel diameter	21.6 m
Bucket wheel drive	4 × 840 kW
Output per hour	19,000 m³ (loose)
Service mass	13,265 t
Motor power installed	16,900 kW

Bottom

Bucket wheel diameter	5.0 m
Bucket wheel drive	74 kW
Output per hour	750 m³ (loose)
Service mass	352 t
Motor power installed	300 kW

Such crawler travel gear had already been known in connection with open-cut equipment since 1933. The first big machine travelling on two crawler chains was a Dragline built by Bucyrus & Co., Milwaukee. German manufacturers, too, already had bucket wheel excavators in open pits travelling on two crawlers. However, in these machines the crawler frames were rigidly connected to the substructure of the equipment, and the individual travel wheels in the crawler frame which transmit the load to the track plates were not equalized. Owing to this, the vertical gravity load was unevenly distributed over the travel wheels. This travel gear was known from tanks.

In open-cut operations, however, this travel gear proved to be unsuitable for the application, as due to unevenness in the track level, large movements of the whole machine are induced. Here, crawler travel gear was required which could adapt itself to the unevenness of the ground without having serious effects on the whole machine.

It was also necessary to transfer the increasingly growing vertical load in a statically defined way on to the travel wheels to avoid severe overloading of the different travel wheels.

The requirements first resulted in one of the two crawlers being fastened to the substructure in pendulum fashion, and the second crawler frame rigidly connected to the substructure as before. The travel wheels were supported in equalizers and in this way provided statically determined support of the vertical load (Fig. 2).

This design, however, still had several disadvantages in operation. The inclination of the whole machine was still determined by the inclination of the fixed crawler, so that as a result of unevenness of the track level on the fixed crawler side, the whole machine could still make big movements, although due to the equalized travel wheels an improvement over the completely rigid travel gear had been achieved.

Fig. 2: Two-crawler travel gear with one pendular crawler.

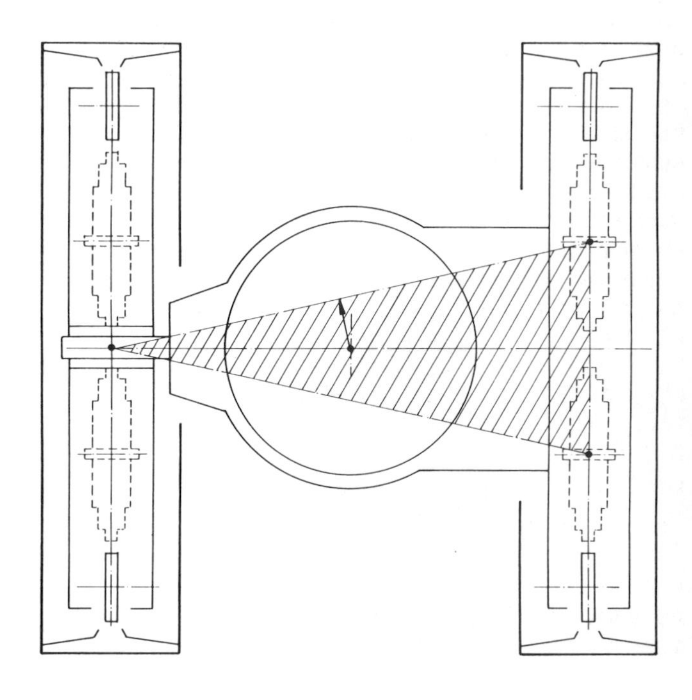

With this crawler design, the overturning edge of the crawler travel gear was very near the slewing axis of the superstructure, so that it became problematic to maintain safety against overturning. The movements of the superstructure, induced by the fixed crawler additionally resulted in further unfavourable shifting of the centre of gravity and reduction of the overturning safety. Consequently, a new form of fastening the crawlers to the substructure was developed, i. e., both crawlers were fastened to the substructure in a way to allow self-alignment, and the third supporting point for the substructure was supported on a transverse girder arranged between the two crawler frames (Fig. 3).

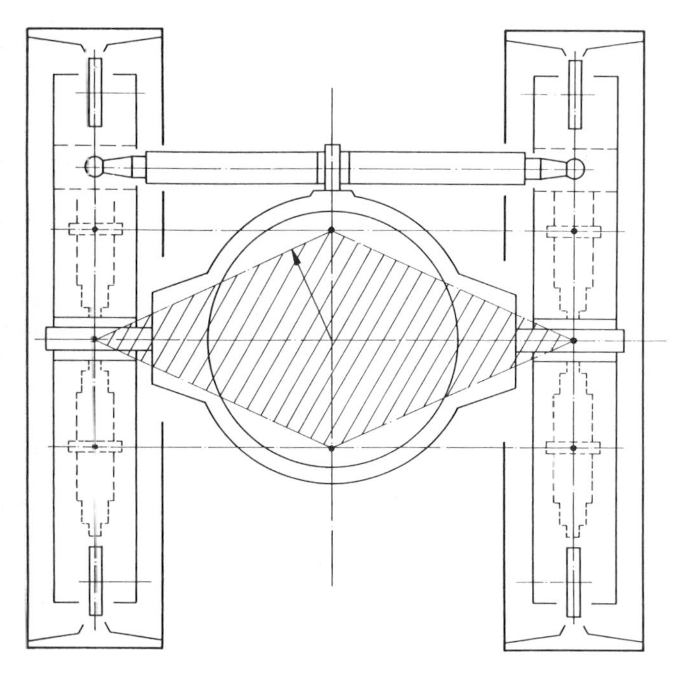

Fig. 3: Two-crawler travel gear with two pendular crawlers.

Owing to this type of support, the distance of the overturning edge of the crawler travel gear from the slewing centre of the machine was increased. In this way, the overturning safety of the machine relative to the crawler travel gear increased as compared with the design in Fig. 2. In addition, the effect of the inclinations of the crawler frames on the whole machine due to unevenness of the ground is halved, and the reduced centre of gravity shifting of the superstructure resulting from this further increased the overturning safety of the superstructure. For this arrangement of crawlers, a patent was applied for by O&K at the German patent office in 1953. The patent was granted under the German patent No. 1002696. In addition, this patent right was applied for and granted in various other countries of the world. Later, several two-crawler machines were also equipped by other makers with this crawler arrangement which proved satisfactory in operation and on bigger machines entirely replaced the crawler arrangement (Fig. 2).

When using bucket wheel excavators in open-cuts, it was realized at an early stage that travel gears of such machines must be capable of maintaining a certain preselected curve radius.

As early as 1936, O&K developed and built the first steerable three-crawler travel gear where the two crawlers on the two-supporting-point side could be slewed by means of a spindle in such a way that the travel movement developed along a given curve radius (Fig. 4).

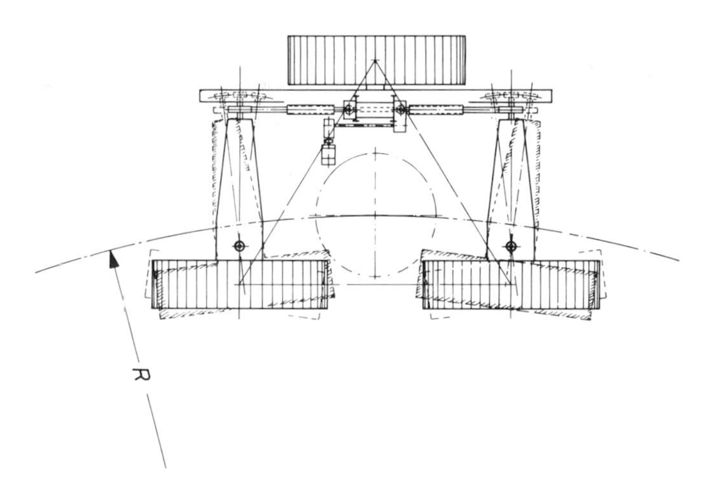

Fig. 4: Steerable three-crawler travel gear.

The development of larger and larger units (dimensions and outputs) resulted also in big increases in weight. Owing to this, the gravity load on the crawler travel gear increased as well. Experience with machines already in service showed that the average ground bearing pressure under the crawlers should not exceed the value of 150 kPa (15 N cm^{-2}). In order to maintain this value, more than three crawler chains were required. Initially 6, later 12 crawler chains were required to transfer the large loads to the ground. Track plates with a width of up to 3.70 m were built. Recently, however, a travel gear has been developed for a giant machine with a daily output of 130,000 m^3 (bank) whose service mass corresponds approximately to that of the machines with a daily output of 110,000 m^3 (bank), and this has only six instead of the twelve crawler chains normally used for such machines until now. For the first time, track plates with a width of 4.5 cm are used. This is a further step in the development of crawler travel gear.

With the increasing loads and the greater number of crawler chains, the forces for crawler steering also increased greatly and, therefore, the design of the steering spindle became more problematic, in particular with regard to lubrication and wear. O&K were the first makers of bucket wheel excavators to find new ways here. They used a hydraulic cylinder as the steering element instead of the conventional steering spindle. This hydraulic cylinder has essential advantages in providing the necessary forces and their limitation but also with regard to wear and maintenance. Fig. 5 shows the design of the steering of a six-crawler travel gear with a steering spindle for a steering force of 3,820 kN. The weight of this steering spindle is 16.4 t.

Fig. 6 shows the same travel gear with a hydraulic cylinder as the steering element. The total weight of this steering mechanism is no more than 9.5 t. Hydraulic cylinders were used for the bucket wheel excavators winning oil sands in the Canadian oil sand area operating under very severe climatic conditions where they have proved their worth for many years in operation and also on the biggest bucket wheel excavators built in recent years which travel on twelve crawler chains. A force of 10,000 kN is required to be exerted by this cylinder. The greatest length in the ex-

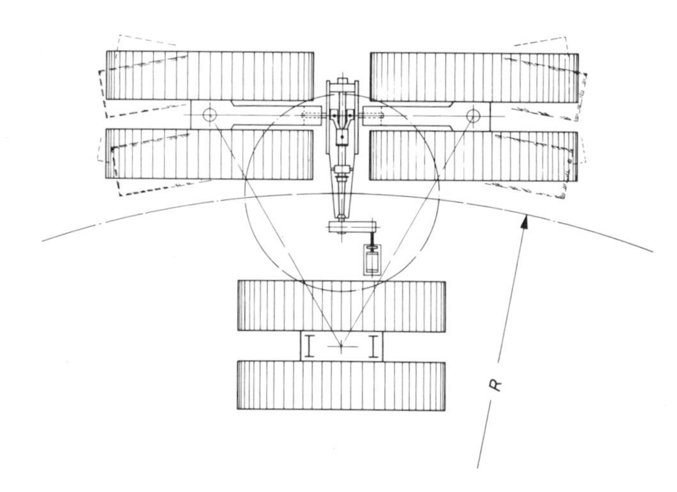

Fig. 5: Six-crawler travel gear with steering.

Fig. 6: Hydraulic steering cylinder for a six-crawler travel gear.

tended condition is 13.6 m, the highest pressure in the cylinder is 17 MPa (170 bar). Cylinders of this size, too, have been working satisfactorily now for several years. Owing to the good experience with hydraulic crawler steering, the hydraulic steering cylinder for the crawler travel gear is a component which is given preference.

3. Slewing Gear

During operation, the slewable superstructure of a bucket wheel excavator is slewed continually relative to the substructure. Therefore, an element has to be provided between the two parts of the machine which allows easy slewing without being subject to heavy wear.

For this purpose, a ball bearing slewing rim is used which is capable of satisfactorily transferring the vertical loads as well as the horizontal forces occurring during operating from the superstructure to the substructure.

With increasing machine size, the diameter of these ball bearings also increased. The largest machines in operation have ball bearing slewing rims with an average diameter of 20 m. The required minimum diameter of such a ball bearing results from the requirement of the overturning safety of the superstructure relative to the ball race still having to be ensured even under the most unfavourable loads. Another reason for selecting the smallest possible ball bearing diameter is that bigger ball bearing slewing rim diameters have unfavourable effects on other dimensions of the machine and its total weight. From the vertical load and its centre of gravity relative to the ball bearing, the biggest load acting on a ball in the ball bearing can be determined.

To get information on the behaviour of ball bearings under operating conditions, rolling tests were already carried out at O & K in the years 1957 to 1963 for which the dimensions of the balls and ball races were so selected that they were close to the dimensions of the ball races built up to that time.

In the course of development of the largest bucket wheel excavators with ball bearing diameters of 20 m and ball diameters of 320 mm, further rolling tests were carried out at the Technical University Hannover and at O & K, which in particular, rendered information on the material to be used for the ball races. These tests were carried out during the years 1971 and 1972. In both test series, various material qualities for the races were tested. In both cases, the tests showed that high-alloy steels and materials of very high natural hardness failed earlier than low-alloy and tough materials which have greater adaptability and higher resistance against local and frequently recurring stress peaks. From these test series, important information for the manufacture of ball races was obtained which is valuable for the practical operation of all bucket wheel excavators. From the rolling tests as well as practical experience obtained from ball races already built, it becomes apparent that the maximum ball gravity load during operation must not exceed a certain value if the service life of a ball race is to reach an adequate value. The resulting reference value was:

maximum gravity load of a ball — P_{max} (N)
ball diameter — d (mm)

$$k = P_{max} \times d^{-2} = 4\ \text{MPa}$$

It was no longer possible to maintain this maximum permissible ball load with the heavy superstructures of giant machines with a single-race ball bearing, Therefore, as long as twenty years ago, a start was made in building such machines with double races (Fig. 7).

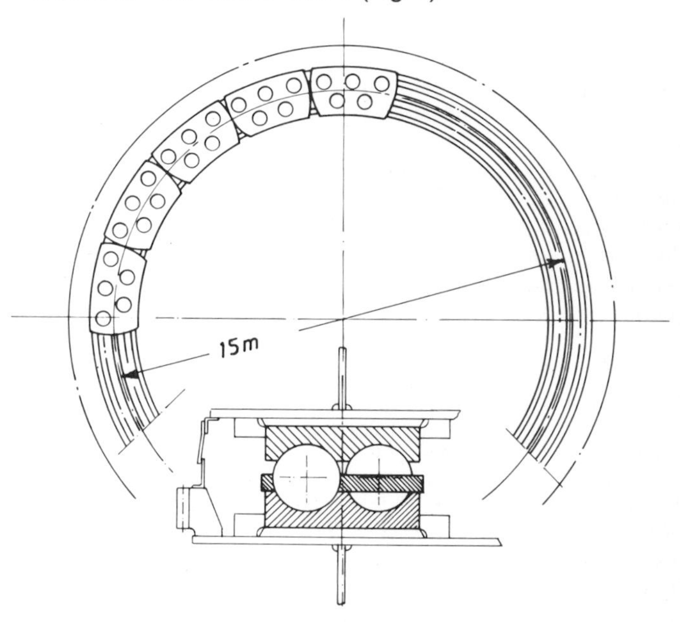

Fig. 7: Double-race ball bearing.

As these first double-row ball races gave good results in operation, the first series of giant machines was equipped with these. These double-row ball races with a diameter of 15 m were installed in two giant machines which went into service in 1959.

4. Bucket Wheel

The development of the cell-less bucket wheel was also partly carried out by O&K. As far back as 1954, the first small bucket wheel excavator with a daily output of 8,000 m³ (bank) with a cell-less bucket wheel went into operation, and in 1958 the first giant machine with a daily output of 110,000 m³ bank and a cell-less bucket wheel started work in the Rhineland brown coal area (Fig. 8).

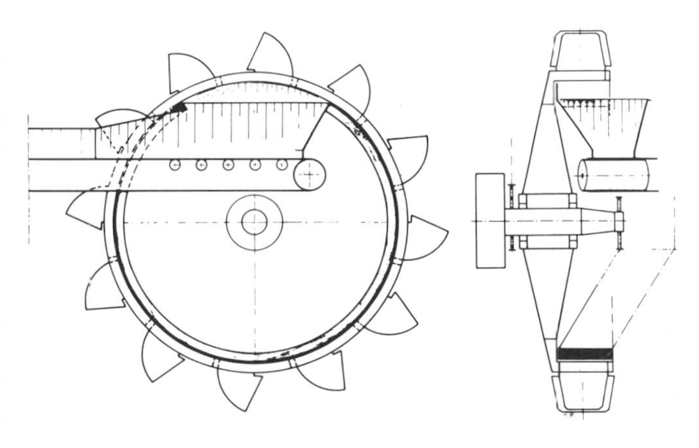

Fig. 8: Cell-less bucket wheel.

To enable the cut material in a cell-less bucket wheel to be properly transferred from the buckets on to the belt in the bucket wheel boom, the bucket wheel has to be suitably arranged. From its optimum cutting position, i. e., its position in a plane which goes through the slewing axis of the superstructure, it has to be slewed horizontally as well as vertically. This results, however, in different cutting conditions for both slewing directions.

As long as the material to be cut is soft and homogeneous, this does not involve any serious disadvantages. With hard materials, however, these differences in the cutting conditions have a very disadvantageous effect. Therefore, for applications of bucket wheel excavators in hard soils, the bucket wheel has been so arranged that the plane of the bucket wheel goes through the slewing axis of the slewable superstructure. In this way, the most favourable position for cutting is achieved. However, with this arrangement the belt in the bucket wheel boom can no longer be taken into the bucket wheel. For transfer to the belt of the material cut by the buckets an additional conveying element, the so-called rotary plate, has to be installed (Fig. 9).

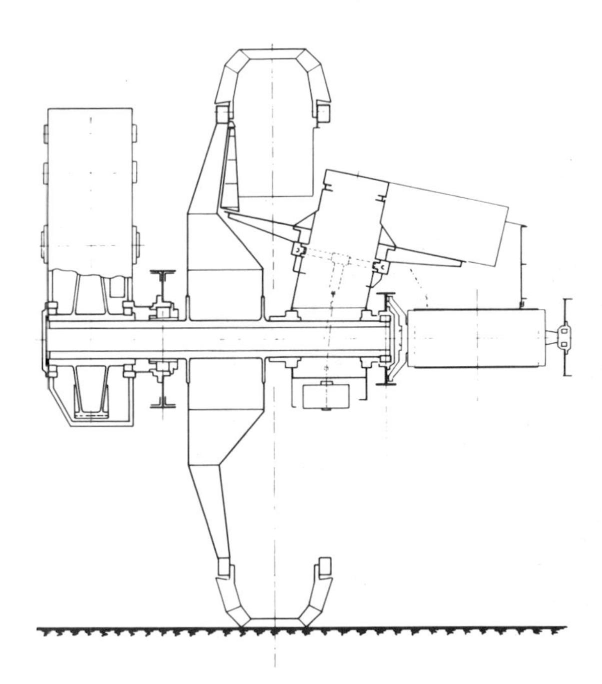

Fig. 9: Cell-less bucket wheel rotary plate.

This design solution has given satisfactory service for many years in machines operating in the open-cut Neyveli, India, in hard sandstone and in the two bucket wheel excavators operating in oil sand in Canada. When developing the giant machines for the Rhineland brown coal area, the ring space under the buckets was enlarged, i. e., so-called half cell provided; in 1963 a giant machine built according to this principle with a daily output of 110,000 m³ (bank) was put into service.

5. Conveying Path

Experience in operation with elevating belt conveyors had shown that the inclination of a conveyor should not be steeper than 18° if trouble in operation was to be avoided. However, with this belt inclination, it was almost impossible to reach substantial cutting depths below track level. Nevertheless, during a certain development phase, it was hoped that greater cutting depths below track level would bring advantages in equipment application and also for operational planning.

In an endeavour to find suitable solutions which would permit the material to be transported even along steeper belts, a second belt, the so-called cover belt, was arranged above the conveyor. The material was held by means of this cover belt and could, thus, be held even on inclinations up to 30°. As far back as 1954 O&K ascertained the feasibility of such "elevating belt conveyors" in a test set-up (Fig. 10). The results of these tests formed the basis for the designs used on bucket wheel excavators later on.

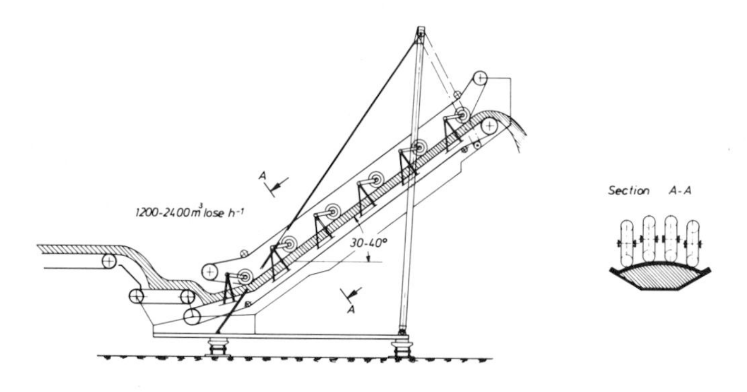

Fig. 10: Material transport on elevating belt with cover belt.

A giant machine with 110,000 m³ (bank) daily output, equipped with elevating belt conveyor equipment, was put into service in the open-cut Fortuna in 1958. The elevating belt conveyor installed in this machine met the demands placed upon it. However, as the open-cut technique advanced still further, big deep-cuts were not used any more and, therefore, elevating belt conveyors were no longer necessary. They were taken off one after the other.

6. Safety Equipment

To make it possible to work with bucket wheel excavators without any greater downtimes, protection devices against overloads are needed for certain drives. For a long time, the LMG catch coupling was an effective overload protection on the different drives. For this design, a German patent was granted in 1938 (Fig. 11). With this coupling, the input step was fixed against the power take-off step by means of a catch which is pressed into a recess by spring load. If the circumferential force at the catch rose above a certain value, the spring load which pressed the catch into the recess was overcome and the catch was forced out. Owing to this, power input and take-off steps were separated. Disengagement of the catch could then be used as signal for switching the drive off.

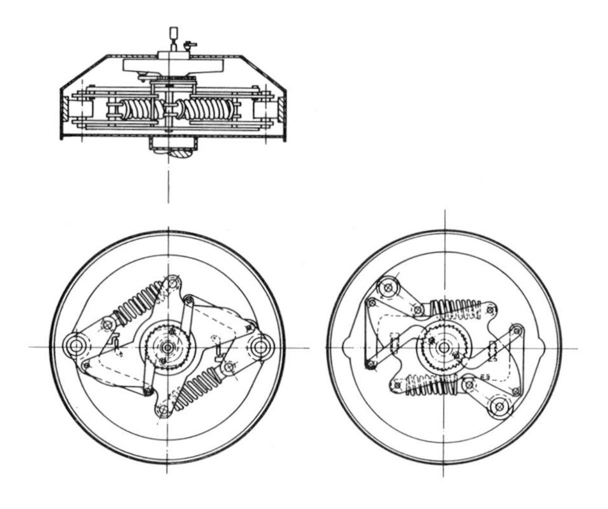

Fig. 11: Overload protection, LMG catch coupling.

When the design of the first giant excavator with a daily output of 100,000 m³ (bank) was developed during 1953 and 1954, the first genuine 2-rope suspension for the bucket wheel boom was engineered and built (Fig. 12). For the first time, rope tension measuring devices were used for the free rope end suspensions which measured the force acting upon the respective hoisting rope, so that the hoisting forces of the bucket wheel boom could be supervised and by suitable tensioning devices could be distributed equally between both hoisting ropes (Fig. 13). At the same time, the development of this machine provided for the support of the superstructure above the turntable in rocker joints (Fig. 14). Owing to this type of support, ex-

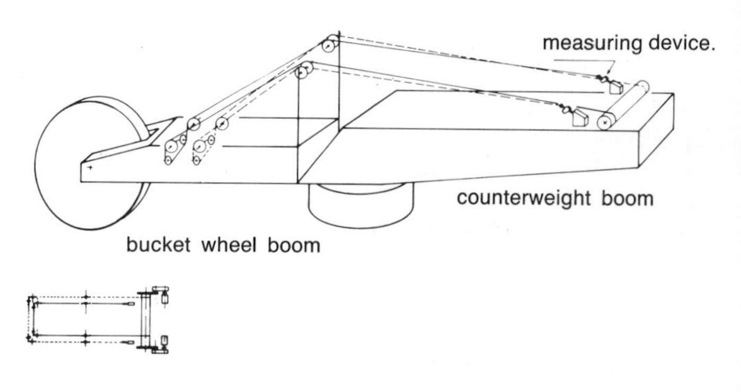

Fig. 12: Two-rope suspension for bucket wheel boom

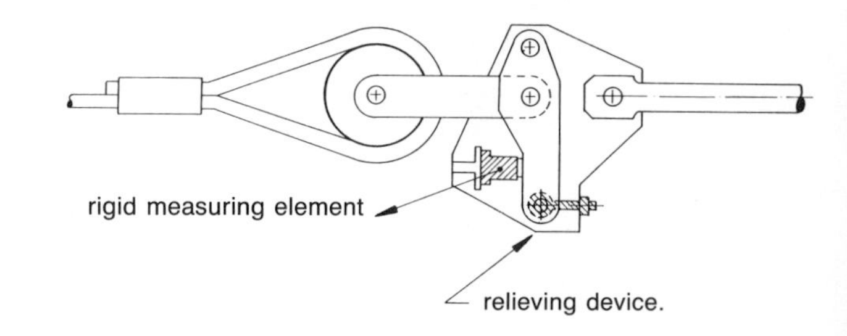

Fig. 13: Rope tension measuring device

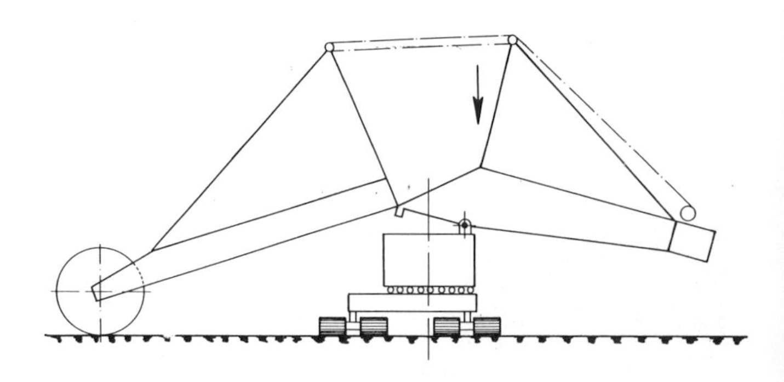

Fig. 14: Superstructure support in the tilting joints.

traordinary forces which can result from burying the bucket wheel or putting it down on the batter, can be intercepted without any overdue stresses on the supporting structure of the superstructure. When the bucket wheel is buried, the superstructure lifts itself from the joints on the ballast side and inclines so far that the bucket wheel rests on the batter. If the bucket wheel lies down on the batter, the supersturcture lifts off the rocking joints on the bucket wheel side. In both cases, lifting out of the rocker joints is indicated. At the same time, the hoisting winch and other drives are switched off. Suitable measures then have to be taken by the responsible supervising person to lower the superstructure again on to the joints.

Nearly all giant machines in the Rhineland brown coal area are equipped with these rocker joints which on several occasions have already contributed towards preventing more serious damage to the machine. Due to the good experience with the rocker joints, giant machines outside Germany were also equipped in this way.

Fig. 15: Bucket wheel excavator in a Polish open-cut.

7. Development of the Machines Themselves

In 1933, O&K built their first bucket wheel excavator with a service mass of 360 t. This machine was put into service in the German open-cut Bitterfeld in 1934. In 1936, a bucket wheel excavator with a service mass of 1,270 t and 800 kW total installed motor power was delivered to a Polish open-pit. This machine is still working today (Fig. 15).

At the end of the thirties, O&K started the development of the first giant excavator with a daily output of 30,000 m³ (bank) for the central German brown coal area. This unit was designed for an hourly output of 1,500 m³ (bank) and was intended to have a service mass of 5,900 t at a total motor power installed of 5,400 kW. Erection was started at the middle of the forties, however, by the end of the war, erection had only got as far as the turntable.

At the end of the forties, it was possible to foresee that the seams near the surface would soon be exhausted in the Rhineland brown coal area and, therefore, the coal at greater depths would have to be exploited. This required handling of huge masses of overburden. Mining planning envisaged machines for this capable of mining 100,000 m³ (bank) per day. In 1952, O&K were given the contract by Rheinische Aktiengesellschaft für Braunkohlenbergbau und Brikettfabrikation (as they were called at that time) to build such a machine.

This unit which was the first of a series of giant machines with a daily output of 100,000 m³ (bank) went into service in the open-cut Garsdorf as early as 1955 (Fig. 16). Right from the beginning, this machine met the requirements stipulated in the contract and is still in operation today.

Fig. 16: Bucket wheel excavator with 100,000 m³ (bank) daily output in the open-pit Fortuna, Garsdorf

Bucket wheel diameter	16 m
Bucket wheel drive	2 × 525 kW
Output per hour	5,830 m³ (loose)
Service mass	5,726 t
Motor power installed	8,900 kW

Until the early seventies, O&K supplied a total of seven giant bucket wheel excavators to the Rhineland brown coal area including one machine with a bucket wheel boom of 100 m length. Participation in the development of a new giant bucket wheel excavator with a daily output of 240,000 ³ (bank) was a matter of course. Rather remarkable is the gear for the bucket wheel of 21.6 m diameter and a drive power of 4 × 840 = 3,360 kW which is shown in Fig. 17.

Fig. 17: Bucket wheel of 21.6 m diameter

Bucket wheel drive	4 × 840 kW
Output per hour	19,000 m³ (loose)
Service mass	13,265 t
Motor power installed	16,900 kW

Apart from these giant machines, O&K also built a series of bucket wheel excavators with a daily output of 60,000 m³ (bank) and many machines with lower output for various applications all over the world.

Based on the experience with bucket wheel excavators for the most varied applications, these machines were also used for mining hard materials such as, for example, in the brown coal open-cut Neyveli in South India.

There, the first bucket wheel excavators went into operation during the years 1958 to 1965 (Fig. 18). Three further machines which started operating in the years 1979 and 1980 are designed for a hourly output of 2,250 m³ (bank). In practical operation, they have often considerably exceeded this output. Maximum daily outputs of up to 80,000 m³ (bank) were reached.

First contacts with Sunoil with a view to exploiting the oil sand fields in Northern Canada go back to the year 1955. After negotiations with the subsidiary of Sunoil, Great Canadian Oilsands Ltd. (GCOS), a decision was taken to use bucket wheel excavators for opening-up the open-pit to the north of Fort McMurray. The contract for the supply of two bucket wheel excavators and two beltwagons was awarded to O&K in 1965. After a building time of less than two years, the machines went into service in 1967. For this

Fig. 18: Bucket wheel excavator in the brown coal area Neyveli/India

Bucket wheel diameter	8.0 m
Bucket wheel drive	650/135 kW
Output per hour	3,570 m³ (loose)
Service mass	1,336 t
Motor power installed	1,730 kW

equipment, tests were carried out to find suitable structural steel to withstand the high and frequently changing loads at the low temperatures prevailing at the site and which would also allow satisfactory welding. At that time, no so-called cold weather steels were available on the German market. Therefore, the maker of this equipment had to acquire the fundamentals in this field first. These machines have now been operating for about 15 years and have met the demands made on them in every way. This is also due to the thorough preparatory work when selecting materials.

After initial difficulties when working with these machines in oil sand frozen to a depth of about 4 m, it was possible — in cooperation with the user's engineers — to bring both machines up to the specified output also during the cold winter months and to supply the extraction plant with a sufficient quantity of oil sand the year round, so that the specified amount of crude oil of 45,000 barrels a day equalling about 7,500 m³ could be delivered. Based on the good experience with these machines when mining large masses, GCOS ordered a bucket wheel excavator for stripping the overburden above the oil sand. As this overburden is to be dumped in different places which are far apart, a conveyor system cannot be used for transport. The material mined by the excavator has to be loaded on heavy trucks with a capacity of about 130 t each. Therefore, a comparatively long discharge conveyor with a heavy discharge chute is attached to the machine (Fig. 19).

Owing to the long discharge belt with the great mass of the discharge chute attached to its end, the position of the centre of gravity and, therefore, the stability of the machine is considerably affected. Special measures had to be taken concerning the slewing rim between superstructure and substructure as well as the travel gears.

Fig. 19: Bucket wheel excavator in the Canadian oil sand open-cut with long discharge belt and discharge chute

Bucket wheel diameter	12.5 m
Bucket wheel drive	2 × 500 kW
Output per hour	7,865 m³ (loose)
Service mass	1,813 t
Motor power installed	3,000 kW

An ordinary ball bearing which cannot transmit any tensional forces between supersturcture and substructure could not be considered for this machine. A roller bearing had to be chosen, which was capable of transferring the tensional forces occurring between superstructure and substructure (Fig. 20).

As under extraordinary operating conditions, the position of the centre of gravity is situated outside the tilting edge of the crawler travel gear, the machine then tilts over this tilting edge. Quick tilting-over is prevented by hydraulic presses which dampen the tilting motion and straighten the machine up again.

The travel gear of this machine has four crawlers, two of which are rigidly connected to the substructure, and the other two are connected by means of a balance beam with the third supporting point of the substructure being arranged in its centre. This balance beam supports the two cylinders contributing to the stability of the machine (Fig. 21).

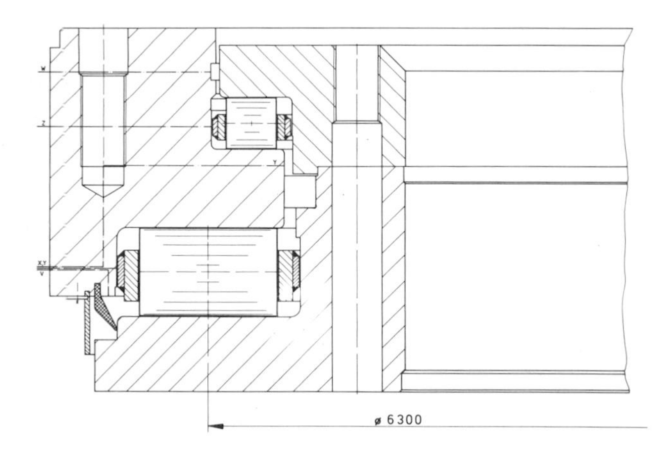

Fig. 20: Roller bearing for bucket wheel excavator in the oil sand open-cut.

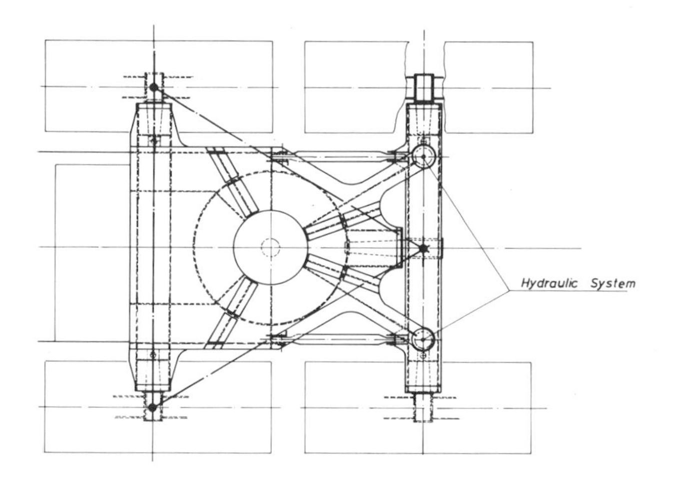

Fig. 21: Travel gear with 4 crawler chains and hydraulic tilting prevention system.

The lower support of the discharge conveyor which is subjected to considerable stress by horizontal forces on the one hand and vertical forces on the other, consists of a ball bearing whose design, in particular with reference to the material used, was newly developed. The surface of the race for the balls, consisting of high-class welding deposit, still has the elasticity required for such a running surface in spite of its hardness. This very expensive ball race is only used where conditions demand its application. On this machine, the restricted assembly space available for this ball bearing necessitated this special design (Fig. 22).

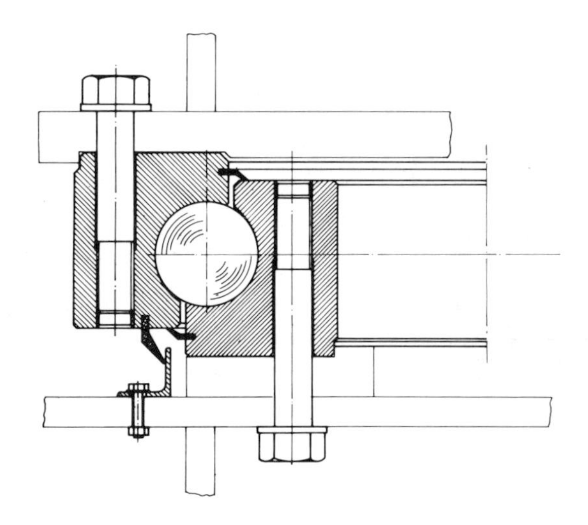

Fig. 22: Special design of a ball bearing.

Applications on considerably inclined grades require that a bucket wheel excavator is capable of standing or being moved even on inclinations of 1 : 6. As early as 1955, O&K built such a machine type for a brown coal mine in Italy. On this machine, the whole slewable superstructure can be levelled by means of an intermediate table fitted above the substructure.

Therefore, all parts of the superstructure are unaffected by the effects of inclinations of the track level. The intermediate table is rigidly supported in one point on the substructure, the other two supporting points can be adjusted in height by means of spindles. In this way, the inclination of the substructure can be equalized by means of the intermediate table, so that the slewable superstructure remains approximately level (Fig. 23).

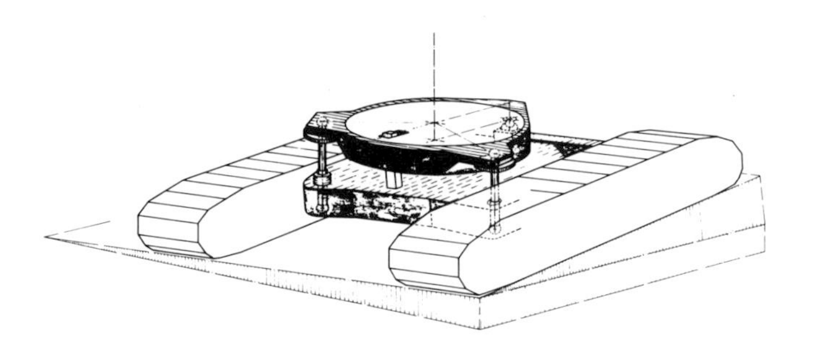

Fig. 23: Levelling equipment for the slewable superstructure.

This design, however, has the disadvantage that a certain overal height is required for the intermediate table and that, therefore, the pivoting point of the bucket wheel boom on the slewable frame must be arranged high up. This requires a long bucket wheel boom and great structural weight.

For this reason, new ways were found for a bigger machine in 1964. Levelling of the whole slewable superstructure was abandoned, only the conveyor belts were kept level. With this design, the whole machine adapted itself to the respective inclination of the grade, and this was appropriately taken into consideration when designing the different components. As, however, conveyors no longer run concentrically in case of too great an inclination transverse to their running direction, the horizontal position of the belts transverse to their running direction on this machine must be ensured even with big inclinations of the track level.

Control of the belt levelling devices is effected in such a way that the transfer conditions at the transfer points are not affected. The discharge belt is, for instance, supported in a cradle and when being levelled by a hydraulic cylinder moves in such a way that the material discharged from the belt in the bucket wheel boom always drops centrally on to the discharge belt (Fig. 24).

Fig. 24: Levelling equipment for the discharge belt.

At the beginning of the seventies — building on experience gained with hydraulically operated shovels — O&K developed the first all-hydraulically operated bucket wheel excavator with a nominal bucket capacity of 0.4 m^3, an output of 500 m^3 (bank) to 1,000 m^3 (bank), depending on the material, and a drive power of the three drive motors of the bucket wheel of 160 kW. The first machine was put into operation in the chalk pit Hemmoor in Northern Germany in 1972. It is being used successfully for stripping the overburden above the chalk. Since it was first put into operation, this machine has met all expectations (Fig. 25). On the basis of this unit, several machines with lower or higher capacities were built. The power can either be supplied via cable from outside or by built-in diesel engines. Several machines of both versions have been operating for years.

In 1967, Compagnie Française d'Entreprise placed a contract for the supply of a bucket wheel excavator system for construction of the Chasma-Jehlum Canal in Pakistan.

Fig. 25: Bucket wheel excavator in the chalk pit Hemmoor

Bucket wheel diameter	6.3 m
Bucket wheel drive	160 kW
Output per hour	1,800 m^3 (loose)
Service mass	184 t
Motor power installed	495 kW

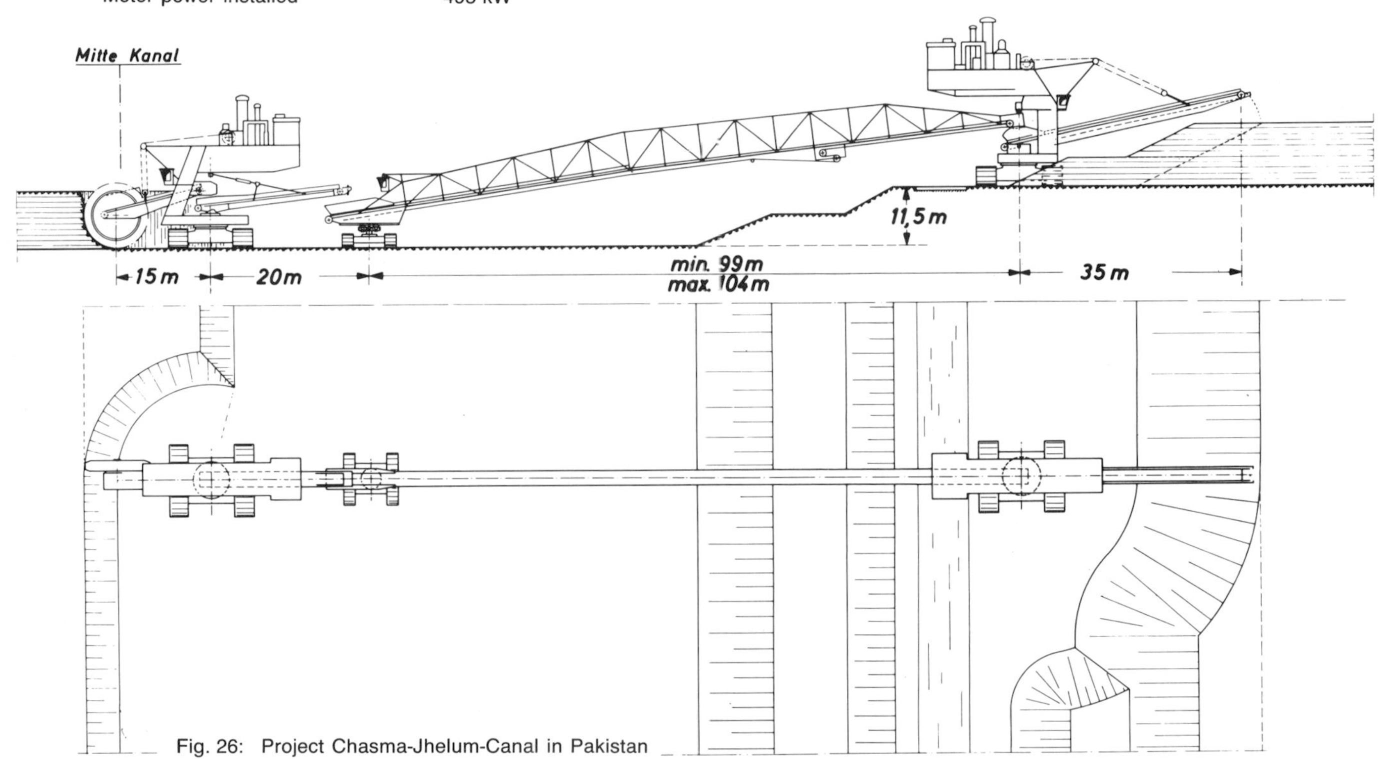

Fig. 26: Project Chasma-Jhelum-Canal in Pakistan

Fig. 27: Bucket wheel excavator during construction of the Chasma-Jhelum Canal, Pakistan

Bucket wheel diameter	10 m
Bucket wheel drive	2 × 360 kW
Output per hour	8,880 m^3 (loose)
Service mass	952 t
Motor power installed	1,900 kW

Total system: excavator + conveyor bridge + spreader

Service mass	2,110 t
Motor power installed	3,900 kW

This system was ready for operation in fourteen months, the components were made in Germany, shipped to the site, and erection was carried out there. Operation of the machine was started in mid-1968 (Fig. 26).

The whole system, a bucket wheel excavator followed by conveyor bridge and spreader, was designed for an average output of 3,750 m^3 (bank) an hour, but had to be designed for a maximum output of 8,800 m^3 (loose) per hour. Along its operating distance, the machine had to travel through zones of soils with little bearing capacity. The average bearing pressure was, therefore, not permitted to exceed 130 kPa (13 N cm^{-2}). The power is generated on the system itself by four diesel engines. Two engines each are accommodated on the excavator and on the spreader. As the machine had to operate under very unfavourable climatic conditions at ambient temperatures of up to 55°C and sometimes during heavy sandstorms, the air required for the diesel engines as well as for the generators had to be thoroughly cleaned.

That was not only done by filters but additionally by precleaning by means of mechanical equipment. All these devices are attached to the housing in which the diesel and generator equipment is accommodated (Fig. 27).

With this bucket wheel excavator system, the company commissioned with building the Canal succeeded in moving masses of 46×10^6 m^3 (bank) along a Canal length of 97 km by the end of 1970 and, thus, completed the canal six months earlier than planned. During this time, the system has mined 4,800 m^3 (bank) maximum per hour and produced a monthly peak output of 2.3×10^6 m^3 (bank). The total travel distance of the system for digging the canal was 1,250 km. After completing the work on the Chasma-Jehlum Canal, the system was bought by another French Company for use in the construction of the Jonglei-Canal in the Sudan.

For operation in the Sudan which placed higher demands on the bucket wheel excavator as regards the material to be mined, several modifications were carried out in connection with the bucket wheel. After these modifications were carried out, the system is working at the moment in the construction of 380 km Jonglei-Canal also under very adverse climatic conditions. The system has an average output of 70,000 m^3 (bank) per day.

Acknowledgements

The author would like to thank the management of O&K Orenstein&Koppel, Werk Lübeck, Federal Republic of Germany, for permission to publish this article and for making available illustrations and photographs. The author is also indebted to the many mining companies which use the bucket wheel excavators described in this paper.

This paper is an edited version of a paper first published in "Braunkohle" E I/82.

Volume 3, Number 1, March 1983

Output and Availability Factors of Bucket Wheel Excavators under Actual Mining Conditions

Joachim F. Rodenberg, Germany

Summary

The author defines the term "theoretical output" [bm³/h] as basic value for the assessment of the short and long time effective output [bm³/h] of bucket wheel excavators.

Approximately 40 bucket wheel excavators operating in overburden on four continents, are analyzed on the basis of outputs actually obtained in performance tests and long term operation.

Efficiency and availability are determined for all these machines.

Finally, a comparison is made with so-called "mobile equipment" i.e., shovels and draglines.

1. Introduction

Such terms as "theoretical output", "effective output", "daily output" and "average output" are generally quoted without appropriate and realistic consideration of the particular mining conditions and the time factor. The output for the equipment is often estimated too optimistically. This applies especially to new equipment.

Information published on the subject matter is often vague and inaccurate.

Theoretical output calculations for bucket wheel excavators (BWEs), based on well established formulae, are of course necessary and are of value in determining the average output of the equipment. Such calculations have, however, no resemblance to actual "real life" operating factors.

In the following an analysis of the output, operating factors and availability of 40 bucket wheel excavators is presented without priority to size and location of the machines.

All of the BWEs were taken into service between 1960 and 1980 and are still operating. They are excavating unconsolidated and cemented soils.

The machines work in four different continents and their output reflects:

— type of material excavated
— mine management
— climatic conditions and other environmental forces
— various material handling systems.

2. Theoretical Output

The starting point for the determination of the output of a BWE is the "theoretical output", where:

a) $$Q_{th} = I_N \times s \times 60$$

Q_{th} = theoretical output in loose m³/hour
s = bucket discharges per minute
I_N = nominal bucket capacity in m³
I_N = usually given as the volume of the bucket plus 50 % of the volume of the ringspace, the space forming part of the bucket but being located within the wheel body. The ringspace has usually a volume of 50 % of the actual bucket.

Therefore, I_N is usually given as:
Bucket volume 1.0 + 0.5 x 0.5 = 1.25 of the actual bucket volume. For all excavators considered here, this interpretation of "I_N" is used.

The term "bucket volume" is sometimes considered without the volume of the cutting edge or cutting lip, but often manufacturers include this volume in "I_N". In some instances, where cutting teeth are used, even the volume delineated by the teeth is included in the bucket volume. Due to the question of bucket volume definition a better determination of Q_{th} would be:

b) $$Q_{th} = H \times V_s \times t_m \times f \times 60$$

where:

H = height of slice in metres
V_s = slewing velocity in the deepest cut at a slew angle $\varphi = 0$ measured in metre per minute
t_m = maximum depth of cut taken by the bucket in metres
f = swell factor.

The product of $V_s \times t$ stays constant as long as the depth of cut 't', which decreases with increasing slew angle φ, can be compensated for by increased slew speed 'V_s' (Fig. 1).

Experience has shown that V_{max}, the maximum slew speed, should not exceed 30 m/min for kinetic reasons — forces encountered at slew reversal and impact forces when hitting obstacles.

The function $\frac{1}{\cos\varphi}$ drops drastically between $\varphi = 60°$ and $\varphi = 90°$. It is not economical to increase the slew speed to compensate for an angle greater than 60°.

c) Determination of the theoretical output of a BWE, using the installed conveyor capacity, or the diameter of the bucket wheel, is inaccurate. (± 30 % variation).

Lumpsize, stickiness of the material and free-cutting conditions of the wheel are only three of many reasons rendering such formulae useless.

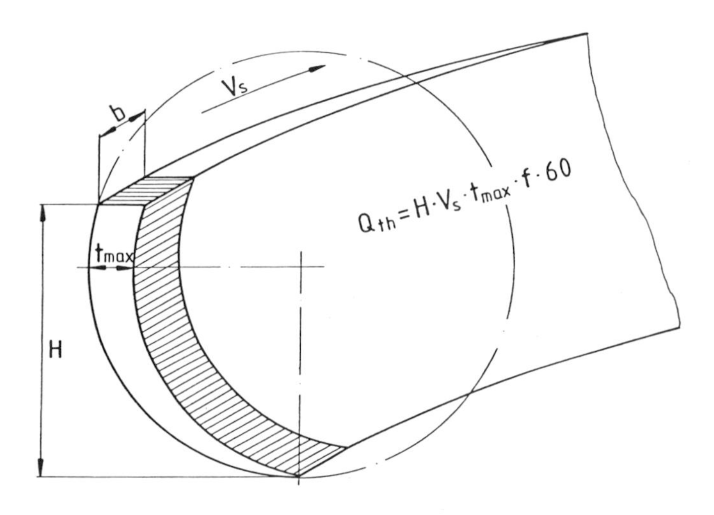

Fig. 1

d) To have a basis for comparison of the output of the excavators, the following approach is taken:

Since all mine planning is done in bank m³, the Q_{th} defined in 1a) and 1b) as loose m³ must be changed by the swell factor 'f' to bank m³ (bm³). This factor varies between 1.2 in loose sands to 1.7 in cemented materials.

$$Q_{th}\,[\text{bm}^3/\text{h}] = \frac{Q_{th}\,[\text{loose m}^3/\text{h}]}{f}$$

This definition will be compared to the effective output Q_{ff} expressed in bank m³.

e) The output efficiency factor is therefore derived as

$$\eta\, L_{eff} = \frac{Q_{eff}\,[\text{bm}^3/\text{h}]}{Q_{th}\,[\text{bm}^3/\text{h}]}$$

This factor varies with the timespan of the observation period. Depending on the time period, the factor may or may not account for the following:

η Soil — Influence of the soil to be excavated (hardness, lumpsize, cementation, consolidation, stickiness).

η Mining — Deviation from the optimum operating conditions, such as height of mining face, width of block excavated, trimming of the mine floor.

η Maint. — Type of maintenance, such as availability of parts and labour, maintenance of sharp teeth and cutting edges etc.

η Oper. — Operator efficiency: Ability of the operator to run at maximum capacity and optimum cut configuration.

η_{T_t} — Time losses due to raising and lowering of the boom, slew reversal and travelling time of the excavator.

η_R — Restriction caused by the transportation system behind the BWE, such as capacity limitation of the conveyor system, stacker or storage bin, or extraction plant.

The efficiency factor of a BWE mining system can be expressed as:

$\eta\, L_{eff.} = \eta$ soil x η mining x η maint. x η oper. x η_{T_t} x η_R

assuming that the time factor is considered in the foregoing factors.

η_{T_t} alone can be approximated mathematically. The time losses are dependent on the block dimensions to be excavated, such as height, width, number of terraces, slope angles of the highwall and the excavation face as well as the dimensions and capabilities of the excavator — such as boom length, wheel diameter, bucket size, slew speed, hoist speed, travel speed. η_{T_t} is calculated as follows:

$$\eta_{T_t} = \frac{t_b}{t_b + t_o}$$

where

t_b = actual excavating time

t_o = lost time due to hoisting and travelling, as well as slew reversal.

3. Test Output and Test Output Efficiency Factor

To demonstrate the capability of a BWE or a whole mining system, it has become customary to conduct a performance test. The mining company and the manufacturer enter into a contract. The performance test becomes an important part of the commercial undertaking. If the test output is achieved, the manufacturer will not have to pay penalties for non-performance or even take back the equipment if a certain minimum output is not achieved. Performance tests were conducted for periods ranging between eight and 1,000 hours.

In addition to the actual test operating time, certain allowances must be made for maintenance and service of the machine. The test is therefore usually conducted over a predetermined calendar period. A reduction of time is granted for delays caused by the mining company. The delays caused by the mining company seem to increase disproportionally with increasing test period times.

The value of a test over a long test period becomes therefore questionable. Tests are costly for the mining company and the manufacturer and require a good deal of management, planning and personnel.

Test periods normally ranging between eight and 150 hours, and in rare instances up to 500 hours, appear appropriate and indicate to the expert the capability of the BWE.

The test efficiency factor — η_L test — of an excavator depends on the length of the test period, the cutting resistance of the soil and the block dimensions excavated.

Table 1: Performance test results

Continent	Type SchRs	Material	Eff. Test Time [h]	Average Output in bank m³/h Q_{eff}	Test Output Efficiency Factor η_L Test	Theoretical Machine Factor η_{T_t}
Europe	$\frac{4000}{20}$ x 50	overburden > 50 % clay f = 1.5	144.5	3,395	0.70	0.97
	$\frac{4500}{12-14}$ x 41	sandy overburden f = 1.3	95	5,227	0.77	0.88
	$\frac{700}{9.4}$ x 29	overburden with clay content f = 1.4	179	1,500	0.83	0.92
	$\frac{1900}{5}$ x 30	overburden f = 1.3	60.5	3,318	0.86	0.87
	$\frac{270}{7}$ x 13	overburden f = 1.3	5.25	526	0.92	0.90
	$\frac{250}{7}$ x 13	overburden with clay content f = 1.4	48	475	0.88	0.85
	$\frac{1500}{6}$ x 31	overburden with high clay content f = 1.55	430	2,725	0.77	0.88
	$\frac{900}{6}$ x 25	sandy overburden f = 1.3	88	2,358	0.75	0.85
America	$\frac{1000}{1.5}$ x 26	tar sand f = 1.39	122	2,730	0.68	0.80
	$\frac{2450}{1.5}$ x 18	overburden f = 1.39	133	4,050	0.72	0.67
Africa	$\frac{560}{1}$ x 12.5	sand f = 1.25	410	1,770	0.70	0.67
	$\frac{350}{5}$ x 12.8	overburden f = 1.3	195	893	0.92	0.79
	$\frac{350}{5}$ x 12.8	overburden f = 1.3	164.5	920	0.95	0.79
	$\frac{150}{0.5}$ x 10.5	phosphate with calcium layers f = 1.4	69	358	0.85	0.86
	$\frac{2300}{1.5}$ x 12.5	overburden with high clay content f = 1.5	83	3,471	0.61	0.66
Asia	$\frac{2000}{1}$ x 12	sand f = 1.3	280	4,887	0.72	0.57
	$\frac{250}{1}$ x 12.5	weathered granite and clay f = 1.55	15	594	0.72	0.75
	$\frac{630}{1.2}$ x 15	weathered granite and clay f = 1.5—1.7	9	1,051	0.58	0.72
	$\frac{1500}{2}$ x 26	pre-blasted sand stone f = 1.45	8	3,550	0.76	0.81
	$\frac{1500}{2}$ x 26	pre-blasted sand stone f = 1.45	1000	2,515	0.53	0.81

3.1 Short Performance Tests

Short performance tests of between five and 70 hours resulted in an average η_L test of 0.79 for 9 study cases, that is 79% of the theoretical output expressed in bank m³ were achieved.

Only one BWE deviates to η_L test = 0.58 over a 15-hour test period. The cutting force required for the test material consisting of clay and weathered granite was 110 N/cm².

The test outputs indicate that the installed bucket wheel power, as calculated on the basis of lab test results, is sufficient for the actual field conditions.

The reasons for a high test efficiency factor can be recognized as:

— Low wear and few plug-ups for the digging head and the conveyors.
— High concentration and efficiency of the operator and the maintenance personnel.
— Constant supervision and positive influence by the manufacturer's representative.
— Few interruptions due to delays caused by the mining company.
— Availability of auxiliary equipment for mine floor clean-up etc.
— With BWEs having short booms, the hoisting, travelling and slewing required to start a new terrace, are quite often done simultaneously while productivity is maintained. This technique leads in cases to the elimination of the factor η_{T_t}.
— In one case, the nominal bucket capacity is exceeded by a factor of 1.3 over short periods of time.
— Continental influences are negligible (rain, temperature, management).
— Influence of downstream delays is minimal (conveyor system — full bin etc.).

3.2 Long Performance Tests

Long performance tests over periods of 70 to 1,000 hours result in an average output efficiency factor of η_L test = 0.71 for 15 machines surveyed.

The lowest value is η_L test = 0.53 for three machines in Asia, tested over a 1,000 hour-period.

The highest value is η_L test = 0.95.

For the long test periods the influence of weather factors and operational restraints result in a reduction of the efficiency factor η_L test compared to the short test periods. Significant is that in both cases the long-term efficiency factors for the normal mining operation sinks below the test period factor.

Interesting is that for nine out of 24 test results the factor η_L test is approximately equal or higher than the theoretical factor η_{T_t} which reflects only the operating losses due to hoisting and walking etc.

Only ten excavators show an influence of η_{T_t} on the effective total efficiency factor η_L. These machines have bucket wheel booms in the range of 30 to 62 m and cannot be operated in an unorthodox way.

4. Long-term Output Efficiency Factor η_L for Mining Operations

After the performance test, the BWE operates under normal mining conditions.

It is unavoidable that unfavourable influences encountered during the performance test only in a limited way, or not at all, start to have a greater influence.

— A greater variety of soil types may have to be excavated.
— Selective mining may be required.
— Rocks must be removed by special excavating techniques.
— Auxiliary equipment is not used for pit floor clean-up and is left to the BWE.
— The block height changes to unfavourable conditions.
— Worn-out bucket teeth are not changed when required.
— The excavator is not operated at full motor load.
— Operation continues under unfavourable conditions. (Frost, rain etc.).
— The operator does not work in a concentrated way but operates routinely and in a more relaxed fashion.
— Under-dimensioned mine conveyor systems lead the operator to operate at lower rates to avoid conveyor tripouts.

It appear that after a period of one to three years the yearly output stabilizes, if mining conditions do not change substantially. Q_{eff} in bank m³/hour stays reasonably constant.

While a mine with experienced personnel may reach a constant effective output only after two or three years of operation, it can be observed that an improvement at least of the output is evident, even in mines with inexperienced personnel after the first year of operation; although the final Q_{eff} is not reached until a later date.

Optimum conditions for favourable efficiency factors η_L exist in central European mines.

40 years of experience with bucket wheel excavators, the tool of the German open-pit mines, sand and clay overburden, favourable climatic conditions and well organized maintenance and operating crews result in long-term output efficiency factors η_L = 0.60—0.85.

The survey shows that 12 BWEs of all sizes, working in three German lignite pits, have a factor η_L = 0.69 over 29 total bucket wheel operating years.

One large German open-pit shows an efficiency factor of η_L = 0.76 for 95 BWE operating years for machines of the 60,000 to 200,000 bank m³ per day class.

The same open-pit shows a factor η_L = 0.85 over a 12-month period during 1976/1977 for one of the 200,000 bank m³ machines. 13% of the BWE production was coal and during 20% of the 4,400 yearly operating hours the machine was cutting below track level.

The open-pit, where the three largest BWEs of the world with a machine weight of 13,000 t are working, records a η_L = 0.8 after initial opening-up cuts were completed. This mine expects to have the three machines of the 240,000 bank m³ class operating at a η_L = 0.85 in the near future.

One especially good efficiency factor is recorded for the first BWE with an output of 100,000 bank m³. This machine was designed in 1951, started to operate in 1955 and worked in several pits, excavating mainly overburden. Over a period of 25 years, this machine excavated 469 million bank m³ with an efficiency factor $\eta_L = 0.857$. The machine is in first class mechanical condition due to good maintenance practices and will shortly be loading a conveyor system. Previously it had been used to load rail cars.

The efficiency factors of 24 BWEs surveyed in other areas of the world are much lower. The machines removing overburden for mines located on other continents have a $\eta_L = 0.45$ over 115 BWE operating years.

7 BWEs working in Africa in sandy soils have a factor $\eta_L = 0.58$ for 26 BWE operating years. This may be a result of the good digging conditions and the extensive experience with BWEs in these mines.

The machines working in North and South America have to battle with adverse climatic conditions. In South America five months of rainy season, and in Canada four months of very cold temperatures, may have an influence on the efficiency factor $\eta_L = 0.45$.

The machines working in Asia have encountered the hardest digging conditions (Fig. 2).

Fig. 2: Bucket Wheel Excavator SchRs $\frac{1500}{2}$ x 26 (top) and SchRs $\frac{700}{3}$ x 20 operating in Cudalore sandstone in South Asia.

Blasted sandstone with a swell factor of 1.45 and weathered granite with a swell of 1.7 combine with adverse weather conditions during the long rainy season. This may have an influence on the resultant factor $\eta_L = 0.4$. This factor may even be considered excellent when taking into account these adverse conditions.

The foregoing does not imply that certain efficiency factors are resultant from operations on certain continents, such as Asia $\eta_L = 0.4$, America $\eta_L = 0.45$, Africa $\eta_L = 0.58$ and Central Europe $\eta_L = 0.69$.

It rather indicates that hard digging conditions and climatic influences result in $\eta_L = 0.4$ over a very long period and $\eta_L = 0.5$ over a short survey period (Fig. 3).

Easy digging conditions result even under tropic, subtropic and arctic conditions in efficiency factors $\eta_L = 0.45—0.65$.

Optimum conditions in Europe result in efficiency factors $\eta_L = 0.6—0.85$. Over short periods of time (1—5 years), the factors are even $\eta_L = 0.75—0.85$ (Fig. 4).

5. Formulae in Literature

General formulae for calculation of effective output hardly ever meet reality.

— DIN 22266 shows the following formula for effective output of BWEs:

$$Q_{eff.}\,(\text{bm}^3/\text{h}) = \frac{0.8}{f} \times Q_{th}$$

— The recommended value for f (swell factor) is 1.3

— The combination of an efficiency factor of 0.8 as well as a swell factor of 1.3 are rarely experienced in mines and, when applicable, then only during short performance periods and under optimal conditions.

— Fully automatic BWEs are supposed to result in:

$$Q_{eff.}\,(\text{bm}^3/\text{h}) = \frac{0.96}{f} \times Q_{th}$$

Under the best conditions such results may be possible for a short performance test but not for long-term average mining conditions.

— If a BWE works in material of different hardness, the output is supposed to vary with the square of the cutting resistance of the material k, expressed in [N/cm].

$$\left(\frac{Q_{eff\,1}}{Q_{eff\,2}}\right) \cong \left(\frac{k_{l_2}}{k_{l_1}}\right)^2.$$

This statement might also be at best true for short time periods of an hour or a day only.

— Another formula states that:

$$Q_{eff}\,[\text{bm}^3/\text{h}] = \frac{Q_{th}\,[\text{loose m}^3\,\text{h}]}{f\,(l + t_e/t_b)}$$

where t_e = the sum of all delays and
t_b = the actual digging time
f = the swell factor.

Experience shows that reference to operating and delay times alone does not give a realistic estimate of BWE output.

— Operating factors resulting from machine dimensions may be calculated or approximated reasonably well. The influence of human factors, climatic and mining conditions, in relation to the digging machine, cannot be determined mathematically. One must unfortunately rely on experience for these factors.

— The relationship of performance test output and long-range output is to be calculated by:

$$\frac{\eta_L \text{ test}}{\eta_L \text{ long-term}} = 1.2 \text{ to } 1.6$$

This, of course, is dependent on the length of test and the number of years of BWE experience. It is relatively independent of the varying digging conditions, changing climatic conditions or management influences.

Fig. 3: Bucket Wheel Excavator SchRs $\frac{2450}{1.5}$ x 18 in overburden above the tar sand fields in Canada.

Fig. 4: The first "Giant Bucket Wheel Excavator" SchRs $\frac{3600}{5}$ x 48 with a daily output of 100,000 bm³; put into operation in 1955.

Table 2: Effective operating time

		Open Pit X		Open Pit Y
Calendar time		8,760 h		8,760 h
./. statutory holidays	5 days:	120 h	10 days:	240 h
./. expected weather delays	4 days:	96 h	10 days:	240 h
./. general overhaul	3 weeks:	504 h	3 weeks:	504 h
./. operational delays — equipment moves conveyor moves, bin full etc.	10 days:	240 h	12 days:	288 h
	325 days:	7800 h	312 days:	7488 h
./. lost time due to 2 shift instead of 3 shift operation		—		2496 h
./. daily preventative maintenance	3 h/day:	975 h	2 h/day:	624 h
scheduled operating time		6,825 h		4,368 h
./. unscheduled downtime approx. 15 % of scheduled operating time		1,025 h		658 h
Effective operating time T_{eff}		5,800 h		3,710 h

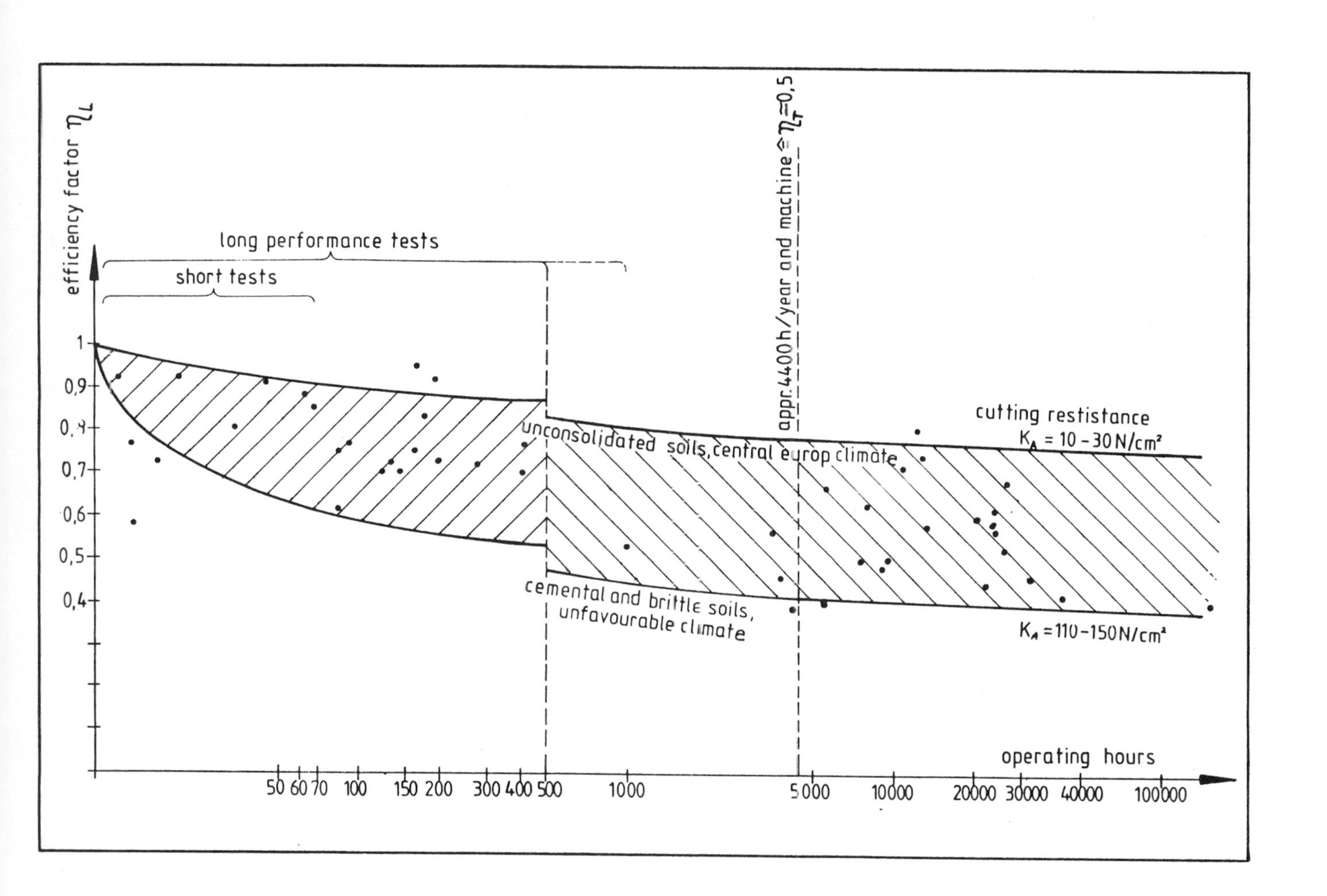

— Almost every BWE can repeat the performance test output; even after many years of operation.

The long-term output efficiency factor η_L does not indicate primarily the capability of the machine, but rather indicates how the machine is applied and what positive or negative influences are experienced under real-life operating conditions (Fig. 2).

6. Operating Time Factor

This factor is calculated as follows:

$$T_{eff} = T_{cal} - T_R - T_{PM} - T_D$$

Where:

T_{eff} = Effective operating time
T_{cal} = Calendar time
T_R = Time at rest (Sundays etc.)
T_{PM} = Preventative maintenance
T_D = Unscheduled down time.

T_R — Can be planned reasonably well, as it constitutes operating policies as to work times or rest times during legal holdidays and expected down time due to adverse weather conditions.

T_{PM} — Can also be planned considering the total mining operation including associated extraction plant, power plant or other plant connected with the particular mining system.

The preventative maintenance is usually carried out during one shift per week or two hours per day — or both. A one to three week outage per year is usually planned for the overall inspection of the machine or a major overhaul. Usually, conveyor moves are planned during the latter outage, or a move of the stacker or BWE from one working area to another may take place during the extended outage.

T_D — includes all down time exceeding several minutes caused by electrical, mechanical or operational problems, as well as plug-ups of the conveyor system, pit wall instability etc.

T_{eff} — is the actual operating (digging) time. It often includes down time of several seconds or minutes that are not registered as down time. Such short outages may be caused by overload trips of the BWE motor, slew motor, conveyor.

In addition it includes lowering, hoisting, travelling, also short discussions of the operators during changes of positions. Since the bucket wheel is rotating in an idling mode during such operational changes, the time is registered as digging time.

The effective operating time T_{eff} and the unscheduled down time T_D make up the operation time.

The operation time factor is

$$\eta_{T1} = \frac{T_{eff}}{T_{eff} + T_D}$$

$T_R + T_{PM}$ is the planned idle time

The planned idle time factor results from the differences between calendar time and operation time and is expressed as:

$$\eta_{T2} = \frac{T_{eff} + T_D}{T_{cal}}$$

Calendar time is taken as 8,760 hours/year.

η_T, the overall time factor is $\eta_T = \eta_{T1} \times \eta_{T2}$

$$\eta_T = \frac{T_{eff}}{T_{eff} + T_D} \times \frac{T_{eff} + T_D}{T_{cal}} = \frac{T_{eff}}{T_{cal}}$$

This says that the time factor of a BWE coupled with a mining system is governed by two time definitions.

T_{eff} = effective digging time per year
T_{cal} = calendar time = 8,760 h/year

The calender time is well defined. The digging time varies between 3,000 and 6,000 hours per year depending on governing conditions in various open pits.

Two examples will show what effective digging times can be achieved (Table 2).

The overall factor for pit "X" is

$$\eta_T = \frac{5{,}800}{8{,}760} = 0.66$$

Time factor η_T for pit "Y" is

$$\eta_T = \frac{3{,}710}{8{,}760} = 0.42$$

Mines overseas usually work three shifts per day for seven days per week so that worldwide time factors of

$\eta_T \approx 0.5$ = 4,380 hours are achieved.

German mines, with years of bucket wheel experience, do not achieve average time factors much better than that.

A BWE system — under Japanese management — achieved a time factor of 0.68 over a period of four years. The time factor for two bucket wheels was 0.71 ≙ 6,240 hours in this mine in Singapore for one particular year.

The assessment of the time factor in terms of calendar time gives a definite term of reference. It must, however, be considered in the light of influence such as climatic conditions, legal holidays etc. governing a BWE application.

In North America the term "availability" of a mining system is used. The "mechanical availability" is sometimes expressed in terms of effective or actual digging time and unscheduled down time

$$\eta_{T1} = \frac{T_{eff}}{T_{eff} + T_D}$$

resulting in time factors of approx. 0.85. Sometimes it is expressed in terms of effective digging time and repair regardless of repair time being scheduled or unscheduled.

Such interpretation of mechanical availability results in factors much lower than 0.85.

Table 3: Long-term efficiency factors

Continent	Machine Type		Overburden Material Excavated	Climate	No. of Machines	Operating Years	Time Efficiency Factor η_T	Output Efficiency Factor η_L
Europe	**SchRs**							
Central Europe	$\frac{4500}{12—14}$	x 41	sandy overburden f = 1.3—(1.4)	continental	1	5	0.54	0.62
Central Europe	$\frac{4500}{12}$	x 44	sandy overburden f = 1.3—(1.4)	continental	1	5	0.6	0.68
Central Europe	$\frac{6300}{9—17}$	x 51	sandy overburden f = 1.3—(1.4)	continental	3	3	0.48	0.8
Central Europe	$\frac{6300}{9—17}$	x 51	sandy overburden f = 1.3—(1.4)	continental	1	1	0.51	0.85
Central Europe	$\frac{700}{9.4}$	x 29	with high clay content f = 1.4	continental	1	3	0.47	0.72
Central Europe	$\frac{450}{10}$	x 20	with high clay content f = 1.4	continental	1	3	0.52	0.62—0.83
Central Europe	$\frac{200}{7}$	x 12	with high clay content f = 1.4	continental	1	4	0.37	0.74
Central Europe	$\frac{250}{7}$	x 13	with high clay content f = 1.4	continental	2	2	0.46	0.62
Central Europe	$\frac{300}{4.5}$	x 14	with high clay content f = 1.4	continental	1	3	0.5	0.58
	Average for η_T and η_L				12	29	0.5	0.69
East Europe	$\frac{4600}{2.5}$	x 30	clayey sand f = 1.25—(1.4)	continental, partly cold winters	1	2	0.47	0.79
East Europe	$\frac{4600}{14}$	x 50	clayey sand f = 1.25—(1.4)	continental, partly cold winters	1	2	0.41	0.56
	Average for η_T and η_L				2	4	0.44	0.68
Africa	**SchRs**							
South Africa	$\frac{400}{0.6}$	x 11	Sand f = 1.3	tropical	1	6	0.45	0.59
East Africa	$\frac{2300}{1.5}$	x 12.5	high clay content f = 1.5	subtropical	1	2	0.44	0.50
West Africa	$\frac{560}{1}$	x 12.5	sandy overburden f = 1.25	tropical	1	6	0.49	0.53
West Africa	$\frac{350}{5}$	x 12.3	sandy overburden f = 1.25	subtropical	2	2	0.59	0.72
West Africa	$\frac{350}{1}$	x 14	high clay content f = 1.4	subtropical	1	5	0.54	0.57
West Africa	$\frac{560}{1}$	x 12.5	sandy overburden f = 1.25	tropical	1	5	0.48	0.60
	Average for η_T and η_L				7	26	0.49	0.58

Continent	Machine	Type	Overburden Material Excavated	Climate	No. of Machines	Operating Years	Time Efficiency Factor η_T	Output Efficiency Factor η_L
America	**SchRs**							
North America	$\frac{1000}{1.5}$	x 26	tar sand $f = 1.4$	continental partly arctic	2	12	0.41	0.42
South America	$\frac{200}{2}$	x 19	sand with clay content $f = 1.3$	subtropical	1	1	0.65	0.5—0.84
South America	$\frac{250}{2}$	x 19	sand with clay content $f = 1.3$	subtropical	1	2	0.53	0.43—0.55
South America	$\frac{150}{0.5}$	x 10.5	sand with clay content $f = 1.3$	subtropical	1	1	0.63	0.37—0.43
South America	$\frac{600}{2}$	x 20	sand with clay content $f = 1.3$	subtropical	1	1	0.43	0.4—0.53
South America	$\frac{400}{2}$	x 15	swampy clay $f = 1.4$	subtropical	1	1	0.50	0.35—0.43
South America	$\frac{400}{3}$	x 20	swampy clay $f = 1.4$	subtropical	1	2	0.55	0.5
	Average for η_T and η_L				8	20	0.46	0.45

Continent	Machine	Type	Overburden Material Excavated	Climate	No. of Machines	Operating Years	Time Efficiency Factor η_T	Output Efficiency Factor η_L
Asia	**SchRs**							
South Asia	$\frac{1500}{2}$	x 26	pre-blasted sand-stone $f = 1.45$	subtropical	3	7	0.53	0.46
South Asia	$\frac{700}{3}$	x 20	pre-blasted sand-stone $f = 1.45$	subtropical	4	58	0.50	0.39
South East Asia	$\frac{630}{1.2}$	x 15	weathered granite $f = 1.5—1.7$	subtropical	2	4	0.68	0.48
	Average for η_T and η_L				9	69	0.51	0.40

The mining system coupled to the BWE has the greatest influence on the time factor. Every component of the system has an influence on the availability of the other components. The bucket wheel could be delivering directly to a conveyor bridge for a direct overcasting operation or could be coupled to an conventional around-the-pit-system with bucket wheel, beltwagon, face conveyor, connecting conveyor, dump conveyor, tripper and stacker.

As long as the components following the BWE are designed for appropriate maximum output, the BWE time factor is affected to a greater degree than the output factor Q_{eff}.

The system should be designed so that:

Q_{max_1} bucket wheel < Q_{max_2} bucket wheel conveyor < Q_{max_3} mine conveyors < Q_{max_4} dump equipment conveyors.

An appropriate design factor for these components could be recommended as follows:

$$Q_{max_2} \cong 1.2 \text{ to } 1.25 \times Q_{max_1}$$
$$Q_{max_3} \cong 1.1 \times Q_{max_2}$$
$$Q_{max_4} \cong 1.05 \times Q_{max_3}$$

7. Comparison of Mining Systems

The time factors $\eta_T \cong 0.5$ and effective output factors η_L = 0.4 to 0.85 may not look too attractive to the uninitiated observer.

However, when one considers the terms of reference, that is, "calendar time" and "theoretical output", the picture looks much more attractive. The term "theoretical output" for mobile equipment and shovels is usually never calculated and would result in factors η_L of surprisingly low magnitude.

Comparisons of mobile equipment application (shovels and trucks or dozers and scrapers) with bucket wheel application (BWE, conveyors and stacker) or comparison of dragline versus bucket wheels and conveyor bridges (cross-pit conveyors), show that the bucket wheel systems are often more economical.

Due to the limited reach of draglines and stripping shovels, double handling is often required. More than half the operating time for such equipment consists of swing time and not excavating time.

BWE systems show generally better economics in soils with a cutting resistance of 1,500 (to 2,000) N/cm or 150 (to 200) N/cm².

A well-known North American manufacturer of draglines and shovels, who has also built a limited number of BWEs, published a paper in 1980 (Ref. No. 6) regarding removal of deep overburden. The economics on a comparable basis are quoted as follows:

Long boomed dragline	—	83 ¢/bank cu.yd.
Shovel and cross-pit conveyor	—	85 ¢/bank cu.yd.
Dragline and cross-pit conveyor	—	73 ¢/bank cu.yd.
BWE and cross-pit conveyor	—	68 ¢/bank cu.yd.

This comparison includes operating and capital costs for loosening, excavating and transporting materials on a comparable basis. The economics are greatly influenced by the output and time efficiency factors.

References

[1] Rasper, L., "Der Schaufelradbagger als Gewinnungsgerät", Trans Tech Publications Clausthal, 1973; "The Bucket Wheel Excavator — Development, Design, Application" Trans Tech Publications, Clausthal, 1975

[2] Krumrey, A., "Schaufelradbagger für feste Erdstoffe", Deutsche Hebe- und Fördertechnik (1981), No. 10

[3] Lubrich, W., "Zur Beurteilung von Geräten über den Zeit- und Lastgrad in Tagebausystemen".

[4] Henning, D., "Erste Betriebserfahrung beim Aufschluß des Tagebaus Hambach", BRAUNKOHLE, (1981) No. 7

[5] Schönfeld, G., "Der Tagebau Fortuna von 1955 bis 1990", BRAUNKOHLE, (1977), No. 9

[6] Learmont, T., Chare, H.B., "Area stripping productivity solutions in deep overburden", Australian Coal Miner, (1980) June

[7] Durst, W., "Der Schaufelradbagger"(unpublished)

Acknowledgement

The author would like to thank the management of O&K Orenstein & Koppel, Werk Lübeck, Federal Republic of Germany, for permission to publish this article and for making available illustrations and photographs.

 Volume 3, Number 3, September 1983

Continuous Mining Technology — Its Impact on the Largest Opencast Lignite Mines in India — Neyveli

T. S. Kasturi, India

1. Introduction

Lignite, the so-called "Brown Diamond", has transformed the featureless, sleepy hamlet "Neyveli" into a modern township. Neyveli, located in the State of Tamil Nadu, about 200 km from the well-known port city of Madras, unknown to many people in India three decades ago, suddenly gained national importance in August 1961 when the first seam of lignite was exposed. Now, Neyveli is humming with industrial activities with more than 19,000 people working in the various units of the Neyveli Project.

The existence of lignite in this area was known as early as 1928, but systematic geological investigations were undertaken in 1943 to 1954 by the Geological Survey of India and the State of Tamil Nadu and later on by the Neyveli Lignite Corporation which established the existence of a deposit with about 3,300 million tonnes of lignite covering an area of 480 km², known as the Neyveli Lignite Field (Fig. 1).

The Neyveli Lignite Field accounts for 95% of the total explored lignite reserves in India. Lignite availability at other locations like Rajasthan, Gujavat and Kashmir accounts for only 5% of the country's total reserves (Fig. 2).

The occurrence of lignite in the State of Tamil Nadu has a special significance in view of the non-availability of coal or other fossil fuels in the Southern Region. The nearest coal field is located about 700 km from Madras. The Bengal and Bihar coal fields are more than 1,500 km away. Hence not only is the transport of bulk coal cumbersome but it also makes coal an expensive fuel.

The hydro-electric sources in the state have almost been tapped, and therefore lignite fuel seems to be the only answer to more thermal power generation if the Southern Region and particularly the State of Tamil Nadu is to progress industrially (Fig. 2).

The Neyveli Lignite Corporation was incorporated as a Government of India project in 1956 with the aim to construct and operate an integrated operation for mining initially 3.5 million tonnes of lignite by opencast mining methods using the "continuous mining technology" and in order to operate a 250 MW Power Plant (later on expanded to 600 MW), a Fertilizer Plant for the production of 152,000 tonnes of urea annually (later the feedstock was changed to fuel oil), a Briquetting and Carbonizing Plant for the production of 329,000 tonnes of domestic briquette fuel called LECO and a Clay Washing Plant for producing 7,000 tonnes of White Clay for the ceramic and other industries.

Mr. T. S. Kasturi, B.E., F.I.E., is General Manager, Neyveli Lignite Corporation, Neyveli, South India.

2. The Neyveli Deposit

The overburden strata of the deposit area belong to the Cuddalore series of the Upper Miocene age of the Tertiary era. The overburden is 65 to 70 m thick. The main types of soil encountered are top soil, laterite sandstone, argillaceous sandstone and clay. The sandstone mainly contains angular to sub-angular grains embedded in a clayey matrix. When the excavated mine faces are exposed for a long time, the strata becomes hard.

The soil condition varies very much from bench to bench. The top layer of 2 m overburden consists of laterite sandstone and loam. The soil between 2—5 m is fairly soft. Below 5—7 m there is a hard sandstone and below 25—30 m fire clay is found. The comparative strength of the overburden ranges from 15—20 kg/cm² up to a depth of 5—25 m to 20—25 kg/cm² from 25—45 m and 10—20 kg/cm² below 50 m. The bulk density is 2 t/m³.

The hardness of the material does not show any specific pattern but at some locations it lies well above the marking range of the Spring Scale Cone Hardness Tester.

The characteristics of the lignite are:

Calorific value, kcal/kg	2,500 to 2,700
Moisture content, %	52.00
Ash content, %	2.4
Bulk density in-situ, g/cm³	1.155
Fixed carbon content, %	20.60
Volatile matter, %	25.00

Mining of lignite at Neyveli is associated with many peculiar problems, among them:

— Presence of very hard Cuddalore sandstone formations in the overburden: For the effective operation of bucket wheel excavators a systematic forward preparation by drilling and blasting of nearly 45—50% of the overburden is necessary as opposed to about 5—10% assumed in the original mining plans. The consumption of explosives is roughly 1,800 t/annum.

Volume 3, Number 3, September 1983

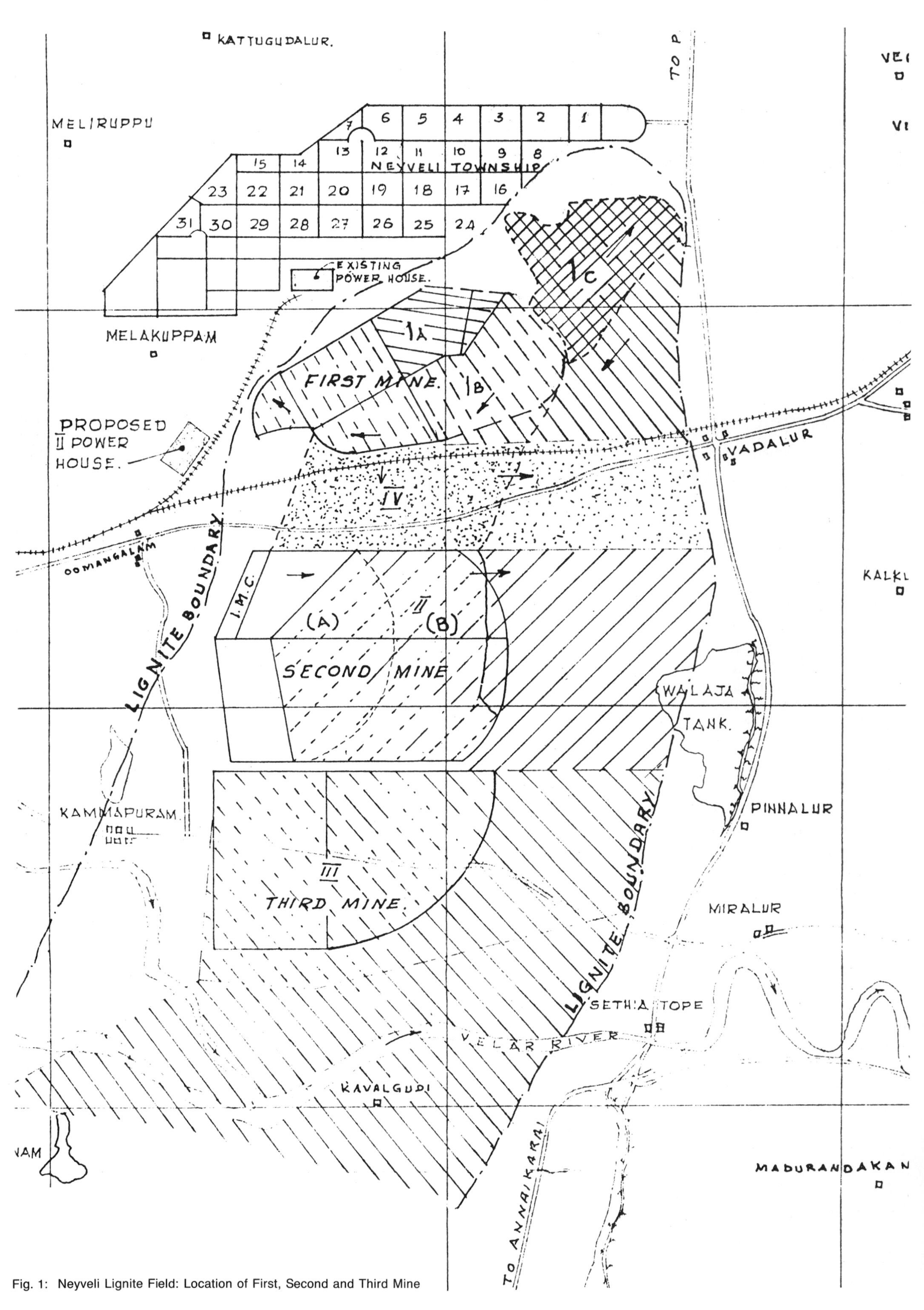

Fig. 1: Neyveli Lignite Field: Location of First, Second and Third Mine

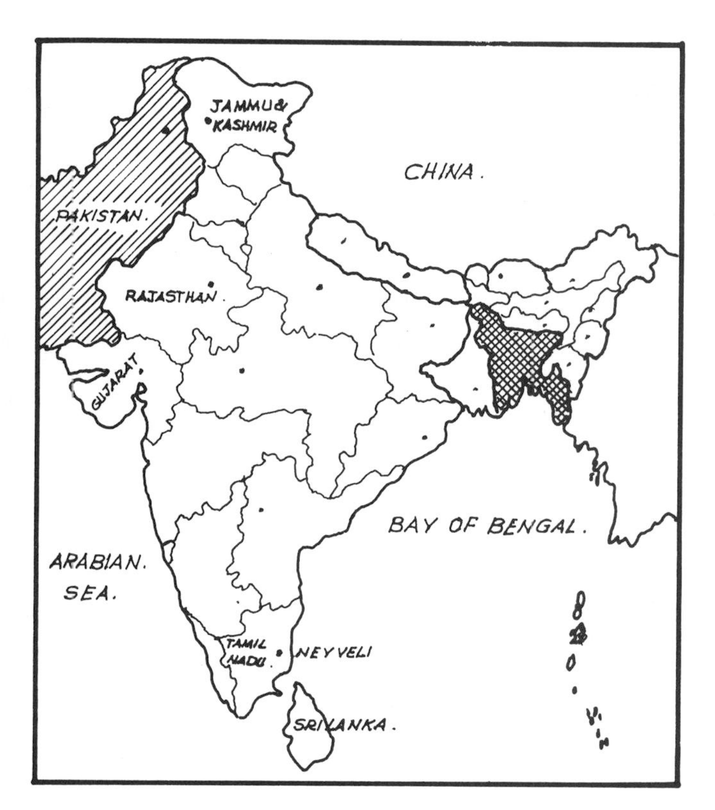

Fig. 2: Map of India showing the location of lignite deposits

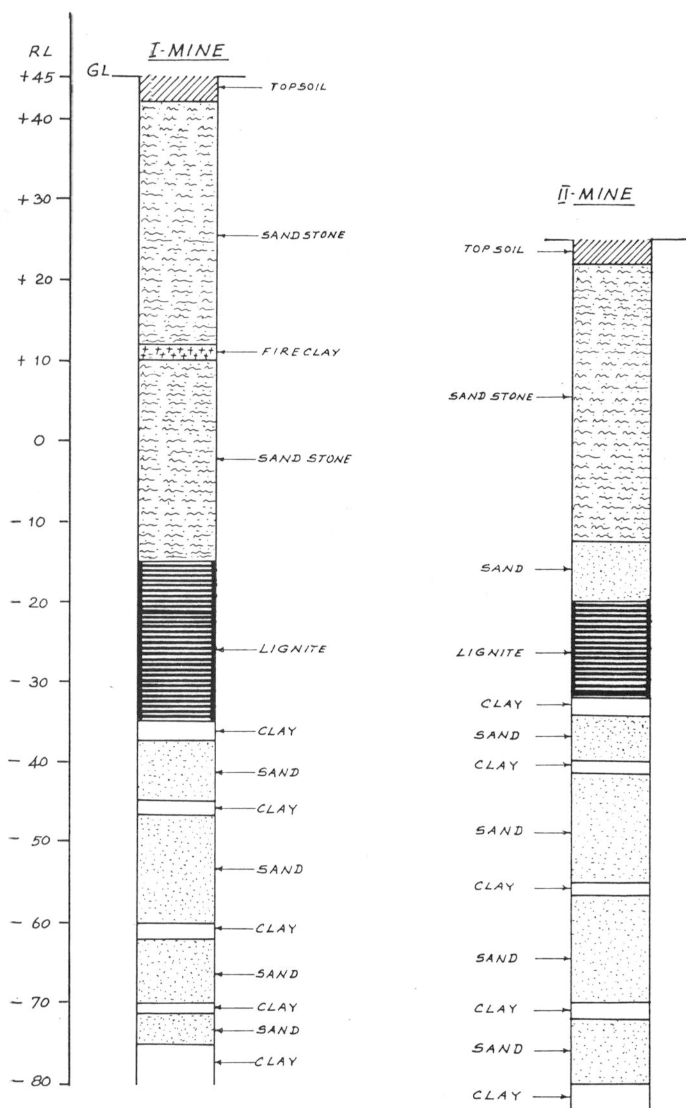

Fig. 3: Typical vertical section of strata to be found at Mine I and Mine II

— Below the lignite seam there is confined ground water in aquifer sand formations. This exerts an upward pressure of 5 to 8 kg/cm^2. In order to mine the lignite successfully, depressurisation of this aquifer is essential and the pressure surface is constantly maintained below the lignite seam by large scale pumping (about 145 m^3/min). For mining one tonne of lignite it is necessary to pump 20 tonnes of water from a depth of 90—100 m. A stand-by Diesel Power Station of 6 MW capacity has been installed near the mine to come into immediate service in case of a main power supply failure, so that pumping operations are not affected.

— Pumping of surface water especially during the monsoon season is another problem. As much as 3.5 million m^3 of water have to be pumped in addition to the aquifer water during the monsoon months of September, October and November.

— Neyveli lies in the cyclonic region where gales of up to 130 km/h have been encountered. The design of the equipment and the layout of the mine have to take this problem into consideration.

— In addition to the above, the problem of managing a sophisticated mechanised mine with the available talent was a major challenge. As in all developing countries where the population is large, an optimum blending of men and machines is a very sensitive subject.

— In developing countries where sophisticated equipment and spare parts have to be imported, foreign exchange is always a major constraint.

3. Mining Plan and Equipment for the First Mine

To excavate the overburden and lignite, a number of alternatives were studied in the early planning stage:

a) Shovel, dragline, dumper combination
b) Dragline, dumper, cross-filling combination
c) Bucket wheel excavator, dumper combination
d) Bucket wheel excavator, conveyor, spreader combination etc.

For reasons of higher productivity, economy, earlier exposure of lignite and also because of the hard strata, ground water problems, monsoonic and cyclonic conditions etc., a decision was taken to adopt the "continuous mining technology" with bucket wheel excavators for excavation and belt conveyor and spreader for the transport system as used in the lignite mines of Germany.

For the initial opening of the mine, approximately 5.6 million m^3 of overburden were excavated with the help of conventional mining equipment such as Euclid loaders; Caterpillar dozers (D8H, D9, D8K), dumpers (630 B/DW 20) and motor graders (112 F), pipe layers (583 H/583 K); Haulpak dumpers; Tata/P&H draglines and shovels; tractors; cranes, etc., altogether about 100 machines.

The mine was planned to have 7 Blocks viz. A, B, C, D, E, F, and G covering the selected mining area (Fig. 1). After completing the excavation of Blocks A and B mining will continue on the next lower level with Block C. The excavated overburden is refilled into the mined-out area. It is expected that the deposits in these blocks will last for 40—45 years.

The 65—70 m of overburden are to be removed in 4 benches of 15—18 m in height and 35 m in width. The 16—20 m of lignite are to be excavated in one or two benches. Each bench will have a bucket wheel excavator (BWE), a set of conveyors (5—10) and a spreader. For the transport of overburden and lignite, belt conveyors with a width of 1,000; 1,200 and 1,500 mm were used but these conveyors are progressively changed into larger belt widths.

For carrying out the mining operation, specialized mining equipment was ordered and put into service (Table 1).

The conveyors used are mostly of the "track-shiftable" type unlike permanent stationary conveyors installed in power plants, cement plants etc.

The progressive increase of conveyor widths and lengths from 1960 to 1983 is shown in Table 2.

The position of the conveyors as of January 1, 1983 is shown in Fig. 4.

The cumulative production of overburden and lignite since the beginning in 1957 until the period ending April 30, 1983 is shown in Fig. 5.

4. First Mine Expansion Scheme

In 1975 it was found that the capacity of the mining equipment originally purchased was not sufficient any longer due to ageing of the equipment, the increased ratio of overburden to lignite from approximately 4.5 to 5.8 and the increased hardness of the overburden, etc.

Therefore, additional equipment to increase the capacity of the mine from 3.5 to 6.5 million tonnes was purchased for the First Mine Expansion Scheme, replacing the 700 litre bucket

Table 1: Equipment used in the First Mine

Type of equipment	Number of units	Manufacturer	Year of Commissioning
350 litre BWE*	2	O & K Orenstein & Koppel West Germany	6/1959 & 1/1960
700 litre BWE	4	O & K Orenstein & Koppel West Germany	10/1960; 5/1961; 2/1966 & 7/1966
500 litre BWE	2	Friedrich Krupp West Germany	5/1969 & 6/1969
Speader 5,000—8,000 t/h	4	O & K Orenstein & Koppel West Germany	8/1959; 4/1960; 3/1962 & 6/1973
Mobile Transfer Conveyor 2,700 t/h	2	Gerlach Werke West Germany	1/1963 & 4/1963
Bucket Chain Excavator 1,000 t/h	1	Buckau-Wolf West Germany	12/1967
Slewing Belt Tripper	1	Buckau-Wolf West Germany	12/1967
Conveyors	~ 20 km		6/1960
1,000 mm belt width		Friedrich Krupp, Germany	
1,200 mm		Friedrich Krupp, Germany	
1,500 mm		Hewitt-Robins, USA Tata Robins, India	
1,800 mm		Demag, Germany Garden Ship Builders & Engrs., India	7/1976

* bucket wheel excavator

Table 2: Increase of conveyor belt widths and lengths

Width of conveyor (mm)	Length of conveyor (m)						
	1960	1978	1979	1980	1981	6/1982	1/1983
1,000	—	1,838	1,838	1,838	1,838	1,838	1,838
1,200	2,700	5,341	5,151	4,211	1,950	1,915	1,915
1,500	—	8,612	8,348	5,190	3,200	4,980	6,593
1,800	—	2,638	2,643	—	4,000	2,678	2,934
2,000	—	—	2,987	7,050	11,000	11,000	12,540
Total	2,700	18,429	20,967	18,289	21,988	22,411	25,820

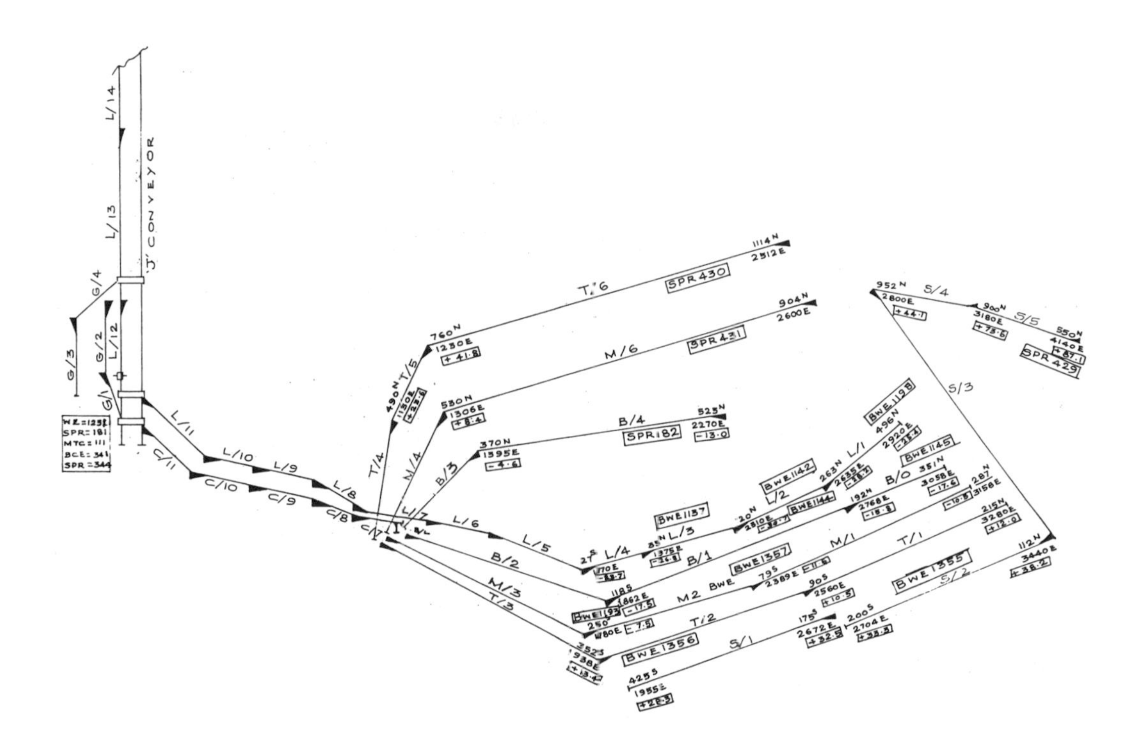

Fig. 4: Location of conveyors in the Neyveli mines as of January 1983

wheel excavators and the 1,500/1,800 mm belt conveyors, thus strengthening the 3 overburden benches (surface, top and middle) by 1,400 litre bucket wheel excavators; 2,000 mm wide conveyors (Steel Cord); 11,000 t/h spreaders etc. The bottom bench 1,500 mm conveyors were also replaced by 1,800 mm conveyors. Table 3 lists this new equipment.

Table 3: Equipment ordered for the First Mine Expansion Scheme

Type of equipment	Number	Manufacturer
1,400 litre BWE*	3	O & K Orenstein & Koppel, West Germany and Southern Structurals, Madras
Spreader	3	Buckau-Wolf, West Germany
Conveyors (2,000/1,800 mm)	15 km	Weserhütte, West Germany
Tripper Car (11,000 t/h)	3	Elecon, India Weserhütte, West Germany
Walking Pads	2 sets	Weserhütte, West Germany
Steel Cord Belting (St 2000/2250)	~ 36 km	Clouth Gummiwerke, West Germany
Electrical Systems		Siemens, West Germany & India NGEF, India (in collaboration with AEG, West Germany)

* bucket wheel excavator

The first system consisting of

— one 1,400 litre bucket wheel excavator (Fig. 6)
— one 11,000 t/h spreader
— one 1,000 t/h tripper and
— a system of 2,000 mm conveyors with a length of 4.5 km (Fig. 7)

was commissioned on August 30, 1978.

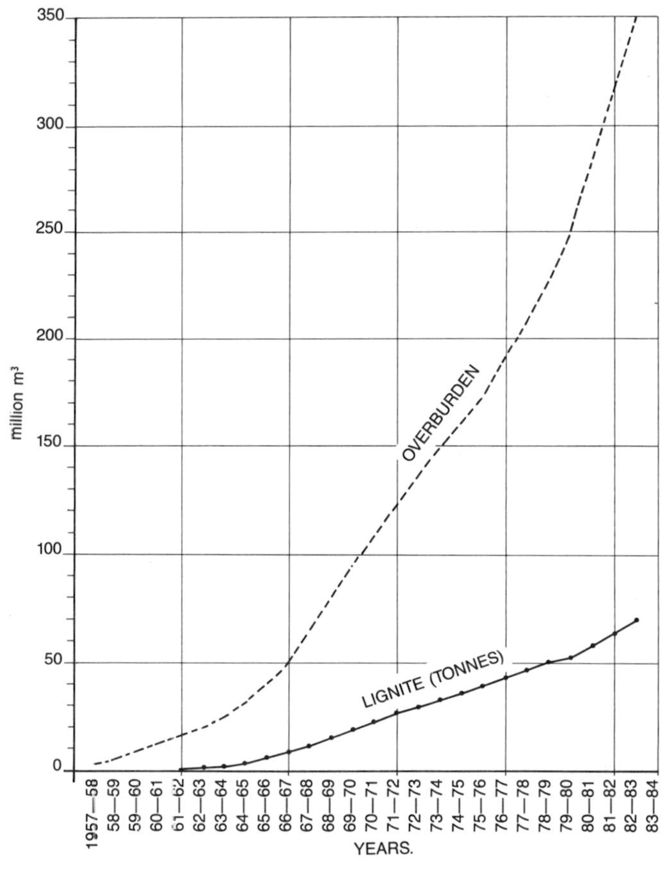

Fig. 5: Cumulative production of overburden and lignite from 1957 until April 30, 1983

Fig. 6: 1,400 litre O & K bucket wheel excavator in operation

Fig. 8: 2,000 mm garland-type overburden conveyor with belt scale for energy consumption studies at various mass flows (top) and the mobile field station housing the monitoring and recording equipment (bottom). (Institut für Fördertechnik und Bergwerksmaschinen, University of Hannover, West Germany)

Fig. 7: 2,000 mm conveyor system with steel cord belting

The second system consisting of

- one 1,400 litre bucket wheel excavator
- one 11,000 t/h spreader
- one 11,000 t/h tripper and
- a system of 2,000 mm conveyor with a length of 4.8 km

was commissioned on May 21, 1979 and the third set of similar equipment with 5.8 km of conveyors on March 29, 1980.

The 1,800 mm conveyors with a length of 2.8 km were commissioned in September 1980 to form a system with two 700 litre bucket wheel excavators and a tripper car procured earlier. By having one more flight of lignite conveyors, continuous lignite supply is ensured to the consuming power plant units. This completed the expansion of the First Mine and at the end of 1981 all units were working to the required capacity.

In order to arrive at optimum operating conditions for the conveyor systems, experience and know-how gained by West German lignite mines, equipment and belt manufacturers as well as by research institutions are systematically used. In this context the investigations and studies undertaken by the Insitute of Materials Handling and Mining Machines (Institut für Fördertechnik und Bergwerksmaschinen) at the University of Hannover, Federal Republic of Germany, should be mentioned. These studies are conducted to investigate the specific energy consumption of the garland-type overburden conveyors with a belt width of 2,000 mm for various mass flows at ambient temperatures of up to 40°C. Fig. 8 shows a section of the investigated conveyor line with belt scale arrangement (top) and the mobile field station containing the monitoring, recording and evaluation devices.

5. Belt Reconditioning Plant

When action was taken to increase the capacity of the conveyor system it was felt that a plant for reconditioning the worn out belts is a must to reduce the cost of belting and to save valuable foreign exchange.

To get the maximum life out of the belting, a Belt Reconditioning Plant was commissioned at Neyveli in 1976. The know-how for this plant was obtained from Clouth Gummiwerke AG, Cologne, West Germany and the plant itself was supplied and erected by Wagener & Co, Schwelm, West Germany.

This plant with a reconditioning capacity of 10,000 to 20,000 m/annum is the only one of its kind in Asia where textile as well as Steel Cord belts up to a width of 2,000 mm are reconditioned. Up to now, already 30,000 m of belting, both textile and Steel Cord, have been repaired. In addition to meeting Neyveli's own demand, large quantities of old belting from other mining projects, ports, cement plants etc. are being reconditioned and rejuvenated. The cost of reconditioning is about 35 to 45% of the original cost belting and most of the reconditioned belts have a life of 60 to 80% of the original life.

In this plant, more than 100 different trials in reconditioning methods, use of various chemicals etc. have been carried out. The deficiency, defects and the quality aspects of belts supplied by different manufacturers are analysed here. Based on observations, Neyveli has now changed the specification of the belting to suit its working conditions and is in a position to suggest suitable optimum specifications to other users, based on the work done in this plant.

6. Power Station

80% of the total lignite production is consumed by the Thermal Power Station. The main technical data of this station are as follows:

Installed capacity	600 MW
Number of turbo generators	6 of 50 MW 3 of 100 MW
Number of lignite fired boilers (outdoor type)	12
Steam generating capacity	220 t/h
Feed water temperature	218°C
Pressure of superheated steam	100 kg/cm²
Temperature of superheated steam	540°C
Lignite consumption per boiler	1,500 t/d
Boiler efficiency	77% on GCV basis

The equipment for the power station was supplied by Technopromexport, USSR and was commissioned from 1962 to 1970.

The power station at Neyveli has many "firsts" to its credit:

- It is the first and only lignite fired thermal power station in India
- It is the first power station in India close to the pit head, demonstrating the advantages of pit head power stations especially when fossil fuels of low calorific value are being used.
- It is the first large modern power station in India which was erected and commissioned by Indian engineers and technicians on their own responsibility under technical guidance by Soviet specialists.

The performance of the Neyveli Thermal Power Station is considered to be the best in the country.

7. Second Lignite Mine

Based on the success of the First Mine, a second lignite mine with an annual capacity of 4.7 million tonnes, about 5 km south west of the existing mine (Fig. 1) was planned and work commenced in April 1981.

Table 4 lists the equipment ordered for Mine II after international tenders had been issued and evaluated.

Table 4: Mining equipment for Mine II

Type of equipment	Number	Manufacturer
1,400 litre BWE*	2	M.A.N., West Germany W.M.I., India
700 litre BWE	2	Buckau-Wolf, West Germany Buckau-Wolf, India
Spreader (11,000 t/h)	2	Buckau-Wolf, West Germany Buckau-Wolf, India
Mobile Transfer Conveyor (4,700 t/h)	2	M.A.N., West Germany W.M.I., India
2,000 mm Conveyor	10.65 km	Elecon, India Weserhütte, West Germany
Crawler Mounted Tripper for 2,000 mm Conveyor	2	Elecon, India Weserhütte, West Germany
1,500 mm Conveyor	5.4 km	Elecon, India Weserhütte, West Germany
Rail Mounted Tripper for 1,500 mm Conveyor	2	Elecon, India Weserhütte, West Germany
1,800 mm Conveyor for Lignite	3.7 km	Southern Structurals, India 2. O & K, West Germany
2,000/1,800/1,500 mm Steel Cord belting	50 km (approx.)	Clouth, West Germany Phoenix, West Germany

* bucket wheel excavator

The first 1,400 litre bucket wheel excavator, manufactured by M.A.N. and WMI Cranes, India, moved out to Mine II on March 28, 1983 and was commissioned on April 14, 1983 (Fig. 9).

Fig. 9: 1,400 litre M.A.N. bucket wheel excavator moving to Mine II

About 2.2 million m³ of overburden have already been removed from the initial mine cut area with the help of conventional equipment such as shovels, dumpers etc. A quantity of 22.3 million m³ of overburden is to be removed initially to expose the lignite in the new mine. The mine is expected to reach the full production stage of 4.7 million t/annum in 1987/1988.

The lignite is expected to be exposed by the end of 1984 to commence supply to the first 210 MW unit of the 630 MW Second Thermal Power Station now under construction as a pit head power plant.

8. Expansion of Mine II

To meet the increasing power demand of the State of Tamil Nadu in particular and of the Southern Region in general, it is proposed to expand the Second Thermal Power Station to 1,470 MW capacity. To meet the lignite requirements of the expanded power plant, it is proposed to increase the second mine's capacity to 10.5 million t/annum. The above expansion has already been approved in principle by the Government of India. Equipment for this expansion scheme is in the tendering stage at present.

9. Third Lignite Mine

Plans exist for the opening up of a Third Lignite Mine (see Fig. 1) at Neyveli for an annual capacity of 14.0 million tonnes to cater for a Super Thermal Power Station of 1,500 MW capacity and a few other ancillary industries to be erected at Neyveli.

10. Conclusion

The development of the Neyveli lignite fields using the continuous mining technology have been described. Due to space restrictions only a few of the salient features of this project could be covered, just to prove that an enormous potential lies in the Neyveli Project — for 300 years more!

References

[1] Kasturi, T. S. and Lachmann, H. P., "Correctives for Conveyor Belts in Operation in Developing Countries", Braunkohle (1979), January/February

[2] Kasturi, T. S., "Computerization in the Neyveli Lignite Mines on the Expansion Scheme, Equipment Erection", Braunkohle (1980), December

[3] Kasturi, T. S., "Development of Bucket Wheel Teeth in the Neyveli Lignite Mines to Meet the Hard Overburden Condition", Braunkohle (1980), August

[4] Kasturi, T. S., "Neyveli — Yesterday and Today", Braunkohle (1981,) August

[5] Kasturi, T. S., "General Development of the Neyveli Lignite Corporation — Tomorrow", Braunkohle (1982), August

[6] Kasturi, T. S., "The Most Modern and Largest Conveyor Transport System in Asia", Mine Tech (India) Vol. 5, No. 1, (1981), March

New Excavators for a New Open-Cast Mine at Neyveli

H. Grathoff, Germany

1. Introduction

The only fossil fuel resources in the southern Indian state of Tamil Nadu are located at Neyveli, roughly 200 km south of Madras, and are in the form of large lignite deposits, operated by the Neyveli Lignite Corp. (NLC). So far, an aggregate power station output of approx. 3,300 MW has been installed in this state, 600 MW of which is from Neyveli. Noteworthy in this connection is the Neyveli Power Station which, because of its high availability of over 90%, makes an over-proportionate contribution to the annual energy production.

However, because of the ever-increasing energy demands of steadily growing industry, particularly in the Madras conurbation, the gap between the demand for and supply of power has been widening. This prompted the Indian Government to have a second power station for 630 MW built at Neyveli. The foundation stone was laid in 1981. To supply this power station, plans were made for a second open-cast mine to be situated near the first one and having a capacity of 4.7 million tonnes.

For the 2 bucket wheel excavators and 2 crawler-mounted transfer conveyors needed to open up this mine, the NLC placed an order in 1980 with a consortium of M.A.N. Nürnberg and WMI Bombay who shared the work as follows.

M.A.N. accepted the total responsibility for the planning, design and basic engineering of the machines and supplied such components as the bucket wheel head, including the complete bucket wheel, its shaft, mounting and drive unit, the slew motions with ring gear and large-diameter ball race, the hoist drive, the crawler drives and the steering mechanism.

The detail engineering of the WMI content was largely done in Madras in an office established for this purpose with the assistance of a M.A.N. engineer. Engineers were also delegated to transfer the necessary know-how for fabrication of the large Indian scope of supply, which involved the crawler chains, crawler beams, hoist motion components, integral conveyors and the complete steelwork. M.A.N. engineers and chief erectors took care of supervising the erection work as well.

Dipl.-Ing. Hartmut Grathoff is Department Manager, M.A.N. Maschinenfabrik Augsburg-Nürnberg AG, D-8500 Nürnberg, Federal Republic of Germany

Although the approx. 2,400-tonne unit weight of the individual NLC excavator does not come anywhere near the size of the excavators working in the German coal fields (M.A.N.'s largest weighs 13,000 tonnes), it is nevertheless possible to point to Neyveli with pride and say that there is none larger in all of Asia.

2. Mining Conditions at Neyveli

As beneficial as the lignite deposit in Neyveli is, it is all the more difficult to get at. The overburden is without doubt the toughest there is in comparable mines anywhere in the world. The major part is made up of the "Cuddalore" sandstone which combines several extremely adverse properties. It is above all very hard, which results in unusually high digging forces. In general, it can only be excavated economically if it is loosened beforehand by blasting. Besides that, the "Cuddalore" sandstone is very abrasive and therefore demands specially designed teeth and hard-facing on the buckets and bow cutters [1].

The third disadvantage of this overburden is that it is exceptionally sticky. The fine particles are very adhesive and steadily build up in the buckets and ring spaces during operation, even if the material appears to be completely dry. For this reason, it is necessary every day, even under relatively favourable conditions, to shut down the entire system, consisting of excavators, conveyors and stackers, in order to replace worn teeth and remove the encrustations building up on the buckets and bucket wheel.

The unusual conditions attributable to the Neyveli overburden also made it necessary to apply special design features in the excavators. Compared with other bucket wheel excavators, those in Neyveli are particularly rugged in design. Since the digging forces are very high and, on top of that, occur very suddenly, it was imperative that the bucket wheel be arranged on the centreline of the boom system and not toed in or tilted as is customary practice (Fig. 1). The result of this exactly concentric arrangement of the bucket wheel is that sudden digging forces are distributed uniformly over the entire supporting system. The stress range, i.e., the quotient from maximum and minimum stresses, is thus very low in the overall load bearing system, thereby ensuring good fatigue strength of the steel structure even if the loading pattern, i.e. particularly the frequency and amplitude of the peak loads, is unfavourable and uncertain.

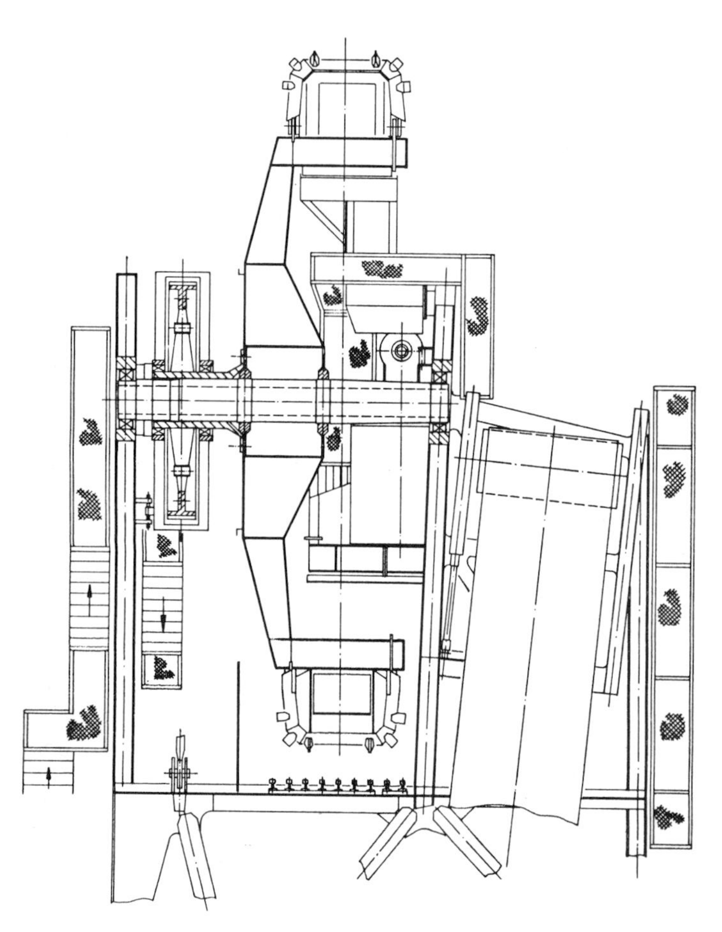

Fig. 1: Bucket wheel arranged on centreline of boom system

The ideal bucket wheel arrangement on the boom centreline can only be realized with a rotary plate feeder. This feeder takes the material from the bucket wheel centre with low loss of height and transfers it to the belt conveyor provided at the side of the bucket wheel (Fig. 2).

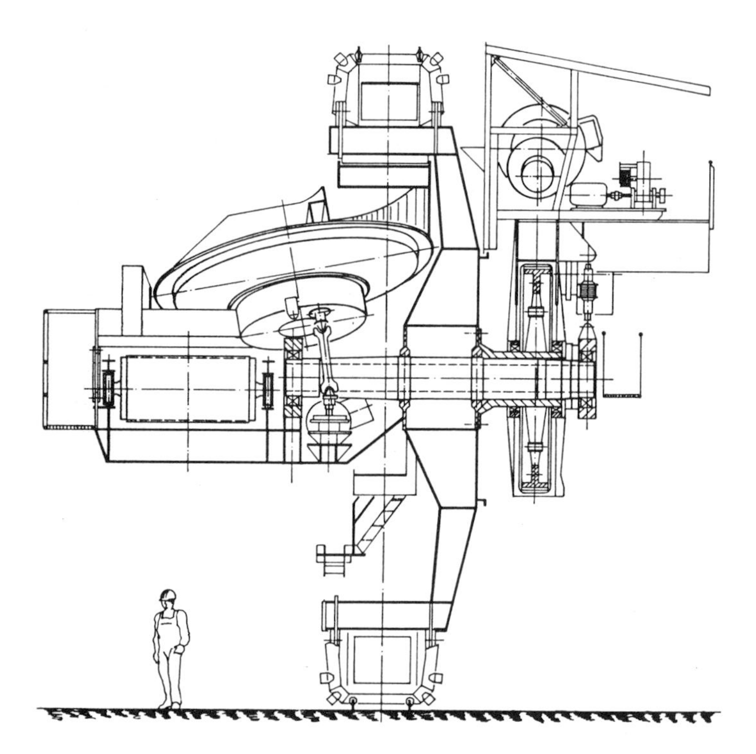

Fig. 2: Cross section of bucket wheel boom head

3. Bucket Wheel Drive

Customarily, there are two drive arrangements for excavators of this type and size, i.e. with the drive either "overhung" or "riding" on the bucket wheel shaft. For Neyveli, M.A.N. selected the second drive arrangement. The main advantage of an overhung bucket wheel drive is that it can be installed and removed as a complete unit. However, with a drive of this size the advantage of such an arrangement is marginal because replacement of a complete drive unit is out of the question. Besides, there is hardly any possibility of using a sufficiently large crane and transport vehicle in the open-cast mine for this purpose.

The riding arrangement, where the drive is located between the two bearings of the bucket wheel shaft (see Fig. 2), ensures approximately equal loads on the bucket wheel bearings, thereby making it possible to have bearings of identical size. It also affords the largest possible bearing centre distance. The especially dangerous and frequent peak loads for the bucket wheel are uniformly distributed between the two bearings and transmitted to the steel structure.

The gearwheel transfers the drive torque directly to the bucket wheel so that the bucket wheel shaft is subjected only to bending stresses. The ring gear of the gearwheel can be replaced without having to remove the gearbox. It is hardened and tempered for high strength and has large reserves against forced fracture. The ring gear is driven by 4 pinions which are connected in pairs to a reduction gear with a differential to ensure that there is complete load balance between the two pinions. The reduction gears can be removed individually and are interchangeable. In cases of emergency, the excavator can continue operation at 50% load on only one reduction gear.

The bucket wheel is powered by two squirrel-cage motors, each with a capacity of 750 kW, via turbo-couplings which primarily serve as overload trip devices, as opposed to the turbo-couplings on the conveyor drives which essentially serve to facilitate starting. In both cases the continuous slip of the turbo-couplings provides torque compensation because the squirrel-cage motors with their steep characteristic are coupled mechanically.

For a bucket wheel excavator of this size, an input power of 1,500 kW is extraordinarily high. In terms of kW per m³ of material, it is double the connected load for a large excavator operating in German lignite mines.

A primary consideration in designing the bucket wheel drive was to keep the rotating masses between the bucket wheel and the overload trip as low as possible and thereby minimize the impact resulting from sudden blockage of the bucket wheel. Because the masses rotating at high speeds play such a decisive part, no brake wheel was provided on the gear input shaft. The brake for the bucket wheel drive is used only for erection and maintenance purposes. For this reason, a brake was provided only on the bucket wheel auxiliary drive which is not connected to the bucket wheel under normal operating conditions.

4. Crawler Travel Motion

The excavator travels on 6 crawlers which are arranged in pairs at each portal leg. One 2-crawler assembly is rigidly connected, whereas the other 2-crawler assemblies are arranged one behind the other and have interconnected steering. Both steering bars are connected mechanically according to German Patent 2949279 and are supported together in the vertical plane by a wheel in a channel. In the horizontal plane they are held and moved by a hydraulic cylinder. The hydraulic power pack is equipped with 2 independently driven pumps. If one pump system fails, the cylinder continues to move with the same force but at half the speed. Each of the 6 crawlers is powered by a drive unit with a compact planetary worm gearing.

5. Slew Motion

The slew motion for the revolving superstructure consists of two compact units, each having two pinions which engage the large-diameter ring gear (Fig. 3). This ring gear is 11.4 m in diameter and is made up of bolted segments. Each segment is fastened to the steel structure by means of three large pins (Fig. 4). The holes in the segments are larger than the pins themselves. The space between the edge of the hole and the pin is filled with a grout. In the case of Neyveli, final alignment of the ring gear and subsequent grouting of the pin holes were performed only after the excavator had been completely erected. This made it possible to compensate for the radial deformation of the steelwork due to the excavator's deadweight. If this had not been done, the deviation from the ideal circular form would have been several millimetres.

The slew motion drives are units consisting of motor, brake, overload coupling, planetary gearing with differential and two drives, as well as the two drive pinions. These slew motions can be installed and removed with relatively little difficulty and, for example, a replacement unit can be installed without the need for any matching work (see Figs. 3 and 4). This type of slew motion was developed by M.A.N. for the large-scale excavators in Germany [2].

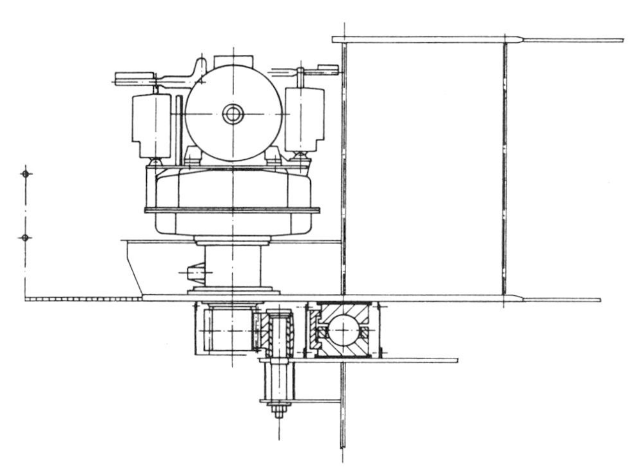

Fig. 3: Large-diameter ring gear and slew motion of revolving superstructure

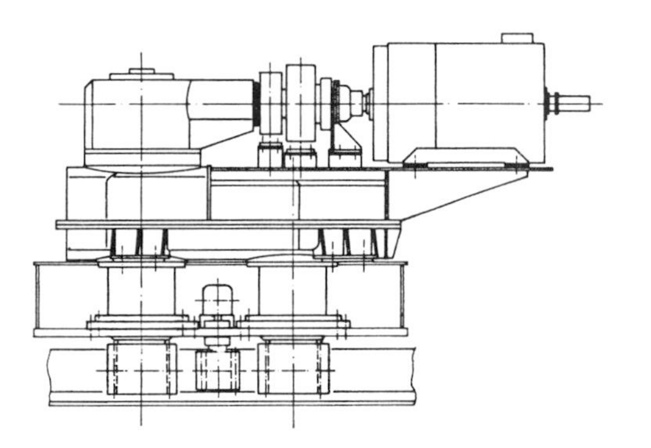

Fig. 4: Slew drive

The slew bearing for the superstructure is a M.A.N.-made single-row ball bearing with a mean diameter of 10 m and ball diameter of 250 mm. Instead of latches, this excavator has a retainer ring on the ball race to prevent the superstructure from tipping over under extraordinary conditions. However, before this retainer ring becomes effective, a limit switch trips all major excavator motions. The ball race is embedded in a 3-component grout. Before it is embedded, the ball race is not made completely level but given a camber which is determined by computer. This causes the peak loads to be reduced at the so-called "hard spots" — which are, in particular, the portal legs for the undercarriage and the pylon connections for the platform. This arrangement results in a significant extension of the life of the ball race.

The discharge boom is moved by a separate slew motion which is arranged under the portal. It is completely independent of the main slew motion and operates only when the discharge boom has to be moved.

6. Hoist Motion

As is customary practice, the bucket wheel boom has a two-rope hoist motion to make sure that the boom is held with sufficient safety in case one system fails. The discharge boom is operated by a hydraulic hoist motion. As in the case of the hydraulic steering motion, the power pack is equipped with two pump systems so that in the event that one system fails, it is possible to continue operation at half speed (Figs. 5 and 6).

7. Actual Operation and Performance

The start of operation for the second open-cast mine posed considerable problems for NLC. The properties of the overburden proved to be quite different from those in the old mine. The soft layer lying above the infamous "Cuddalore" sandstone is substantially thicker and extremely sticky. Attempts to dry out this layer by drilling drainage wells proved fruitless. The uppermost layer of approx. 5 m thickness could be excavated fairly easily. It was relatively dry — no doubt due to the drought that has been affecting southern India for years. However, operation below that layer was incredibly difficult. The bucket wheel became clogged after a very short time. In a matter of hours the 1,400-l buckets were completely packed with deposits so that they could no longer carry any material. M.A.N. first tried to line the buckets with wear-resistant 2-component plastic liners and glued-on rubber mats. Then an attempt was made to use cage-like buckets made only of narrower cutting strips and rods instead of closed sidewalls. Success was finally achieved by lining the buckets with polyethylene panels approx. 10 mm thick. This material is distinguished by its marked cross-linked molecular structure. The molecular weight is between one and five million which means the material is very resistant to wear. The most important factor, however, is that it extensively prevents deposits from sticking in the buckets. If material does start to build up now and then, it either dislodges itself quite easily or at least can be removed without much difficulty. In this sticky overburden it was soon found out that the bucket wheel worked better without bow cutters.

The second excavator which was put into operation on the second level had to excavate mostly "Cuddalore" sandstone

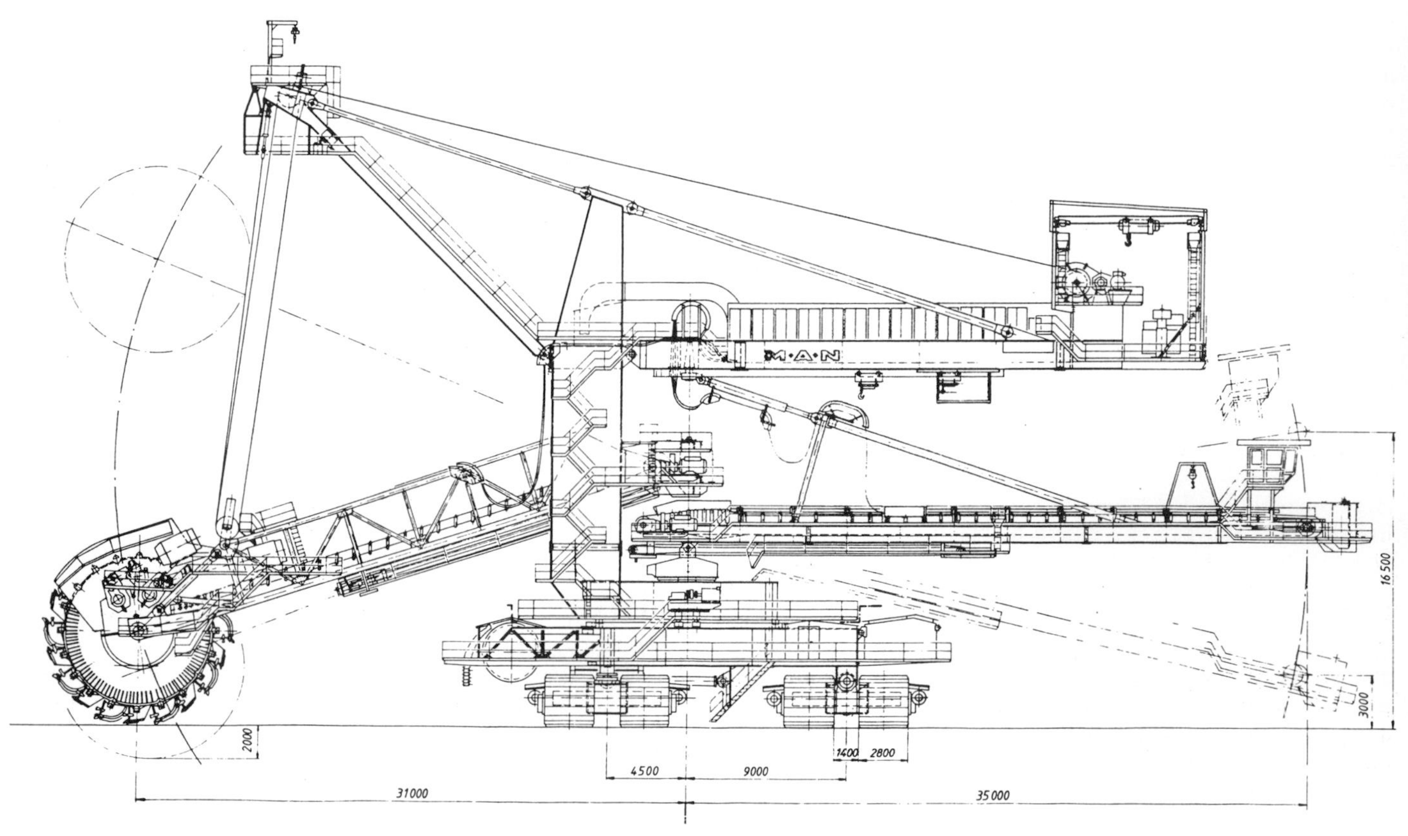

Fig. 5

Fig. 6

which posed no problems — after all the machines are designed to cope with this material.

One of the conditions of the contract was to prove that the excavators could develop a peak capacity of 2,980 m^3/h with solid overburden in an 8-h test. In addition, a mean handling rate of 2,250 m^3/h solid overburden had to be attained in a test, together with the conveyor systems and stacker, lasting 1,000 operating hours.

Despite enormous difficulties with the unexpected properties of the material, both tests proved successful right off the bat.

References

[1] Kasturi, T.S., "Development of Bucket Wheel Teeth in the Neyveli Lignite Mines to Meet the Hard Overburden Condition", *Braunkohle*, Special Issue, pp. E14—E20, 1980

[2] Grathoff, H., "Neue Antriebe für Großbagger (New Drives for Large Excavators)", *fördern und heben*, 26, No. 2, pp. 122—126, February, 1976

All articles published in this book have appeared originally in the bi-monthly journal **bulk solids handling** — The International Journal of Storing and Handling Bulk Materials, the leading technical journal in this field throughout the world. If you are interested in receiving a free sample copy or information on subscription rates, please write to:

Trans Tech Publications, P.O. Box 1254, D-3392 Clausthal-Zellerfeld, Federal Republic of Germany.
Telex: 9 53713 ttp d

 Volume 4, Number 3, September 1984

Bucket Wheel Excavators for Hard Mining Operations at Neyveli

M. Mildt and R. Trümper, Germany

Summary

Almost three decades ago, the first bucket wheel excavators were ordered for mining the lignite deposits at Neyveli in South India. In spite of the well known difficult mining conditions, caused by heavy monsoon rains and very hard overburden which in some parts is also sticky, the mining technique with continuously operating bucket wheel excavators has proved to be economic and advantageous. This article reviews the development of the bucket wheel excavator technique at the Neyveli Lignite Corporation from the beginning until today and presents an outlook on further possibilities for development in the future.

1. Introduction

In 1937, the first lignite deposit was discovered by well sinkers near the present township of Neyveli in the state of Tamil Nadu in the south of India. Towards the end of World War II, it was thoroughly investigated over an area of 50 km² and initially a mining field of 14 km² was ascertained. Today, a field extending over 250 km² and containing over 1.5 billion t of coal, including 1 billion t which is considered to be economically mineable by the open-cut method, is known. The coal is covered by alternating layers of clay and sand, with the sand in some parts being consolidated to an extent that sandstone consistency is reached. Therefore, the digging forces required average about 3—4 times those needed in German overburden mining operations.

As materials of this consistency had not previously been cut by bucket wheel excavators, new measuring methods had to be developed which made it possible to calculate the digging forces of the bucket wheel drives for this application. For this purpose LMG developed a method of determining the cutting pressure by which hard and consolidated soils could be classified.

In 1956, instructions were given to a British engineering firm. The project provided envisaged opening up the mine by conventional, mobile equipment such as scrapers, Euclid loaders and bulldozers.

Dipl.-Ing. Martin Mildt is Project Manager and Dipl.-Ing. Reinhard Trümper is Project Engineer, both with O & K Orenstein & Koppel AG, D-2400 Lübeck, Federal Republic of Germany

Even at that time, it became apparent that the loaders were hardly able to cope with their task. Not only were they much inferior to the bucket wheel excavators as far as digging force and outputs obtainable were concerned, but also with regard to economy.

2. Mine I

2.1 Equipment with 350 l BWEs and 700 l BWEs

In September 1956, Orenstein & Koppel AG (then LMG Lübecker Maschinenbau AG) received a contract for all the equipment on the winning and dumping sides. On 3 overburden digging faces, two 700 l BWEs and one 350 l BWE were to be provided.

The overburden from the upper and central mining benches was transported to the dump by means of conveyors. The overburden of the lower bench was planned to be dumped by direct overthrow via a mobile transfer conveyor and a discharge conveyor unit. The pre-dump built up in this way was followed by 2 benches; from the lower bench a low dump was built up by a spreader and from the top bench a high and a deep dump also by means of a spreader. For mining the coal, a second 350 l BWE was provided (Fig. 1), capable of winning the coal by high and deep cutting and delivering it onto a coal conveyor.

For the first 5 years a production of 3.5 million t of coal had been planned per year, while later (from 1963) production was to be stepped up to 6 million t/year. In addition to a fertilizer and a briquetting plant, a 250 MW power station was to be supplied.

In 1964, Orenstein & Koppel delivered two further 700 l excavators (Fig. 2), so that a total of four 700 l BWEs were put into operation and the 350 l BWE which until then had been working in overburden, was also mining coal. With this equipment combination Mine I was operated for another 12 years until 1976.

Owing to the difficult conditions in the mine caused by hard and abrasive overburden and adverse climatic conditions (maximum precipitation in the monsoon season is approx. 230 mm/day compared with values in Germany of 40—50 mm/day, i.e. 4—5 times as high as in the traditional German mining areas where bucket wheel excavators are

Fig. 1: Bucket wheel excavator SchRs $\frac{350}{5} \cdot 12$

Technical data:

Theoretical capacity	1,500 m³/h
Outreach of bucket wheel boom	17.2 m
Outreach of discharge boom	20.0 m
Belt width	1,400/1,200 mm
Digging height	12 m
Service weight	approx. 470 t

Fig. 2: Bucket wheel excavator SchRs $\frac{700}{3} \cdot 20$

Technical data:

Theoretical capacity	3,570 m³/h
Outreach of bucket wheel boom	25.5 m
Outreach of discharge boom	35.0 m
Belt width	1,800 mm
Digging height	20 m
Service weight	approx. 1,300 t

used), and due to a deteriorating overburden: coal ratio, the coal production of 6 million t/a originally planned could not be reached, but varied between 3 and 4 million t/a. To meet the increasing demand for energy, it was decided in 1974 to extend the open-cut by the Mine I extension.

2.2 Development of Digging Elements

As the mine was initially opened up with the 350 l BWEs which were to be used in coal later, it was possible to carry out tests with these machines aimed at developing the best possible design for the cutting elements.

On the basis of the cutting resistance value obtained in laboratory tests, it appeared from the soil samples taken at Neyveli that the values of hard clay known until then and which could be cut by bucket wheel excavators were far exceeded. In addition to this, the overburden — known as Cuddalore Sandstone — is extremely abrasive.

2.2.1 Design of the Teeth

In the beginning, teeth consisting of two parts were used to allow quick replacement of the tooth points. The cast steel tooth adapter was riveted to the bucket lip and the tooth point with conical internal seating was fastened on the tooth adapter by means of a wedge. However, before long it became apparent that due to the hard cutting conditions, the conical seating face became worn and the wedge connection loose. This resulted in premature loss of the tooth tip and damage to the cone of the tooth adapter as well. These disadvantages made it necessary to introduce a tooth in one piece which could be changed quickly and fastened by means of a torque wrench to the bucket lip by means of high tensile bolts with a two-shear connection. This solution has to date proved its worth (Fig. 3).

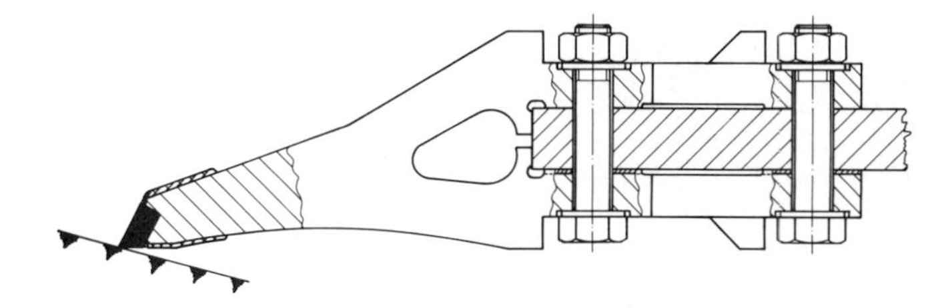

Fig. 3: Typical Neyveli tooth for hard cutting conditions

2.2.2 Tooth Shapes

The so-called spade shape of the teeth (Fig. 4), known until that time with cutting edges parallel to the lip edge, made penetration into the digging face more difficult and resulted in low production. In addition, the life of the tooth point (consisting of cast steel with 2% manganese and armouring of the cutting edge with wear resistant electrode metal with a proportion of approx. 35% chrome) lasted only about one shift and was altogether unsatisfactory.

A further test was made with teeth having cutting edges positioned at right angles to the bucket lip and, consequently, also at right angles to the digging face, but did not result in any significant improvement in production or tooth life. Although the so-called diamond teeth (Fig. 5) had the advantage of reducing the specific power required from the bucket wheel drive motors, they increased the loads on the slewing gear motor.

From the knowledge gained from tests with the two tooth shapes, wooden models on the scale 1:1 were developed in the carpenters' shop of the maker O & K — a so-called harp model (Fig. 6). The strings represent the cutting direction

Fig. 4: Bucket with spade shape teeth

Fig. 5: Bucket with diamond shape teeth

Fig. 6: Wooden tooth models in "harp model"

resulting from circumferential and slewing speeds with the buckets slewing either to the right or left.

Taking the free-cutting, cutting and lip angles — familiar from turning tools — into consideration, the tooth outfit which is still giving good service has been developed for the Cuddalore Sandstone after various tests (Fig. 7).

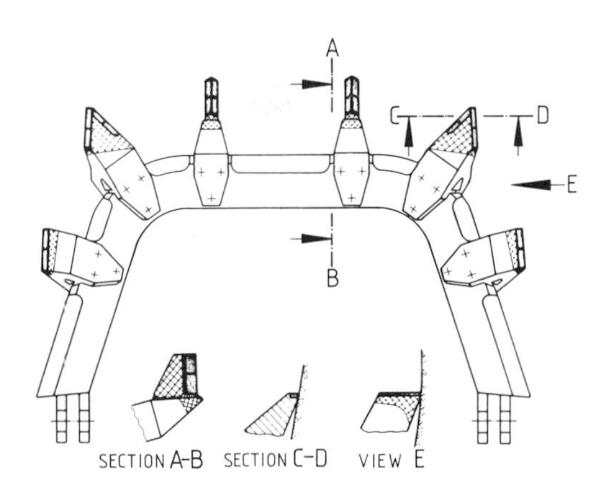

Fig. 7: Teeth arrangement on buckets for hard soil

The outfit consists of two side teeth, a specially shaped corner tooth on the right and left sides and two ripper teeth in the centre which are provided to facilitate penetration into the hard digging face.

The results of the tests with the 350 l BWE whose bucket wheel was not vertical relative to the digging face, showed further that this inclined position of the bucket wheel had a rather negative effect on the operation of the bucket wheel excavator when mining hard material. For this reason, the 700 l and 1,400 l BWEs delivered later by O & K subsequently were supplied with their bucket wheels absolutely vertical to the face. This was possible owing to the design with a rotary plate as the element discharging the material from the bucket wheel to the belt on the bucket wheel boom. The design of the rotary plate allowed a relatively small bucket wheel and thus a relatively low total weight of the machine.

With this design of the teeth, tooth position and additional shock blasting of the bench in hard material and the vertical position of the bucket wheel relative to the face, the highest output (production) was obtained with the minimum power required from bucket wheel and slewing gear motors.

2.2.3 Protection against Wear

The life of the teeth was considerably increased by soldering tungsten carbide inserts onto the cutting edges and also a wear protection (fastened to the face by autogenous welding) consisting of a highly wear resistant wolfram alloy in which wolfram carbides of 0.5 mm grain size are embedded.

In the Cuddalore sandstone of Neyveli, tooth life of approx. 150 h (ripper teeth) to approx. 500 h (corner teeth) is obtained and even then such worn teeth are reconditioned in the mine's own workshop.

To find out the most suitable quality of tungsten carbide inserts, a number of tests was carried out.

It is a known fact that with increasing hardness and therefore life, the toughness of tungsten carbides decreases. A medium quality (Vicker's hardness 1,320—1,360 HV_{30}, bending strength 220—190 kp/mm²) for mining purposes has prevailed at Neyveli which still has adequate toughness to prevent chipping off when hard layers in the digging face are penetrated.

All the trials to obtain the best possible tooth shape were carried out from the beginning when the open-pit at Neyveli was opened up with 350 l BWEs at the end of the fifties in

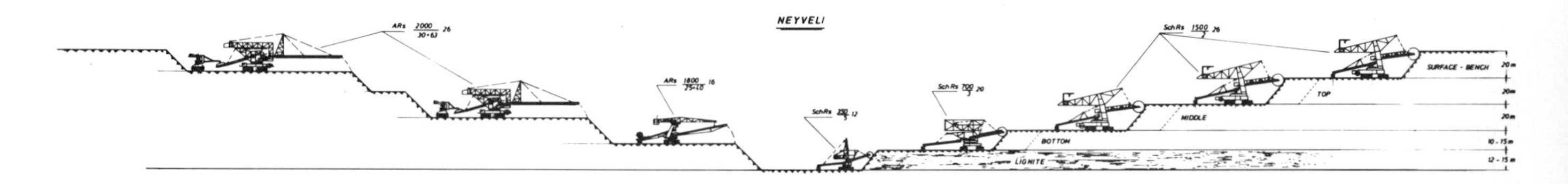

Fig. 8: Mining scheme of Mine I extension

cooperation with the supplier of the bucket wheel excavators O & K and the engineers of NLC. The knowledge thus gained was used later as a basis for designing the digging elements of the 700 l and 1,400 l BWEs.

These days, the replacement teeth are manufactured in India and the workshop of NLC which has been suitably equipped to carry out all work in connection with reconditioning of worn bucket teeth.

3. Mine I Extension

3.1 Equipment with 1,400 l BWEs

For the Mine I extension it had been planned to mine the overburden on 4 benches and the lignite on 1 bench (Fig. 8). For each of the three upper benches a new bucket wheel excavator with an effective output of 2,250 bank m³/h (1,400 l BWE) was required (Fig. 9).

Fig. 9: 3 O & K 1,400 l BWEs

The mined overburden is delivered via 3 conveyors with 2,000 mm belt width to the dump site where it is dumped by 3 spreaders. Each spreader has a capacity of 11,000 t/h. On the lower overburden bench, 2 old bucket wheel excavators with an effective output of approx. 1,100 bank m³/h are used (700 l BWEs). This overburden is transported to the dump via a conveyor system with a 1,800 mm belt width and it is dumped there by a spreader with a capacity of 7,000 t/h. On the bottom bench, the coal is mined with the remaining old equipment, with existing 1,500 mm conveyor systems being used instead of 1,200 mm conveyors.

3.2 1,400 l Bucket Wheel Excavator

The bucket wheel excavator is designed for an effective output of 2,980 bank m³/h which was to be proved in an 8-hour test. The theoretical capacity is 5,460 loose m³/h based on a bucket capacity of 1,400 l net (Fig. 10).

Fig. 10: Bucket wheel excavator SchRs $\frac{1400}{2} \cdot 26$

Technical data:

Capacity	5,460 m³/h
Outreach of bucket wheel boom	31.3 m
Outreach of discharge boom	30 m
Belt width	2,000 mm
Digging height	26 m
Service weight	approx. 2,200 t

The whole machine has a service weight of approx. 2,200 t.

As the mining conditions for this excavator far exceeded the degree of difficulty known up to that time, in particular as regards the hardness of the material to be mined, a concept had to be chosen for the excavator which would cope with these conditions.

In addition to the buckets, the bucket wheel was equipped with pre-cutters to obtain adequate lump sizes. Inside the bucket wheel, the material was transferred to the belt in the bucket wheel boom via a rotary plate. As already mentioned, due to a rotary plate, a bucket wheel of fairly small diameter converting the relatively high torque at the bucket wheel shaft into great digging forces could remain vertical in a vertical position and always at right angles to the cutting edge. The bucket wheel gear was driven with 2 x 750 kW motors.

At the centre of the machine the material is transferred to the discharge belt which is slewable by 95° to either side.

The bucket wheel boom is held by a 2-rope winch installed on the upper platform of the counterweight girder.

The complete superstructure of the machine is slewable on a ball bearing through 360°. The slewing gear consists of two slewing gearboxes arranged exactly opposite each other. The discharge boom slewing mechanism is also connected to the main slewing mechanism and ensures a constant position of the discharge belt boom, independent of the respective position of the superstructure, by means of a superimposed drive. The sub-structure of the machine is supported on a 6-crawler travel gear. Each crawler has its own travel drive. The travel gearboxes are designed as planetary gears with worm input step and are interchangeable.

3.3 Long-Term Performance Test

To prove the long-term performance with regard to output and availability of bucket wheel excavators in combination with a conveyor system, tripper car and spreader, the three 1,400 l BWEs ordered in 1975 by NLC from O & K for the First Mine extension were subjected to an 8-hour peak load test and a 1,000 h long-term performance test.

With a theoretical capacity of 5,460 m³ (loose material) — only the net bucket content is calculated — the following were specified in the contract:

Peak load test	minimum 2,980 bank m³/h ≙ 4,320 loose m³h
Long-term performance test	minimum 2,250 bank m³/h with an effective operating time of 1,000 h in a time of 3 months with 3-shift operation and 7 working days per week.

The following describes the experience gained with the first system:

The peak load test resulted in a production of 3,329 bank m³/h, i.e. 12% extra production.

The long-term performance test was carried out during the period 22.2.79—5.5.79 and the production achieved was 2,650 bank m³/h, i.e. 18% extra production.

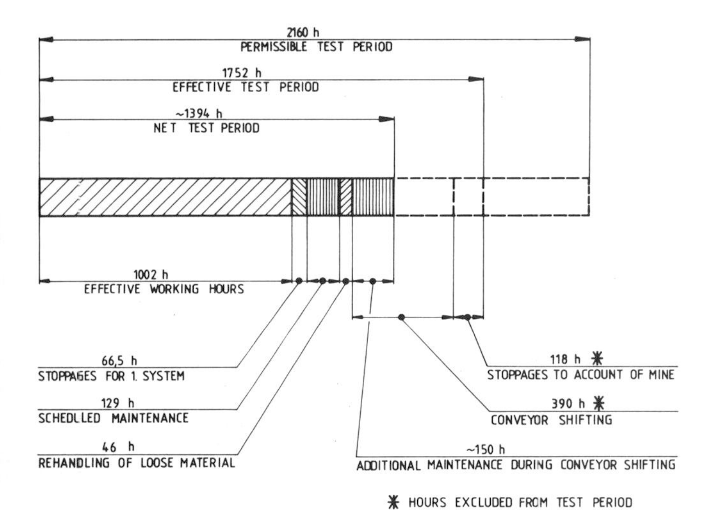

Fig. 11: Bar chart test period

According to the contract, the times of shifting the conveyor system as well as downtimes for which the mine was responsible were not to be counted in with the total testing time. If it is further considered that during shifting periods some maintenance work was carried out, the testing time amounted to approx. 65% of the specified minimum time.

The bar chart (Fig. 11) shows that, in addition to the effective working time of the system of *1,002 hours*, no more than *66.5 h* downtime have occurred due to mechanical and electrical faults, i.e. approx. 6.2% of the total operating time of the system which was available.

The non-operating times of the equipment of the system were made up as shown in Table 1.

Table 1: Non-operating times of the system equipment

	Mechanical		Electrical		Total	
	h	%	h	%	h	%
Bucket Wheel Excavator (O & K Orenstein & Koppel)	15.63	1.46	10.88	1.02	26.51	2.48
Conveyor System incl. Tripper Car (Elecon-Weserhütte)	7.52	0.70	9.29	0.87	16.81	1.57
Spreader (Buckau-Wolf)	19.74	1.85	3.48	0.32	23.22	2.17
Totals	42.89	4.01	23.65	2.21	66.54	6.22

These are percentages which refer to the effective operating time, plus breakdown periods.

The electrical equipment was delivered by Siemens India and Siemens Germany.

Fig. 12 shows from bottom to top the effective operating hours per day and from top to bottom the actual maintenance time per day, so that the daily breakdown times appear between the bottom and top curves.

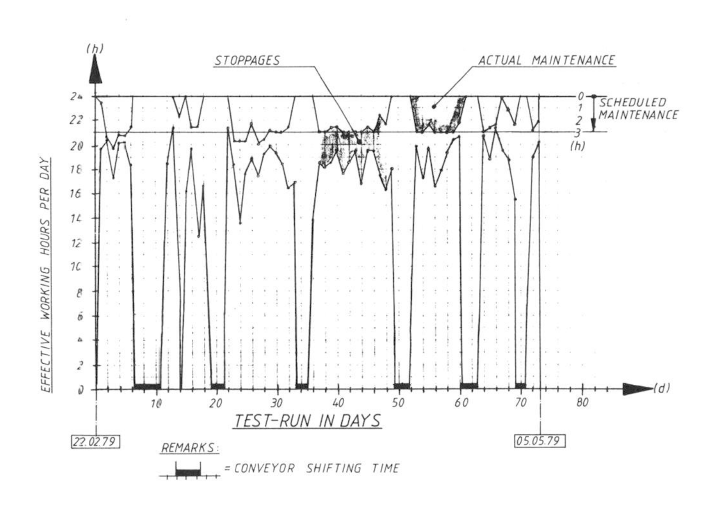

Fig. 12: Time diagram, long-term performance test

The long-term performance test was carried out by operating personnel of NLC. The suppliers of the equipment machines and equipment for this system, associated in a consortium for the purposes of this test, delegated supervisory personnel, so that the machines whose efficiency the suppliers had to guarantee were supervised around the clock as far as mechanical and electrical equipment was concerned.

Volume 4, Number 3, September 1984 bulk solids handling

The performance tests of the second and third system were concluded with similarly good results, so that the contractual obligations towards NLC were fully met.

4. Mine II

In 1977, the Indian Government decided to develop a second mine to supply a 630 MW power station with coal. With an overburden: coal ratio of 4.8:1, 187 million t of lignite were available in the second mine to be won at a rate of 4.7 million t/a. The overburden had a max. thickness of 75 m, while the max. thickness of lignite was 18 m.

In 1980, the Neyveli Lignite Corporation ordered the equipment required for Mine II. The overburden was to be mined on three levels. One 1,400 l BWE each was operating on the two upper benches. The excavators had mainly the same characteristics as the 1,400 l machines of the Mine I extension.

Via two conveyor systems with a 2,000 mm belt width, the overburden was transported to the respective spreader designed for an output of 11,000 t/h. On the bottom overburden bench a 700 l BWE was used, delivering via a mobile transfer conveyor and 1,500 mm wide conveyor system onto a spreader already available from Mine I.

On the coal face a new 700 l BWE and an old 350 l BWE still available from Mine I were used. The coal was transported via an MTC and a 1,500 mm wide conveyor system rated for 4,700 t/h.

5. Further Possibilities for Development

5.1 Application of a BWE with Conveyor Bridge

With increasing size of equipment, the investment costs also rise considerably. To ensure economic operation of a system, it is necessary to minimize losses in output as far as possible. If, in the case of 3 small systems for instance, one system should fail, total production will only be reduced by 1/3; however, in the case of a large system, all the follow-on equipment is also forced to stand still and production drops by 100 %.

An important point for reducing non-operating times is to avoid freely moving transfer points in the conveying path.

The outreach of small excavators is increased by using mobile transfer conveyors, whereas conveyor bridges are normally used in conjunction with giant equipment.

The size of bucket wheel excavators for which a conveyor bridge can be used with advantage fluctuates, but can be assumed to apply to bucket wheel excavators with long booms with a range between 1,600 mm and 2,200 mm belt width. An example of the economy of a bucket wheel excavator with a bridge, even with a narrow belt width, is the BWE type SchRs $\frac{900}{7} \cdot 30$ which was delivered by O & K to the People's Republic of Poland in 1981 (Fig. 13).

5.2 Application of a BWE for 100,000 bank m³/day

Owing to the steadily rising demand for energy on the Indian continent, it is to be expected that the projects which are about to materialize at present do not by any means represent the final extension stage for the coal deposit at Neyveli. With the increasing size of the open-cuts, an increase in the size of the equipment also appears advantageous. Whereas the first machines were equipped with 350 l buckets, these data were doubled for each of the following machines. The 700 l BWEs were followed by 1,400 l BWEs.

Assuming a continuous development, bucket wheel excavators in the future could be machines with 2,800 l buckets equal to 100,000 bank m³/day capacity (see also, for example, Fig. 14).

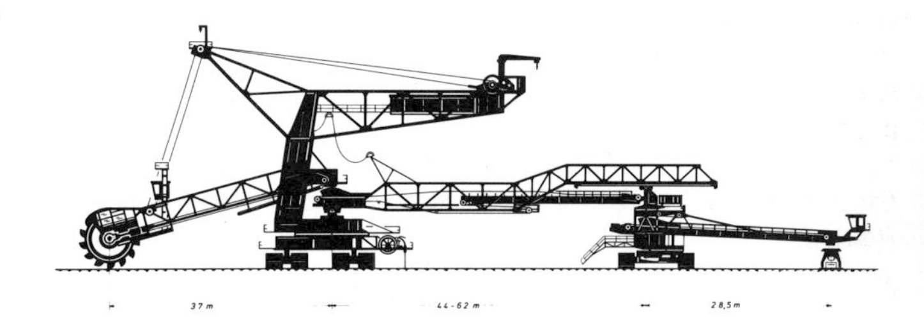

Fig. 13: Bucket wheel excavator SchRs $\frac{900}{7} \cdot 30$

Technical data:

Bucket capacity	900 l
Bucket wheel speed	2.8 m/s
Outreach of the bucket wheel	37 m
Length of intermediate bridge	53 ± 9 m
Length of discharge boom	25 m
Belt width	1,600 mm
Service weight	approx. 2,000 t

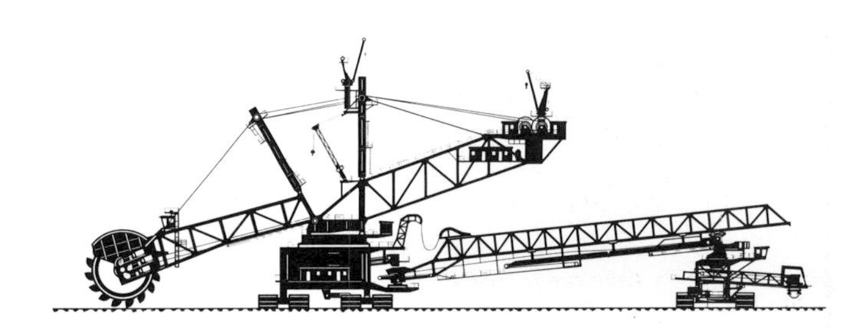

Fig. 14: Bucket wheel excavator SchRs $\frac{4,000}{2.5} \cdot 37.5$

Technical data:

Bucket capacity	4,000 l
Bucket wheel speed	2.6 m/s
Bucket wheel outreach	46.5 m
Length of intermediate bridge	85 ± 12 m
Length of discharge boom	17 m
Belt width	2,600 mm
Digging height	37.5 m
Service weight	approx. 6,500 t

Several different types of machines of the size have already been built by O & K. In addition to the equipment built for the Rhenish brown coal district in the Federal Republic of Germany, O & K also supplied very modern machines of this type, which had comparatively high bucket wheel drive capacities of 3 x 710 kW each, to the People's Republic of Poland.

Excavator components whose reliability has been proved over many years ensure, in connection with up-to-date control techniques e.g. automation for slewing, luffing/hoisting and preselection of advance travel, optimum utilization of the equipment.

6. Conclusions

Bucket wheel excavators have been working at Neyveli in hard mining operations for many years.

The bucket wheel technique at Neyveli started with 350 l and 700 l BWEs about 25 years ago and has taken into operation the next generation of 1,400 l BWEs, which have been running now for several years. The first problems regarding wear and lifetime of the digging elements have been overcome by a good cooperation between the supplier O & K and the engineers of NLC.

Although the mining conditions at Neyveli can be considered to be extremely hard — no other mine has material with a comparably high digging resistance to Cuddalore Sandstone — the bucket wheel technique has proved its economy for many years. Regarding further development caused by the high demand for energy in India, big bucket wheel excavators will also be needed in the future to supply lignite to the power stations.

References

[1] Rasper, L. and Rittner, H., "Opening-up of the Lignite Mine of the Neyveli Lignite Corporation and Experience with Bucket Wheels in Hard Overburden", Braunkohle 10/61

[2] Durst, W., "Schaufelradbagger für Tagebaue in der Welt", Braunkohle 4/79

[3] Vogt, W., "Stand und Entwicklung des südindischen Braunkohleprojektes Neyveli", Braunkohle 5/75

[4] Kasturi, T.S. and Streck, W., "Expansion of Neyveli Lignite Mine, India" Braunkohle

[5] Rodenberg, J., "Großschaufelradbagger für polnische Braunkohlentagebaue", Braunkohle 6/81

 Volume 1, Number 4, December 1981

Basic Design of the Bucket Wheel Excavator for the Goonyella Mine

H.C.G. Rodgers and J.R. Brett, Australia

Grundlegende Konstruktionsmerkmale des Schaufelradbaggers für den Goonyella Tagebau
Conception de base de l'excavateur à roue à augets pour la mine de Goonyella
Diseño básico de la excavadora de cangilones para la mina de Goonyella

グーニエラ炭坑のバケットホイール掘削機の基本設計

贡耶拉矿场杓轮挖掘机的基本设计

التصميم الأساسي لحفارة الدواليب ذات القواديس الخاصة بمنجم جونييلا

Summary

Overburden stripping at Goonyella Mine has already reached the stage where draglines alone cannot handle the depth or achieve the production targets required without the aid of additional (pre-stripping) equipment. Currently, a fleet of twelve scrapers is being used for this purpose but long-term planning indicates an ever increasing need for pre-stripping, to such an extent that more sophisticated, more efficient systems will be required early into the 1980s.

Investigations of options available for this pre-stripping showed a continuous system, i.e. Excavator-Conveyor-Spreader, would be the most attractive. This would mean introduction of entirely new plant types and related operating methods to the Central Queensland mines. Similar systems have operated for many years in Victoria and overseas, however, these existing systems are in deposits much less consolidated than those at Goonyella. Soil investigations suggest a continuous system will meet some of the most arduous conditions yet encountered by this type of plant. Consequently, extensive investigations have been made to judge whether a suitably designed Bucket Wheel Excavator (BWE) could be obtained for Goonyella, and, this being so, in drawing up the specification of the machine and associated transport and disposal units.

1. Introduction

The open cut coal mines managed by Utah Development Company (UDC) in Central Queensland use the dragline stripping system, as developed in North America, to expose the coal seams. The coal is excavated by a heavy duty shovel and loaded into wheeled transport. To date, typical overburden (O/B) depths are 35—45 m and even with the high rehandle rate necessary at these depths, stripping with draglines is highly efficient (Runge, 1979). However, the coal seam dips at some 5° and the corresponding increase in O/B depth as mining progresses necessitates increased rehandle by the draglines. Stripping by draglines alone would then become inefficient and at depths over 45 m a tandem system for stripping is required.

The options open to UDC for pre-stripping included:

scrapers; shovels and trucks; a semi-continuous system using shovels, crushers and conveyors; and a fully continuous system.

Scrapers are being used at present for auxilliary stripping, but the greater the depth of O/B the further the spoil has to be moved in order to maintain stable slopes, and wheeled transport begins to lose its advantage in favour of conveyors. The semi-continuous system would have the advantage of conveyors and would be applicable if extensive blasting and crushing at the operating face were necessary. However, much of the upper layers of O/B at Goonyella can be very sticky, which would prejudice shovel and crusher productivity.

Investigations to date suggest less than 25 % of the O/B to be pre-stripped will be too hard for a BWE, but could be blasted to a lump size suitable for handling by conveyors. The majority of the O/B appears suitable for BWEs, although harder than that in most of the open cuts where BWEs have been applied satisfactorily. Its occasional high stickiness poses severe design and operating problems.

UDC were aware there had been some costly mis-applications of BWEs in hard deposits, and therefore undertook digging resistance tests to confirm suitability for the BWE application. The results of these and other judgements suggested a suitably designed BWE could be applied and the decision was made to proceed on the basis of a continuous system. Specifications for the main plant units were issued late in 1977 and orders placed progressively in 1978. The system is expected to be in service in 1981.

2. Present Goonyella Operations

The Goonyella Mine became operational early in 1971 and is designed to produce $4.5 \cdot 10^6$ tonnes of coking coal annually. The Goonyella Middle seam is being mined at present. Its thickness ranges from 6—10 m and it yields a high quality, hard, medium volatile coal. Present operations use the extended bench method of dragline stripping; shovel and trucks are used for the mining and hauling of coal. Both coal

H.C.G. Rodgers, Consultant, Huen Rodgers & Associates, 938 Toorak Road, Camberwell, Vic. 3124, Australia and J.R. Brett, Senior Mining Engineer, Surface, Utah Development Company, Australia.

Correspondence to be addresses to Mr. H.C.G. Rodgers.

Paper presented at the International Conference on Mining and Machinery, July 2—6 1979, Brisbane, Australia.

and O/B are blasted prior to digging. The coal is hauled out of the pits up ramps and roadways to a crushing and screening plant. The ramps are established in gaps left in the spoil heaps. The present overburden stripping fleet consists of:

2 Bucyrus Erie 1370W draglines (45 m³ bucket)
2 Bucyrus Erie 1350W draglines (34 m³ bucket)
1 Marion 8050 dragline (45 m³ bucket)
5 Caterpillar D9 bulldozers (one to each dragline)
6 Caterpillar 657B scrapers
2 Caterpillar 633C scrapers
1 Caterpillar 633 scraper
3 Caterpillar 631B scraper
2 Caterpillar D9 bulldozers

It has been estimated that the existing dragline fleet can economically strip 45 m of O/B and maintain sufficient reserves of uncovered coal to meet current production targets. At depths greater than 45 m the operating cost of the dragline system rises and the rate of uncovering coal drops because of the high rate of rehandling necessary and generally lower productivity (Runge). Eventually the cost of rehandling would be at least as high as a pre-stripping system. This situation has led to the decision to pre-strip in those areas where the O/B depth exceeds 45 m.

3. Future Operations

As referred to in 1 above, of the options open to UDC for the pre-stripping system the continuous type was selected. The dragline system will remain for main stripping for the time being and hence the new system will have to integrate satisfactorily with the existing one.

A wide range of continuous systems is available. For instance BWEs are built from small standard units of only 50 tonnes mass to gigantic units of 13,000 tonnes and daily outputs of 240,000 m³ (bank). Machines in the large ranges are capable of digging up to 50 m above and 15 m below transport level and even greater vertical ranges are possible by use of multiple bench operations (Rasper, 1975).

Planning for the pre-stripping system at Goonyella was based on continuing production to the year 2000. By this time the O/B depth will average more than 90 m, and if the lower 45 m of this were allocated to draglines, about the same depth would eventually have to be taken out by the pre-stripping system. The total pre-stripping requirement builds up to some $22.5 \cdot 10^6$ m³ (bank) per year during the 1980s. However, this assumes final (steep) highwall slopes being reached in 1999/2000. If production beyond 2000 were contemplated, higher outputs would be needed, not only from inherently deeper overburden, but also because flatter overall batter angles would be needed to ensure pit stability, that is, the pre-stripping benches would have to be advanced further ahead of the dragline benches.

Theoretically one BWE alone with a daily output of 80,000 m³ (bank) would be capable of $22.5 \cdot 10^6$ m³ (bank) annually. Such a large unit would of course necessitate conveyors with widths of 2.4 m or more and a correspondingly large spreader. The system would represent a large capital investment and a high operational risk for several reasons. Firstly, the only comparable heavy duty machines in this size range are in Ekibastus, Siberia and are not readily available for inspection. Others have not been proved in hard digging. Secondly, the six pits at Goonyella stretch over 18 kilometres and the use of one BWE alone would necessitate frequent changes of site to balance the respective dragline operations in the various pits. A correspondingly complex conveyor and disposal system would be involved. Thirdly, the intention is to superimpose pre-stripped overburden on the dragline spoil heaps. Such spoil piles do not have the capacity to receive efficiently pre-strip spoil at such a high rate over a limited area.

When initial planning was undertaken by Rheinbraun Consulting (RC) in 1975 (Peretti, 1976) it appeared two "standard" heavy duty machines with digging heights of up to 15 m and masses of 500 tonnes each would suffice. However, subsequent information on the O/B indicated the "standard" machines might not be robust enough and the low digging height could necessitate additional conveyors. Also, the parameters set for the RC feasibility limited highwall depth to only 75 m. After consideration of the various alternatives UDC opted for three individual systems, each with a nominal annual output of $7.5 \cdot 10^6$ m³ (bank). The excavator would have an overall digging range of 33 m. The programme is such that if the first unit were installed by early 1981 sufficient operating experience could be gained before it would be necessary for the second and third systems to be ordered. Three systems would also give a desirable degree of flexibility, and procurement of the first system has proceeded on this basis.

The disposition of the existing dragline system and the new continuous system is diagrammatically shown in Fig. 1. The basic BWE unit would have a digging range of up to 25 m above and 11 m below the transport level, and some variations in these heights would accommodate the nominal 33 m strip. The conveyor system would initially have three simple flights and the O/B would be transferred by a crawler-mounted tripper with 20 m transfer boom to a crawler-mounted spreader with 35 m connecting conveyor and 60 m discharge boom. The spreader would be capable of high dumping to a height of 23 m as well as deep dumping. The details for these three basic plant units and the main parameters which were considered in specifying and developing their designs were developed following the completion of a detailed study of the Goonyella O/B in 1977.

4. Nature of the Goonyella Overburden

4.1 Basic Details

Typically the coal seam (the Goonyella Middle) sub-outcrops some 15 to 30 m below surface. The seam is 6 to 10 m thick and dips to the east at some 5°. Weathering up-dip of this location renders the coal unsuitable for coking. Strike length of the mine is about 18 km.

Within the zone for pre-stripping the O/B consists of Tertiary sands and clays and minor gravels, with occasional boulder beds between 5 to 15 m deep overlying unconformably Permian sandstone, siltstone, claystone, and occasionally thin coal beds with associated carbonaceous claystone.

In the Tertiaries very stiff to hard, moist clays dominate, with minor dense beds of clayey sand, river wash with hard rocks up to head-size, occasional weathered boulders of sandstone, siltstone and claystone.

The Permian rock types include weathered rock fragments forming the coarse fraction and clay minerals forming the rock matrix and cement. Most sandstones contain between 30 % to 50 % clay minerals. The important exception to this

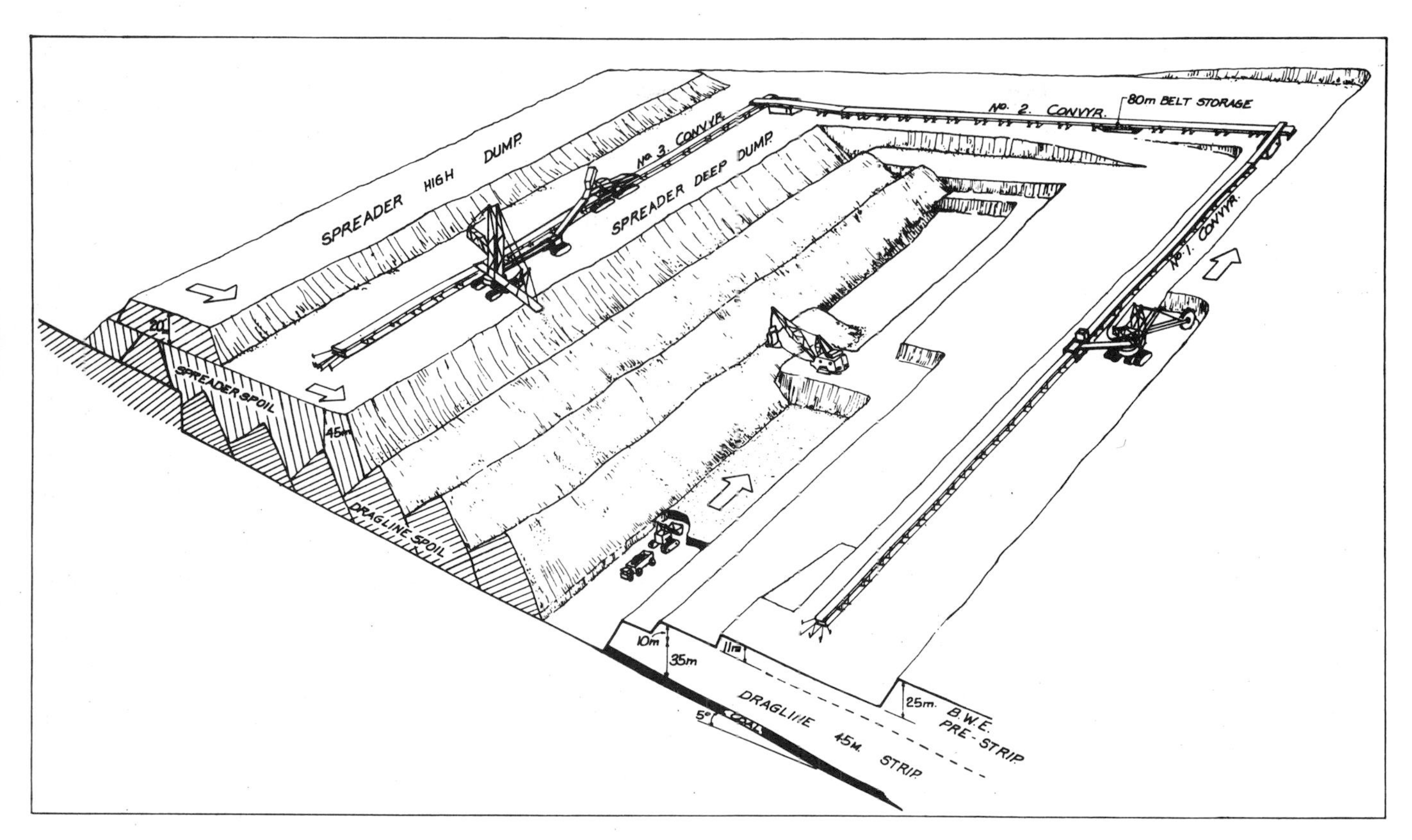

Fig. 1: General arrangement of pre-stripping system in relation to existing operations

is where carbonate minerals form the cement, the effect of which is to increase rock hardness and strength significantly.

4.2 Batter Stability

The mix of the various materials which form the Goonyella O/B necessitates blasting for excavation by draglines. Blasting aggravates the deposit's natural tendency to instability and highwall and spoil dump failures have been a continuing problem. This has increased the need for pre-stripping in order to lower dragline operating faces.

Highwall failures are mostly caused by slip planes dipping into the pit; spoil failures are primarily caused by weakened spoil material. In the mid 1970s a joint study was undertaken by CSIRO and UDC (Boyd, Komdeur and Richards, 1978) to determine the relationship between depth of rock weathering and stability. This study involved extensive large core drilling and the opportunity was taken concurrent with this (in 1976—77) to carry out digging resistance tests on the various rock types.

4.3 Digging Resistance

In any O/B system the behaviour of the material at each point in its movement from the undisturbed operating face to the disposal point, where it must become a stable O/B dump, strongly influences the design of each plant unit within the system. While this may be axiomatic many of the misapplications of BWEs have occurred through an ignorance of, or an ignoring of, the need to relate closely the nature of the deposit to the design of the plant (Winzer 1976, Rodgers 1976).

For the BWE the first consideration is design of the wheel and its drive to ensure it will be able to separate the hardest of materials from the operating face at a steady rate of output, with high reliability and without undue wear and tear. For this the designer needs to know the digging resistance and characteristics of the O/B. Various methods of determining this have evolved and usually design of the drive power and digging parts of the wheel are based on a number of factors. In an open cut such as Goonyella which has long existing operating faces it is possible to cut pit-wet samples from beneath the dry surfaces and make laboratory tests on these. However, it is also necessary to determine characteristics of the material remote from the operating faces which will be dug some years hence. Core drilling gives a ready means of such checks.

Among the methods frequently used for measuring digging resistance are the O & K Wedge Test, unconfined compressive strength tests, and the Protodjakanov method.

The Protodjakanov method is used in East Germany but did not appear to be of sufficient value to the Goonyella investigation, although some samples were sent to Freiberg for testing. Drilling and testing was carried out as follows:

Initial — 50 mm diameter open-hole drilling with rock type and weathering recorded from cuttings. "Hardness" was related to drilling performance.

Intermediate — a grid of 50 mm diameter core holes with samples taken at 1.5 m intervals for compressive strength testing.

Final — based on the results of the above a wider spaced grid of 150 mm diameter core holes was drilled. Holes were continuously cored in the pre-stripping range and 30 cm samples taken at 3 m intervals, or whenever a change of rock type or weathering occurred. These samples were subjected to the O & K Wedge Test.

The O & K method was developed by Orenstein & Koppel of Lübeck, West Germany (O & K). The procedure is illustrated in Fig. 2. This is drawn to show the test carried out on the 150 mm drill cores at Goonyella. With larger samples taken close to the operating face it is usual and possible for them to be shaped to 150 mm cubes, but the principle of the testing on cylindrical corings is the same. For laboratory purposes the standard wedge, with dimensions as shown at A of Fig. 2, is mounted in a compression testing machine. For field testing at Goonyella a portable rig as shown at B of Fig. 2 was used.

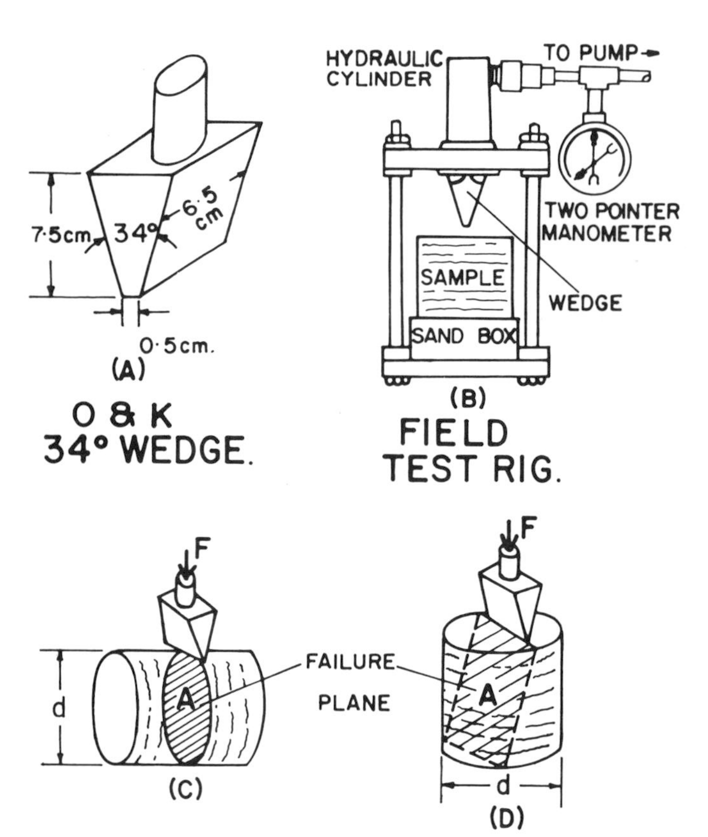

Fig. 2: Test rig and details: O & K wedge digging resistance tests

The wedge is first pressd into the centre of the sample as shown in C of Fig. 2 and parallel to the bedding planes. The test as shown in C yielded two further samples each approximately 15 cm dia. by 15 cm long. Each of these was then subjected to the test as shown in D of Fig. 2 in which the wedge was applied normal to the bedding planes. For each test the force F at which splitting occurred was recorded on a pressure gauge and the depth of penetration of the wedge and the area of the surface as split were measured.

The tests yielded two sets of results for Specific Cutting Force. One of these was related to the area of splitting and was expressed as Ncm^{-2} with the symbol K_A. The other related to the length of the wedge and expressed as Ncm^{-1} with the symbol K_U.

In addition to the wedge test on the 150 mm core samples, specific compressive strength normal to the bedding planes was also measured and compared with the results of similar tests on the 50 mm sample. They tended to be lower, showing the effect of sample size. The results in these cases were also expressed as Ncm^{-2} with symbol C_A.

UDC classified the rocks according to their increasing degrees of hardness into six classes:

(i) Clays — extremely weathered rock
(ii) Highly weathered rock
(iii) Moderately weathered rock
(iv) Slightly weathered rock
(v) Fresh rock
(vi) Calcareous material (slightly weathered to fresh)

The O & K wedge test appeared to be the most applicable and results of these tests were used in evaluation of digging resistance and as a basis for determining productive performance of the proposed BWE.

It should be noted the method of application of these measurements to dimensioning of buckets and teeth and drive of a wheel is largely empirical. Correlation between wedge test measurements and compressive strength test results and actual cross-section cut by the bucket is usually qualified by a multiplication factor applied by the designer and dependent on his judgement of likely problems in digging the various materials. This judgement is based not only on the test results but from a qualitative comparison between the deposit under consideration and others where BWEs have been successfully and unsuccessfully applied. Whilst the task for the designer for an open cut mine such as Goonyella having long, exposed faces and a large diameter programme, is less difficult than for a deposit in virgin ground with only a limited number of small diameter drill cores available, many preliminary enquiries were answered with the suggestion that the only way to determine "diggability" of the O/B was to install a BWE for test purposes or airfreight 300 kg lumps of overburden to Europe for testing.

This was not possible and the evaluation of likely behaviour of the various rock types and application of the digging resistance tests results had a dominant influence on plant design philosophy and detailed specifications as explained below.

5. Basic Design Philosophy

The specifications for the units associated with the pre-stripping system were the first UDC had issued for this type of plant and a prime objective was to ensure the highest degree of man safety and plant reliability. A vital consideration for many Australian mining projects is that they are in remote locations with limited mechanical and electrical engineering workforce and maintenance personnel. These people must of necessity be versatile and are frequently masters of improvisation but they cannot be expected to be completely up to date with some of the maintenance techniques required for sophisticated electrical equipment nor can they be expected to keep in service mechanical items which are not of first-rate quality and simplicity. Where mining operations are in densely populated areas, such as the Rhineland, West Germany, specialists can be obtained in a matter of hours, but in Australia it may be several days before a specialist can be brought from, say, the Pilbara region to the Bowen Basin. These considerations, and the very arduous duty expected from the studies of the deposit, caused the specifications to be such as to give emphasis to the need for safe, simple, robust, high-quality plant components. The following examples are typical of the way this attitude permeated drafting of the specifications and the subsequent negotiations with tenderers to work up safe and reliable designs.

5.1 Structure

It was anticipated the machines, i.e., BWE, spreader and tripper would be of East or West German design. Over nearly 50 years these countries jointly, and then separately, have developed advanced and comprehensive codes for the design and construction of large machines. By the use of a wide range of Load Assumptions for different degrees of operational severity of loading, and sophisticated strength and stability analyses, their machines can work safely to high permissible stresses. As a control of this practice, it is mandatory in both countries for the design to be audited progressively by a government accredited Proof Engineer (Independent Expert). The UDC specifications provided for such an appointment and an eminent West German engineer was engaged to check the designs.

The material used and the actual manufacture also are subject to approval of the Independent Expert (I.E.) and he normally makes a personal inspection before the plant is cleared for operation.

5.2 Mechanical Items

The life rating to be expected from plant of this nature is equivalent to 100,000 hours of full load operation of the BW drive. The aim of those who drafted the UDC specifications was to call for life ratings where appropriate with the object of some 20 calendar years without significant "wear and tear" mechanical breakdowns. A typical example of this is that specified for gears. All bearings are to have a life rating of 75,000 hours at input to the gearbox of the most arduous combination of torque and speed when the drive motor is operating at full output. Tenderers were given the option for gears to be designed to AS61 (i.e. BS436) and kindred standards or the AGMA standard. For the former case the expected running hours with full load rated output of the drive motor applied at each side of teeth of each gear wheel were specified. These hours were selected according to the frequency of operation of the drive and, for instance, 100,000 hours for both wear and strength were given for the BW, whereas for the BW hoist only 26,000 hours was stated. For variable elevation conveyors 26,000 hours was nominated, whereas for fixed elevation conveyors (and this also applied to the main, long conveyors) the figure was 52,000 hours. Other qualifications were applied to ensure reductions in these hours were not introduced by application of zone factors and other means.

Where the tenderer opted for AGMA he was required to use the curves of flank life C_L and root life K_L which provided for a diminution of these factors up to not less than 10^9 cycles of tooth contact. The gears were to be classified for reliability at "Fewer than one Failure in 100" and the appropriate value of C_R was, in any case, to be not less than 1.0. Also, appropriate overload factors C_0 were to be applied to deal with such conditions as starting torque and shock and in any case are to be not less than 1.25. Tenderers were also required to show that provisions would be taken to have the elasto-hydro-dynamic (EHD) oil film thickness maintained under full load in the gearboxes and were to state what the minimum values of film thickness would be.

Table 1: Comparison of heavy duty BWEs

		O & K 1340	O & K 1355	O & K 1367
		Athabasca	Neyveli	Goonyella
BW Diameter	m	12.5	10.5	12.25
Bucket Capacity l_1		1900	1400	1300
Ring Space Capacity l_2		550	800	1000
Number of Buckets	S	14	10	10
Number of Pre-cutters		0	10	10
Discharge Range	buckets/min			
HIGH		53	65	48
LOW		26.5	65	16
Cutting Speed Range	m/s			
HIGH		2.48	3.57	3
LOW		1.24	3.57	1
Bucket Wheel Drive	kW	2 x 500	2 x 750	2 x 600
Basic Digging Force	tonnes	35.0	36.4	34.7
Belt Width	mm	2200	2000	1800
Belt Speeds	m/s	4.7	4.5	2.5/4.2
Ground Pressure	kPa	160	131	127
a. BW outreach	m	21	31.3	36
b. Discharge boom outreach	m	35	30	40
A. Max. height to centre of BW	m	17	26	25
T. Max. range below transport level	m	10	8	11
Ballast	tonnes	220	270	270
Mass Ready for Service	tonnes	1725	2170	2230
Total Installed Power	kW	3000	3750	3200

5.3 Electrical Equipment

Electrical equipment for the system will be to the same high standard as the structural and mechanical components. Electrical supply is the one common denominator in the whole system, particularly from the point of view of control. Interlocking of conveyor drives and other operational functions are essential.

The conveyors must be started and stopped in sequence and under controlled conditions, otherwise flooding and blocking of chutes would occur. Such interlocking and sequential control are commonplace in similar installations throughout the world. Where the Goonyella installation differs electrically from most other operations, is as follows:

5.3.1 Bucket wheel drive — 2 x 600 kW DC motors

DC motors provide an appropriate means of speed and torque control to a degree not possible with the simple AC alternatives. "Cascade" and other AC speed control systems lack the range of speed control and efficiency of the DC motor control and also require provision of heat dissipation and additional equipment for start-up.

Weight and cost of DC motors, with no real need for a wide speed range, have not favoured DC in most BWE applications. However, use of aluminium piping and remote ventilation have enabled DC motors of comparable weight to the AC motors to be designed for the Goonyella machine.

5.3.2 Pole-changing motors

Full and half-speeds have been specified for all main machine and trunk conveyors. Variable speed couplings permit a range of speeds to be used for a conveyor installation but the range of speeds is limited by, again, considerations of efficiency and heat dissipation. Pole-changing motors, although not in common usage, are considered to present the best solution to the speed range problem. On board the machines the motors will be squirrel-cage AC motors and on the conveyors resistor controlled, wound rotor motors.

5.3.3 22 kV power supply

Draglines, drills and shovel already in use at Goonyella are fed at 6.6 kV. The larger motors on the pre-stripping equipment will be 3.3 kV. However, due to the length of trailing cable that would be required and with the necessary derating for layers on cable drums, the supply to the mobile machinery has been selected at 22 kV. The 22 kV trailing cables will be designed for a rated capacity of 3 MVA to accommodate possible, future machinery such as mobile belt conveyors.

5.4 Manufacture

As a follow-up to quality design, specifications called for stringent and well-controlled manufacturing processes and these are to be subject to inspection by UDC as well as by the contractor.

5.5 Testing

Goonyella lies some 200 km north of the Tropic of Capricorn and its shade temperature ranges from —7 °C to +45 °C. This and other climatic details were given in the specifications and the main objective of workshop and field testing was to ensure that plant components could give sustained operation under these conditions. This meant that for units such as gearboxes and electric motors, works testing was required to be carried out in "hot boxes". Furthermore the practice often followed by some manufacturers of extrapolating results from smaller units was to be avoided and as far as practicable components were to be given tests before delivery equivalent to full load.

Particular attention was given to the quality of metals both by chemical, physical and detailed ultrasonic testing.

6. Basic Plant Unit

6.1 Bucket Wheel

As explained in Section 3 the annual output for the initial system was planned at $7.5 \cdot 10^6$ m^3 (bank) and the BWE is to have an overall range of 33 m. From these main parameters and the batter stability characteristics of the deposit the basic hourly outputs were developed.

In some mining circles much emphasis is placed on plant availability rather than the more realistic aspect of utilisation. While individual units in a system may have high availability, the overall system might not have high utilisation. A major influence is the total number of hours worked annually and obviously a system working one shift 5 days weekly will have higher inherent utilisation than one on 3 shifts 7-day continuous operation. The latter is planned for the Goonyella operations and of the 8,760 hours available annually the planned hours of full operation was taken at 5,000. This utilisation of some 57 % is consistent with continuous systems throughout the world and it has been explained (Winzer, 1976) that utilisation of a continuous system is really no worse than a remote dumping system with wheeled transport.

The outcome of this consideration was a nominal 1,500 m^3 (bank) hourly. The UDC specification called for this output to be guaranteed over a test period of one month. Tenders were invited from major BWE manufacturers most of whom had built BWEs for heavy digging. Most of the offers were capable of being developed into what could be expected to be a suitably designed machine for Goonyella. However, the machine finally selected was the O & K 1367 and it is dealt with here as a typical example of how a tenderer and client can develop a design prior to the contract ad without shifting the responsibility for the performance from the tenderer (contractor).

At the time the Goonyella tenders were under consideration a system with a similar hourly and annual capacity had commenced operation at Yallourn (Rodgers & Mitchell, 1978). The BWE for this sytem is a well engineered and versatile machine with high output potential and should prove to be well suited to the Yallourn conditions. However, the digging at Goonyella will be much harder than in the Latrobe Valley; hence the BWE, and in particular the BW drive, will need to be more powerful. The latest heavy duty machine readily available for comparison is the O & K No. 1340 which commenced operation at the Fort McMurray Mine (Athabasca) of Great Canadian Oil Sands Ltd. (GCOS) in Canada late in 1976 (Winzer, 1976). This is a custom-built unit designed to dig glacial drift in temperatures ranging from +40 °C to —40 °C and to load to 150 tonne back-dump trucks. It is shown in Fig. 3. As well as 1340, a group of three heavy duty O & K machines Nos. 1355-56-57 were under construction at Neyveli, India and the first of these went into operation late in 1978.

Fig. 3: O & K 1340 BWE at Athabasca

Neither 1340 nor 1355 were readily applicable to Goonyella — because of the inherent differences between the three deposits — but they served some useful comparisons in working up the design for No. 1367.

The design ordered for No. 1367 is illustrated in Fig. 4 and the main dimensions are set out in Table 1, which also gives some basic comparisons between the three other machines referred to above.

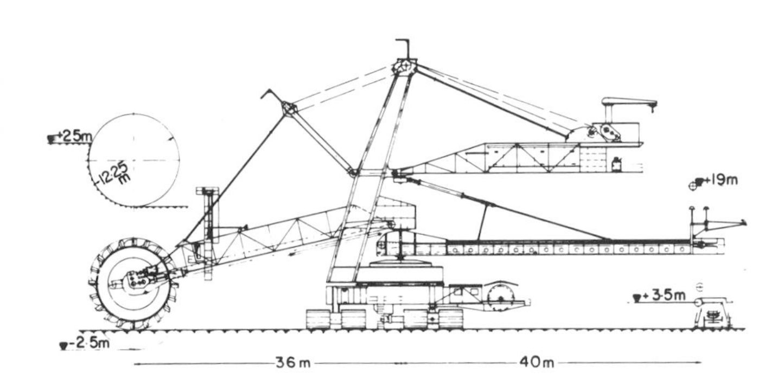

Fig. 4: O & K 1367 BWE for Goonyella

6.1.1 Bucket wheel head and bucket wheel drive

The nominal hourly output of 1,500 m³ (bank) is to be maintained in the weathered clays, which at times can be extremely sticky, and it was realised and agreed that lower guarantees would suffice in the harder and more abrasive types of O/B. At the extreme, where material will have to be blasted, lumps up to 0.5 m³ would have to be handled by the whole system. For this condition belt speeds will be lowered.

In the Goonyella system in order to attain the annual output the wheel must be proportioned to handle the sticky O/B and in the case of 1367 this is attained by large, roomy buckets (Volume l_1) and by having the ring space (Volume l_2) twice the size of the normal German convention (Rodgers & Mitchell, 1978). With these proportions and a speed variation control, the digging rate can be varied so that O/B is not packed firmly into the bucket but can "float" in the bucket and ring space and so discharge more readily. Also, with these proportions, when free-flowing material is met, both l_1 and l_2 can be filled to overflowing and high production can be maintained. However, an upper limit needs to be set on production rate to minimise the risk of excessive spill and blockages at the conveyor transfers of the system. This was set at 6,600 m³ (loose)/h. At an assumed swell factor of 1.5 this is equivalent to 4,400 m³ (bank) or nearly three times the guaranteed output, and is a measure of the conservative rating used in the Goonyella planning. It also indicates how a wheel of the 1367 proportions has a good deal of reserve capacity and can cope with a wide spectrum of digging conditions and O/B variations.

The close relationship between the batter slopes and block widths has been explained (Rodgers, 1976) and for Goonyella the stability investigations suggested that 45° at 25 m would be the norm for the standing side batter and for the same height 50° could be tolerated for the front batter. The zone allocated for No. 1367 will vary in heights between 20 and 25 m above the conveyor transport level. It was also realised that to attain the hourly and annual outputs the wheel diameter would need to be in the order of 11 m and in order to optimise the frequency of conveyor shifting the block width should be a nominal 40 m at the 25 m face height. From these considerations it appeared that the BW outreach should be between 36 and 38 m. Once the order of the BW outreach and the block width were known, it was then apparent the discharge boom would be of the order of 40 m. Its hoisting equipment and clearances to the BWE top structure should enable it to load to conveyors standing up to 11 m above the BWE operating level. With these basic proportions, variation in face conveyor grading would enable No. 1367 to command its nominal 33 m strip. It and the associated transport and disposal equipment would also have sufficient electrical reserve capacity to enable it to operate with a mobile slewing conveyor (bandwagon).

Having determined the basic proportions of the BWE the next main consideration was the BW drive. The digging resistance tests and comparison with other heavy duty machines indicated that a cutting speed of 2 m/s and a force at the bucket cutting circle of the order of 35 tonnes, would be needed for digging the more difficult material. An overall drive efficiency of 85 % was assumed and from this the following simple formula was derived for the relation between the cutting force P and the kilowatts output of the BW drive motor *s*, viz:

$$P = 0.0867 \times \mathrm{kW} \div V.$$

where kW = kilowatts input to bucket wheel drive when drive motor(s) are at maximum torque without overload, V = cutting speed in m/s at the bucket wheel cutting circle and 0.0867 = 85 % x 0.102 (the conversion factor of kW to tonnes). (For wheels with multiple speeds the largest value of kW ÷ V is to be used).

With the nominal 35 tonnes for P this gave 800 kW at 2 m/s and as it was desired to dig at 3 m/s in soft and free-flowing regions, 2—600 kW (i.e., 1200 kW total) motors were nominated. It was also considered the wheel speed would need to be reduced to 1 m/s to prevent excessive loading and impact on the transport system when handling blasted material. This would be readily obtainable with a DC drive or with an extended version of an AC "Cascade" system. The capacity of the drive for 1367 thus became comparable with those of 1340 and 1355 and of course resulted in a very heavy bucket wheel head. DC motors were chosen in preference of AC for the BW drive.

A major influence on the design of the BWE as a whole is the moment of the BW and its boom about the boom pivots. The greater this moment, the stronger and, therefore, the heavier the machine must be. The outreach from the slew axis to the outer lip of the cutting circle influence the dimensions of the top cut and consequently block width and batter slopes. Also, consideration must be given to the slope of the BW boom conveyor in the highest and lowest digging position and to the angle of the transfer chute from the bucket to the BW boom conveyor. Thus the BW outreach and hence the length of the BW boom has to be integrated with the wheel diameter, and in this case rather than choose a 37 m boom with 11 m wheel O & K opted for a 12.25 m wheel and 36 m boom, the larger wheel diameter giving better clearance angles. Permissible "foreland", or advance per terrace cut, needs to be considered also (Rodgers, 1976).

6.1.2 BWE conveyors

1.8 m belt width was selected as the appropriate balance with wheel proportions. The belts will be supported by 5-roll impact idlers in the loading zone and 3-roll 40° idlers in the main length of the conveyors. In the loading zones 5-roll plain steel impact idlers are to be used spaced at 400 mm; but there are some reservations as to whether these will prove to be as satisfactory under Goonyella conditions as 3-roll steel or rubber disc idlers. For this reason the frame and the chutes at the tail ends of conveyors are to be designed for ready modification to suit the 3-roll idler configuration.

The normal belt speed will be 4.5 m/s but in order to cope with the sharp and heavy blasted material this can be reduced to the half speed of 2.25 m/s, a feature not only in the BWE but throughout the whole system. Also, to prevent belt damage by sharp, heavy lumps impact tables are to be installed at transfer points. Naturally these would very quickly cause blockages when handling sticky materials and hence the impact tables will be retractable.

In order to improve belt tracking, even with steel cord belts, drive pulleys are to be crowned.

6.1.3 Ground pressure

Because of the high clay content of the Goonyella deposit and its inherent instability, it was considered that ground pressure on the BWE should be low enough to minimise the risk of bogging of the crawlers and was set at an average based on the operating weight of < 125 kPa, with a maximum of not more than 200 kPa. These ground pressure measurements are taken on the convention of distance between the crawler end tumblers with the take-up tumbler in the mid position. Even so, the risk of bogging cannot be overlooked and the crawlers steering system and underframe are to be designed so that the machine can travel with full steering with the crawlers on one side bogged to a depth of 0.8 m. The crawlers also have to be powered and proportioned for a grade of 1:20 in operation and 1:10 when changing sites.

6.1.4 Power supply

Whereas 1340 and some machines in Europe have a separate cablecar, No. 1367 has followed the Australian practice of having a large cable drum mounted on the BWE underframe. This helps to make the BWE fully independent of other plant and tends to minimise manning requirements. For the Goonyella system, a central feed point is contemplated on the 2.5 km face conveyor and the drum on the BWE itself will accommodate 1,500 m at a supply voltage of 22 kV.

6.1.5 Load assumptions

The basic design code for 1367 will be the West German Basis for Calculation of Large Open Cut Machines (the B.G.)

and related codes and standards. The current version of this code is 1960. UDC became aware that revisions and upgrading are in progress and provision has been made to incorporate these revisions into the 1367 design. For example, under the BG the theoretical live load is based on the bucket contents I_1 and half the ring space contents I_2, but for 1367 a higher figure, from the maximum normal cross-section of load on the BWE conveyors, has been taken. This results in a nominal live load of 14,000 tonnes/h to which the structure are to be designed, compared with some 10,000 tonnes/h at the nominal peak load of 6,600 m³ (loose).

7. Australian Manufacture

The highest practicable Australian participation in its projects has been an objective of UDC and was applied again in this case. The result was that only some specialised mechanical and electrical items and those which the contractor insists on making overseas will need to be imported. A high percentage of the machines and conveyors will be made in Australia. However, as indicated in Section 5, the design parameters for the machines are higher than hitherto and much of the structures are of plate with maximum use of welding. This type of steelwork is expected to be of a higher standard than the highest quality of bridge steelwork and the contractors' detailed welding instructions have been audited by UDC and local welding authorities to ensure a close integration of the best of German and Australian welding practices. This approach has been carried back to the design stage, where the drawings are required to have full size details of welds rather than symbols only, so that the designers intentions can be made clear to the welder on the shop floor or in the field.

Close cooperation has been achieved between the local steel suppliers (BHP), UDC and the contractor to ensure that plates of the required quality will be produced by local mills. As with earlier machines of this type steel to AS1204-350-L15 will be used in lieu of the German St52-3 for plates up to 30 mm thickness and above this size the local equivalent of the low sulphur German OK400 will be used. AS1204-250-L15 is to be the equivalent of the German steel St37.

In terms of contract value the percentage of Australian manufacture will be 56 for the BWE, 67 for the Spreader, 72 for the Tripper and 80 for the conveyors. Of necessity, most of the design has to be carried out overseas and this reduces the Australian content on a contract basis, but as far as work in Australian shops is concerned the percentages are higher than the figures stated above.

8. Conclusions

The Goonyella pre-stripping system marks a new venture into heavy duty O/B stripping by a BWE system but only broad details have been given here because of the length limitations on this paper. As with most new systems teething troubles can be anticipated in the early stages of this operation. However, as explained herein a systematic approach has been made, firstly to try to understand the nature of the deposit and then to develop plant design parameters to suit the deposit.

This was followed by the issue of performance and engineering standards and specifications, and then discussion with each of the competitive tenderers; and finally close discussions with selected tenderers prior to placing contracts. It is also intended close and continuous communication be maintained with the contractors until the end of warranty periods. This will not shift any responsibility to the client but will rather help the contractors to develop the optimum plant for this application. Such user input is worthy of application in most mining ventures and readers are commended to the following quotations:

"No mining operation is a piece of cake"... and... "Good mining equipment is only constructed in cooperation between manufacturer and user". (Winzer, 1976).

9. Acknowledgements

The authors extend their thanks to Utah Development Company for permission to publish this paper and officers of that Company who have been so helpful in its preparation. The views expressed are those of the authors and not necessarily those of UDC.

References

[1] Beale, R.F. (1978), "Better Design through Problem Solving and Vendor Liaison". Presented at Australian Mining Industry Council, 1978, Maintenance Conference (unpublished).

[2] Boyd, G.L., Komdeur, W. and Richards, B.G. (1978), "Open Strip Pitwall Instability at Goonyella Mine, Causes and Effects". Aus. I.M.M. 1978. North Qld. Conf. papers. pp. 139—157.

[3] Peretti, K. (1976), "The Development of the German Open Cast Mining Technology in the World (selected examples)". Braunkohle F 20353E. July 1976. Special Export Edition pages 66 to 77.

[4] Rasper, L. (1975), "The Bucket Wheel Excavator: Development. Design. Application". Trans Tech Publications, Clausthal, Germany, 1975

[5] Rasper, L. and Rittner (1961), "Opening up of the Lignite Mine of the Neyveli Lignite Corporation and Experience with Bucket Wheels in Hard Overburden". Braunkohle Wärme und Energie Vol. 10, 1961.

[6] Rodgers, H.C.G. (1976), "Experience with BWEs in Australian Brown Coal". Presented at a Symposium on Bucket Wheel Technology. Canada. Oct. 1976 (unpublished).

[7] Rodgers, H.C.G. and Mitchell, D.M. (1978), "The Yallourn Southern Overburden System". Inst. Eng. Aust. 1978 Annual Conf. Papers. pp. 248—256.

[8] Runge, I.C. (1979), "The Next Generation of Australian Surface Coal Mines". International Conference on Mining Machinery, July 1979.

[9] Winzer, S.R. (1976), "Experience with Bucket Wheel Excavators in the Tar Sands". Presented at a Symposium on Bucket Wheel Technology. Canada Oct. 1976. (unpublished).

Volume 1, Number 4, December 1981

A New Bucket Wheel Excavator Complex for the Goonyella Mine

W. Fleischhaker, Germany

Ein neuer Schaufelradbagger für den Goonyella Tagebau
Un nouveau complexe excavateur à roue à augets pour la mine de Goonyella
Un nuevo complejo de excavadora de cangilones para la mina de Goonyella

グーニエラ炭坑の新型バケットホイール掘削機コンプレックス

贡耶拉矿场的新型杓轮挖掘设施

مجمع حفارة الدواليب ذات القواديس الجديد الخاص بمنجم جونيلا

Summary

A bucket wheel excavator — conveyor system — spreader complex will be put into operation in Australia for the first time in particularly difficult winning conditions. These difficulties were taken into account by the exceptional stipulations laid down for the layout of the machines. The complex will be put into operation at the beginning of 1982.

1. Introduction

The Utah Development Company (UDC) operates a coal open cast mine at Goonyella, Bowen Basin in Central Queensland. This is coking coal situated at a depth of approx. 45 m. The overburden has been removed to date using American draglines (Fig. 1).

As the seam sinks deeper, the dragline operation becomes uneconomic. As a result UDC decided to remove the upper overburden layer of approx. 30 m depth with the help of a pre-stripping system and the rest as before with draglines.

The overburden consists mainly of hard clay with gravel and boulder beds. Strata of sandstone, siltstone and argillaceous rock occur which have to pre-blasted. Chunks with 80 cm edge length must be conveyed.

O & K Lübeck received the order for the design and delivery of the pre-stripping system consisting of excavator, tripper car and spreader in the summer of 1978. The conveyor system is being delivered by PHB Weserhütte, Australia.

2. Special Conditions

Comparable machines in Australia have only been put into operation by the SEC Victoria. These machines mainly win brown coal and small amounts of clayey overburden. In contrast to this in Goonyella there is an extremely hard overburden, which alternates with sharp edged pre-blasted stone. In addition to this, the overburden is extremely sticky during the rainy season.

In talks between UDC and the vendor it was therefore agreed that not only the German "Basis for Calculation of Large Open Cut Machines" should be observed, but also additional exceptional loading conditions were defined. The gearboxes were designed according to AGMA for an operating life of 100,000 h under full load.

3. Excavator Type SchRs $\frac{1800}{2.5} \cdot 25$

The excavator has a height of 25 m in high cut and a depth of 2.5 m in deep cut. The bucket wheel boom has a length of 36 m with a wheel diameter of 12.25 m. The discharge boom has a length of 40 m. The discharge boom can be raised and lowered between 3.5 and 19 m above crawler level and conveys by means of a hopper car onto the conveyor system. The belt width is 1,800 mm. The machine travels on six crawlers (Fig. 2).

The bucket wheel moves a theoretical output of 5,200 m^3/h loose with a specific weight of the masses of 1.8 tonnes/m^3.

The bucket wheel is driven by two direct current motors each of 600 kW. The wheel has a maximum circumferential speed of 3.1 m/s at the tooth circle. A digging force of 350 kN works at the buckets at 100 % motor output.

For the dimensioning of the machine for fatigue 150 % motor output or a digging force of 500 kN has been assumed. For operating conditions a digging force of 590 kN has been assumed. The 150 % motor output corresponds to the switching value of the safety coupling. This unit is an oil-cooled disk coupling.

The bucket wheel is conceived as a single-disk-wheel in the form of a conical shell. A particularly good emptying characteristic for sticky material is thereby achieved with this wheel design.

In addition the wheel has no hollow body which could fill with dirt, thereby worsening the position of the centre of gravity of the machine. The bucket wheel has been designed for a digging force of 350 kN. In addition an impact force of 1,280 kN has been taken into account.

The machine slewing gear delivers 120 kW for a slewing speed at the wheel of 30 m/min. The lateral force acting on the wheel for the determination of the fatigue strength has been derived from 150 % motor output here, too. This is 300 kN. This is considerably larger than the inertia forces occuring when the wheel is slewed at full speed against the

Dr.-Ing. W. Fleischhaker, O & K Orenstein & Koppel AG, Werk Lübeck, P.O. Box 1601, D-2400 Lübeck 1, Federal Republic of Germany

Volume 1, Number 4, December 1981

Fig. 1: General arrangement of pre-stripping system

bank face and is decelerated over a distance of 30 cm. The loading of the belts has been determined alone from the cross-section of the belts independent of the wheel output. A load of 8.6 kN/m is derived from the normal cross-section which corresponds to a conveyor output of 7,740 m^3/h or 14,000 tonnes/h.

In addition an exceptional cross-section of 19.5 kN/m is required which is only possible when the space between the guide plates above the conveyor belt is filled. The guide plates have been substituted for by ropes in order to prevent the occurance of this loading condition. In spite of this the load bearing structure has been designed to carry this exceptional payload.

Fig. 2: Bucket wheel excavator

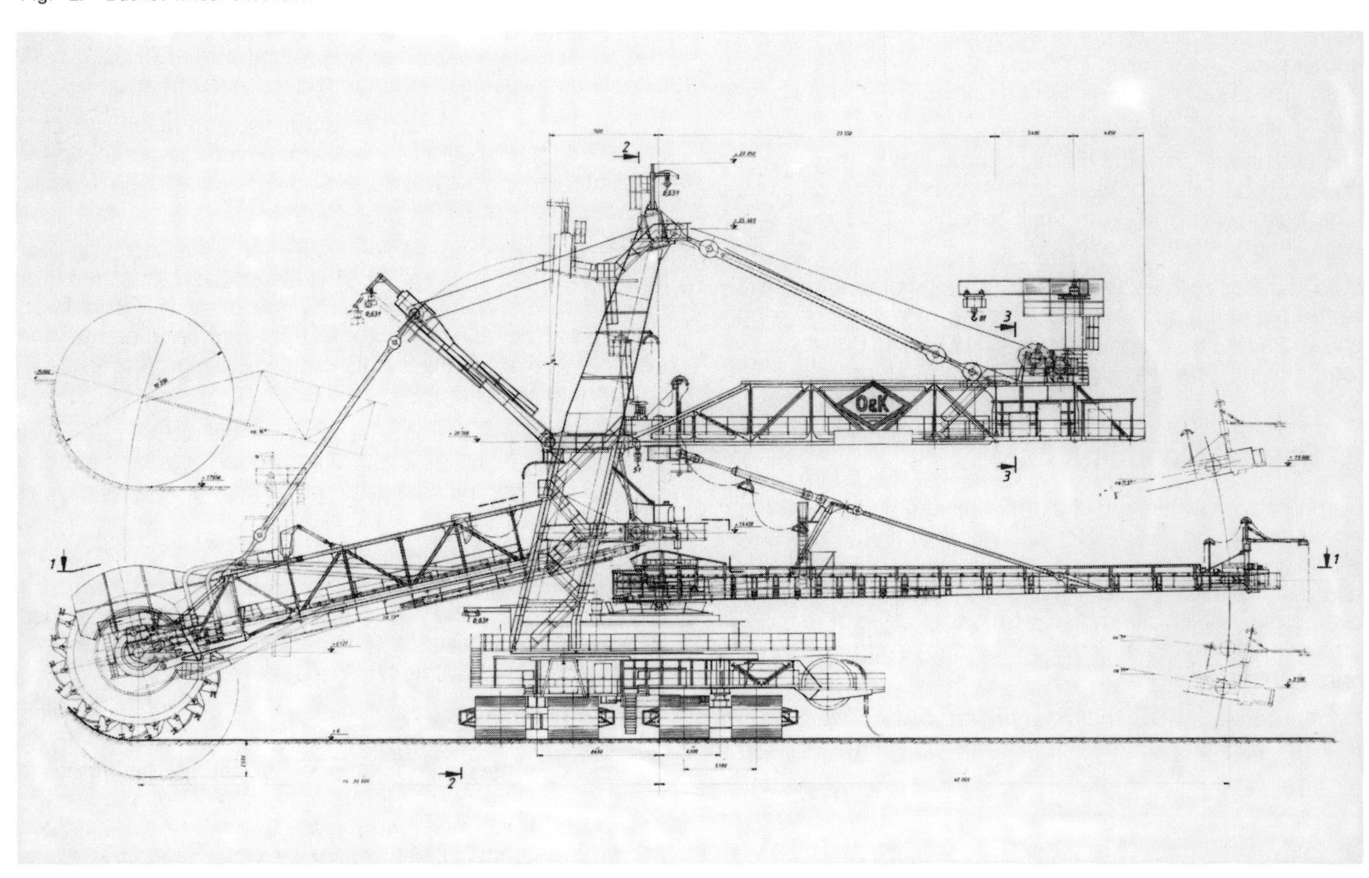

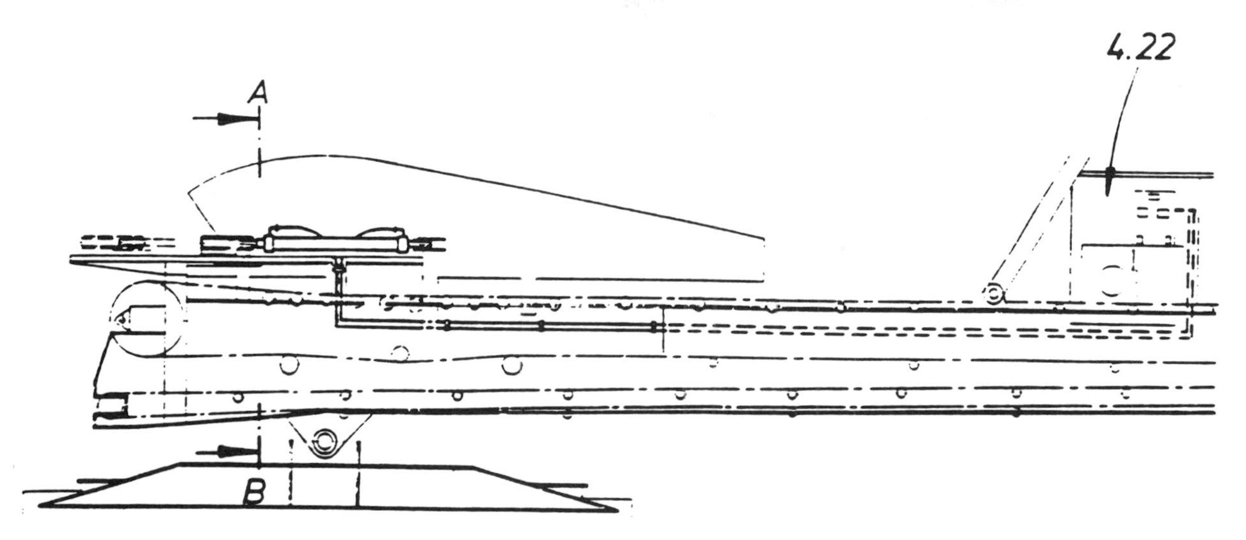

Fig. 3: Transfer chute with rock box

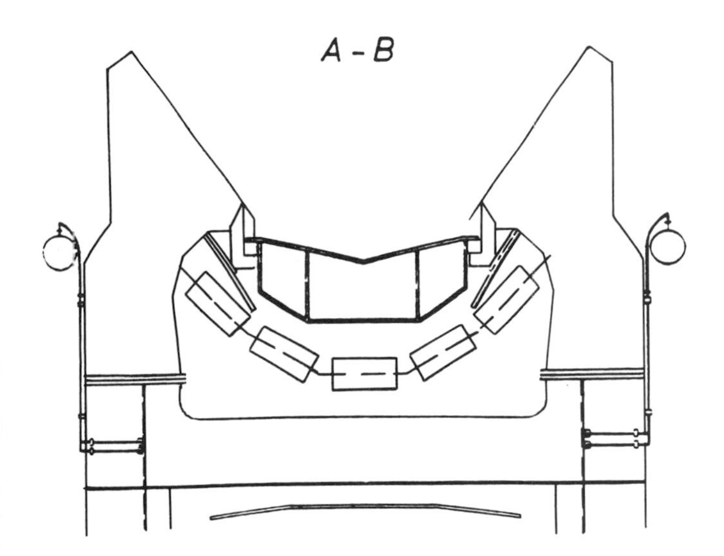

Fig. 4: Transfer chute with rock box, Section A—B

A belt speed of 4.5 m/sec is provided for the conveyance of clay and sand. If, however, sharp edged chunks (blasted rock) are conveyed the belt speed can be reduced to 2.25 m/sec. For this purpose change-pole motors are provided on the belt drives.

In order to prevent damage to the belts at the transfer points, by sharp edged chunks, a movable baffle plate (rock box) can be hydraulically positioned over the belts. When sticky material is being conveyed the baffle must be retracted (see Figs. 3 & 4).

The wind loads are also exceptional at Goonyella. For the load case "out of operation" a wind speed of 160 km/h must be taken into consideration. This corresponds to a static pressure of 1,230 kPa.

The excavator carries a cable drum, with a feed length of 1,500 m, on the undercarriage.

The installed electric power is 3,400 kW. The machine has a service weight of 2,600 tonnes including 300 tonnes ballast.

Fig. 5: Tripper

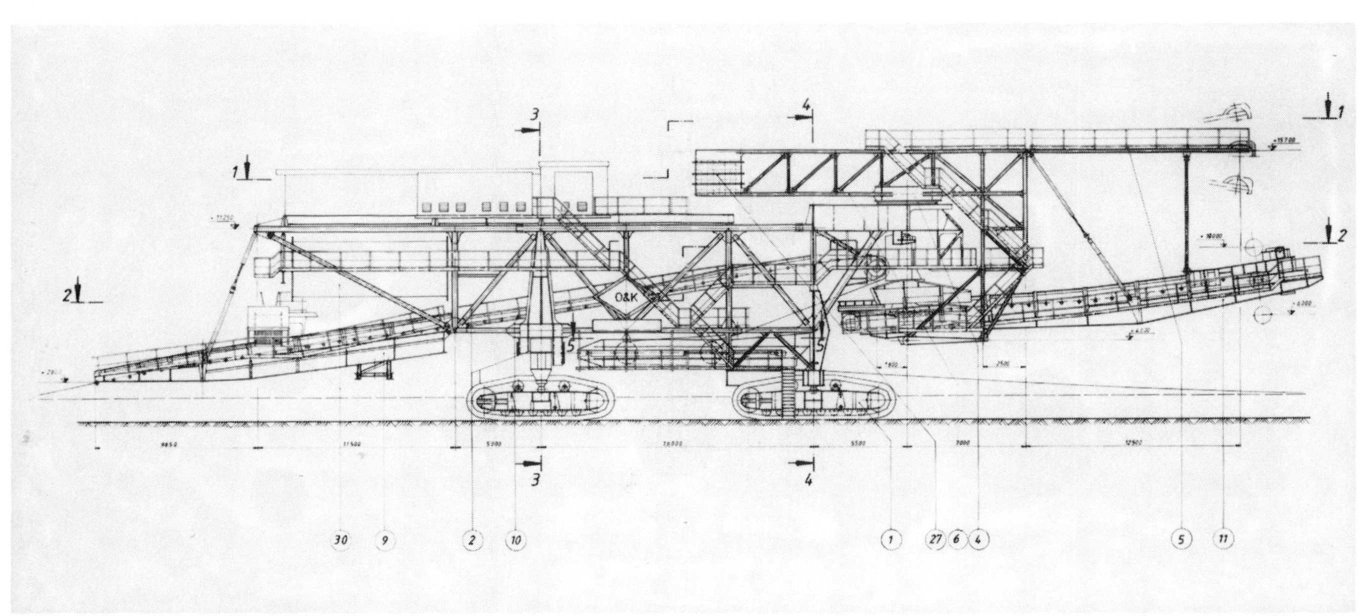

4. Tripper Car ÜR 1800 x 20

The conveyor belt has a width of 1,800 mm and runs at a speed of 5.2 m/sec. The speed may be reduced to 2.25 m/sec.

The tripper car carries the belt drive with a rated capacity of 2 x 700 kW and conveys the masses over the 20 m long boom with a belt loop onto the spreader. The belt width on the slewable boom is 2,200 mm. Both boom and receiving section are raisable and lowerable (Fig. 5).

The tripper car moves on four steerable crawler units. In addition the tripper car carries a large cable drum with a length of 9 m and a diameter of 3.2 m. The wound length of cable is 2,000 m. The spreader and tripper car are electrically connected with each other and interlocked. The installed power of the tripper car is 3,800 kW, and the service weight is 770 tonnes.

5. Spreader ARS $\frac{1800}{40 + 60}$ x 25

The last member of the transport chain is the spreader with receiving and discharge boom (Fig. 6). The belts have a width of 1,800 mm. Their speed may be either 4.5 m/sec or 2.25 m/sec. The receiving boom rests on a separate movable support car. The support length is 40 ± 2 m. The 60 m long discharge boom can be raised from 6 m to 25 m above crawler level. The spreader travels on three crawlers, two of which are steerable. 500 m of cable are wound on a cable drum. A hydraulically retractable baffle plate is installed as is on the excavator. The payload laid down for the normal operating conditions is 14,000 tonnes/h, for the dimensioning of the machine, however, 32,000 tonnes/h.

In addition to the load cases defined in the "Basis of Calculation for Large Open Cut Machines" the following loads for dimensioning the machine were defined: For the load case "out of operation" a static wind pressure of 1,230 kPa has been assumed.

The discharge boom may be laid on the bank slope with a load of 260 kN at its extreme and simultaneously collide laterally with the slope with a force of 110 kN. This lateral force corresponds with a slew gear output of 150 %.

In addition it has been assumed that the boom may dig into the slope at the discharge pulley with a vertical force of 170 kN. This force corresponds to a lifting winch power of 110 %.

These load assumptions, which are very unpleasant for the dimensioning of the machine, led to an exceptionally robust construction with a service weight of 1,340 tonnes including 100 tonnes of ballast. The installed power is 1,400 kW.

6. Production and Erection

All structural members and a large proportion of the mechanical components were produced in Australia. The material is a particularly low sulphur content rolling, better than AS 1204 — 350 — L15. Welded connections were chosen as far as was technically possible. Particular attention was paid to the production of knotch-free and close toleranced weld seams.

Remaining erection connections are carried out as high tensile friction joints according to German standards. The erection was carried out by an Australian firm with the help of a German erection engineer.

The machines will be mechanically operational at the end of 1981. They will be weighed in December under the supervision of an independent expert.

It is to be determined, by means of the experimental evaluation of the position of the centre of gravity, whether or not the calculated amount of ballast must be corrected, in order to guarantee sufficient stability of the machines under all conditions. Figs. 7—12 show the different machines during erection.

7. Conclusion

The three machines will be put into operation at the beginning of 1982. According to the experience that the vendor has made with hard overburden and pre-blasted rock in mines in Canada and India, it is expected that the new

Fig. 6: Spreader

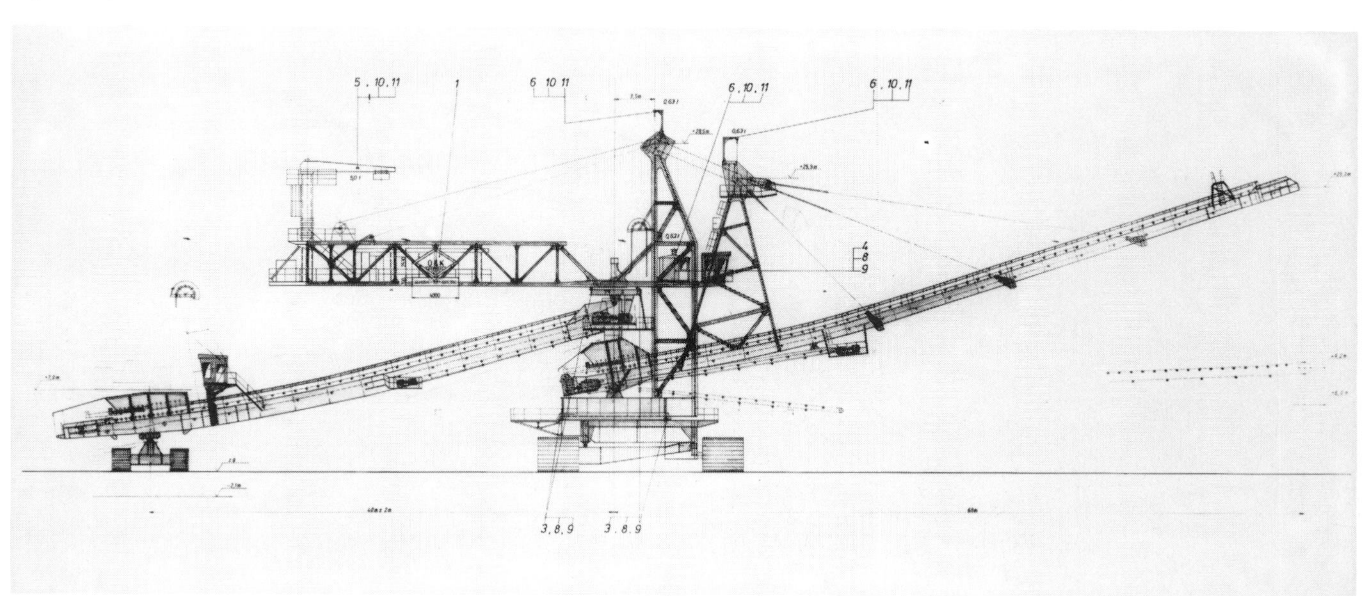

Fig. 7: Bucket wheel excavator during erection

Fig 8: Bucket wheel excavator during erection

Fig. 9: Crawler

Fig. 10: Steering girder

Fig. 11: Undercarriage of BWE

Fig. 12: Tripper

complex will accommodate the described difficult mining conditions prevalent in Goonyella. This confidence is supported by the exceptionally heavy dimensioning of the machines and by the great care spent on their design and construction.

References

[1] Rodgers, H.C.J. and Brett, J.R., "Basic Design of the Bucket Wheel Excavator for the Goonyella Mine", Int. Conf. Min. Mach., 1979, Brisbane

[2] Hanemann, D.K., "Combined Bucket Wheel Excavator/Dragline Application: Goonyella Mine Australia", Braunkohle 12, 1980

[3] Durst, W., "Earth Moving Machines for Larger Open-Cuts — in Particular Bucket Wheel Excavators and Spreaders", Inst. Eng. Australia, Paper M1065, 1980

bulk solids handling Volume 5, Number 6, December 1985

Experience with the Goonyella Bucket Wheel Excavator Working under Extreme Conditions

W. Fleischhacker, Germany

Summary

A bucket wheel excavator has been working in extremely hard overburden in Queensland, Australia, for the past 30 months. The operator has gained experience in preblasted and unblasted material. Both working methods are possible. The limits for the implementation of a bucket wheel excavator have not been reached in spite of a specific cutting resistance of K_A = 200 N/cm².

1. Introduction

The Utah Development Co. Ltd. (UDCL) operates the Goonyella opencast coal mine in the Bowen Basin of Central Queensland, Australia. There coking coal is mined which is situated at a depth of approx. 45 m. The overburden has been removed in the past using draglines. As the seam deepened, the dragline operation became uneconomic. Therefore, UDCL decided in 1978 to remove the upper overburden layer of approx. 30 m depth with the help of a prestripping system and the rest, as before, with draglines.

2. The Material

The overburden consists mainly of hard clay with gravel and boulder beds. Strata of sandstone, siltstone and argillaceous rock occur. Trunks of petrified wood are inbedded. These trunks are extremely hard, their diameters vary between a few decimeter and about 3 m.

The digging resistance of the overburden increases with depth. In the highest layers the cutting resistance K_A is 40 to 60 N/cm², going downwards there are layers of 60 to 80 N/cm² and 100 to 150 N/cm², and in the lowest one K_A is 100 to 200 N/cm² (Fig. 1).

3. The Prestripping System

UDCL ordered a system of a bucket wheel excavator, tripper car and spreader from O&K Lübeck. The system is described in Ref. [4]. The nub of the system is a bucket wheel excavator (BWE), type Sch Rs $\frac{1800}{2.5} \cdot 25$ (Fig. 2). The

Fig. 2: BWE Sch Rs $\frac{1800}{2.5} \cdot 25$ in Goonyella

bucket wheel is driven by DC motors of 1,200 kW. The wheel diameter is 12,250 mm, the outreaches of the BW boom and the discharge boom are 36,000 mm and 40,000 mm, respectively, and the belt width is 1,800 mm. The wheel has ten buckets and ten precutters, each with eight teeth. (Figs. 3 and 4).

Commissioning commenced end of May 1982. The system went into two-shift operation in July 1982, and in February 1983 three-shift, seven-day continuous operation commenced.

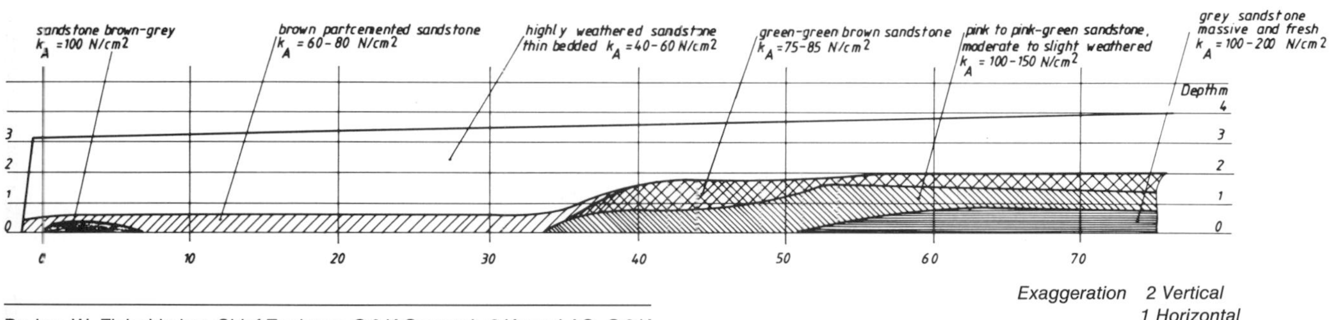

Fig. 1: Digging resistance in different layers

Dr.-Ing. W. Fleischhaker, Chief Engineer, O & K Orenstein & Koppel AG, O & K Tagebau und Schiffstechnik, P.O. Box 1601, D-2400 Lübeck 1, Federal Republic of Germany

Fig. 3: Bucket wheel

Fig. 5: Preblasted material

Fig. 4: Buckets and teeth

Fig. 6: Preblasted material

4. Experiences in Hard Material

At the onset, UDCL intended to preblast the very hard layers of unweathered sandstone. In August 1982, the first trials were carried out. Digging of the preblasted material has been successful. The main problem remained to obtain a maximum size of the blasted material of 50 cm (Figs. 5 to 11).

In the next step, the operation returned to unblasted material, even in hard sandstone (Figs. 12 and 13).

Fig. 7: Preblasted material

Fig. 8: Preblasted material

Fig. 9: Preblasted material

Fig. 10: Preblasted material

Fig. 11: Preblasted material

Fig. 12: Unblasted material

Fig. 13: Unblasted material

The results, reported over a period of ten months, were:

Maximum output	— 4,000 m^3(bank)/h
Average output during one month	— 2,800 m^3(bank)/h
Average output during ten months	— 2,300 m^3(bank)/h

The theoretical output of the machine is 4,400 m^3(bank)/h, the contractual figure is 1,500 m^3(bank)/h.

The availability of the system was:

Maximum digging time per month — 570 h,

$$\text{i.e., } 100 \cdot \frac{570}{19.2 \cdot 30} = 99\,\%$$

Average digging time per month — 385 h,

$$\text{i.e., } 100 \cdot \frac{385}{19.2 \cdot 30} = 67\,\%$$

As can be seen from these figures, there is no major problem in getting a high output, even in unblasted material.

5. Lifetime of Teeth

Looking at the experience of about 30 months, it seems to be a question of economics whether or not extremely hard material should be preblased.

UDCL has conducted some experiments with various teeth. They have used:

O&K teeth	— cast steel GS with hard metal
ESCO teeth	— hardened cast steel
Hensley teeth	— tungsten carbide mounted on cast steel

The reported data show the following:

ESCO teeth	— 65—80 h average lifetime
Hensley/O&K teeth	— 160—240 h average lifetime

The lifetime depends on the hardness of the material (ratio of minimum to maximum lifetime about 1:3) and on the position of the teeth in the bucket (ratio of minimum to maximum lifetime about 1:8).

6. Outlook

UDCL intends to dig a preblasted bench in the near future. The proposal is to obtain better information about the most economic operation in hard material. UDCL is also trying to replace the steel wear plates by ceramic material. The wear of this kind of material is less than a tenth of that of high tensile steel. The main problem is to prevent the plates from breaking (Fig. 14).

Fig. 14: Ceramic wear plates

Acknowledgement

The author thanks UDCL for its support and the permission to publish its results.

References

[1] Rodgers, H.C.J., and J.R. Brett: Basic Design of the Bucket Wheel Excavator for the Goonyella Mine. Int. Conf. Min. Mach., Brisbane, 1979.

[2] Hanemann, D.K.: Combined Bucket Wheel Excavator/Dragline Application: Goonyella Mine Australia; Braunkohle (1980) No. 12.

[3] Durst, W.: Earth Moving Machines for Larger Open-Cuts — in Particular Bucket Wheel Excavators and Spreaders. Inst. Eng. Australia, Paper M 1065, 1980.

[4] Fleischhaker, W.: A New Bucket Wheel Excavator Complex for the Goonyella Mine; bulk solids handling Vol. 1 (1981) No. 4, pp. 629-633.

Volume 5, Number 5, October 1985

Application of Bucket Wheel Excavators in a Philippine Continuously Operated Open Pit Mine

K. Wimmer, Austria, and H. Goergen, Germany

1. Introduction

Based on a feasibility study of Austromineral GmbH, Vienna, Austria, carried out in cooperation with the Technical University Aachen, Federal Republic of Germany, in May 1981, Semirara Coal Corp., Manila, Philippines, awarded Voest-Alpine AG, Austria, a contract for the delivery of the complete equipment, erection, start-up of operation as well as technical management of an open pit mine to be developed in Semirara, Philippines.

This mine is the largest open pit mine in South East Asia, with a production of 1 Mt/a of coal at an average overburden-to-coal ratio of 8.5 m^3(solid) : 1 t. The coal serves as fuel for a 300-MW power plant to be erected at Batangas, south of Manila, as well as for different branches of industry.

The present paper describes the geological and hydrological conditions of the Unong deposit situated in the Southeast of the Semirara island, approx. 270 km south of Manila. Also, the planning of the open pit mine will be dealt with in detail and the technical data of the equipment will be given.

All equipment has been installed during an erection period of twelve months. Production — in three-shift operation — was started in mid 1984.

2. Geological and Hydrological Conditions

The coal bearing formations of the Unong deposit, going back to the Miocene, are mainly composed of soft clayish sandstones and silts. In the utmost western part, this mine is limited by coralline limestones.

To prevent the pit from being flooded by tidal springs and high waves caused by typhoons, spring tides etc., a 6 m high breakwater dam has been erected in the critical part in the southeast.

Dr. Klaus Wimmer, Senior Vice President, Voest-Alpine AG, Finished Products Division, P.O. Box 2, A-4010 Linz, Austria, and Prof. Dr. H. Goergen, Technical University Aachen, Mining Department III, D-5100 Aachen, Federal Republic of Germany

The first data of the hydrological conditions have been registered by means of three test wells during the feasibility study. These test data showed a low permeability of the cap rocks with k-values in the range of $0.55 \cdot 10^{-7}$ m/s up to $3.8 \cdot 10^{-7}$ m/s, corresponding to a flow velocity of 0.2 to 1.4 m/h. These values are in accordance with a series of laboratory tests.

During the phase of development, a further seventeen test wells with a total of 4,000 drilling meters and an average depth of 150 m have been drilled for the further well layout. Due to the high salinity of the groundwater, the well has been installed with set up corrosion-resistant PVC pipes, specially designed and screwable, with an internal diameter of 175 mm. In areas of hydraulic contact with the sea, the groundwater is controlled by the application of corrosion-resistant immersion motor pumps.

In addition to the hydrological investigations, the usual soil mechanics tests have been carried out on undisturbed soil samples of the drill cores.

Tests to determine the required cutting force of the bucket wheel excavator were performed with a suitable specially equipped wheel loader. For successful excavator operation, all layers have to be won with a compressive strength of up to 1 kN/cm^2. The circumferential force at the bucket wheel therefore required is 120 kN to max. 180 kN at a theoretical winning capacity of the bucket wheel excavator of 1,050 m^3 (solid)/h.

The layout of the Unong open pit mine is influenced by a large number of negative factors:

1. 95% of the existing coal reserves down to a depth of 150 m are below sea level.
2. Being 1.6 km^2 in size, the field is relatively small.
3. With 3,000 mm/a on average, the quantity of rainfall is high.

Due to the existing geological conditions, the coal reserves of the Unong deposit are won in open pit mining, the field being relatively small in size, but deep. Three possible mining methods were tested, excluding strip mining with dragline as uneconomic because of the deposit conditions (Table 1).

Table 1: Alternative mining methods

Mining Method	Equipment Cutting	Loading	Hauling	Dumping
I discontinuous	dozer/ ripper	shovel excavator	trucks	dozer/ grader
II discontinuous/ continuous	dozer/ ripper	shovel excavator	belt conveyor	spreader
III continuous	bucket wheel excavator		belt conveyor	spreader

The soft, plastic overburden material renders impossible the application of rubber-wheeled vehicles for several months of the year, especially during the rain seasons (alternative I).

In the technical and economic evaluation, it has been considered, too, that the initial investment costs of alternative I are lower compared with alternatives II and III and relevant experience in conventional open pit mining on the Philippines is available.

Alternative II requires a large number of benches (max. 12 m high) and, therefore, a large number of expensive belt conveyor systems and belt waggons.

Finally, continuous mining has been decided for as the specific costs with the bucket wheel excavator are the lowest because of the simultaneous cutting and loading process.

Further advantages are a high operational availability, minimization of the number of haulage seams due to a possibly large excavation height, independent of diesel fuel, and the favourable total costs of production.

3. Mine Layout and Technical Data of Equipment

In order to secure coal production during the phase of development, the opening-up ditch was cut in parallel to the solid main seam. After having finished the opening-up, the open pit was further developed by slewing mining operation.

Coal and overburden are excavated on seven main haulage levels. The standard block width of the bucket wheel excavator amounts to 26 m. The average slope is 26.5 ° (1:2) with corrections to 12° to 18° in the southern and southeastern parts of the mine due to unfavourable soil mechanics conditions. The size and depth of the open pit, the petrography and geology as well as economic considerations require a bucket wheel excavator featuring a low weight and low soil pressure, but a high cutting force for cutting possibly existing hard sandstone strata. By an extremely long discharge boom, application of the excavator shall be made possible with a possibly large excavation height at high and low stages as well as a high cut without using a belt waggon.

3.1 Bucket Wheel Excavator

To meet the requirements, Voest-Alpine has developed and furnished an SR 400-type excavator with a total weight of 430 t. This excavator reaches a max. total excavation height of 33 m, i.e., from one position of the face conveyor it is possible to cut in three levels, 15 m, 8 m and 10 m (Figs. 1 and 2). This possibility is provided by:

- max. excavation height above subgrade of 15 m
- long discharge boom
- good buckling ability of the discharge boom
- small folding angle, i.e., an angle between bucket wheel boom and discharger of approx. 30 °.

Fig. 1: SR 400-type bucket wheel excavators

Fig. 2: Front of the bucket wheel of the SR 400-type bucket wheel excavator

Normally, the aforementioned total excavation height of standard equipment can only be achieved by means of a belt waggon. The advantages of having dispensed with the belt waggon are, on the one hand, reduction of the investment and operating costs, respectively, and, on the other hand, reduction of the number of belt transfer points, i.e., decrease in wear on transfer idlers and belts.

Technical data of the bucket wheel excavator are:

Diameter of bucket wheel	6.5 m
Number of buckets	10
Bucket capacity	400 l
Theoretical output at 75 discharges/min and 100% bucket filling	1,800 m³/h
Effective output	17,000—20,000 m³(solid)/d
Circumferential force	120—180 kN
Slewability of bucket wheel boom	150°
Discharges/min	75
Soil pressure	8.2 N/cm²
Length of bucket wheel boom	18.3 m
Length of discharge boom	39.0 m
Max. excavation height above subgrade	15 m
Installed power	1,020 kN

The coal is transported from the belt junction via an overland conveyor over approx. 5 km across the island to the blending stockpile with a capacity of 150,000 t. After homogenizing, the coal is moved by a rail-mounted reclaimer to a new loading pier at Dap Dap in the wind-protected part of the island. The loading pier at Dap Dap allows the loading of colliers of capacities of up to 20,000 t of coal. The capacity of the shiploader is 1,000 t/h.

3.2 Spreader

The overburden is dumped directly into the sea via a mobile spreader with crawler track assembly.

Technical data of the spreader are:

Capacity	5,400 m³/h 9,700 t/h
Belt width	1,600 mm
Belt speed	5.5 m/s
Installed power	520 kW

The spreader is connected with a tripper carriage and a shiftable belt conveyor with a conveying capacity of 5,400 m³/h, a belt width of 1,600 mm, a belt speed of 5.5 m/s and an installed power of 4,250 kW. The total service weight of the spreader including tripper carriage is 125 t.

3.3 Coal Stacker

For stacking and blending the coal, a rail-mounted mobile coal stacker is operated.

Its technical data are:

Capacity	3,060 t/h
Belt width	1,400 mm
Belt speed	5.2 m/s
Boom length	29.6 m
Max. stacking height	16 m
Total weight including tripper carriage	250 t

3.4 Reclaimer

The reclaiming of coal is carried out by a rail-mounted mobile reclaimer with the following technical data:

Capacity	1,000 t/h
Diameter of bucket wheel	6.3 m
Number of buckets	8
Length of bucket wheel boom	22.8 m
Belt width	1,200 mm
Belt speed	4.2 m/s
Total weight	177 t

3.5 Shiploader

The coal is transported from the reclaimer via a 2.5 km long belt conveyor system to the shiploader (Fig. 3). The ship loading can be done, independent of the size of the collier, via the so-called trimmer with a belt velocity of 15 m/s or via a loading chute with an integrated dust exhaust unit (air quantity for dust exhaust 1,800 m³/h).

Technical data of the shiploader are:

Capacity	1,000 t
Belt width	1,200 mm
Belt speed	5.2 m/s
Length of discharge boom	20 m
Total weight including tripper carriage	196 t

Fig. 3: Shiploader

3.6 Belt Conveyor Systems

The total length of the installed belt conveyor systems is 15 km with belt widths from 1,000 mm to 1,600 mm. The longest single belt conveyor with a length of 4,069 m and a width of 1,000 mm transports the coal from the mine across the island to the blending stockpile at Dap Dap.

The conveying capacity is 1,020 t/h at an installed driving power of 3,280 kW.

Volume 3, Number 1, March 1983

Reclaiming Gold and Uranium in South Africa

Benno Woelkie, South Africa

Summary

Bucket wheel excavators coupled with mobile conveyor systems are used almost exclusively in open-pit operations for removing overburden and mining minerals.

The purpose of this article is to describe the technically and economically successful application in a new field, viz. reclaiming slimes dams containing gold and uranium.

1. Background

Since gold was discovered in 1886 on the Witwatersrand which extends from Springs 50 km East of Johannesburg to the West as far as the Klerksdorp goldfields and the Orange Free State goldfields at Welkom, slimes dams were constructed.

They originate from underground gold mining and are the remainder of a gold containing pyritic conglomerate, found extensively in the Witwatersrand basin, which was crushed and pulped before extraction of the gold, and then dumped on the surface of the mining area.

Apart from gold "left-overs" significant quantities of uranium minerals are contained in this waste product. The dam construction consisted of an initial dike frame of compacted dry material. These were filled with slurry and were allowed to dry before the next layer of slime was deposited.

Incidentally, the slimes dams surrounding Johannesburg became a most striking landmark with their golden yellow colouring and their desert-like dust clouds over them on windy days, until grassing of the old mine dumps began some years ago.

A few decades back gold mining processing had not developed far enough to make close to 100 % yields economically possible. It was easier to mine the gold containing conglomerate in the traditional fashion with abundant available labour than to use a sophisticated chemical extraction method whose cost would have been prohibitive anyway regarding the relatively low gold price in those days.

Consequently the waste material was dumped with a considerable amount of gold remaining in it together with significant quantities of uranium, a mineral whose importance and values were not recognised.

Since gold mining rapidly became costlier over recent years, especially through payment of higher wages, and because the demand for uranium increased heavily on the world market, the viability of re-mining the previously uneconomic old mine dumps became an important issue in South Africa.

A case in point is the Stilfontein Gold Mine. It was established that the mine's slimes dams contain a certain amount of gold together with significant values of uranium representing considerable value in foreign exchange and strategical terms.

2. Reason for the Application of a Bucket Wheel Excavator Together with a Shiftable Conveyor Belt

In March 1978 Stilfontein Gold Mine's parent company "General Mining and Finance Corporation" (now Gencor) asked Weserhütte South Africa (now PHB Weserhütte S.A. or PWH S.A.) to develop an overall plan with detailed design for a continuous reclaiming system of the gold mine's slimes dams.

To transport the sand from the slimes dams to the extraction plant a conventional hydraulic pressure system is normally used which consists basically of a high pressure water jet generator (water cannon), pumps, repulpers and pipelines.

The Stilfontein Gold Mine decided against the application of this traditional equipment and a bucket wheel excavator coupled with a shiftable conveyor belt was selected instead (Fig. 1). The criteria finally deciding the method to be

Fig. 1: Slimes dams reclamation

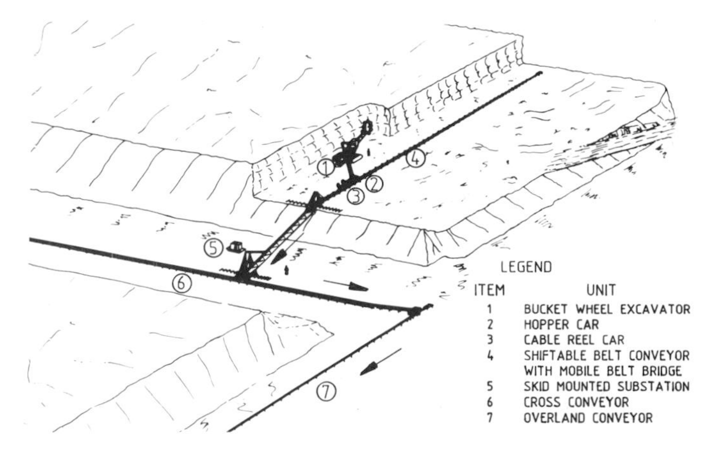

employed were: water and energy consumption, wear and tear cost, flexibility of the equipment and the requirements of the material's subsequent metallurgical treatment.

The water used for the conventional hydraulic pressure system as transport medium can only partly be re-utilised which results in a significant waste of a commodity which is rather scarce in South Africa and the energy required to maintain adequate water pressure in the conveying pipes represents a substantial cost item in a mine's profit calculation.

Water is also not an ideal transport medium for sand because its density degree in relation to sand causes the latter to sink very quickly unless a strong impetus is prevalent to keep it flowing. Consequently, the water has to be agitated continually through turbulence in the pipes; or, alternatively, a flotation agent must be introduced into the water which then carries the sand along. Either way additional costs occur.

Pipes, repulpers and pumps are all subject to heavy wear and tear because there are many points of metal friction which cannot be eliminated. The extensive pipework needed for conveyance must be dismantled for shifting. For the subsequent extraction treatment, viz., leaching and filtering, one half of dry material reclaimed and transported by the bucket wheel and conveyor system and one half of diluted material were needed. The excessive water in the pulp would have presented a distinct disadvantage, because it would have had to be removed again. (In the case of Stilfontein Gold Mine the diluted material is supplied by a neighbouring mine).

3. Time Schedule and Production Program

The plan designed by PWH South Africa for reclaiming the slimes dams at Stilfontein Gold Mine provided for a "crawler mounted bucket wheel excavator" together with a "shiftable face conveyor coupled with a mobile belt bridge".

The contract, concluded in March 1978, specified delivery times too short for the construction of such a specialised plant but by sharing the work load with PWH's parent company in Germany completion was achieved in time. The design of the bucket wheel excavator was carried out in Germany. The shiftable conveyor belt with hopper and cable reel car as well as the conveyor bridge were designed in South Africa.

The time schedule adhered to was as follows:

Completion of all engineering work	end of May 1978
Delivery of the most important imported parts for the bucket wheel excavator	December 1978
Supply of the locally manufactured components	February 1979
Erection completed	June 1979
Commissioning period	end of July 1979

4. Geometric Requirements

The hourly processing requirements for this project were relatively small and called for a compact type of bucket wheel excavator. (For the first working period 250 t/h based on a specified annual production of 1,800,000 t). A compact type of excavator is designed mainly for work where high-cuts are required because the short bucket wheel boom cannot make extensively deep cuts. Cutting can be done without difficulty up to 9 m digging height which allows for one block up to 10.5 m wide if the slope inclination is not below 45 degrees and provided the slope does not collapse (Fig. 2).

The geometric dimensions of the bucket wheel excavator were a result of the specified reclaiming capacity which determined a length of 8.8 m for the excavator boom and a length of 16 m for the discharge boom. With these dimensions a well balanced load combination of the slewable super structure and the load receiving undercarriage was achieved.

Reclaiming operations were started at slimes dam No. 3 and the downward operating conveyor bridge was designed especially for this type of dam operation.

The height of the dam varies and contains approximately $1475 \times 15 \times 325 = 7.2 \times 10^6$ m³ mining material which equals approximately 10.0×10^6 t with a bulk density in situ of 1.4 t/m³.

The effective excavator capacity (according to contractually laid down capacity increases at later stages) determined the volumetric dimensions of the plant as a whole and the static dimensions of the bucket wheel excavator and of the conveyor bridge were calculated in accordance with the European regulations for the construction of mobile open pit equipment (FEM).

5. System Description

PWH South Africa received the order from Stilfontein Gold Mine for a complete reclaimer plant consisting of:

- One crawler mounted bucket wheel excavator
- One rail mounted hopper car
- One rail mounted cable reel car
- One shiftable face conveyor coupled with a rail mounted mobile belt bridge
- One mobile substation on skids
- One bulldozer with side arm attachment and shifting head
- Electrical fittings and complete installation for the reclaiming plant (Figs. 3 and 4).

Detailed descriptions of the bucket wheel excavator, technical data and basic working method have been published and are available from the relevant literature.

Main data are contained in Table 1 providing general information. For the application of the excavator at the Stilfontein Gold Mine certain modifications had to be made and special features had to be incorporated in the various components. The special cutting conditions at this mine and the cutting

Table 1: Technical data of the bucket wheel excavator SR 250 (Fig. 5)

Theoretical digging capacity	750 m³/h
Bucket content	250 liter
Number of discharges	50 per minute
Bucket wheel diameter	4.8 m
Belt width	1050 mm
Belt speed	2.62 m/s
Digging height	9.0 m
Bucket wheel boom length	8.8 m
Discharge boom length	16.0 m
Average ground pressure	65 kPa

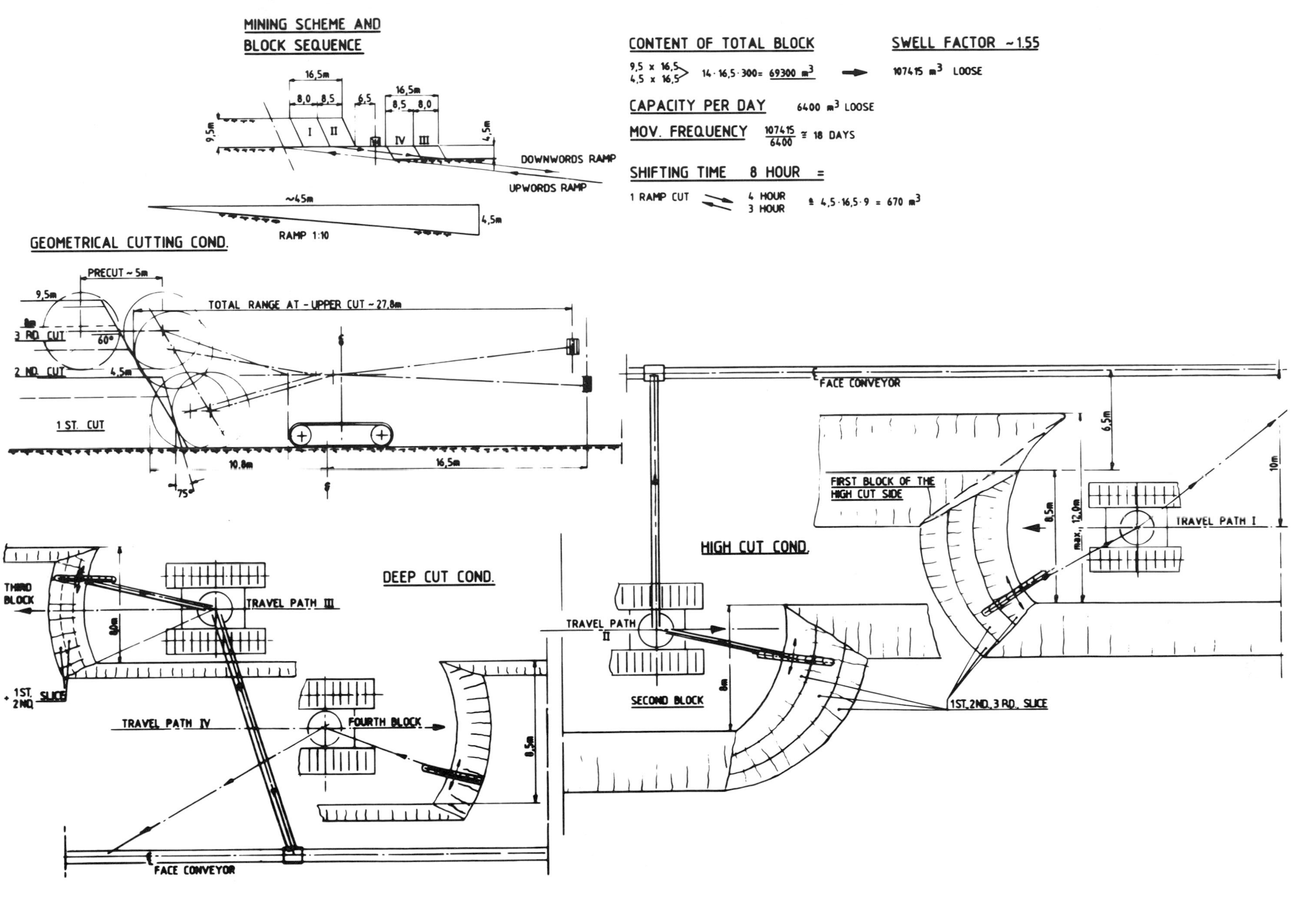

Fig. 2: Cutting configuration

Fig. 3: Plain view after opening cut

Fig. 4: View of the plant at the second cut

Fig. 5: Bucket wheel excavator while digging

method applied are demonstrated in Fig. 2. Buckets, ring body, ring and discharge chute had to be adapted to the wet and sticky properties of the conveyed material. All points coming into contact with the material were coated with HDPE (High Density Polyethylene). The internal gearing of the slewing device was made completely spillage-free and the foot-plates of the crawlers were broadened to compensate for the comparatively low bearing capacity of the ground which can contain up to 40 % water.

The foot-plates were also coated with an anti-caking layer for safety reasons. Friction moments between foot-plates and ground made it necessary to enlarge travel drives. The driver's cabin and the electrical room were insulated and pressurised.

The reclaiming method designed by PWH South Africa is shown in Fig. 6 and demonstrates the various working phases from formation of the travel ramp up to the block mining method at the slimes dam No. 3. A detailed description of the working stages would take up too much space in this article, but a short description outlining the major concepts of the design follows (Fig. 6).

The conveyor belt system was designed for a two-cut mining method because the total mining height ascended from 15 m to 20 m (slimes dam No. 3) and because of the order in which the belt was arranged. For the upper cut the shiftable conveyor belt is positioned on the cutting level of the bucket wheel excavator but later re-located to the natural surface level when making the lower cuts. The shiftable conveyor and the belt bridge are a coupled unit. The belt forming the link with the shiftable conveyor is designed to be established on any working level of an inclination between Nil and 13 m transporting the material downwards or horizontally between the excavator and the overland conveyor. The head pulley of the bridge drives the conveyor belt. The bridge is hinged on two supporting frames which are wheel mounted on rails and move independently of each other. Conveyor bridge and shiftable face conveyor move simultaneously to their new position.

A mobile hopper is installed between bridge and overland conveyor. Contrary to original apprehensions due to the convex guide line of the belt at the transfer points between the shiftable part and the wheel mounted bridge, the steel cord belt performs well and retains its troughing shape. To achieve this performance the belt is pretensioned to the limit of the transportable force able to be transmitted.

Table 2: Technical data of the shiftable face conveyor (Figs. 7 and 8)

Conveying capacity	750 m³/h (1,150 t/h)
Belt width	900 mm
Belt speed	2.62 m/s
Troughing angle	35°
Lift distance	from —13 m to + 4 m
Belt quality	ST 500 (5 + 5)
Length between the pulleys	350 m

Fig. 7: Bucket wheel excavator in operation

Fig. 8: View of the hopper and cable reel car

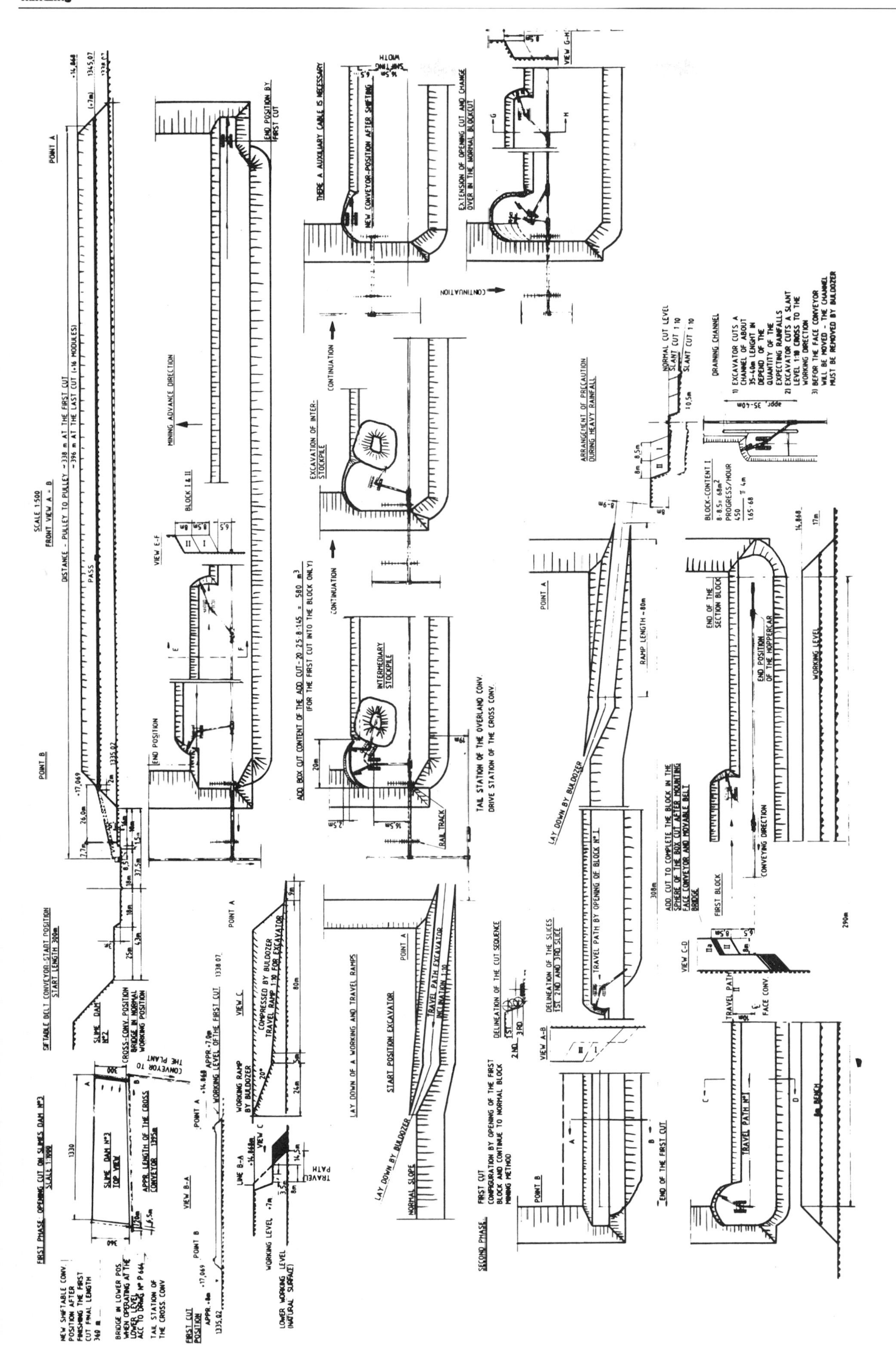

Fig. 6: Working plan

6. Operating Experience and Evaluation of the System

After a short time of operation and training of personnel the plant performed reliably according to the designed capacity. The results even exceeded expectations from the beginning. During the first months after commissioning, approximately 400 t/h on average were processed and these quantities were intermittently increased to 650 t/h, compared with 250 t/h budgeted and peak performances of 1,050 t/h were achieved from time to time.

The quantities above were, however, produced sporadically and were not within the scope of the average requirements of the plant.

During the initial years of operation at the lower reclaiming rate of 135,000 tons per month (approx. 4,500 t/day) the operating costs excluding depreciation and interest have been around 0.39 R/t. When the reclaiming rate has reached the 270,000 ton per month level planned for later years the operating cost will be around 0.20 R/t.

Loss of production time due to shifting the face conveyor equipment was recorded at 18 hours in the beginning and reduced to 16 hours during the later production periods reaching 8 hours under extremely favourable conditions after each 380 working hours, during which time approximately 107,000 m³ bulk were mined.

The useful life of the hard metal faced welded bucket cutters and teeth was extended up to 9 months or approximately 3,500 working hours, respectively.

Wearing chute liners and plates made from HDPE lasted for approximately 12 to 14 months. The break-down time of all conveyor idlers, after a running-in period, was reduced to 4 % of the total numbers per year.

The efficiency of the bucket wheel excavator increased from approximately 65 % during the first year to approximately 80 %. During the following years a further improvement to approximately 86 % was achieved. The availability of the whole plant was increased from approximately 68 % to 74 %. The comparatively high capital outlay required for the plant had been justified by the high degree of utilisation as well as the low operating cost achieved.

Bucket wheel excavator coupled with conveyor belt equipment used in a continuous system of reclaiming the slimes dams has certainly distinct advantages over other transportation systems and the application at the Stilfontein Gold Mine has proved that the changeover to this new technology can be very successful.

Acknowledgement

The author wishes to thank the management and staff of General Mining Union Corporation Limited (Gencor) for their contribution towards this article and the management of PHB Weserhütte (S.A.) for permission to publish this paper.

Photographs and drawings:
PHB Weserhütte (S.A.) (Pty.) Ltd.

 Volume 3, Number 2, June 1983

Box and Ramp Cuts with Bucket Wheel Excavators in Opencast Mines and Mining Bench Working Schedules

Horst V. Birkheuer, Germany

Summary

The author reviews extensively procedures for continuous mining systems using bucket wheel excavators and shiftable conveyor systems. General information is presented with respect to open cast mining technology including detailed analysis of box cuts and ramp cuts. Working schedules are presented for a variety of equipment combinations and the need for adequate consideration of the mining strategy to be adopted, in the mine planning stage, is stressed.

1. Introduction and Definitions

Where economic reasons call for the exploitation of a deposit by opencast mining, the special mining techniques of using bucket wheel excavators, shiftable belt conveyor systems and spreaders is continuously gaining in importance throughout the world.

When adopting this principle, the bucket wheel excavator removes the top overburden of a deposit at a certain width over the entire pit length, thus exposing a strip of the deposit.

The pay mineral is exploited by bucket wheel excavators or similar equipment, transported via shiftable belt conveyor systems (bench conveyors) towards one of the mine perimeter sides for subsequent haulage out of the pit by means of an elevating conveyor.

The excavators operating in the overburden also feed onto shiftable belt conveyors. Belt conveyors installed on the second face (face conveyors) ensure transportation of material to the mine perimeter (dump). Shiftable dump bench conveyors take the overburden for subsequent feed onto belt type spreaders, which in turn permit backfilling of the waste into the mined out pit.

The pit advances over the deposit by adopting a parallel and/or slewing operation, always maintaining, however, the initial width. The dump area constructed by the spreaders can subsequently be recultivated.

The mining method described has been successful because it permits maximum continuity of mining operations. The worldwide trend towards steadily increasing efficiency and production rates contributed considerably to the development of this mining technique.

Mining with bucket wheel excavators and belt conveyor systems requires repeated initial cuts (box cuts) when changing from one strip (block) to the next.

When mining on several faces (benches), there exists the need to construct inclined travelling faces (ramp cuts).

Since these procedures have to be continuously repeated during the mining cycle, they should be considered during operational planning and not be decided during running operations. Any non-consideration of these procedures might cause high costs due to avoidable shutdown times.

The conditions imposed on bucket wheel excavators and belt wagons during box and ramp cuts have to be considered *prior to the investment,* because they influence the dimensioning of the machines required.

In the past, the above-mentioned points were often not taken into account. This caused frequent interference during the running operations.

Because the following description requires a fundamental knowledge of the design and working method of bucket wheel excavators and shiftable belt conveyor systems, some basic information is given here for a better understanding.

2. General Information on Machines and Belt Conveyor Systems

2.1 Bucket Wheel Excavators

Figs. 1—4 show the three main construction types adopted for bucket wheel excavators.

Normally, mining operations with bucket wheel excavators develop parallel to the bench conveyor, i.e., by the so-called block operation. In most cases, the excavators move on crawlers. The relevant block cross-section (theoretically a parallelogram) is frequently mined out in several digging faces, through continuous slewing of the bucket wheel with its boom and associated superstructure.

The loading boom or bridge do not make any slewing motions but make matching motions. Depending on the construction type adopted for the bucket wheel excavator, the

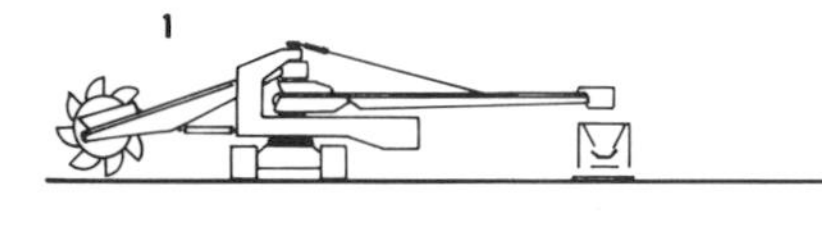

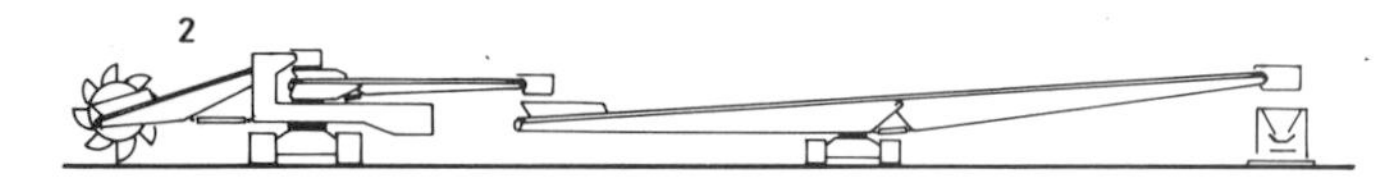

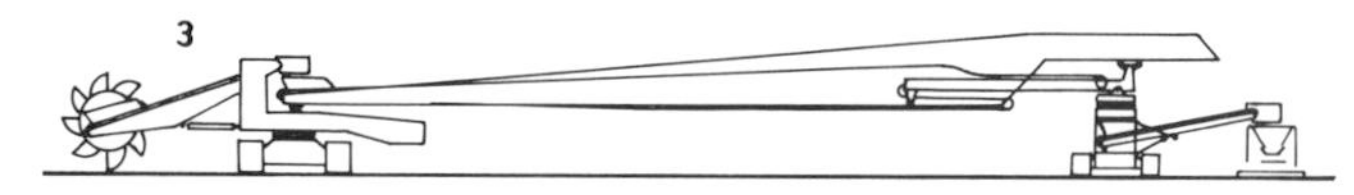

Fig. 1: Bucket wheel excavators: 1) with self-supporting loading boom 2) with belt wagon 3) with bridge and hopper car

Fig. 2: Bucket wheel excavator with self-supporting discharge boom (Fig. 1.1)

Fig. 3: Large scale bucket wheel excavator with bridge and hopper car in an opencast mine of the Rheinische Braunkohlenwerke, output = 90,000 m³/day overburden or coal (Fig. 1.2)

Fig. 4: Bucket wheel excavator with belt wagon in an open cast mine in the USSR (Fig. 1.3)

material taken by the bucket wheel is transmitted through the slewing center onto the belt of the loading boom or that of the bridge. Subsequently, the material is fed onto the shiftable belt conveyor system or an intermediate belt wagon.

The loading boom or conveyor bridge are also slewable and in most cases possess a luffing function.

The advantage of a bridge over a belt wagon lies naturally in the field of large-scale machines, since in most cases distances (> 80 m) have to be overcome which are no longer within an economical range for a belt wagon. Also, a matching of the feeding points can be made more easily with bridge and hopper car than with a belt wagon.

2.2 Belt Wagon

In principle, a belt wagon represents a belt conveyor moving on crawlers to overcome the distance between excavator and bench conveyor. The belt conveyor axis is slewable in the horizontal plane to the travel mechanism and in most cases both boom ends are of the luffing type.

The advantage of using belt wagons is found in the mobility of the excavating system. It is in most cases more adaptable to changing operating conditions, especially when mining operations cover several benches.

The advantages and disadvantages of a mining system need, however, to be judged for each specific case, as to which construction type should be adopted.

2.3 Shiftable Belt Conveyor System

In the case of a continuous material flow in an opencast mine, the belt conveyor system represents the best suited means of transport to do the haulage job involved. When this transport method is selected, bench conveyors are located in the pit on the relevant levels (benches). They can be used for both transport of the mineral or haulage of the overburden to the dump.

The layout of a shiftable belt conveyor system is as follows:

The tail station with belt return pulley is located at the rear of the conveyor, the tiedowns of this station being anchored in the ground. This return pulley can however, also be mounted in a car and be used as hoist-rope tractable take-up pulley. There exists also the possibility for accommodation of additional drives. The individual conveyor frames supporting the belt and the material are lined up and connected by shifting rails. Normally, a hopper car with chute and impact table travels on these shifting rails. This arrangement guarantees a perfect and smooth material transfer from the excavator or belt wagon onto the bench conveyor.

The head station (Fig. 5), serving in most cases both as drive and take-up station, is located at the transfer point onto the next conveyor in line. This head station is equipped with an intermediate bridge in the area of the inclined conveyor section.

Fig. 5: Drive station of a shiftable belt conveyor system in a Canadian oil sand mine

Since the geographical location of the mining stope changes — either development of mining operations at right angle to the bench conveyors (parallel operation) or mining around the feed point at the head station (slewing operation) —, the bench conveyors have to be shiftable in a transverse direction. Special crawler-mounted shifting vehicles do this job, pulling the belt conveyor during each longitudinal travel by approximately 1 to 2 m until the new position is reached. Both tail and head station are shifted separately. The heavy head station is shifted by bulldozers, walking pads or transport crawlers.

3. General Information on Box and Ramp Cuts

In an open cast mine, where a bucket wheel excavator has to work onto a belt conveyor system by the "advance block operation", it should be considered that the excavator has to cut its own box for the new block. When adopting the multiple-bench operation, the excavator has to travel to the adjacent benches over ramps situated parallel to these benches. The excavator has to dig these ramps, too.

All benches to which the excavator descends over ramps need no additional box cut, since the ramp cut permits immediate start of the "advance block-operation" (Fig. 6).

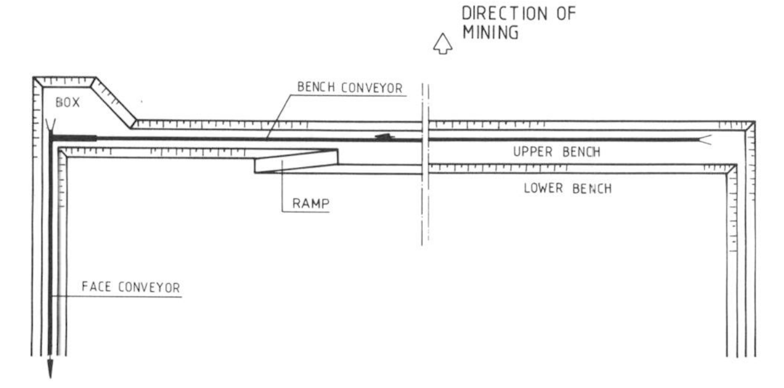

Fig. 6: Excavator bench in an opencast mine

The shape and size of a box cut and its position depend mainly on the selected mining method, the geological conditions and the equipment available.

If the mining operation is not to be repeatedly interrupted by digging box or ramp cuts, it is recommended that the excavator be able to assume this job, too.

The dimensioning of the machine depends on the distance to be overcome by the material flow from slewing center of the excavator up to the feed point onto the bench conveyor.

4. Dimensioning of the Excavator for Box and Ramp Cuts

4.1 Clearance Angle at the Bucket Wheel

When designing an excavator for operation in an open cast mine, special consideration has to be paid to the layout of the bucket wheel head, especially with the goal of obtaining a minimum clearance angle. For a real understanding of this term see Fig. 7.

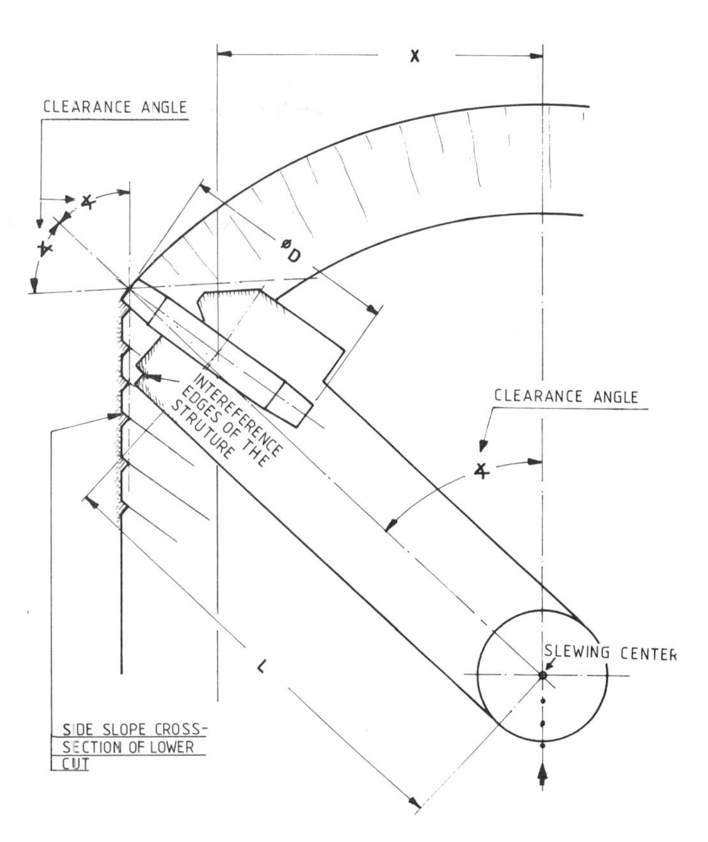

Fig. 7: Clearance angle at the bucket wheel head

The minimum distance (*X*) of the excavator from the slewing center up to the side slope base is determined by both the clearance angle and bucket wheel diameter (*D*), as well as by the boom length (*L*). The dimension (*X*) cannot be smaller than this minimum during a box or ramp cut, without leading to a collision between bucket wheel head contour and side slope. (Exception: Lower cutting height is < 0.4 of wheel diameter).

The economical range for a clearance angle is between 30 and 45 degrees. This angle is not freely selectable, it depends, among others, on the outreach of the bucket wheel boom in the top digging position, considering the general slope angle (Fig. 8).

4.2 Curve Radius of the Crawler Travel Gear

In case of excavator crawler travel gears having more than 2 crawlers, the minimum negotiable curve radius obtained by this configuration restricts the manoeuverability of the excavator during the box cut. A repeated shunting, to reach a defined position in a relatively small box cut, is no longer economical.

5. The Box Cut — Requirement

The cutting of a box on the excavator top bench accomplishes the following three conditions:

1. The box enables the excavator to start digging along the bench conveyor at full block cutting position.
2. The box can be used for parking of the excavator during bench conveyor shifting close to the side slope, in order not to restrict the space required for conveyor shifting.

   Fig. 24, part 2, shows that the box can be used in addition for parking of a belt wagon during the mentioned shifting procedure.

3. In case of multiple-bench operation, the excavator travels around the tail station of the bench conveyor to reach the next operating bench. If a box is cut at the tail station before the excavator leaves the top bench, this box enables the excavator after its return to reach the new digging position around the tail station of the bench conveyor. (In the meantime the bench conveyor having been shifted close to the side slope.)

Normally, box cuts are made at the ends of the excavator operating bench. It might, however, occur, that a box cut needs to be made within a running operating bench.

5.1 Box Cuts at the Head Station

The excavator position illustrated in Fig. 10 shows the bridging distance of the rear excavator section (*A*) — considering the head station length (*KS*) and the minimum angle β of the excavator booms to each other.

In the case of insufficient outreach of the discharge boom, or of the belt wagon or bridge during the last box cut at the head station prior to shifting, one can for the reasons indicated above, cut the box at a position before reaching to the bench end.

The block remaining at the head station can then be removed *subsequent* to shifting of the excavator bench conveyor, since the angle α between the rear bridging section of the excavator unit and the bench conveyor is now more acute and permits overcoming of the distance *KS* (see Fig. 18, part 1).

The usual method of cutting a box, as shown in Fig. 10, is also adopted in case of direct-overcasting operations and represents an economical and practicable solution.

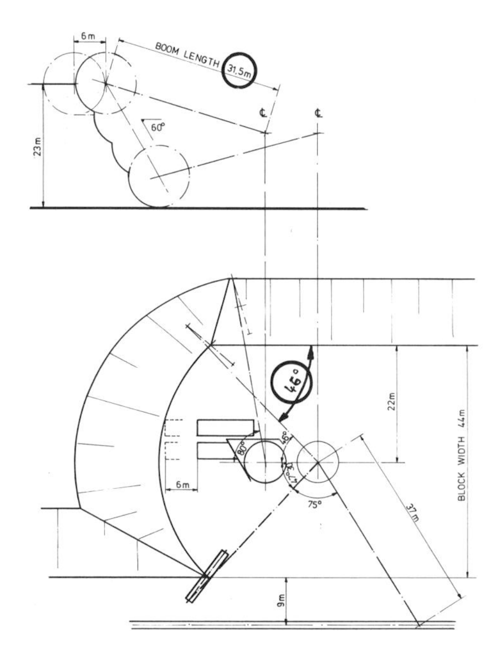

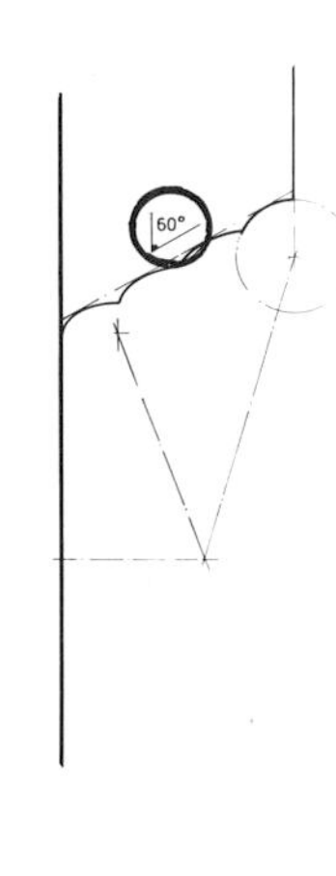

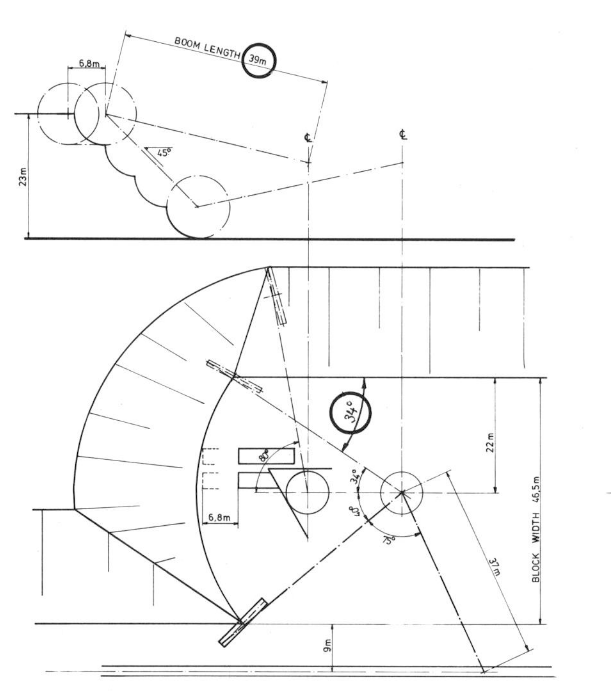

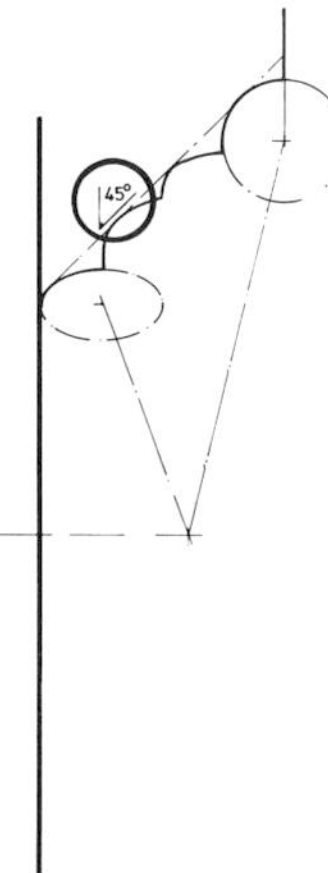

Fig. 8: Example of the dependence of wheel boom length and side slope angle on the clearance angle

Fig. 9: Excavator with bridge and hopper car during a box cut

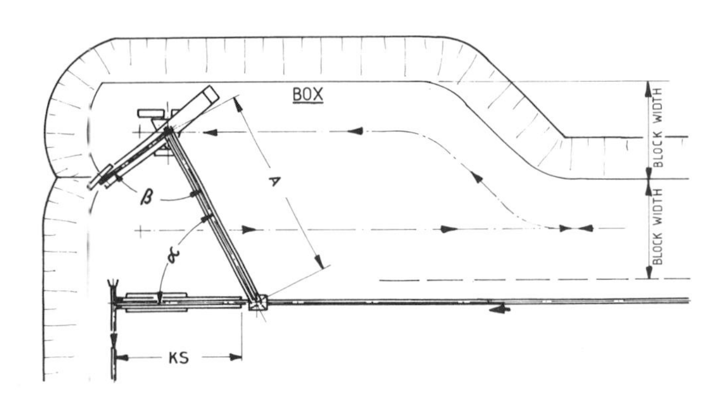

Fig. 10: Typical box cut shape close to the head station

Fig. 11: Box cut during direct-overcasting operation

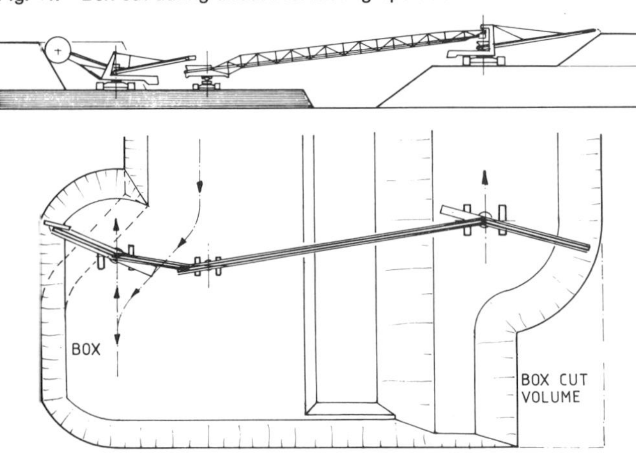

The descending face conveyor is not always situated on the bench conveyor level, but can be positioned on the bench on top of it. Should this be the case, there is frequently the need to use inclined belt conveyor bridges at the headstation-sided end of the excavating bench.

Fig. 12 shows the shunting position of an excavator with belt wagon during a ramp cut at the head station. It can be seen that the transfer points I, II and III are not only feed points onto the face conveyor, during the box cut procedures, but also onto a mobile inclined bridge (see the dotted contour on the face slope).

When making a box at the bench conveyor head station, attention should be paid to the need of having sufficient space for the tail station of the face conveyor and its anchoring. This space requirement can be guaranteed by an additional cut, which should preferably be made by the bucket wheel excavator or if necessary by auxiliary equipment (see Fig. 12).

5.2 Box Cuts at the Tail Station

The length of a box to be cut at the tail station depends primarily on the fact that the excavator — feeding onto the first possible bench conveyor transfer point — can start its new block cut at the slope of the opening cut. Secondly, the box length extending beyond the tail station during multiple-bench operation results from:

Length of tail station + anchoring — crawler travel gear width of the excavator + safety clearance (during position change around the tail station).

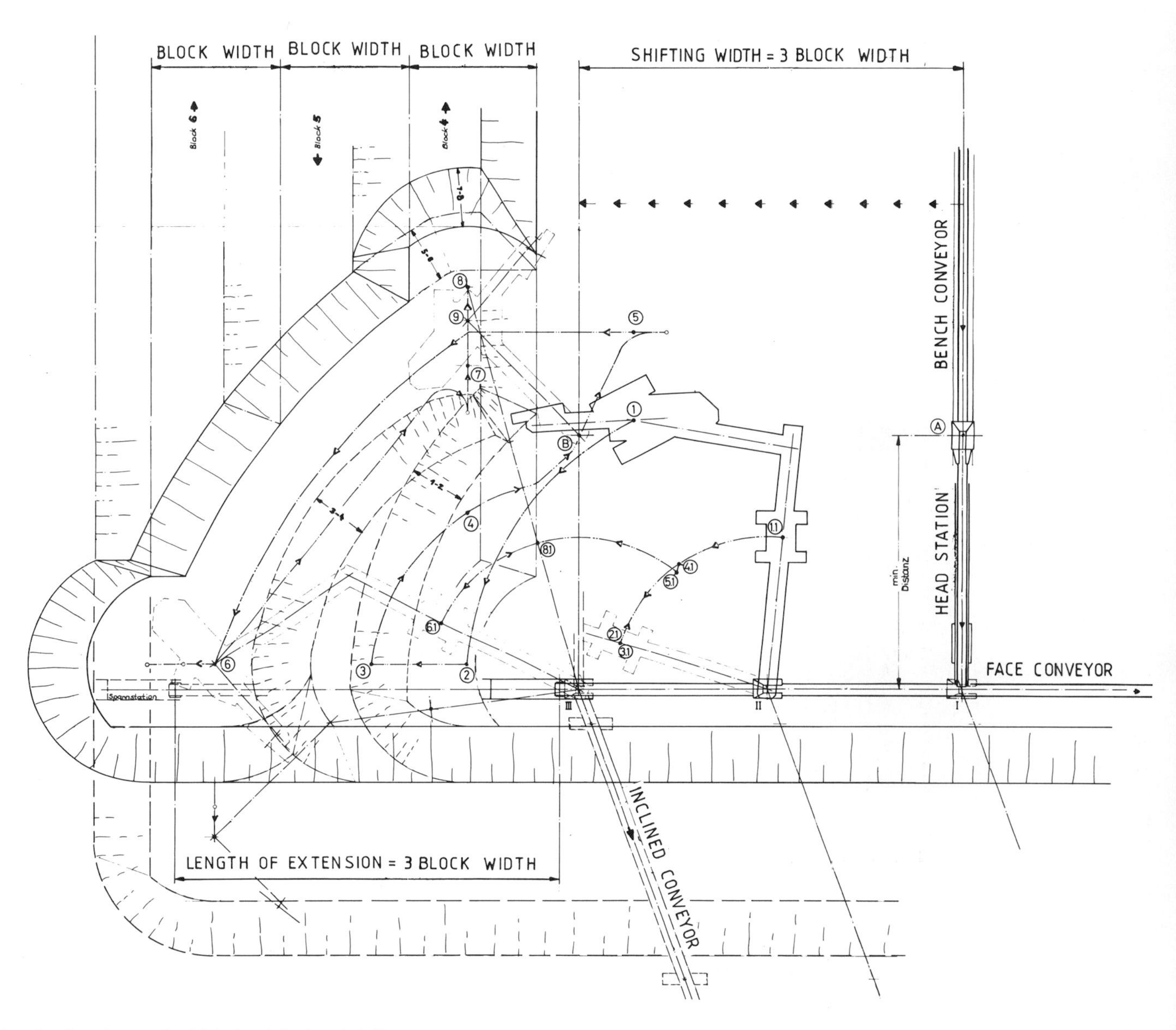

Fig. 12: Opening cut for 3 blocks at the head station

Description of the box cut procedure in Fig. 12:

Face conveyor is extended. Hopper car moved from position I to position II.

Position 1 and 1.1:
Start of the opening cut.

Position 2 and 2.1:
Completion of the first opening cut.

Position 3 and 3.1:
Excavator advances, belt wagon does not move.

Position 4 and 4.1:
Completion of the trailing second cut. Position change of the excavator to position 5. Position change of the belt wagon to position 5.1. Tracking of the hopper car from position II to position III.

Position 5 and 5.1:
Start of the third opening cut. Belt wagon now feeds at position III.

Position 6 and 6.1:
Completion of the third opening cut.

Position 7:
Position change of the excavator to starting position for the fourth opening cut.

Position 8 and 8.1:
Completion of the fourth and last opening cut.

Position 9 and B:
Start of block 4, conveyor has been shifted in the meantime. Material feed is made at point B without the belt wagon. Inactive belt wagon travels along with the excavator.

This type of box cut is applied and shown during the mining method in Fig. 21, parts 10 and 11.

These factors may be an important criterion in the dimensioning of the rear bridging section of the excavator, considering additionally the last possible feeding point onto the bench conveyor (see Fig. 22, Part 1, position 0 and Part 2, position 6).

Normally, the box is cut at the tail station *prior* to shifting of the bench conveyor close to the side slope.

There exist, however, mining methods (as in e.g. Fig. 21) *not* envisaging a so-called preventive box cut at the tail station *prior* to shifting of the bench conveyor.

Fig. 13 shows the possibility to dig — when using a belt wagon — a new block *subsequent* to shifting in the range of the tail station. Despite the somewhat difficult operation of the belt wagon, this method grants the advantage of continuously good digging conditions ahead of the excavator, since any inclined wedge-shaped cuts are unnecessary.

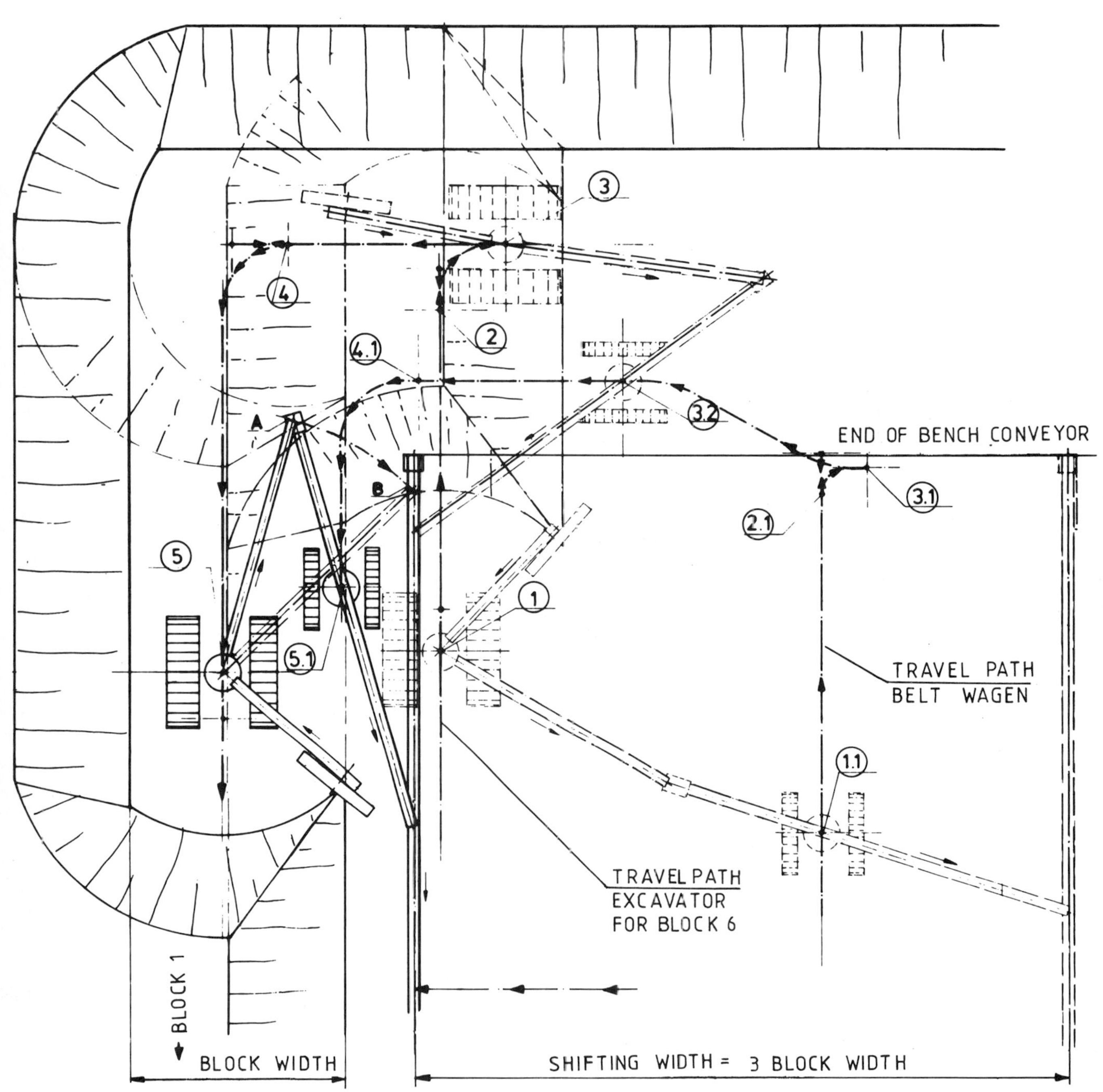

Fig. 13: Opening of a new block at the tail section subsequent to shifting — excluding a box, but using a belt wagon (see also Fig. 21 Parts 1—3)

Description of the box cutting procedure in Fig. 13:

Position 1 and 1.1:
Excavator and belt wagon approach the end of a block.

Position 2 and 2.1:
Excavator and belt wagon at the end of the block.

Position 3 and 3.1:
Position of the excavator and belt wagon at the beginning of the box cut. Shifting of bench conveyor. Belt wagon moves towards position 3.2.

Position 4 and 4.1:
Excavator and belt wagon at the end of the box cut.

Position 5 and 5.1:
In position 5, the excavator slews its discharge boom from position A to B and the belt wagon is put out of operation.

Fig. 14 shows turning of a bucket wheel excavator with conveyor bridge in the box.

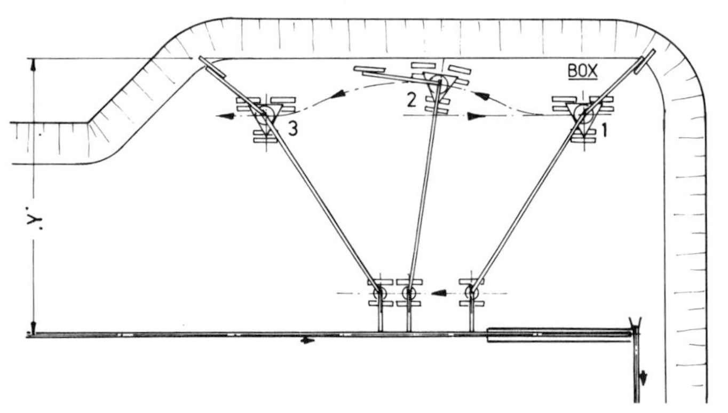

Fig. 14: Dimension "Y" = required bench width for the case of change in direction of excavator with bridge and hopper car

6. Ramp Cut

If a bucket wheel excavator in an opencast mine is intended to operate on more than one digging bench in changing cycle, this change from one bench to another involves the need for the construction of ramps. Their slope has to be adapted to the travelling characteristics of the excavator (normally approx. 1 in 15 to 1 in 20).

Excavators possessing a levelling device have, however, in the past cut ramps with an inclination of 1 in 6. Considering economy, the excavator should cut its own ramps and remove them, too. The masses accumulating should be transported out of the mine by means of the excavator bench conveyor. This requirement can only be guaranteed if the ramp is constructed parallel to the excavator bench conveyor (Fig. 15).

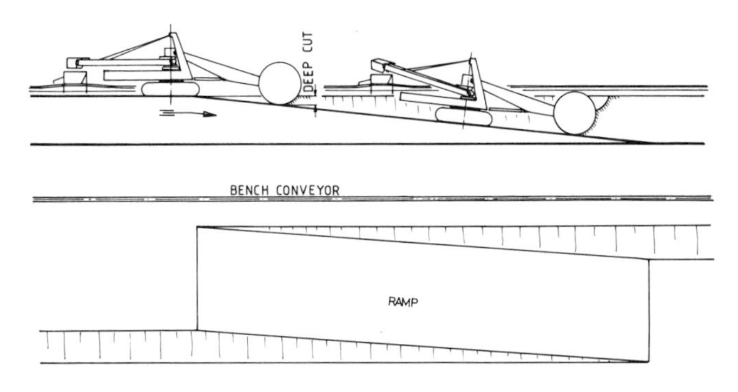

Fig. 15: Bucket wheel excavator during a ramp cut

The mass of a ramp, positioned at right angles to the bench conveyor, could not be transported over the bench conveyor, neither during construction nor during removal of such a ramp. In addition thereto, any non-removal of such a ramp would reduce the useful bench length.

The best position for a parallel ramp is generally to be found at half the relevant bench length, in order to minimize the idle operations of the excavator and thus possible times of non-production.

Normally, ramps are cut from top to bottom. Considering this special feature, the opening phase envisages a deep cut for the excavator.

In special cases, however, the excavator can make an upgrade cut subsequent to a box cut (Fig. 16).

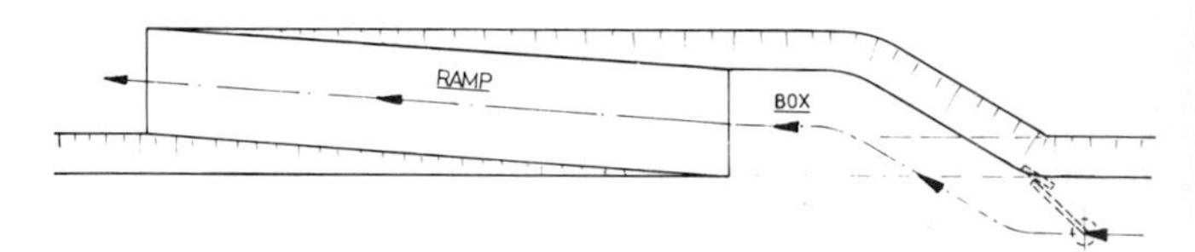

Fig. 16: Upgrade ramp cut with preceding box cut

The width of a ramp to be cut by the excavator, as well as the position of the excavator on this ramp, are mainly determined by strong adherence to the afore-mentioned clearance angle at the bucket wheel head. For economical reasons it seems to be most suitable to use the dimensions of the block width as a guide for the ramp width to be dug, since keeping of this width will permit continuous passage from a ramp cut into a subsequent block operation. (See Fig. 22, part 3).

The ramp maintained after removal of the old ramp has to have a minimum width that grants sufficient space for travel of the excavator crawlers during a change from one bench level to the next. (See e.g. Fig. 22, Parts 3 and 5).

A ramp situated in the center of the excavating bench would lead to a subdivision of the blocks to be excavated, and cause laterally offset blocks. Removing of the so-called old ramp and advancing of the offset block (2) — visible in the figure for development of mining operations in case of multiple benches — will generally be of important influence on the design to be adopted for the rear bridging section of the bucket wheel excavator.

Should the bucket wheel excavator not have this outreach for economical reasons, it is possibile to remove the ramp to the area of the tail station. This, however, will impair the continuity of the system. (See Fig. 17, Parts 1 to 3).

Fig. 17 shows a solution to be adopted for removing an old ramp when the loading boom length of the bucket wheel excavator is not sufficient to permit direct material feed onto the bench conveyor.

As far as the direction of ramp sloping is concerned, the following conclusive information has to be given on the no-load travel of the bucket wheel excavator required to be made during multiple-bench operation:

The ramp side, extending towards the conveyor bench, should point in the direction of the tail station. The mining schedule in Fig. 27 shows that only this arrangement of ramps permits a position change of the excavator around the tail station to higher benches.

7. Several Mining Bench Working Schedules

Starting from the working range of *one* bucket wheel excavator, differentiation is generally made between single-bench and multiple-bench operation. Depending on the selected mining method, the shifting width of the bench conveyor is equal to one, two or three times the block width of the bucket wheel excavator. Every effort should be made to achieve the shifting procedure over several block widths in one operating cycle.

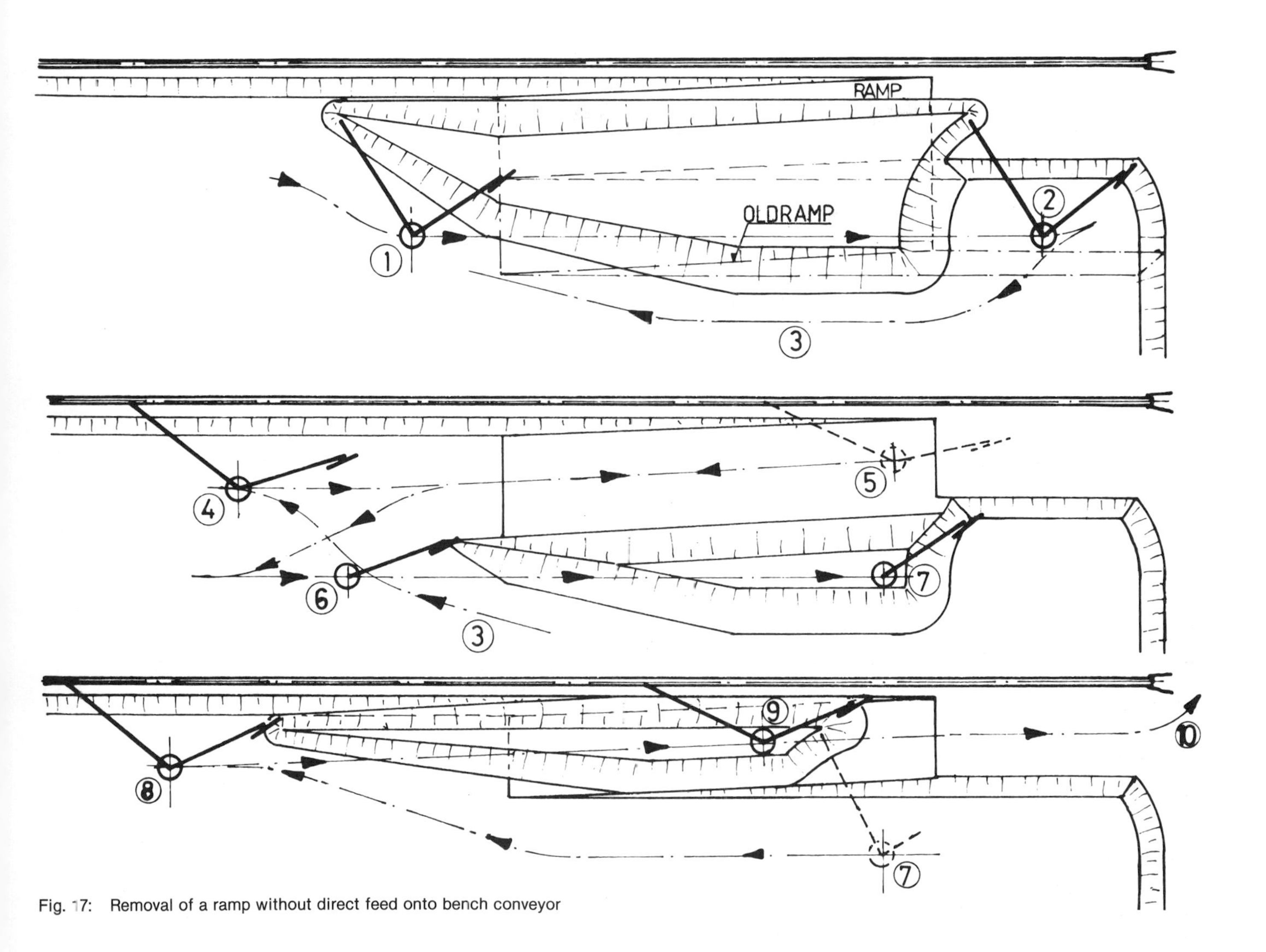

Fig. 17: Removal of a ramp without direct feed onto bench conveyor

Explanation of Fig. 17:

Position 1 to 2:
Removal of the old ramp.

Position 3:
Travelling around the still standing ramp.

Position 4 to 5:
Digging of the ramp to remain (material feed on to bench conveyor), reverse travel.

Position 6 to 7:
Removing of the pile beside the ramp (discharge on to the ramp), reverse travel.

Position 8 to 9:
Repeated clearing of the ramp.

Position 10:
Travel around the tail station.

The mining method to be selected on the excavating bench side depends mainly on geological conditions, and/or on the intended production rate of the mine.

7.1 Working Range of the Excavator: On One Bench

The mining method represented in Fig. 18 is the simplest one existing, i.e., a single-bench operation, using a bucket wheel excavator with self-supporting loading boom. This configuration does not, in most cases, need a belt wagon.

The shifting width of the bench conveyor is equal to the block width of the bucket wheel excavator. Box cuts are made at the head and tail station of the bench conveyor, their depth corresponding to the block width. Subsequent to the box cut, position 8, there would theoretically be the possibility to cut a second block prior to shifting of the bench conveyor. This would, however, lead to uneconomical excavator operations during the next block cut. The loading boom outreach is determined by positions 2 and 7 (visible in parts 1 & 3). Here, the maximum width is required during the box cut or at the end of the block, since the hopper car cannot travel in the area of the inclined section towards the headstation (route a). No material transfer is thus possible.

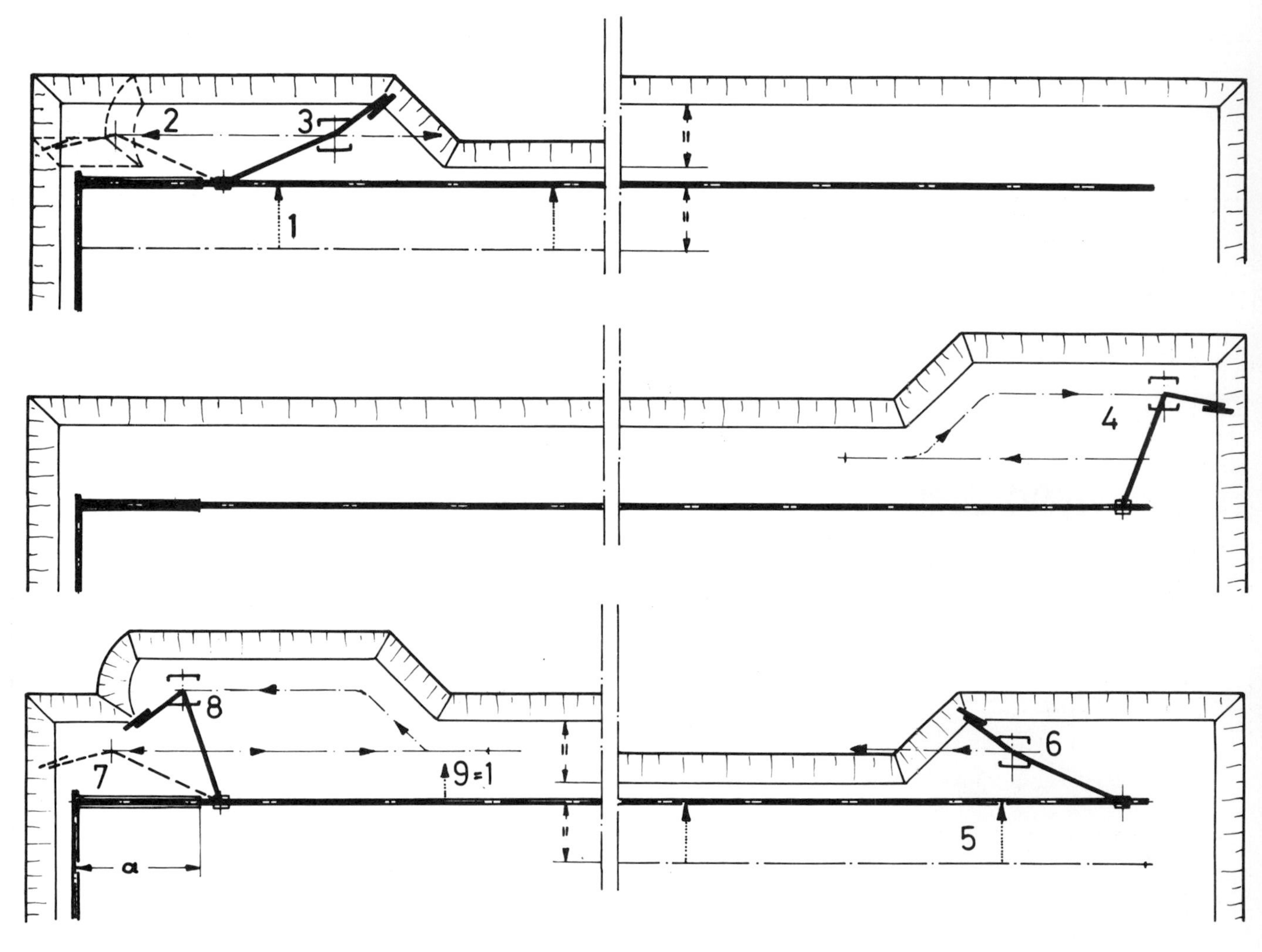

Fig. 18: Working schedule: Single bench operation (shifting width = block width)

Explanation of Fig. 18:

Position 1:
Shifting of the bench conveyor by one block width.

Position 2:
Digging of the remaining box at the head station.

Position 3:
Advancing of the block out of the box.

Position 4:
Digging of the box at the tail station.

Position 5:
Shifting of the bench conveyor by one block width.

Position 6:
Advancing of the block out of the box.

Position 7:
End of block cut.

Position 8:
End of the block cut at the head station.

Position 9:
Shifting of the bench conveyor by one block width.

It is, however, possible to make this section feedable, too, e.g., chuting and equipping this section with closely spaced impact idler sets or use of a separate feeding car on this inclined conveyor route. These possibilities are, however, uneconomical compared to the costs incurred by using a longer loading boom or a belt wagon.

A more highly developed mining method is shown in Fig. 19. During this single-bench operation, the frequency of conveyor shifting procedures is halved by shifting the bench conveyor over a double block width, subsequent to mining out of two blocks. This naturally results in a reduction of down-time. This solution requires, however, the operation of a belt wagon.

The advancing of block 1, position 3, is decisive for dimensioning of the loading boom outreach, since block 1 is mined out without the aid of a belt wagon. The required length of the belt wagon can be taken from position 7, since the maximum distance has to be overcome during this box cut situation.

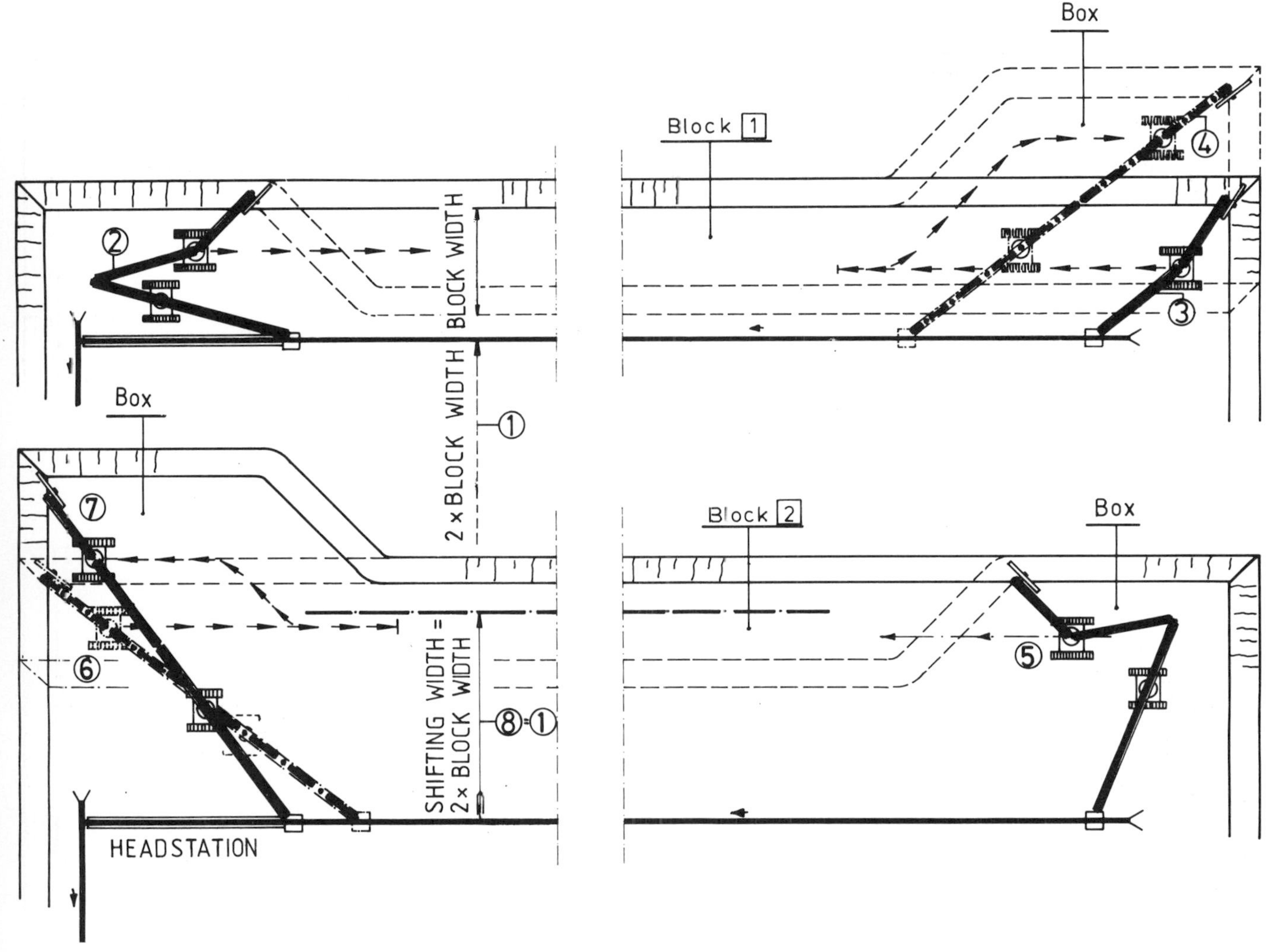

Fig. 19: Working schedule: Single bench operation (shifting width = 2 block widths)

Explanation of Fig. 19:

Position 1:
Shifting of the bench conveyor by two block widths.

Position 2:
Advancing of block 1 out of the box.

Position 3:
Final position of the excavator for block 1.

Position 4:
Final position of the excavator during a box cut at the head station for block 2.

Position 5:
Start for advancing of block 2.

Position 6:
Final position of the excavator for block 2.

Position 7:
Final position of the excavator during box cut at the head station.

Position 8:
Shifting of the bench conveyor by two block widths.

This mining method can be used even without a belt wagon, however, using a bucket wheel excavator having a conveyor bridge with hopper car instead of a self-supporting loading boom.

Figs. 19 and 20 show a mining method incorporating a "3-block-cycle", the first section of the working sequence corresponds to the above described mining method.

Should there be a wish to cut down the bench conveyor shifting procedures to a minimum, it is possible to cut a third block with the same machinery prior to shifting (see positions 8 and 9).

When this solution is adopted, there will be no possibility of shifting the bench conveyor by three block widths, since a previous new box cut has to be made. The outreach of the machines are, however, not sufficient to make a further box cut, at the tail end. The bucket wheel excavator and belt wagon (position 10) have to make a no-load travel towards the head station, to cut a new box.

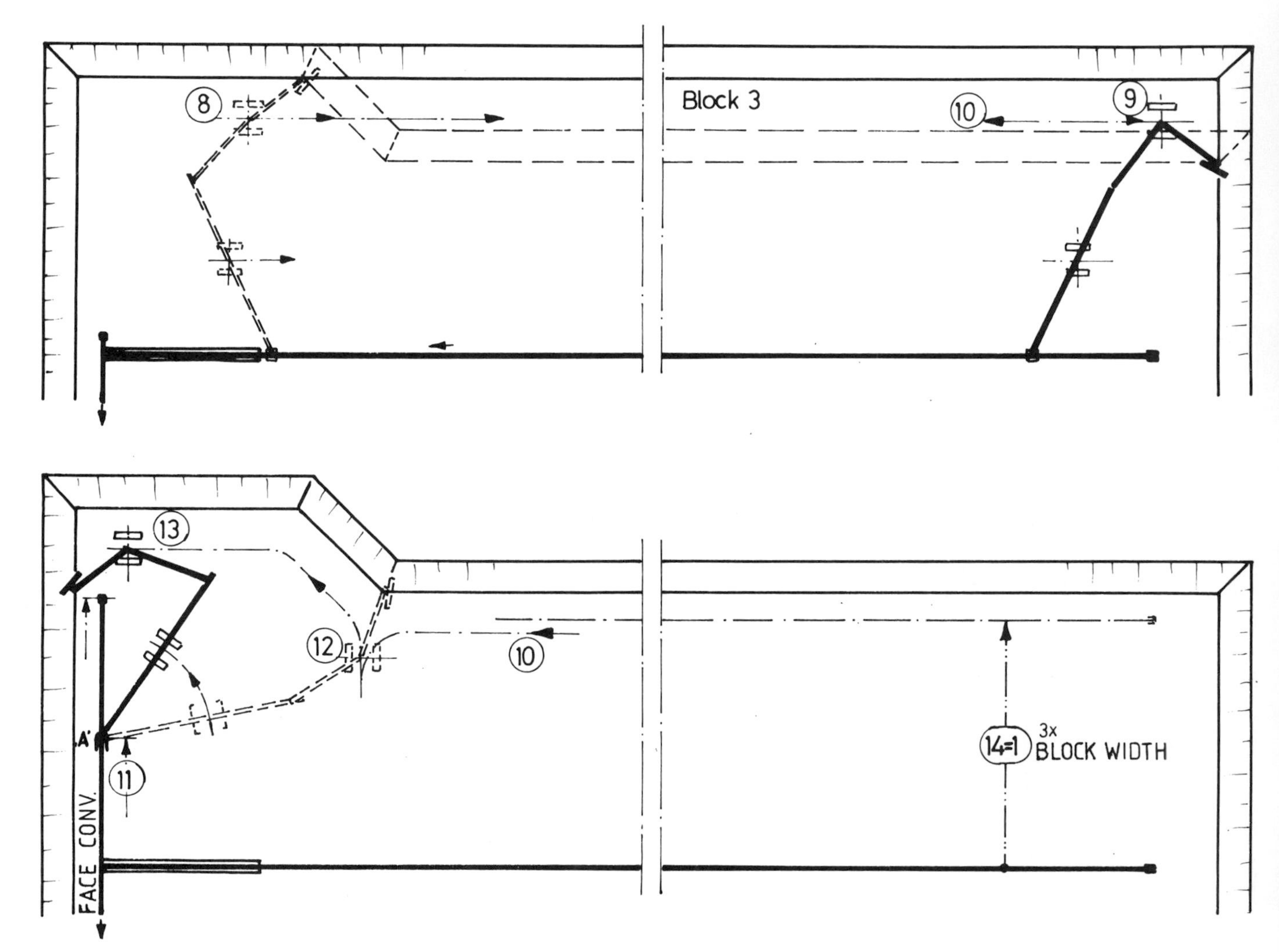

Fig. 20: Extension of working schedule in Fig. 19 (shifting width = 3 block widths)

Explanation of Fig. 20:

Position 8:
Start for advance of block 3.

Position 9:
Final position of the excavator in block 3.

Position 10:
No-load travel of the excavator and belt wagon towards the head station.

Position 11:
Advancing of the face conveyor and moving of the feed table "A".

Position 12:
Starting position of the excavator for box cut of block 1.

Position 13:
Final position of the excavator in the box.

Position 14:
Shifting of the bench conveyor by three block widths.

The resulting time of non-production is, of course, disadvantageous for the described 3-block-system. During this no-load travel, the face conveyor can be extended by the length equal to three block widths (position 11). The feeding table A under the head station of the bench conveyor is turned towards the tail station of the face conveyor. It serves now as feed point during the subsequent box cut (positions 12 and 13). The bench conveyor is out of operation during the above procedures and can be shifted the three block widths. The feeding table of the face conveyor is subsequently re-positioned under the head station (positions 14 and 1).

The described operating method can also be accomplished by a bucket wheel excavator with conveyor bridge and hopper car. Fig. 21 shows a single-bench operation for three blocks, differing from the preceding primarily in the fact that two mining cycles are adopted alternately (positions 1 to 9 and 10 to 15).

This results in the advantage that one idle operation is deleted during a three-block operation (odd number). The box cuts at the head and the tail station are, however, somewhat complicated. The required box depths at the head station amount to 2 and 3 block widths (see positions 5 and 10). To permit this, there is the need — as adopted for the previous method, too — to extend the face conveyor at its tail station by three block widths and to twist the feeding table towards the tail station. In the meantime, shifting of the bench conveyor can be commenced (see positions 9, 10 and Fig. 12).

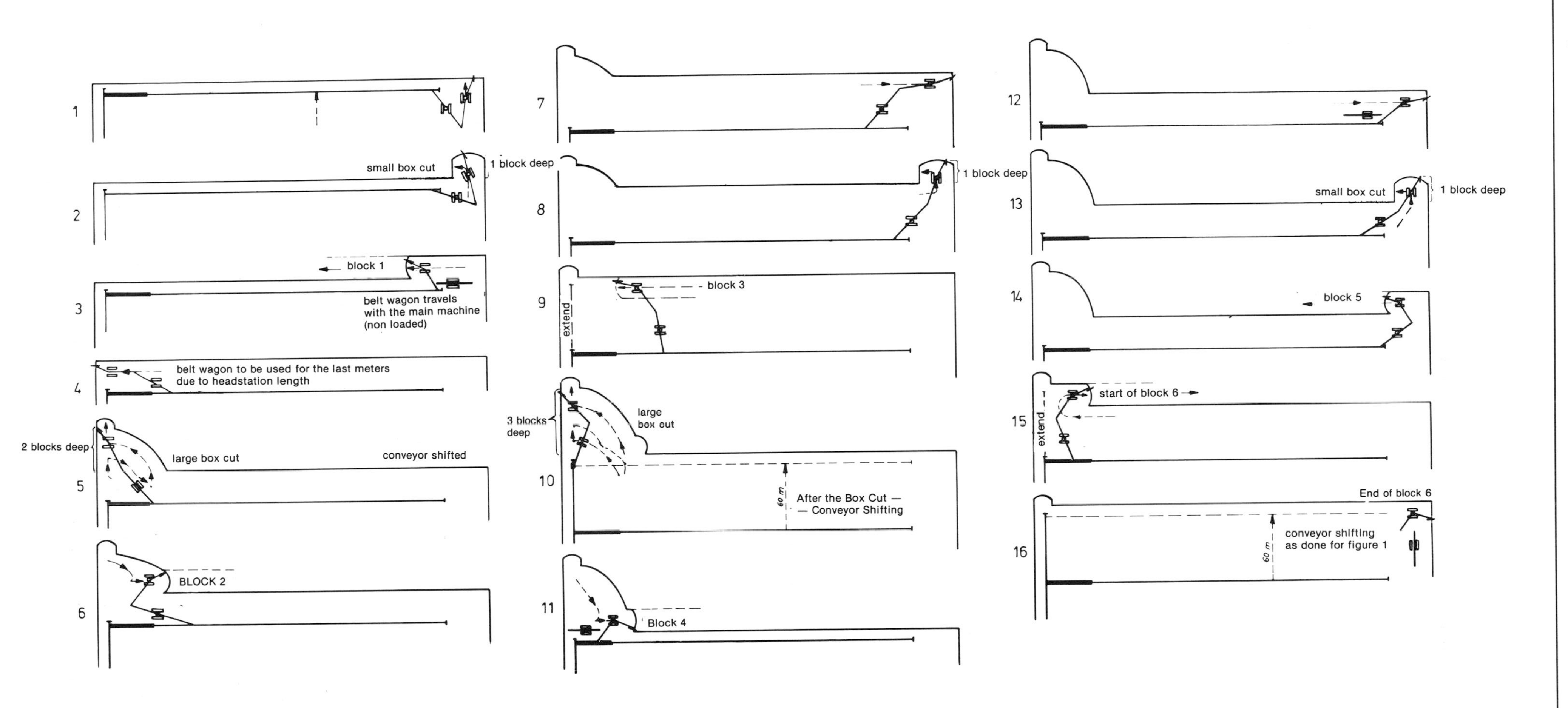

Fig. 21: Working schedule: Single bench operation (shifting width = 3 block widths, alternating cycles)

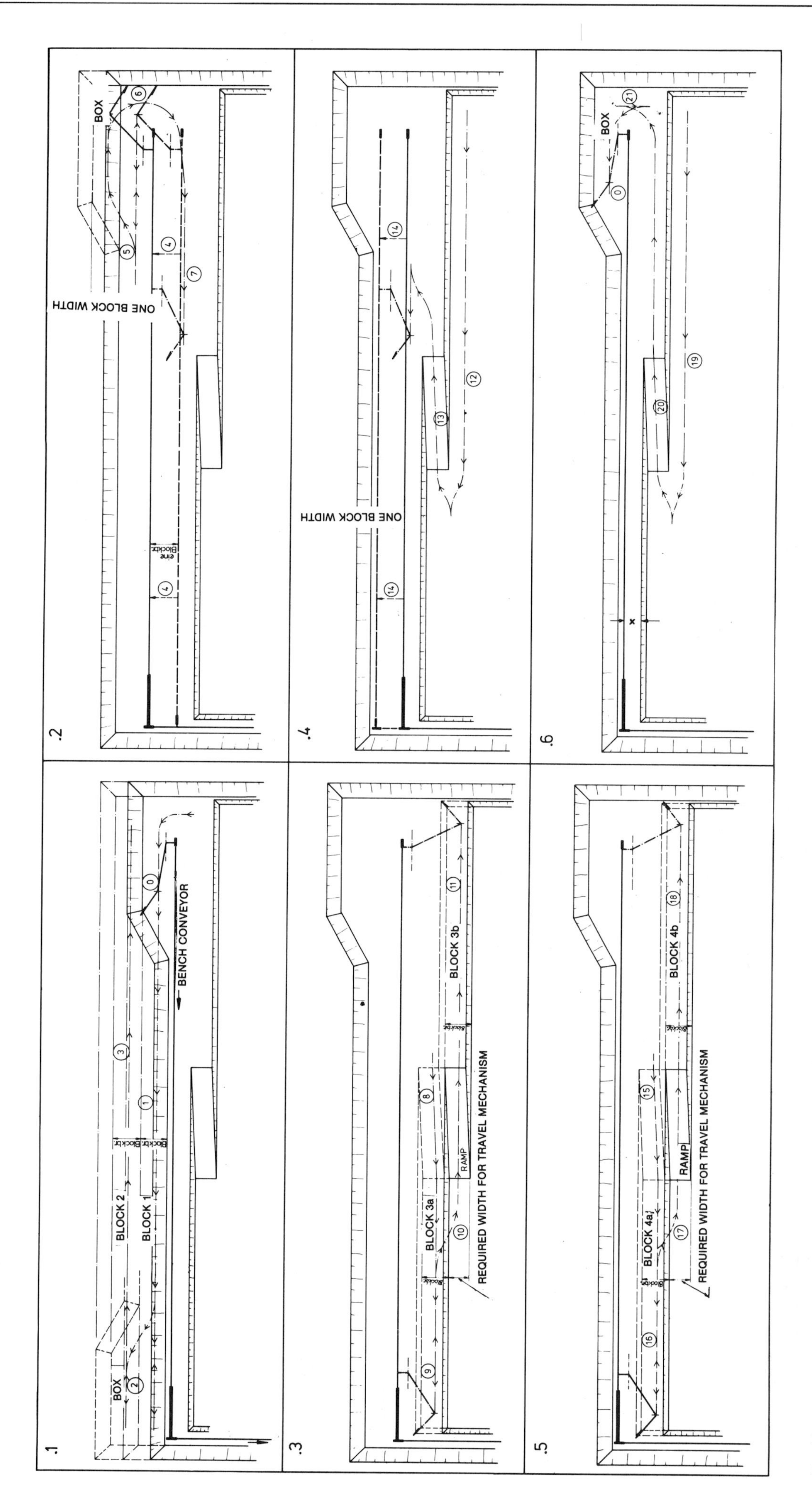

Fig. 22: Working schedule: Double bench, two block operation

Explanation of Fig. 22:

Position 0:
Excavator in starting position for removal of block 1.

Position 1:
Mining of block 1.

Position 2:
Mining of the box cut at the head station.

Position 3:
Mining of block 2 towards direction of head station. After termination of this block cut:

Position 4:
Shifting of the bench conveyor and extension of the face conveyor by one block width.

Position 5:
Starting of the excavator for box cut at the tail station.

Position 6:
Bucket wheel excavator travels around the tail station and then to:

Position 7:
to start down-grade ramp cutting

Position 8:
Down-grade ramp cut and subsequent

Position 9:
advancing of block 3a.

Position 10:
Start for block mining 3b up to:

Position 11:
on the lower excavator bench.

Position 12:
Idle travel back to the ramp and

Position 13:
up-grade over the ramp, in the meantime

Position 14:
shifting of the bench conveyor by one block width.

Position 15:
Repeated down-grade ramp cut and

Position 16:
subsequent advancing of block 4a

Position 17:
Start of block mining

Position 18:
up to the lower excavator bench.

Position 19:
Idle operation of the excavator towards the ramp.

Position 20:
Up-grade travel over the ramp.

Position 21:
Excavator travels around the head station in order to reach the initial position 0 in the existing box.

The boxes at the tail station have to be cut at right angles to the bench (see positions 2, 8, 13 and Fig. 13). The belt wagon may have a length identical to that for a 2-block operation (see position 5). Due to the complicated box cuts, an excavator with bridge and hopper car cannot be used here.

7.2 Working Range of the Excavator: On Two Benches

The two-bench double-block mining method, represented in Figs. 22 and 23, requires one to shift the bench conveyor twice, i.e., for the first time subsequent to mining out of the two top blocks and for the second time between mining out of block 3b and 4a.

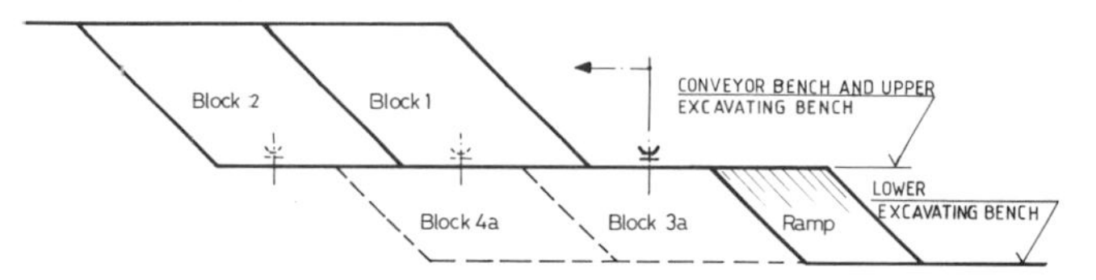

Fig. 23: Cross-section of excavator benches in Fig. 22

This method is, however, only to be recommended if a bucket wheel excavator with bridge and hopper car is to be used.

The space "X" in Part 6 is required to permit hopper car travel during removing of block 3a and 4a. The hopper car can only transfer the material on to the bench conveyor, if both are standing on the *same* grade level.

Quite another situation exists when using a belt wagon as shown in Fig. 24. The greater flexibility of the system makes it advantageous to adopt the somewhat more effective mining method — 4 blocks on two levels — i.e., conveyor shifting by two block widths after the mining of four blocks.

Since the excavator can travel on the lower bench during the block cut IVa and feed onto the higher situated bench conveyor, the dimension "Y" in Fig. 4 can be reduced and be smaller than the dimension "X" in Fig. 22, Part 6. When doing this, the two lower blocks III and IV move closer to the bench conveyor. This close approach of the blocks to the bench conveyor guarantees economic dimensions for the belt wagon and leads to the required operational flexibility.

7.3 Working Range of the Excavator: On Three Benches

The last mining method described here represents the maximum mass which can be reached by one excavator, one belt wagon and one bench conveyor, before shifting of the bench conveyor will be required.

Mining of 6 blocks on 3 levels, with only one shifting procedure, requires one to make several no-load travels of the machine. This must not inevitably lead to a negative effect on the economy of this method. The required belt wagon has to be designed such that it can serve as a connection link between the three operating levels (Figs. 28 and 29).

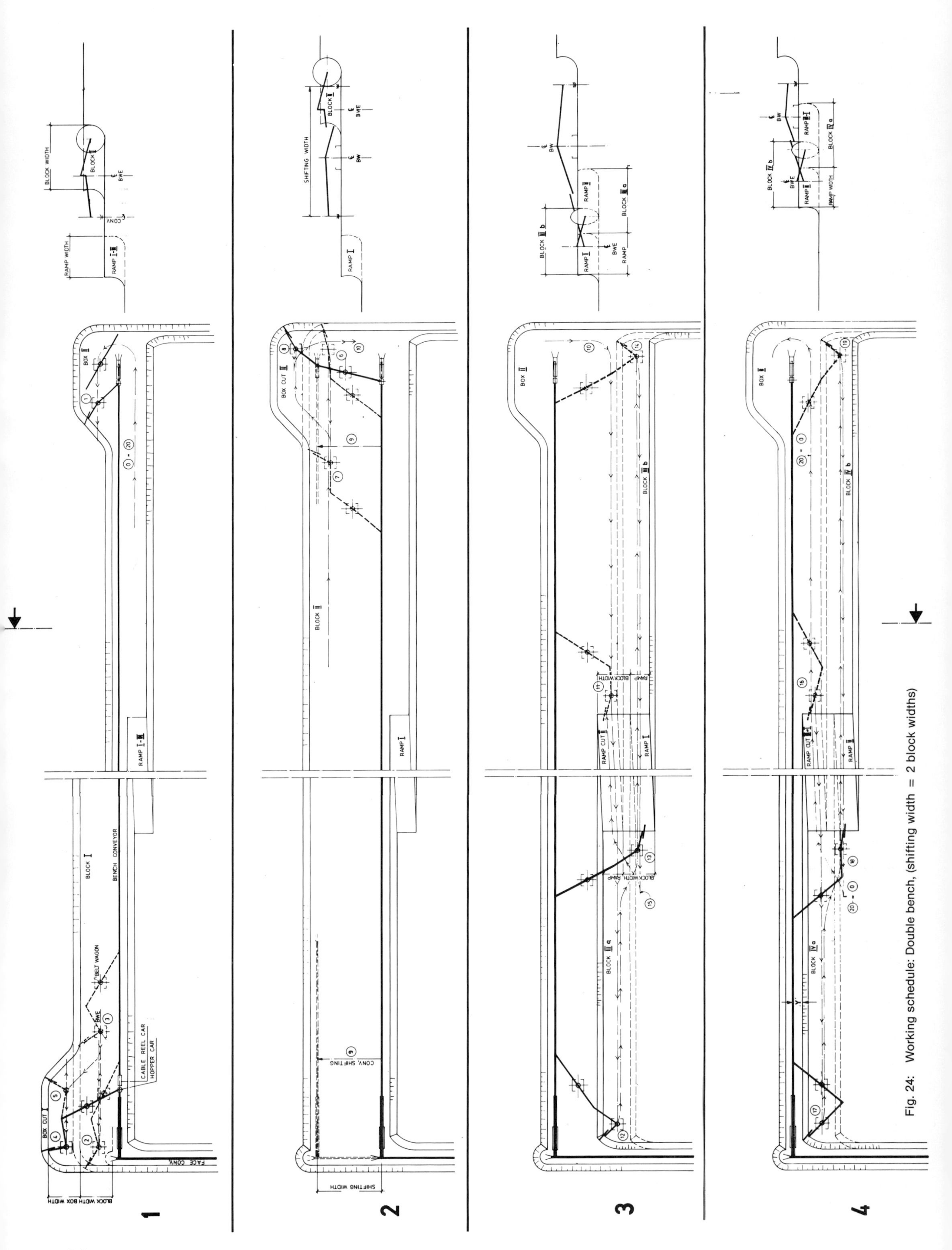

Fig. 24: Working schedule: Double bench, (shifting width = 2 block widths)

Explanation of Fig. 24:

Mining method for four blocks over two levels with one excavator and one belt wagon.

Phase 1

Position 0:
The excavator changes over ramp I—III from the lower bench onto the conveyor bench. Excavator and belt wagon travel around the tail station of the bench conveyor to position 1.

Position 1:
Start of block cut 1. The box was cut previously (phase 2). The block I can be dug up to the corner of the head station, without using the belt wagon.

Position 2:
End of block cut I. The rest of block I close to the head station can only be removed by means of a belt wagon, since no hopper car can travel in the area of the head station.

Position 3:
Start of box cut I.

Position 4:
End of box cut I.

Position 5:
Start of block cut II.

Phase 2

Position 6:
End of block cut II.

Position 7:
Start of box cut II.

Position 8:
End of box cut II.

Position 9:
Extension of the face conveyor and shifting of the bench conveyor by two block widths.

Phase 3:

Position change of excavator and belt wagon around the tail station of the bench conveyor.

Position 11:
Start of ramp cut II and start of block cut II a. Ramp II is cut besides ramp I over one block width down to the lower bench. The belt wagon follows the excavator on the conveyor bench.

Position 12:
End of block cut IIIa.

Position 13:
Start of block cut IIIb. Due to removal of the ramp I by one block width, only a width remains for ramp II that is sufficient to permit travel of the machines.

Position 14:
End of block cut IIIb.

Position 15:
Position change of excavator and belt wagon. The excavator changes from the lower bench over ramp II to the conveyor bench.

Phase 4

Position 16:
Start of ramp cut III = I and start of block cut IVa. The ramp III = I is cut besides ramp II and extends over one block width to the lower bench. The belt wagon follows the excavator over ramp III = I to the lower bench.

Position 17:
End of block cut IVa.

Position 18:
Start of block cut IVb. The belt wagon changes over ramp III = I from the lower bench to the conveyor bench.

Position 19:
End of block cut IVb.

Position 20:
Position 0.

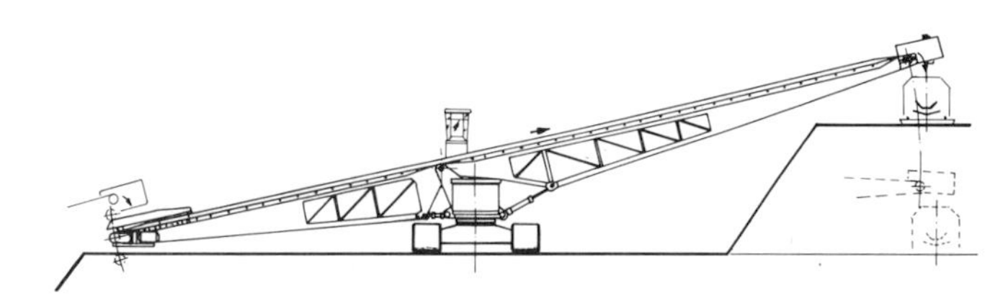

Fig. 25: Type of belt wagon required for the mining method in Fig. 24

Fig. 26: Belt wagon of type shown in Fig. 25 (length 80 m) in an ore mine in the USSR

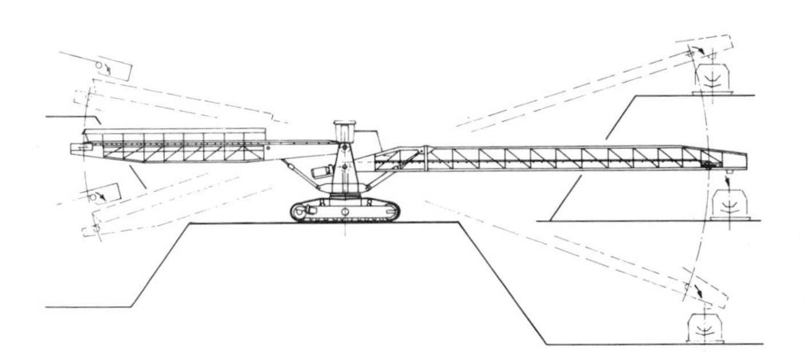

Fig. 28: Type of belt wagon required for mining method shown in Fig. 27

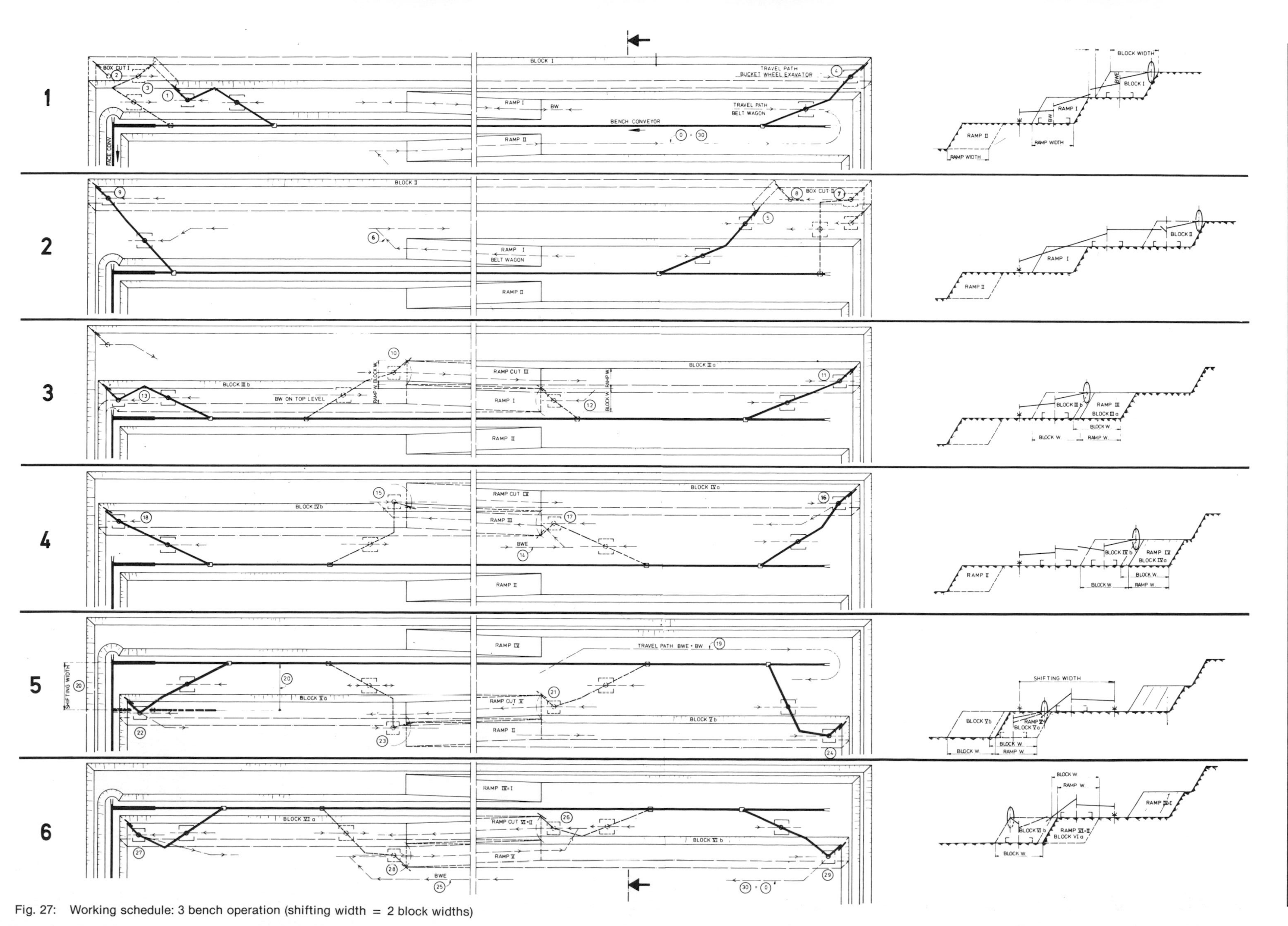

Fig. 27: Working schedule: 3 bench operation (shifting width = 2 block widths)

Explanation of Fig. 27:

Mining method for 6 blocks with one bucket wheel excavator and one belt wagon.

Phase 1

Position 0:
Excavator and belt wagon travel over ramp VI = II around the tail station of the bench conveyor and over ramp IV = I towards position 1.

Position 1:
Start of box cut I.

Position 2:
End of box cut I.

Position 3:
Start of block cut I. The belt wagon changes its travelling face from the upper bench over ramp I to the conveyor bench.

Position 4:
End of block cut I.

Phase 2

Position 5:
Start of box cut II.

Position 6:
Change of face of belt wagon. After the opening cut in box II, the belt wagon travels over ramp I from the conveyor bench to the upper bench for completion of the box cut.

Position 7:
End of box cut II.

Position 8:
Start of block cut II.

Position 9:
End of block cut II.

Phase 3

Position 10:
Start of ramp cut III and start of block cut IIIa. The ramp III is cut beside ramp I over one block width towards the conveyor bench. The belt wagon follows with the excavator from the upper bench over ramp I to the conveyor bench.

Position 11:
End of block cut IIIa.

Position 12:
Start of block cut IIIb. Due to removal of ramp I by one block width, the remaining ramp III has only a width to permit travel of the machines. Block IIIb can be mined out up to the head station without the use of the belt wagon.

Position 13:
End of block cut IIIb. The remaining portion of block IIIb beside the head station can only be removed by the use of a belt wagon, since no hopper car can travel within the area of the head station.

Phase 4

Position 14:
Position change of excavator and belt wagon. The excavator changes from the conveyor bench over ramp III to the upper bench.

Position 15:
Start of ramp cut IV and start of block cut IVa similar to position 10. The belt wagon remains on the conveyor bench.

Position 16:
End of block cut IVa.

Position 17:
Start of block cut IVb.

Position 18:
End of block cut IVb.

Phase 5

Position 19:
Position change of excavator and belt wagon around the tail station of the bench conveyor.

Position 20:
Simultaneous extension of the face conveyor and shifting of the bench conveyor for 2 block widths.

Position 21:
Start of ramp cut V and start of ramp cut Va.

Position 22:
End of block cut Va.

Position 23:
Start of block cut Vb.

Position 24:
End of block cut Vb.

Phase 6

Position 25:
Position change of excavator and belt wagon. The excavator changes from the lower bench over ramp V to the conveyor bench.

Position 26:
Start of ramp cut VI = II and start of block cut VIa. The machines change from the conveyor bench to the lower bench.

Position 27:
End of block cut VIa.

Position 28:
Start of block cut VIb. The belt wagon travels from the lower bench over the ramp VI = II to the conveyor bench.

Position 29:
End of block cut VIb.

Position 30:
Position 0.

Fig. 29: Belt wagon of type shown in Fig. 28 in an Australian coal mine (both booms are capable of luffing independently)

8. Conclusions

We have commented on the following main topics:

- General information with respect to opencast mining technology.
- Basic factors to be considered during box and ramp cuts.
- Several box shapes and methods for their realization.
- Procedures for cutting and removing of ramps.
- Several mining bench working schedules with comments, graduated according to the mass within the reach of one excavator.

The following final conclusions are made:

There exist, naturally, many methods how to make the opening cut into the next block or how to reach the next bench level.

There are mining operations in the world, for which neither during purchase of equipment, nor during selection of the mining method to be adopted have box and ramp cuts been considered in operational planning. It was in most cases left to the mine management to solve this specific problem, partly with the use of auxiliary equipment, such as dozers and shovels, because the excavator clearance angle was insufficient, or the length of discharge boom, bridge or belt wagon too short.

Problems of this type can be avoided by adequate planning. In order to guarantee maximum mining continuity and minimum use of auxiliary equipment, the bucket wheel excavators should dig their own boxes and ramps on the associated benches.

If these requirements are made for a system, the influence on mining method and dimensions of equipment to be selected are large and thus these considerations *must* be made in the planning stage.

Acknowledgements

The author is indebted to the management of Mannesmann Demag Baumaschinen Division Lauchhammer for permission to publish this paper.

Photos: Mannesmann Demag Baumaschinen, Lauchhammer Division, Düsseldorf and author.

 Volume 4, Number 4, December 1984

Soviet Continuous Surface Mining Technology

T.S. Golosinski, Canada

1. Introduction

The history of bucket wheel excavator (BWE) applications in Soviet mines dates back to the mid-1930s. A number of small, home-made excavators was introduced by the construction industry at the time.

A rapid growth of BWE applications took place in the 1960s, when this equipment was introduced to manganese and iron ore mines in the Ukraine. These mines, supervised by the Ministry of Ferrous Metallurgy, presently mine some 120 Mm³ overburden with BWEs each year. The excavators work either with around-the-pit belt conveyors or directly with across-the-pit stackers.

At present, BWE applications escalate further stimulated by their use in coal mining. There were 38 BWEs in Soviet coal mines in 1978, 55 in 1982, and 72 in 1984; over 120 Mt coal were mined by BWEs in 1982. A typical BWE mines coal and loads it into rail cars for transport to a mine-mouth power plant or other coal users. It is planned that the proportion of BWE mined coal will steadily increase as the new mines east of the Ural mountains come to the use of continuous surface mining technology.

2. Continuous Surface Mining Equipment

New BWE applications include both domestic and imported equipment. As experience is gained, the Soviet equipment is upgraded and future mines will rely solemnly on it. At present, the Nikopol manganese ore mines use for example O & K and Takraf equipment, the Kursk iron ore mines use Demag, Takraf, and Czechoslovak equipment, and the Ekibastuz coal mines use Takraf equipment.

As a result of this policy, an impressive range of BWEs, stackers, and belt wagons is manufactured in the USSR with a number of unique features reflecting the variety of applications. The largest equipment presently available has a theoretical productivity of 5,250 m³(loose)/h, with the plans to develop 12,500 m³(loose)/h equipment in the future.

Dr. T. S. Golosinski is Professor at the Department of Mineral Engineering, The University of Alberta, Edmonton, Alberta, Canada

Soviet-made equipment is at average 30 % heavier than German equipment of comparable productivity, reflecting the severity of service under the climatic conditions in the USSR. To cope with this additional weight, a new type of travel mechanism was developed for bigger BWEs and stackers. It is a combination of walking and rail mounted undercarriage. The machines are equipped with an hydraulic walking mechanism, similiar to that of draglines, used to move the machine over distances greater than some metres. Rails are installed inside the walking pads, and the machine also rests on wheels in such a way that it can move over short distances without using the pads. Short maneuvering moves are done fairly easily by moving the machine back and forth on rails. Rotating the pads around the machine permits movement in any direction, thus making maneuvering at a mining face a fairly straight-forward operation. Big pad size assures an acceptable ground pressure, despite a heavy machine weight. This type of walking mechanism is more reliable than conventional crawlers.

2.1 Bucket Wheel Excavators

Selected BWEs manufactured in the USSR are shown in Table 1 (see also Fig. 1). The biggest unit presently available is the ERShRD-5250 (5,250 m³(loose)/h). The 12,500 m³ (loose)/h excavator listed in Table 1 is not available at present; the first unit is to be built within the next years.

The standard description of excavator models used in the USSR consists of a letter symbol for the excavator type, followed by figures detailing its theoretical productivity in m³ (loose)/h, maximum height of cut, and maximum depth. The productivity is normally calculated from the bucket capacity only, without the corresponding ring chute space.

Most of the excavators have a two-speed wheel drive, permitting variation of the rpm of the wheel and its digging force. Maximum digging forces developed at the wheel for lower speeds are very impressive, reflecting the needs of coal mining. For example, the maximum specific digging force of the ERGV-630 type excavator is in the range of 2.1 MPa. In the USSR it is standard practice to relate the specific digging force of the BWEs to the maximum cross-section of the chip cut off the face by a bucket, not to the bucket cutting edge length, hence its value in MPa. 2.1 MPa digging force would likely correspond to over 200 kN/m if related to the cutting edge length.

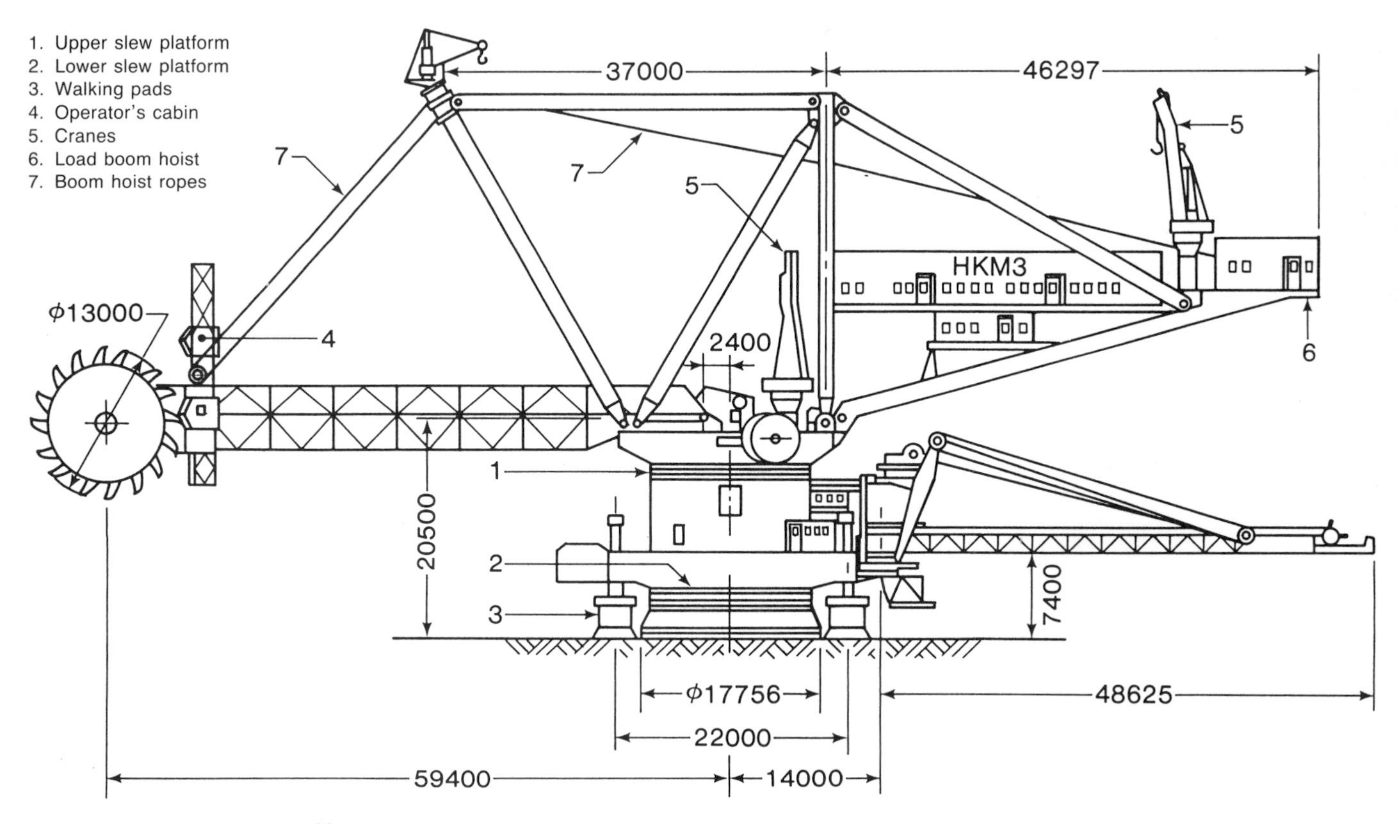

Fig. 1: Excavator OShRD-5000 $\frac{30}{3}$

For the mining of coal, requiring high digging forces, several unique bucket and cutting tooth designs were developed. The bucket cutting edge is usually tilted back (Fig. 2), with six to twelve heavy, self-sharpening teeth bolted to its reinforced rim.

Fig. 2: Wheel of the BWE model ER-1250 D used for coal mines

A very interesting development is the introduction of the so-called centrifugal bucket discharge (ERGV-630 and ER-1250 OTS in Table 1). As shown on Fig. 3, the material is thrown out of buckets by the high centrifugal force resulting from high-speed wheel rotation. Different variations of this design were tested in Soviet mines; one such BWE, the ERG-400 DTS, working in the Ekibastuz mines, reportedly mined 30 Mt coal since its introduction in 1969.

High-speed rotation results in more frequent bucket discharges, thus higher production rates. On the other hand,

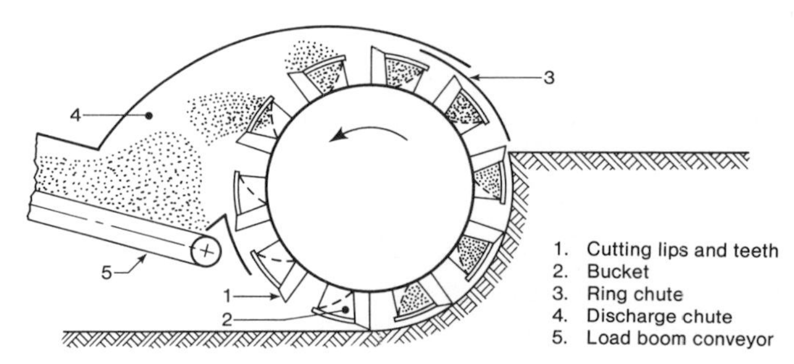

Fig. 3: Centrifugal bucket discharge mechanism [5]

for the same production rate it permits reduction of the BWE front-end weight, thus lenghtening the boom and/or reducing the excavator weight. According to Grindev [3], recent modification of the ERP-2500 by equipping it with a centrifugal discharge mechanism increased its production rate by the factor of 1.36, its digging force by 1.16, the maximum diameter of digging by 1.26, the maximum height of digging by 1.2, and the width of a mined block by 1.23. Similar proportions are true for the ER-1250 D (conventional discharge) and ER-1250 OTS (centrifugal discharge) excavators listed in Table 1. Other advantages of a centrifugal bucket discharge are finer consistency of the mined material and a reduction in the amplitude of the BWE structure vibrations. The use of this mechanism also renders crushers, a standard feature on other Soviet excavators, superfluous.

Experience indicates that this type of bucket discharge cannot be applied indiscriminately. In particular, it is not suitable for mining wet and sticky materials, nor materials with a high internal friction angle. Further, the energy consumption in the process of mining increases by some 30%, and dust generation increases along with material spillage. So far, application of these excavators was sucessful in coal mining only.

Table 1: Specifications of selected Soviet bucket wheel excavators [4]

Excavator Model	ERGV-630 $\frac{9}{0.5}$	ER-1250 $\frac{17}{1}$ OTS	ER-1250 $\frac{16}{1.5}$ D	ERP-1250 $\frac{16}{1}$	ER-2500 $\frac{30}{3}$	ERP-2500 $\frac{21.4}{1}$	ERG-1600 $\frac{40}{10}$ 31	ERShR-5000 $\frac{40*}{7}$	ERShRD-5000 $\frac{30*}{3}$	ERShRD-5250 $\frac{30*}{2.1}$	ERShR-12500 $\frac{32*}{4}$
Min./Max. Theoretical Production/m³(loose) h⁻¹	690/ 1,000	1,250/ 2,100	1,100/ 1,800	1,250/ 2,500	2,500	1,750/ 2,500	2,700/ 4,500	3,500/ 5,000	5,000	5,250	12,500
Max./Min. Specific Digging Force/MPa	2.1/1.1	NA	1.5/0.8	1.5/0.7	0.8	2.0/1.4	0.8/0.5	1.1/0.8	1.4	1.4	0.75
Max. Bench Height/m	9	17	16	16	30	21.4	40	40	30	30	32
Max. Digging Depth/m	0.2	1.0	1.5	1	3	1	10	7	3	2.1	4
Max. Digging Radius/m	16.3	27.6	24.2	24.5	38	38.8	71.7	73.7	65.9	48.1	48.5
Max. Dumping Radius/m	16.3	22.6	22.5	23.4	26	28.4	52.7	59	45	59	38.5
Wheel Diameter/m	3.2	4.0	6.4	6.5	8.5	8	11.5	16	13	11.5	18
Number of Buckets	8	10	9+9	9+9	10	18	10	10	16	22	13
Bucket Capacity/l	140	125	270	400	600	330	1,600	1,600	1,000	600	3,200
Min./Max. Number of Discharges per Minute/min⁻¹	188/272	167/280	76.5	76.5	50/70	90/126	30/50	36/52	56/80	110/143	50/65
Wheel Drive Power/kW	300	500	320	400	500	860	700	1,150	1,720	1,600	2,580
Ground Pressure/kPa	122	NA	135	135	190	125	90	126	140	140	125
Total Power Installed/kW	880	1,063	860	1,500	1,100	3,000	3,940	3,860	5,740	7,850	9,200
Operating Mass/t	290	697	700	1,040	1,350	1,450	3,150	4,265	4,700	3,850	5,700
Costs (Erected and Ready to Operate)/ 10⁶ Rub	0.350	NA	0.960	0.977	1.882**	1.970	NA	7.162	7.303	7.731	NA

* walking mechanisms for excavator travel

** ERP-2500 $\frac{25}{2}$

2.2 Stackers

A range of Soviet-made stackers, shown in Table 2, includes units with 190 m long booms, presently the longest booms world-wide — a Demag stacker under erection in a Texas Utilities mine will have a 205 m long boom. A 220 m long boom stacker, also shown in Table 2, is under development and should be available in the near future.

Stackers with booms of up to 95 m are used for conventional applications, such as the disposal of waste material transported by belt conveyors. 180 m and 190 m boom stackers, in combination with a BWE, are used for direct across-the-pit material transport.

Large stackers, e.g., large BWEs, are mounted on a walking undercarriage with rails, as described above. This assures an acceptable ground pressure; at the same time the stacker becomes highly mobile. It can move in any direction at any time, in particular it can move perpendicular to the dump crest without any delay, should the slope failure be imminent.

The discharge boom of large stackers is non-slewable; the required stacker mobility is assured by a walking mechanism, which can move the stacker in any direction, and a slewable receiving boom. The receiving boom, depending on the stacker model, can be slewn between 45 and 105° in both directions. All the receiving booms are suspended from the stacker structure; they are not supported by a separate undercarriage.

In many Soviet stacker and belt wagon designs, it is a standard feature to run discharge and load boom conveyors at different speeds. The faster discharge boom conveyor is narrower, which allows a lighter boom. Fast belt speed also reduces the weight of material carried on the boom at any time.

2.3 Belt Wagons

A range of belt wagons is manufactured in the USSR to supplement the BWEs and stackers (see Table 3 for specifications). They are used in a conventional way where the BWEs work with belt conveyors. Belt wagons are also used when needed as intermediate units in across-the-pit systems.

All the units listed in Table 3 have two separate and fully slewable booms. As with the stackers, the belt speeds on the longer boom are higher to achieve some weight economy. All the belt wagons are crawler mounted.

The 12,500-m³(loose)/h unit is not available at present. Its design is planned to match the 12,500-class excavators and stackers when they become available.

Table 2: Specifications of selected Soviet stackers [4]

Stacker Model	OSh 1500/ 105	OShR 4500/ 90	OShR 4500/ 180	OShR 5000/ 95	OShR 5000/ 190	OShR 5250/ 190	OShR 12500/ 220
Production Rate/m³(loose) h⁻¹	1,500	4,500	4,500	5,000	5,000	5,250	12,500
Max. Dumping Radius/m	105	88.5	184	95	190	190	220
Max. Dumping Chute Heigth/m	37	30	65	33.3	63	63	90
Length of Receiving Boom/m	46.2	31	61.5	47.4	62.3	62	105
Slew Angle of Receiving Boom/°	±60	±65	±30	±105	±45	±45	±90
Conveyor Belt Width/mm							
— Receiving Belt	1,200	1,800	1,600	1,800	2,500	2,500	2,800
— Discharge Belt		1,600					2,500
Conveyor Belt Speed/m s⁻¹							
— Receiving Belt	4.2	3.7	5.6	5.03	5.0	5.0	5.3
— Discharge Belt	5.2	5		6.25	6.0	6.0	6.5
Travel Mechanism*	W	WOR	WOR	WOR	WOR	WOR	WOR
Travel Speed/m h⁻¹	115	200	120	90	120	120	120
Ground Pressure/kPa							
— Working	82	100	70	100	110	102	125
— Walking	135	100	120	180	157	170	153
Total Power Installed/kW	720	1,840	3,500	2,430	3,900	3,900	7,500
Operating Mass/t	570	1,000	2,220	1,415	2,770	3,000	8,000
Costs (Erected and Ready to Operate)/10³ Rub	NA	NA	NA	2.346	4.221	NA	NA

* W — walking, WOR — walking on rails

Table 3: Specifications of selected Soviet belt wagons [4]

Belt Wagon Model	PG-1950/54	P-1600 $\frac{50}{21}$ M	PG-5000/60	P-5250	PG-12500/85 + 50
Production Rate/m³(loose) h⁻¹	1,600—2,000	1,600	5,000	5,250	12,500
Length of Discharge Boom/m	54	50	60	60	141
Length of Receiving Boom/m	27	23.8	26.5	29	56
Max. Discharge Height/m	21.8	21	15.4	16	27
Conveyor Belt Width/m	1,200	1,200	1,800	1,800	2,500—2,800
Belt Speed/m s⁻¹	4.25	5	5	5	5—6
Ground Pressure /kPa	95	111—140	90	120	125
Total Power Installed/kW	470	550	960	1,000	5,500
Operating Mass/t	287	420—407	437	500	2,610
Costs (Erected and Ready to Operate)/10³ Rub	NA	577*	775	1,496**	NA

Note: All units are crawler mounted.

* P-1600 $\frac{50}{17}$

** PG-5250/120

3. Operating Practices

3.1 Coal Mining

With the exception of one BWE-conveyor-stacker system, the application of BWEs in the coal industry is limited to coal mining. Overburden in coal mines is mined by power shovels in combination with trains or drag lines which sidecast it into the excavated pit. Further, all BWEs load rail cars at present. Rail transportation is considered to be the most economical up to pit depths of around 200 m. Belt conveyors are planned for deeper mines to be constructed in the future, i.e., the Vostochny pit in Ekibastuz.

The digging resistances of all Soviet coals planned for mining are less than 2.1 MPa, thus all coals can be mined by BWEs without blasting. In some mines, however, the coal includes interbedded partings of higher strength (compressive strengths of up to 70 MPa) which cannot be mined selectively. Such coal is preblasted to lower its digging resistance.

The blast is designed in such a way that the digging resistance of mined material is reduced to values lower than the specific digging force of the selected excavator, without mixing the coal and partings. In the Ekibastuz mines, where preblasting is used, the mined material is weakened so that its digging resistance is in the range of 1.2 to 1.3 MPa. It is mined by excavators with specific digging forces of 1.4 to 1.5 MPa. The uniformity and required degree of material weakening is achieved by proper blasthole spacing, blasthole load pattern, and amount of explosives used. To facilitate this design, extensive research precedes the mining, one of its purposes being the finding of relations between the powder factor and the material digging resistance. Typical relations of this type are shown an Fig. 4.

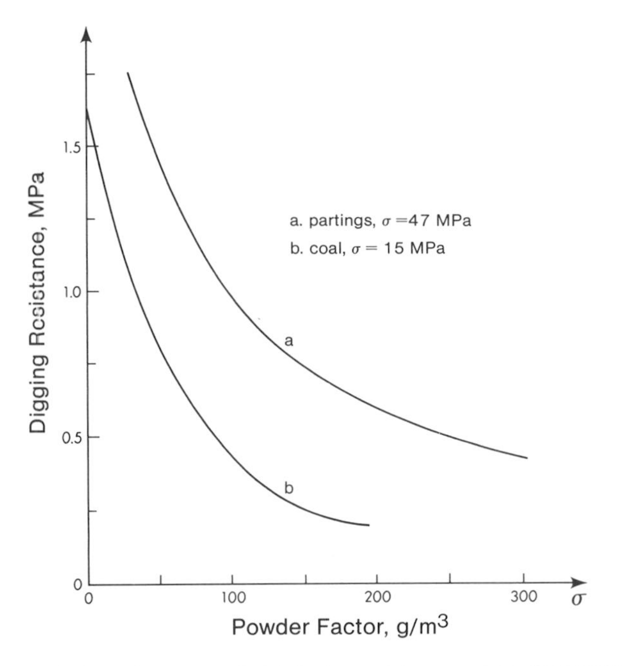

Fig. 4: Relation between material digging resistance and powder factor in the Ekibastuz mines [5]

Detailed understanding of the geology of the mined seam is also needed for a proper design of preblasts. The required information is gathered from both exploration drilling and blasthole drilling.

Preblasting results in a substantial improvement of a BWE's productivity and lowers the maintenance and spare part requirements. These improvements far outweigh the added costs for drilling, blasting, and related engineering work. Table 4 details the improvement in performance of a sample SRs-470 (k) excavator working in the Ekibastuz mines.

Table 4: Sample performance data of a SRs-470 (k) excavator in the Ekibastuz mines [1]

Parameter	Without Preblasting	With Preblasting
Production Rate/t h^{-1}	1,080	1,630
Power Consumption/kWh t^{-1}	0.50	0.19
Tooth Consumption/10^{-3} t^{-1}	0.5	0.005
Excavator Downtime/min 10^{-3} t^{-1}		
— Excavator Breakdowns	18.6	4.6
— Slope Instability	6.4	—
— Scheduled Repairs	19.1	17.6
Total Costs of Mining/Rub t^{-1}	0.162	0.107
Unit Costs/Rub t^{-1}		
— Supplies and Spares	0.038	0.006
— Salaries and Wages	0.027	0.018
— Depreciation	0.079	0.052
— Drilling and Blasting	—	0.019
— Power	0.016	0.010
— Track Shifting	0.002	0.002
Powder Factor/g t^{-1}	—	60

The BWEs in Ekibastuz mine coal with an in-situ compressive strength of up to 25 MPa, containing thick layers (up to 3 m) of shale and sandstone partings with an in-situ strength of up to 75 MPa. Partings thicker than 3 m are mined selectively and rejected with the waste material. Nevertheless, the run-of-mine coal contains up to 45 % ash. It is burned in power plant boilers without further preparation.

The 5000-class and 5250-class excavators discharge coal on a crawler mounted loading hopper, straddling two tracks. The hopper capacity is equal to two car loads so that no interruptions occur when loading switches from one car to another. Two tracks permit two trains to be positioned under the hopper to avoid interruptions in the BWE work when a fully loaded train is pulled out from under the hopper. This arrangement calls for an additional 15 to 20 m bench width to accomodate tracks and hopper, a factor which often complicates the overburden removal scheme.

Fig. 5: ER-1250 D loading railroad cars in the Azeiski mine

Smaller BWEs load rail cars directly, using a small two-way chute suspended from the end of the discharge boom (Fig. 5). Usually one track is used. After the train of 40 to 60 cars is loaded, the excavator stops and waits until the new train with empty cars arrives. This results in a substantial excavator downtime (e.g., 127 to 139 min for a typical 7-h BWE shift [2]).

The excavator crew is usually much larger than in European or American mines using the same technology, ranging from three (for BWEs with 0.3-m³ buckets) to seven (for BWEs with buckets of over 1.0 m³) [2].

3.2 Overburden Mining

In most of the Soviet coal mines the overburden is either too strong to be mined by a BWE or, if relatively soft, it is more economically moved by drag lines. The use of continuous mining systems is limited to several lignite mines in the Ukraine, where bucket chain excavators work with across-the-pit belt bridges. There are plans to use BWEs for mining overburden in the Kansk-Achinsk coal mines, but no details of these plans were available.

Full-scale overburden removal operations with the use of continuous mining technology are limited to the Ukrainian manganese and iron ore mines. In the Nikopol manganese mines, the overburden is either sand or sandy clays with a minor amount of weak shales and/or sandstones and a maximum digging resistance of 0.6 MPa. In the Kursk iron ore mines, similar overburdens often include interbedded layers of harder shales or sandstones (compressive strengths of up to 100 MPa) 0.2 to 0.4 m thick.

The high moisture content of this overburden creates severe problems during winter, particulary rapid wear and frequent damage of conveyor belts. In addition, wet material freezes to the belts, creating problems comparable to those faced in the Alberta oil sand mines. As a result of these difficulties, BWE operations are normally shut down during winter, usually from January to March. Major equipment overhauls are carried out during this period.

Whenever possible, cross-pit systems (XPSs), comprising a BWE in combination with stacker and sometimes a belt wagon in between, are used. The maximum overburden thickness mined by XPSs equipped with a 190 m long boom stacker is in the range of 35 m. Across-the-pit systems are highly productive; their mining costs are comparable to that of a drag lines (Table 5). Consequently, the use of such systems is planned for several future coal mines.

Table 5: Relative costs of mining with different mining systems in Soviet mines

Mining Method	Relative Costs of Mining/%	
	Nikopol Mines [6]	Coal Mining Avg. [4]
Side Casting by Drag Lines	100	100
BWE-Stacker or Belt Bridge Across-the-Pit System	108	122.2
BWE-Conveyor-Stacker Around-the-Pit System	209	NA
Power Shovels and Trains	259	220.3
Shovels and Off-Highway Trucks	310	

The operating techniques of BWE systems in overburden operations are not different from those used elsewhere, and therefore they are not discussed here in detail.

The size of the crew for each excavator depends on its type and size. For stackers it depends entirely on the duty. More operators are involved in loading material on conveyors (total crew size seven) than for a dump construction (total crew size four), because more accuracy is needed in the loading operation. Almost all crews include an electro-mechanic (i.e., a journeyman able to make both electrical and mechanical repairs). His duty is to continuously monitor the equipment performance and to conduct inspections, adjustments, and minor repairs.

4. Performance of Soviet Continuous Mining Systems

In all recent BWE applications, between three and five years were needed to bring the production of the BWE systems to the planned levels. Modifications of equipment components were frequently necessary during the start-up period, as well as the development of suitable operating, mine planning, and maintenance procedures. As an example, it took four years to achieve the acceptable production at the Ekibastuz mines. Original SRs series excavators were modified, material preblasting techniques developed, new BWE operating procedures developed, etc. As the informations are transferred to subsequent applications, new BWE applications of the same type are usually fully productive within one year from the start-up date.

4.1 Coal Mining

Typical performance indicators for BWE systems working in selected Soviet coal mines are given in Table 6. The load, time, and utilization factors were calcuated as follows:

load factor = actual hourly production rate/theoretical hourly production rate

time factor = number of working hours/calendar hours available

utilization factor = load factor x time factor

Distinct performance differences between the Azeiski and the Nazarovski mines, both using the same equipment and working under the similar conditions, reflect differences in the performance of transportation systems. In the Azeiski mine, BWEs load the railroad-owned cars to transport coal some 300 to 500 km. The cars have to be loaded accurately to assure the even load distribution required by a railroad. The complex loading process slows the BWE down. Further, the railroad does not make a sufficient number of cars available to the mine. Resulting BWE downtime may be as high as 20 % to 30 % of the calendar time.

In the Nazarovski mine, the transportation system, owned and operated by the mine, is more efficient, thus improving the performance of the BWE system.

Another interesting conclusion from Table 6 is that the small BWEs in the Ekibastuz mines are more efficient compared to the big ones (SRs 2000); this is believed to be the opposite in Germany. It is likely that inefficiencies of the rail transportation system become more apparent for bigger BWEs. The good performance of the 400 DTs excavator with centrifugal bucket discharge described before is worth mentioning.

Table 6: Performance indicators of selected BWEs in Soviet coal mines (1978) [5]

Mine — Mined Matl.	Excavator	Production		Load Factor	Time Factor	Utiliz. Factor
		Coal or Lignite/ 10^6 t/a	Partings*/ 10^6 m³/a			
Azeiski —	ER-1250 D					
Lignite,	#32	1.86	0.02	0.36	0.50	0.18
γ^{**} = 1.25 t/m³,	#36	2.15	0.02	0.41	0.51	0.21
f^{***} = 14 Mpa (max.)	#41	2.66	—	0.43	0.60	0.25
	#42	2,93	0.01	0.45	0.63	0.28
Nazarovski —	ER-1250 D					
Lignite,	#38	3.10	—	0.57	0.54	0.31
γ = 1.25 t/m³,	#43	3.63	—	0.56	0.64	0.36
f = 13 Mpa (max.)	#47	3.44	—	0.55	0.62	0.34
Bogatir (Ekibastuz) —	SRs 470 (k)					
Coal and Partings,	#3	3.28	0.33	0.36	0.68	0.25
γ = 1.74 t/m³,	#4	3.26	0.08	0.36	0.60	0.22
f = 70 Mpa (max.)	SRs 2000 (k)					
	#5	5.48	0.27	0.24	0.66	0.16
	#6	5.10	1.10	0.24	0.75	0.18
Tsentralny (Ekibastuz) —	ERG-400 DTs	3.09	0.12	0.37	0.77	0.29
as Bogatir	ERP-1250	2.95	0.19	0.38	0.74	0.28

* mined selectively

** γ — in-situ material density

*** f — unconfined compressive material strength

4.2 Overburden Mining

The range of performance indicators for overburden mining systems working with belt conveyor transportation is shown in Table 7. Lower values of the time factor result from winter shutdowns. There is also a distinct difference between the load factors for manganese and iron ore mines. Overburden in the latter mines is more difficult to dig; it contains layers of hard shales and sandstones which increase the wear and tear of the equipment and decrease its reliability.

Table 7: Performance indicators of Soviet continuous mining systems mining overburden (1978) [5]

Mine	Avg. Load Factor	Avg. Time Factor	Avg. Utiliz. Factor
Nikopol Manganese Ore Mines	0.46—0.65	0.36—0.53	0.17—0.34
Kursk Iron Ore Mines	0.23—0.39	0.26—0.43	0.06—0.17

Low values of the utilization factors indicate that a continuous mining system working under difficult climatic conditions (winter shutdown) and mining difficult material needs a tremendous overcapacity to meet the production goals.

Table 8 details the performance of selected overburden mining systems, including BWE-stacker combinations used for across-the-pit material casting. The superiority of the latter system in terms of total production, working time, manpower requirements, and unit costs of mining is clearly visible.

4.3 BWE Downtime

In general, downtime of the Soviet systems is due to planned maintenance or operational delays. In most cases, only less than 5 % of the total downtime is caused by equipment breakdowns. Credit for such a low percentage of breakdowns can be given to excellent preventive maintenance procedures. It is also likely that the downtime resulting from inefficiencies of transportation systems is partially used to conduct opportune maintenance of the equipment.

5. Maintenance Standards

Preventive maintenance is implemented on all Soviet continuous mining equipment according to fixed standards, though these standards are sometimes changed in response to specific requirements imposed by specific site conditions. A maintenance system was developed based on statistics of equipment component wear. Representative statistics of this type are shown in Table 9.

The maintenance cycle of major equipment (BWEs, stackers, belt wagons) consists of four distinct overhauls of different scope, whose timing depends either on the equipment working time or on the equipment production.

Table 8: Performance of selected Soviet continuous mining systems [2], [4]

Mine	Excavator	Material Transport System — Length/km	Stacker	Production/ 10^6 m³/h	Time*** Factor/ %	Unit Costs	Crew Size/ men	Effective Hourly Production/ m³/h
Shevchenkovski*	ERG 1600 $\frac{40}{10}$ 31	PG 5000/60	A_2RsB-8800/110	9.29	0.48	0.19	55	3,000—4,500
	ERG 1600 $\frac{40}{10}$ 31	Conv.—4.0 km	OShR 4500/90	7.26	0,39	0.39	2.08	3,000—4,500
Grushevski*	ERG 1600 $\frac{40}{10}$ 31	—	OShR 4500/180	8.72	0.47	0.14	55	3,000—4,500
	ERG 400	Conv.—5.45 km	OShR 4500/90	1.83	0.43	0.48	87	2,000—3,000
	K 300	Conv.—NA	OShR 4500/90	2.12	0.42		NA	NA
Zaporozhski*	SchRs 1500 $\frac{24}{6}$	ARs $\frac{1800}{30+23}$	OShR 4500/180	11.13	0.57	0.14	61	3,000—4,500
	ERG 1600 $\frac{40}{10}$ 31	Conv.—4.6 km	ARs $\frac{1800}{60+30}$ 28	8.28	0.47	0.43	177	5,000
Severny*	SRs 2400 $\frac{35}{9}$	—	A_2RsB-8800/110	10.54	0.51	0.20	86	6,600—7,200
	SRs 2400 $\frac{35}{9}$	Conv.—4 km	A_2RsB-8800/110	8.64	0.43	0.47	146	6,600—7,200
Basanski*	ERG 400 $\frac{17}{1.5}$	Conv.—4.8 km	OSh 1500/105	1.30	0.57	0.73	70	1,000—1,500
	ERG 1250	Conv.—NA	ZP 1500	1.68	0.50		NA	NA
Mikhailovski**	SRs 2400 $\frac{35}{9}$	Conv.—7.9 km	A_2RsB-8800/110	4.87	0.25	1.30	189	6,600—7,200
	SRs 2400 $\frac{35}{9}$	Conv.—10.1 km	A_2RsB-8800/110	2.43	0.21	3.35	229	6,600—7,200
	SchRs 500 $\frac{13}{3}$	Conv.—4.2 km	ARs $\frac{1800}{60+30}$ 24	2.29	0.38	0.94	123	3,700—4,000
Chkalovski**	ERShR 1600 $\frac{40}{7}$	Conv.—4.5 km	OShR 5000/95	9.76	0.49	0.45	160	5,000
Stoilenski**	KU 800	Conv.—4.9 km	ZP 5500	2.16	0.16	2.85	101	6,600—8,000
	K 300	Conv.—4 km	ZP 1500	1.45	0.31	1.14	64	700—1,000

* Nikopol manganese ore mines
** Kursk iron ore mines
*** time factor = working hours/calendar time available

A minor overhaul, (A), takes place approximately once a month. It includes an inspection of all major equipment components conducted to define their rate of wear, minor repairs such as replacement of wear plates and worn teeth, adjustments of equipment settings and safety devices, etc.

The first major overhaul, (B), takes place after ten to twelve minor overhauls. Typical maintenance work performed is belt exchanges, gear box repairs, crawler pad repairs, bucket repairs, tooth replacements, etc. It is followed by another ten to twelve minor overhauls conducted in monthly intervals.

The second major overhaul, (C), takes place after approximately two years of equipment work, i.e., after completion of the cycle (A)—(B)—(A). It is broader in scope than the (B)-type overhaul; major maintenance jobs include gearbox repairs, travel mechanism repairs, hoist rope exchanges, structural jobs, wheel repairs, etc.

A general overhaul, (D), completes the maintenance cycle. It takes place after approximately four years of equipment work and completion of the overhauls (A)—(B)—(A)—(C)—(A)—(B)—(A). All the major component exchanges are done at that time, often in connection with modifications or upgrading.

Similar standards as applied by the coal mining industry are used by manganese and iron ore mines. These, however, are based on the equipment working time to define the timing of an overhaul. In general, (A)-type overhauls are conducted each 500 working hours, (B)-type overhauls each 9,000 h, (C)-type overhauls each 18,000 h, and (D)-type overhauls each 36,000 working hours. The following pattern is followed:

17 x (A)—(B)—17 x (A)—(C)—17 x (A)—(B)—17 x (A)—(D)

Table 9: Service life of BWE components [2]

Component	Service Live/a	
	Small/ Medium BWEs	Large BWEs
Buckets	1—1	2—3
Wheel	4—5	5—6
Wheel Rim	4—5	5—6
Wheel Drive Gearbox		
— First Stage	1—2	2—3
— Second Stage	2—3	4
— Following Stages	4	4—5
Crawler Pads	2—3	3—4
Crawler Sprockets	4—5	up to 8
Crawler Rollers	2—3	3—4
Crawler Bogie Pins	3—5	4—6
Crawler Drive Gearbox		
— First Stage	1—2	2—3
— Second Stage	3—4	4—5
— Following Stages	5—6	6—8
Slew Mechanism	4—6	5—10

Acknowledgements

The author thanks the Soviet Ministry of Coal Mining for its assistance in making visits to the following locations possible: Skotchinski Institute in Moscow, Karagandashakhstroi mine and Kostenko mine in Karaganda, Vostsibgiproshakht mine in Irkutsk, Azeiski mine in Tulun, and other Soviet institutions. The hospitality of the people in these institutions is greatly appreciated.

Special thanks are due to Dr. Genadij Vozniuk and Mr. Valerij Popkov of the Ministry of Coal Mining for their assistance during the USSR tour.

References

[1] Beliakov, Yu. I., and Vladimirov, V. M.: "Sovershenstvovanie ekskavatornykh rabot na karerakh", Nedra, Moscow 1974

[2] Gorovoi, A. I.: "Spravochnik po gornotransportnym mashinam nepreryvnogo deistvia", Nedra, Moscow 1982

[3] Grindev, A. P.: "Sovershentvovanie proiektirovania tekhnologii dobychi uglia otkrytym sposobom", Shakhtnoe Stroitelstvo, Vol. 6, June 1983, pp. 1—7

[4] Melnikov, N. V.: "Kratkij spravochnik po otkrytym gornym rabotam", Nedra, Moscow 1982

[5] Vladimirov, V. M., and Trofimov, V. K.: "Povyshenie proizvoditelnosti karernykh mnogokovshovykh ekskavatorov", Nedra, Moscow 1980

[6] Vinnitski, K. E.: "Upravlenie parametrami tekhnologicheskikh processov na otkrytykh razrabotkakh", Nedra, Moscow 1984

 Volume 4, Number 4, December 1984

Surface Miner 3000 SM/3800 SM

A New Machine for Open-Cast Mining

H. Goergen, L. Arnold and H. Tuschhoff, Germany

1. Introduction

In the vast field of open-cast mining equipment machines have attained such a technical standard that the peak of development possibilities appears to have been achieved or even exceeded. The equipment presently used consists, for example, of:

- bucket wheel excavators with 240,000 m³/d output
- overburden conveying bridges of 60 m conveying height and 670 m length
- belt conveyor systems of 3 m width
- transfer units with 205 m boom length
- hydraulic excavators with 30 m³ shovels
- semi-mobile crushers with 6,000 t/h throughput
- dumpers with 318 t payload.

The most typical application of these machines is in soft or hard rock open-cast mines, whereby the first mentioned is usually a continuous process and the latter is usually a discontinuous process. An exact definition of soft and hard rock is not possible as the change is gradual and merges in the boundary area between "soft" and "hard" (Fig. 1 and Table 1).

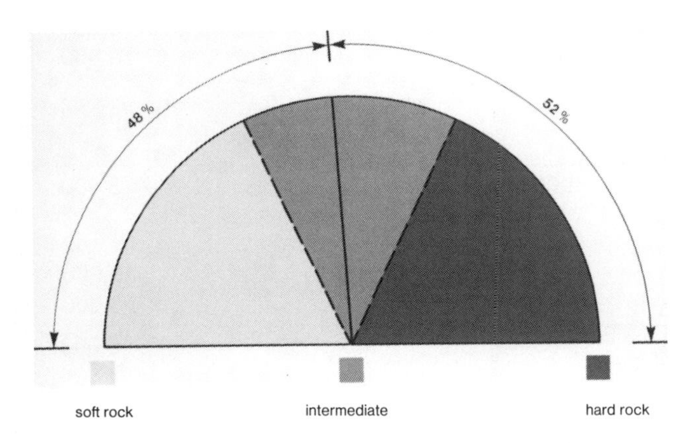

Fig. 1: Soft and hard rock

Prof. Dr.-Ing. Hans Goergen is Director of the Institute for Mining III at the RWTH, Aachen, Dipl.-Ing. Lothar Arnold is from Wiesbaden, and Dipl.-Ing. Hilmar Tuschhoff is from Aachen

The machines have operational limits in, for example, soft rock with solidified seams as a result of their inadequate cutting power, or in hard rock containing thin mineral seams due to their lack of selectivity. This results in high costs when clearing interstratified hard seams by special operations, or alternatively mining losses.

At this point we introduce a newly developed open-cast mining machine, which can assist in the elimination of these disadvantages.

2. Machine Description

In road construction, machines that mill off old road surfaces so that the salvaged material can be partly recycled have been in use for quite some time. With these machines, the process stages "loosening" and "loading" have been incorporated into one machine.

In the case of road surfaces — as opposed to minerals in open-cast mining — road lengths of almost standardized and homogeneously structured materials are involved. Nevertheless, it was just a short step from introducing this continuous system into mining because the principle of continuous cut mining has been practised for numerous years in civil engineering, whether it be a tunnelling machine, a shearer loader, a continuous miner in bituminous coal mining, or a "Marietta Miner" in mineral salt mining operations.

The Surface Miner (SM) from Wirtgen GmbH represents such a machine for open-cast mining.

The framework of the open-cast miner 3000 SM is of rigid, welded-box design with integrated tanks for hydraulic oil, diesel and water. The total length, without conveyor, is approximately 11 m (Figs. 2—4). The machine runs on three (3000 SM) or four (3800 SM) crawler units.

This results in ground pressures which are comparable to other mining machines: 13.8 N/cm² for the 3000 SM and 10.0 N/cm² for the 3800 SM.

Due to the central position of the milling drum, the rear crawlers travel at a level lower than the front crawler(s), which is equal to the cutting depth (h) when mining (Fig. 5).

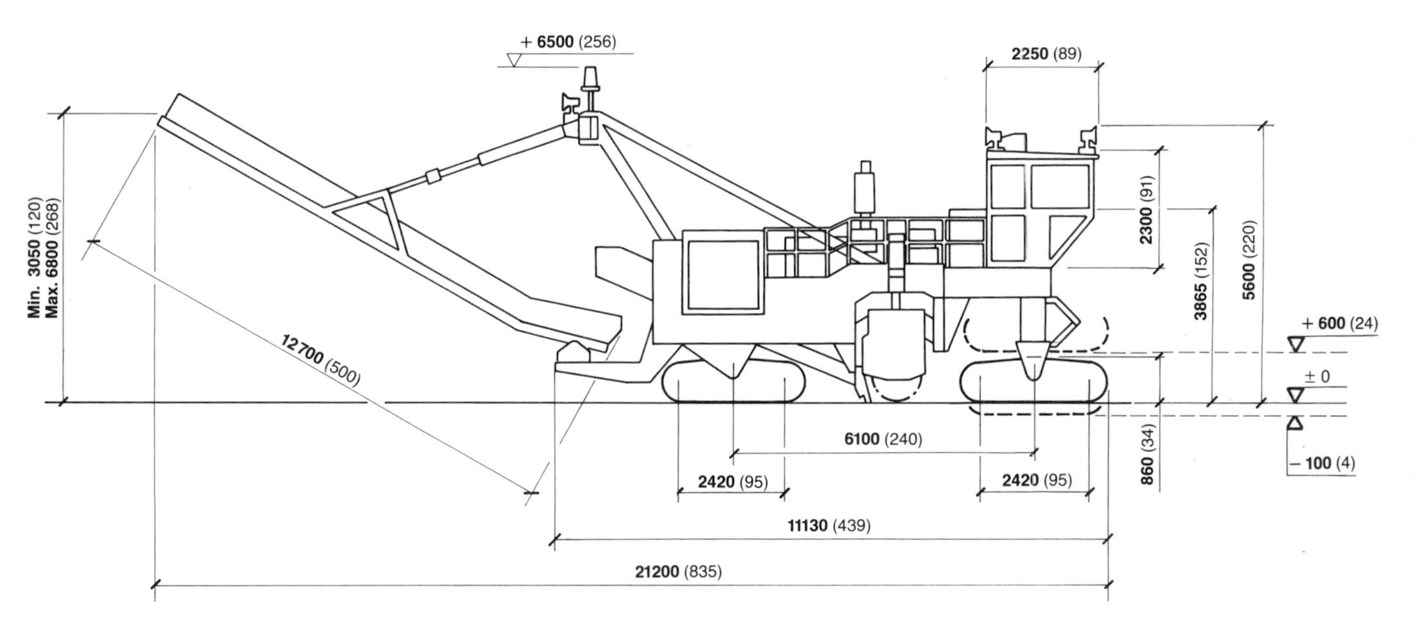

Fig. 2: Dimensions of the Surface Miner 3000 SM

Table 1: Classification of soil and rock according to their specific digging resistance and compressive strength (according to Mannesmann Demag Lauchhammer)

Soil Class	Description of soil/rock		Specific digging resistance		Compressive strength
	General	Examples	k_L (N/cm²)	k_F (N/cm²)	(N/cm²)
0	Bulk material	Ore, coal, etc.			
I	Soft + loose soil, sand	Sand	100—500	4—13	< 300
II	Relatively dense soil	Soft loamy sand; Fine to medium gravel; Damp or loose soft clay.	200—650	12—25	300—800
III	Dense soil	Hard loamy sand; Medium clay; Soft lignite coal; Hard gravel.	250—800	20—38	800—1,000
IV	Extremely dense soil	Hard clay; Clay shale; Soft matrix; Hard coal.	400—1,200	30—50	1,000—1,500
V	Semi-solid rock of low strength; Rock with significant cleavage.	Medium clay shale; Extremely hard clay; Chalk; Soft sandstone; Soft phosphorite; Very soft limestone; Hard coal; Lignite coal; Heavy ore with significant cleavage.	550—1,600	50—70	1,500—2,000 1,500—3,000 <6,000 <8,000
VI	Relatively hard semi-solid rock; Soft frozen rock; Highly cleaved rock.	Soft limestone; Marl, Chalk; Gypsum; Medium sandstone; Hard phosphorite; Shale; Extremely hard coal; Highly cleaved ore.	900—1,950	70—200	2,000—3,000 >3,000 >8,000
VII	Hard semi-solid rock; Medium to hard frozen soil; Rock with average cleavage.	Hard to extremely hard limestone; Marl; Chalk; Gypsum; Hard sandstone; Heavy ore with average cleavage.	1,400—2,600	180—500	3,000—6,000
VIII	Rock with little cleavage	Heavy ore with little cleavage			>8,000
IX	Practically monolithic rock.	Practically monolithic heavy ore.			>8,000

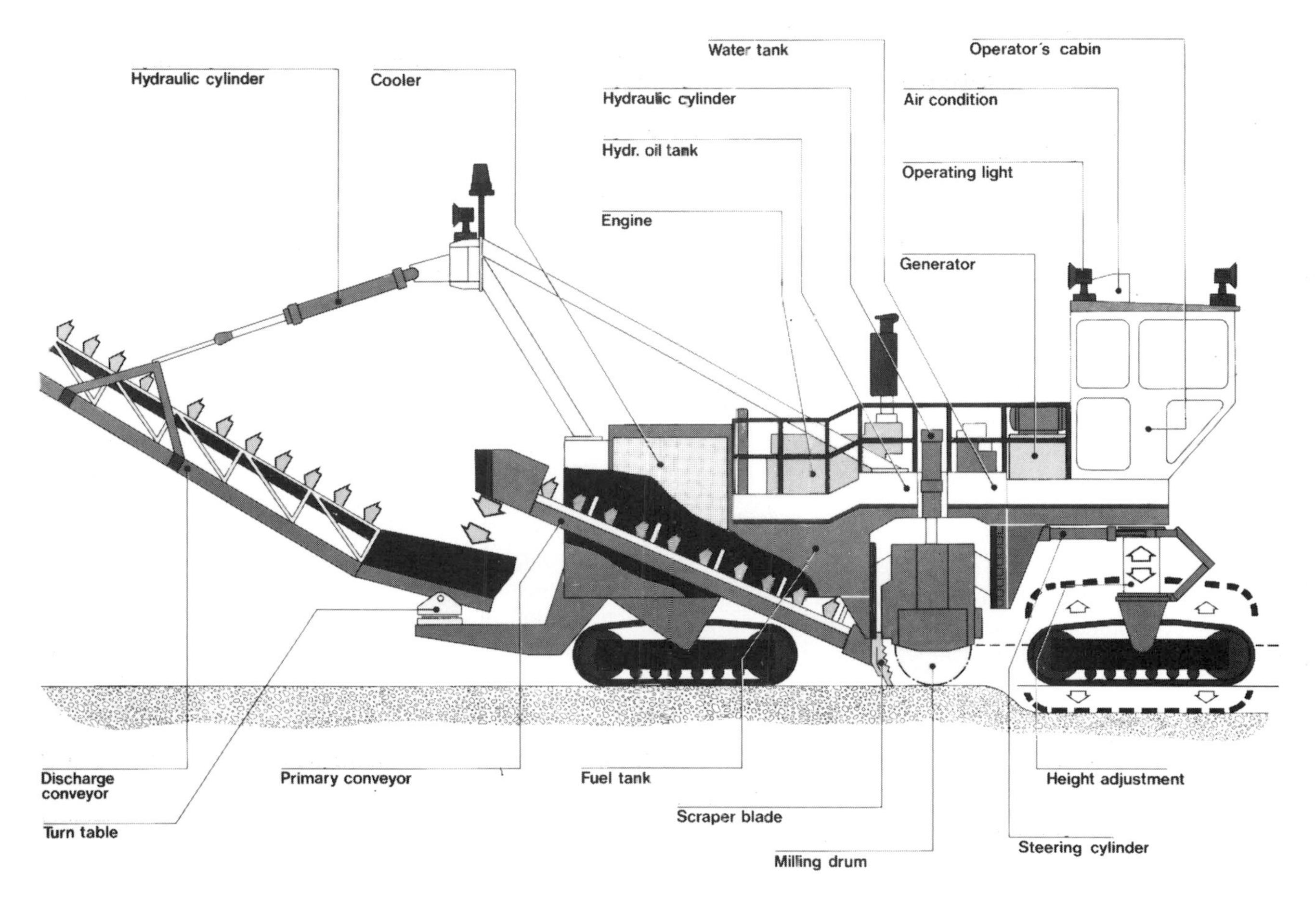

Fig. 3: Operating parts of the Surface Miner

Fig. 4: 3000 SM operating panel

Fig. 5: 3000 SM front track

The relatively high travelling speed of the machine permits to change site rapidly. Its mobile character is emphasized by its independence from external energy sources, such as trailing electrical cables.

The milling drums of 3,000 mm (Fig. 6) or 3,800 mm width (the milling drum width determines the machine designation — 3000 SM or 3800 SM) are the actual mining units of the machine. The conical picks with tungsten-carbide tip inserts presently used on the 3000 SM rotate in tool-holders. The tool-holders are bolted to the outer edge of the drum spiral, which permits the pick flighting to be adapted to the mineral characteristics.

The required milling (mining) depth can be adjusted via 2 hydraulic cylinders which are connected to the drum. The surface slope can be regulated within certain limits by laterally tilting the drum to a maximum of 7° (only 3000 SM). With the 3800 SM, the entire machine is tilted.

The first tests with a gamma-ray sensoring unit to determine exactly the interface of coal/partings were successful, although the development has not been finalized. If the gam-

Fig. 6: 3000 SM milling drum

ma radiation of the mineral is inadequate, or if a difference in the radiation intensity cannot be measured between two separable materials, other sensory methods can possibly be applied. The twin-cutter method [3], radar, infra-red and sonar measurements can be used as control methods to determine differences in hardness.

The material loosened by the almost circular motion of the picks is — similarly to the operation of a spiral feeder — transported in the drum scroll to the drum centre, because the drum is of twin-helix design. Thereafter, the material moves over a sloping scraper blade onto the rigidly installed primary conveyor. It is then fed onto the 12 m discharge conveyor, which can be slewed through an angle of 180° and adjusted to a maximum discharge height of 7.70 m (3000 SM) (Fig. 7).

Fig. 7: 3000 SM discharge conveyor

The discharge conveyor of the machine type 3800 SM has a length of 14 m. It can be slewed through an angle of 210° and the discharge height is 8 m.

3. Working Principle

The working principle of the open-cast miner can best be compared with that of scrapers. Contrary to the block operation of bucket wheel excavators or mechanical shovels, scrapers and the SM are mobile surface mining machines with a relatively high mining speed. While excavators attack the face in head-on operation, and digging and line of advance are therefore horizontal and vertical to each other, mining with the SM is hallmarked by a vertical cutting and horizontal advance. In the following, the two basically possible variations of mining systematics are illustrated:

Fig. 8 clearly illustrates the wide-surface mining method, which is hallmarked by mining the exposed mineral, layer after layer, in strips adjacent to each other.

Fig. 9 illustrates the cut sequence of block mining at the total deposit depth.

The stepped cuts are slightly staggered in order to ensure sufficient bank stability.

While mining method 1 lends itself to dumper transportation systems, method 2 is also suitable for the application of movable conveying units.

The individual cuts should ideally be situated directly next to each other and have the same milling depth in order to achieve a level surface for the supporting transport system and the auxiliary units.

4. Mining Performance

Theoretically the mining performance can be calculated from the machine speed, milling depth, and milling drum width:

$$Q = v_M \times h \times b \times 60 \text{ (m}^3\text{/h)}$$

where Q = mining performance (m³/h)
v_M = machine speed (m/min)
h = milling depth (m)
b = milling drum width (m)

Fig. 10 illustrates the theoretical performance of both milling machines 3000 SM and 3800 SM at constant maximum milling depth for medium-hard bituminous coal.

For the 3000 SM, a theoretical performance of 1,000 t coal/h is given. The 3800 SM is designed for a theoretical performance of 2,000 t coal/h, equal to 15,040 m³/h. The initial test runs will determine whether this theoretical value can actually be achieved.

In order to determine the actual achievable daily performance over a long period of time, the load factor and the time factor must be considered.

The load factor is calculated as a quotient of the effective and the theoretical mining volume per time factor:

$$f = \frac{\text{effective mining volume/time factor}}{\text{theoretical mining volume/time factor}}$$

The load factor is influenced by:

— milling depth
— machine speed
— relation milling time : re-siting time
— mineral hardness
— capabilities of operators.

The load factor drops when the maximum machine speed is not achieved due to high mineral strength or ground frost, or when the maximum milling depth cannot be retained, e.g.,

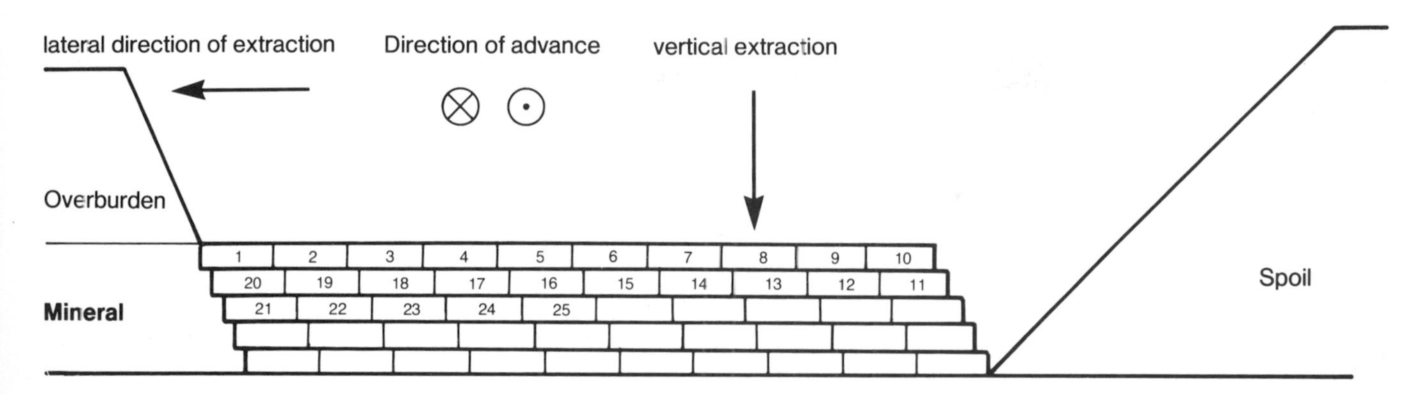

Fig. 8: Wide-surface mining method

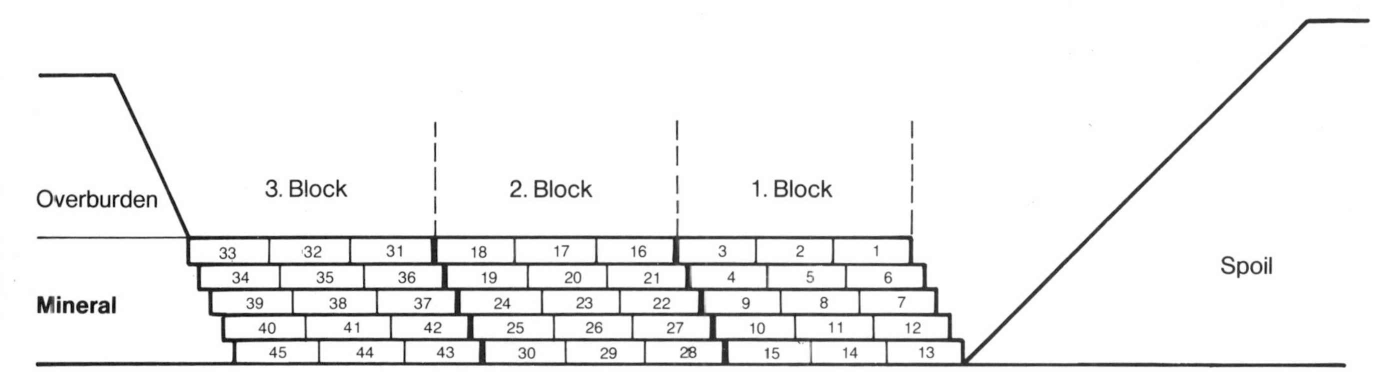

Fig. 9: Cut sequence of block mining

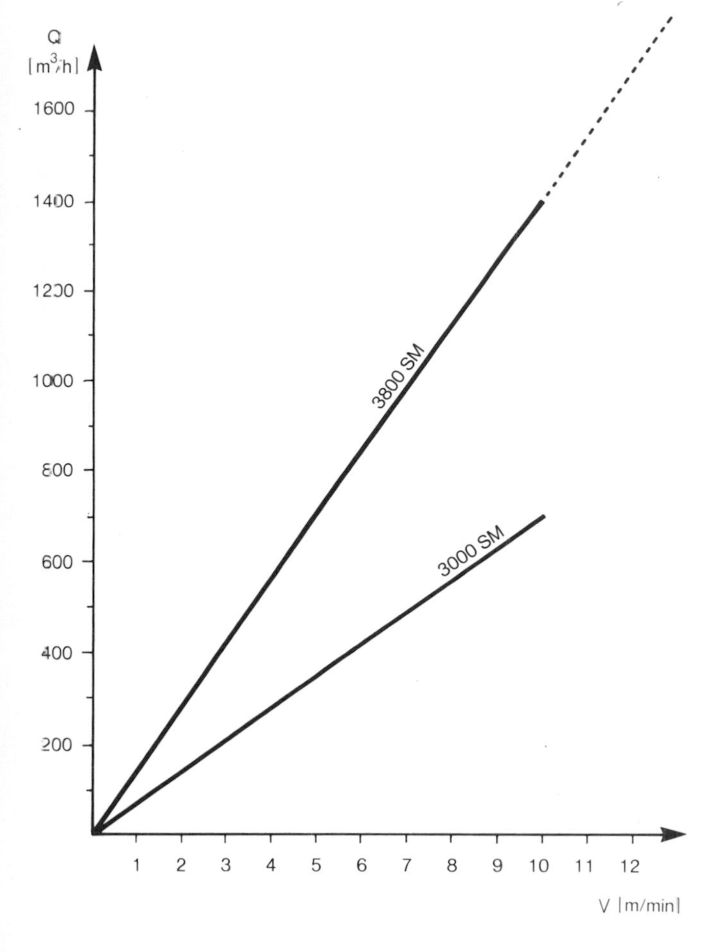

Fig. 10: Comparative theoretical performance of 3000 SM and 3800 SM

during the selective mining of thin layers of partings or the cleaning of hanging walls or base materials.

Unproductive shunting manoeuvres should be as short as possible in relationship to the milling time, which means that long milling sections are desirable.

The time factor (η) represents the actual utilization time achieved per year. It determines the selection of the capacity and performance of mining machines when a yearly volume has been predetermined.

The time factor is calculated on the basis of 8,760 available operational hours per annum:

$$\eta = \frac{\text{actual utilization hours/a}}{\text{8,760 h/a}}$$

The time factor is reduced by:

- — repair and maintenance periods
- — down-time when changing picks
- — availability of supporting transport systems
- — negative climatic conditions
- — unproductive seams.

Assuming a utilization rate of approx. 50% (in a three-shift operation) this results in the following effective annual mining performances, valid for coal at maximum milling depth and a machine speed of 10 m/min:

3000 SM . . . . 3.1 million m^3 (bank)/a
3800 SM . . . . 6.0 million m^3 (bank)/a

As no experience or only inadequate experience of cutting extremely hard rock is available, reductions in performance under difficult conditions must also be expected. At present, no exact figures can be given.

5. Particle Size

At present, no generally valid statements pertaining to the effects of mining by cutting on the particle sizes can be made. Apart from the technical influences, the mechanical influences of the mineral, such as the elasticity modulus or the hardness homogenetics, play an important part.

The demands placed by the mining industry on the broken material are sometimes of a contrary nature. While, for example, the fine particle content in the bituminous coal mining industry should be as low as possible, the limestone industry frequently requires the maximum possible size reduction (of material) during the mining process.

By varying the cutting tool flighting, the material grading can, within limits, be adjusted to the requirements. The machine manufacturer quoted a maximum particle size of 300 mm. Mining of material of a size suitable for belt conveyors is therefore assured.

6. Operations to Date

The Surface Miner was first successfully operated during 1981 in the Frimmersdorf open-cast mine of the Rheinische Braunkohlenwerke AG. The mining performance was up to 1,000 t/h.

Further test runs with the 3000 SM were carried out at:

— Atascosa Mining Co., Texas/USA (lignite coal)
— Palafox Mine, Laredo, Texas/USA (bituminous coal)
— Minsa Mine, Nuevo Rosita, Mexico (clearing).

Using the smaller Surface Miner types 2600 SM and 1900 SM, the mining possibilities were proved by the milling of andalusite (Timeball Mine, Thabazimbi/South Africa), gypsum, anhydrite, limestone, and bituminous shale, whereby the latter was mined underground.

7. Application Possibilities and Mining Methods

Possible future applications of the SM technology are:

— mining of harder minerals which can presently be mined only by blasting techniques (e.g. limestone)
— mining of thin seam deposits
— selective mining of materials with varying mineral content
— creating channels
— digging exploratory trenches
— removal of partings
— mining residue minerals
— removal of consolidated overburden layers, e.g., frozen ground
— road construction and maintenance
— digging drainage ditches
— moving conveyor units with an attached side-boom.

8. Fundamental Requirements of the Deposits

The mining performance, or even the possibility of mining at all, is heavily influenced by the technological characteristics of the rock such as:

— tensile strength
— compressive strength
— shear resistance
— elasticity modulus
— abrasiveness.

These characteristics, on the other hand, are dependent upon mineral contents, particle structure, and weathering conditions.

A level seam deposit without faults is the ideal operational situation because the machine can mine long strips without the utilization rate being substantially influenced by repeated shunting and turns.

If a thin deposit is to be mined, the seam incline/slope must not be steeper than the machinery specification. This limit is dictated by the wear-and-tear rate of the crawler units on side-slopes or by the drive power on gradients. The maximum inclination of the SM is 7° and the climbing capability is limited to 15°.

All common mining methods — bench mining, deep mining, parallel operation, and slew mining operations — are suitable for the application of open-cast Surface Miners. Parallel operation, however, is for the following reasons the most suitable mining method for the SM:

— mining strips which can be many kilometers long ensure a high net milling period
— the long mining faces ensure ideal conditions for the installation of conveying units, as a result of which a systematic complementation of the continuous working method is possible
— the mineral volume transportable by a single conveyor on the same working level is twice as great as that of slew operations.

9. Combination with Supporting Transport Systems

In conventional consolidated rock open-cast mining with its discontinuous stages of "loosening", "loading" and "transport", the "transport" part has with approximately 60% a much higher influence on the total costs of production per ton of mineral [2].

When looking at future cost developments, the selection of the most economical transport system, even when using open-cast Surface Miners, is of great importance.

Possible transportation alternatives in "open-cast mining" are rail, dumpers, and movable or mobile conveyor units.

In principle, the combination of Surface Miner/railway is technically possible, but due to the mobile operational method of the Surface Miner and the resultant frequent movement of the rail-line, the application of this machine combination is not very practical, in particular as the mined volumes per machine are relatively small in comparison to bucket wheel excavators.

Despite its high flexibility, a dumper transportation system in connection with the Surface Miner should be seen as unfavourable because the continuous mining characteristic is partially cancelled by the intermittent operation of the dumpers.

The application of a fleet of dumpers under these circumstances is only justifiable in small or medium sized opencast mines with shallow depth, i.e., when operating with smaller Surface Miners.

The combination of open-cast Surface Miner/conveying unit does not require the intermediate stage of a crusher because the material is reduced to a particle size of < 300 mm during the mining process and is therefore immediately suitable to be conveyed. The resulting cost savings shift the economical limitations of depth of dumpers in favour of conveying operations.

The combination of a Surface Miner and a mobile or semi-mobile conveying unit has not yet been used. With this machine combination, the mobile working characteristics when covering large areas must be given special attention.

The horizontal and vertical distance to the bench conveyor can be bridged by using a mobile conveyor.

With a mobile conveyor of approximately 40 m length and a bench height of approximately 10 m, the total system of mobile conveyor/Surface Miner could mine approximately 800,000 m³ on a bench length of 1,000 m without having to move the conveyor system.

A further possibility is the combination of movable and mobile conveying units.

This system involves a mobile conveying unit on rails, tracks, or wheels as a connection between the Surface Miner and the movable conveyor. The SM mines parallel to the mobile conveyor which is located right-angled to the movable conveyor.

As the mobile conveyor needs to be moved only at intervals, the operational costs of continuously moving the mobile conveyor are eliminated. Nevertheless, higher investment costs than with the other systems must be envisaged because of the larger quantity of tracked vehicles. Conveyor units of this type are presently under development.

10. Comparison of Conventional and Surface Mining Technology

The previously unknown high degree of accuracy in selective mining of overburden or minerals is the most important characteristic of the Surface Miner.

The resulting advantageous aspects are as follows:

- lower mining losses, therefore better exploitation of the available deposit
- transport cost savings with partings because they are tipped away directly without going through the beneficiation process
- lower processing costs because the uncontaminated coal need not pass through the beneficiation plant
- lower processing costs because the total throughput volume is lower and therefore a smaller processing plant can be selected.

The primary crushing effect integrated in the machine has the following advantages:

- down-time caused by oversize aggregates is almost eliminated
- mined material with particle size < 300 mm is suitable for belt conveyors therefore low belt wear-and-tear
- primary crusher in open-cast mines with belt conveying systems not required, therefore elimination of a possible operational bottle-neck and also savings of capital and operational costs
- because cost intensive secondary breaking is no longer necessary, the potential danger of "flying stones" is eliminated
- the primary crushing stage in the beneficiation plant can either be eliminated or dimensioned smaller accordingly.

High mobility when changing sites, flexibility when mining minerals of various hardness grades, production of a clean and level surface and cost reductions achieved by minimizing lifting work are typical characteristics of the surface mining technology.

11. Alternative to Blasting Methods

In as far as the mineral hardness permits the Surface Miner to operate in situations previously the sole domain of blasting technology, when comparing equipment expenditure it must be considered that drilling equipment, explosives and secondary crushing are no longer required. The SM technology therefore represents a considerable simplification of the stages "loosening" and "loading".

It must also be considered that blasting is not possible everywhere. The maximum permissible noise and vibration levels in heavily populated areas, imposed by the environmental protection authorities, can be so restrictive that mining with explosives is either limited, or not possible at all. Roads or utility service systems nearby can also restrict or even prohibit the use of drill and blasting techniques.

Under the above-mentioned conditions, in which mining would otherwise not be possible, the SM technology represents the only possible alternative if the digging power of conventional excavators is inadequate for the material to be mined, but which could still be milled by the Surface Miner.

12. Reducing the Quantity of Auxiliary Machinery

Draglines, hydraulic excavators, bucket wheel excavators, and scrapers generally require assistance from auxiliary machinery in order to operate at high performance. Whilst the above-mentioned excavators require an additional dozer to push up loose material otherwise unreachable, bulldozers are frequently used to increase the traction of scrapers by pushing the actual mining machines in front of them.

Surface Miners, on the other hand, seldom or never require auxiliary machinery to increase their performance, e.g., when clearing over-spills caused by careless loading of trucks.

These savings in auxiliary machinery should not be disregarded when comparing machinery expenditures.

13. Creating Channels

Apart from mining, the Surface Miners can be used to produce canals. The operational conditions are similarly advantageous when digging exploratory ditches. The frequently limited space available does not represent a problem to numerous parallel operating Surface Miners. The investment costs, in comparison with a bucket wheel excavator, are relatively low.

14. Digging Exploratory Ditches

When opening up a new open-cast mine, the surface mining technology can be considered during planning as an alternative to excavation with large bucket wheel excavators.

The elongated dimensions of exploratory trenches are ideal for the operational methods of the Surface Miner. Contrary to bucket wheel excavators, which must first cut an entry ramp and therefore cannot achieve their nominal rated performance during this period, the Surface Miner can start at maximum performance immediately.

If the required mining performance of Surface Miners cannot be achieved as the mining operation advances, the operation can then be taken over by a larger machine such as a bucket wheel or bucket chain excavator. The SM could then start mining thin seams or partings.

The advantage of this process would be much lower start-up and investment costs. This argument is often the decisive criterion of development countries when selecting machines.

The surface mining technology also has certain systematical disadvantages:

— if selectivity in the horizontal direction is required, e.g., with limestone deposits containing dolomite veins, considerable reduction in mining performance is to be expected
— the unproductive turns at the end of each machine pass reduce the performance rate; this disadvantage compared with other machines can be almost eliminated by mining long strips
— due to its mobile operational character, an electrification of the Surface Miner is problematic
— the reduction of the particle size to < 300 mm is not necessary if belt conveyors are not planned; in such cases, energy is used which is actually not required because a material size reduction is unavoidable when operating the Surface Miner
— the mobile characteristic requires that a transport vehicle follows continuously; as a result operational costs with dumpers and mobile conveyors are higher than with a stationary operating machinery combination.

15. Economic Comparison, Using a Limestone Open-Cast Mine as an Example

For the conventional mining of medium to hard limestone, costs of approximately 1.75 DM/t for the operations of loosening, loading, and primary crushing are incurred.

The operational costs of the Surface Miner 3000 SM, which incorporates these three stages in one machine, are ∿ 34% lower than conventional mining for an output of 400 t/h and ∿ 56% lower for an output of 600 t/h.

Other cost advantages, such as reduced transportation costs due to lower lifting height, less tyre wear due to level surface, or savings on auxiliary machinery, cannot be considered in this cost comparison. Nevertheless, the lower capital and operational costs of the SM technology can be clearly seen in this simple example. The cost advantage is particularly obvious if the mining operation with conventional machinery is divided into the three stages of blasting, loading, and primary crushing.

16. Future Fields of Application

The following minerals can be considered for Surface Miner technology in open-cast mines:

— bituminous and lignite coal — maybe peat
— bauxite
— lateritic nickel ore
— oolithic or lateritic iron ore, siderite
— nitrate
— phosphate
— oil sand, oil shale
— uranite
— baryte
— gypsum
— anhydrite
— limestone
— dolomite
— chalk clay
— magnesite
— andalusite
— talc
— clay
— kaolin
— quartz sand and gravel
— detrital deposits of gold, platinum, tin ore, zircon, and magnetite, etc.

17. Summary

The Surface Miner of Messrs. Wirtgen is a new technological achievement in open-cast mining. The possibilities and limitations in its application have been presented, including the possible integration of the milling machines in the total open-cast mining system.

As a continuous and selectively operating mining machine, its future area of application should be mainly at the borderline between soft and hard rock, where the conventional stages of "loosening", "loading", and "primary crushing" can be achieved with one machine when mining with the Surface Miner. The resulting cost advantages compared with conventional mining have been clearly proved in a cost estimation.

If the prerequisites are observed, the Surface Miner technology is an authentic alternative to conventional mining methods from both a technical and an economic standpoint.

References

[1] Tuschhoff, H., "Machinery for Soft and Hard Rock Open-Cast Mining", Braunkohle (1982), Booklet 12

[2] Korak, J., "Technical/Economic Survey of Transport Machinery When Considering the Transport Machinery Combination of Mobile Crushers/Belt Conveyors for Transporting Soft and Hard Rock", Dissertation, Aachen (1978)

[3] Henkel, E.H. and Heck, L., "Regulating the Cutting Horizon of Drum Cutter-Loaders", Glueckauf Research Booklets, Dec., (1977)

[4] Martens, P.N., "Survey of Loading Operations in Hard Rock Open-Cast Mines", Dissertation, Aachen (1977)

[5] Goergen, H., "The Development of Open-Cast Technology Under Consideration of Stone and Earth Open-Cast Mining", Braunkohle (1982), Booklet 8

bulk solids handling Volume 3, Number 4, November 1983

Development Trends in the Construction of Spreaders for Opencast Mine Dumps

Part 1

Spreaders in Around-the-Pit Pre-Stripping Systems

Detlef F. Bartsch, Germany

1. Introduction

Various publications and forecasts for the future have reported on the constantly increasing volumes in the mining of raw materials. Irrespective of these forecast data, today's mining operations already show significant production figures. It is, however, not the intention of the present article to deal with the total production volume, but to highlight the throughput which can be attained by one machine or system within a specific period of time. In this connection it is clear that the total production rate can only be achieved by a limited number of systems, since a quantitative increase in systems operated can only be profitable up to a certain limit. This statement reflects the basic trend which can actually be recognized, i.e. a constant increase in performance of the individual systems. Development trends can be derived from the activities in this special field in the last few years, and hitherto collected knowledge offers a reliable basis for future technologies.

2. Present Status of Spreader Production Volumes

A figure repeatedly mentioned in overburden mining is 240,000 billion m³ per day using a large-scale excavator, a production rate attained in the lignite mines of Rheinbraun. As far as the spreader is concerned, one has to consider that it is the last link in a transport chain and must have a considerably higher capacity, i.e. each link in the materials handling chain must have a higher performance than the preceding one, so as not to produce a critical point where spillage and blockage may occur.

Expressed in figures, this leads to a daily production rate of 375,000 billion m³ for the spreader, with a ratio of 1.56 for the performance increase taking place between the bucket wheel of the excavator and the discharge conveyor of the spreader. Spreader design considers these values.

Fig. 1 shows a spreader capable of achieving the abovementioned performance.

Mr. D. F. Bartsch is Project Manager, Mannesmann Demag Baumaschinen, Division Lauchhammer, Düsseldorf, West Germany.

Fig. 1

3. Parameters Influencing Spreader Operation

3.1 Soil Mechanics

Normally, a spreader is not operating on natural soil, if not used for initial construction of an external dump (out-of-pit dump). Bearing in mind that a natural soil cannot withstand any load imposed, it is clear that for a newly dumped area the admissible ground pressure has to be kept as low as possible. Complicated and especially developed dumping methods offer increased safety, and are in most cases a basic requirement to guarantee a stable dump. Admissible ground pressures of 10 N/cm² or even less have to be kept by the machines.

This requirement can only be met by the use of crawler assemblies with adequate contact surfaces. Rail travel assemblies, which are technically possible, cannot be used because of their reduced flexibility or for operational reasons.

The dump mechanics have considerable influence on the spreader dimensions and especially the boom length. During high dumping operations a defined height has to be guaranteed, but the travelling system must be safe at any time and not be spilled by a possibly sliding pile base. For deep dumping operations, the boom length should at any time per-

mit adequate clearance between the spreader and the dump edge, or if required be extended up to an advanced dump base. The considerable influence on the boom length can be demonstrated by an experiment, during which it was tried to keep this length as short as possible and to compensate the reduced outreach by a significant increase in discharge speed (belt speed) and thus the trajectory of the overburden. These tests revealed — summarised as simply as possible — that the requested high belt speed caused unwarranted high risks during operation, resulting in an increased danger of accidents and hardly controllable vibrations in the steel structure. The relevant values were as follows:

Speed still used on the machine	15 m/s
Speed which seems to be adequate for normal operation	7.5 m/s

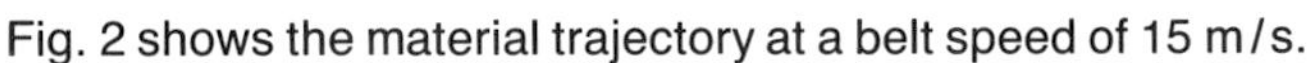

Fig. 2 shows the material trajectory at a belt speed of 15 m/s.

Fig. 2

3.2 Material Characteristics and Throughput

If, as stated above, the belt speed amount to 7.5 m/s, a variation in throughput is only possible by means of the belt width and troughing angle of the belt. Current fabrication possibilities are still facing limits here, but a further development in this special field seems to be possible. It has to be noted in this connection that the currently operated 3.2 m wide belt conveyor systems on spreaders have been tailored to the maximum transport volumes occurring in today's opencast mining operations. But even in case of careful and optimum determination of the handling cross-section and the associated belt speed, the subsequent practical application would not be successful if not preceded by detailed investigations and evaluation of experiences with respect to design and execution of transfer points, chutes and structures for special purposes.

3.3 Availability

The operation of machines will finally depend on the calculated (and possibly proved) profitability. This feature is in close relation to the question of availability, since only a maximum availability factor for both the individual machine and the total system leads to acceptable results in operational cost calculations.

Prior to commenting on the influence of the total system on the design to be adopted for the spreader, a short description follows in Section 3.3.1 centering on the measures that can be adopted to increase machine availability.

3.3.1 Auxiliary devices on the machine

When speaking about the availability of a system, this means the mechanical and electrical operating condition. Any problems caused by weather or other external influences are not taken into consideration. The operational shutdowns reducing the availability of a system are subdivided into scheduled and non-scheduled standstills. The scheduled standstills cover repair and maintenance work at regular intervals. Normally, a long-term working schedule exists, containing all information on type and extent of required activities. A rapid execution of these activities involves the need for provision of efficient "on-board" equipment, in order to avoid long fitting times and delays due to heavy workloads in the central workshop.

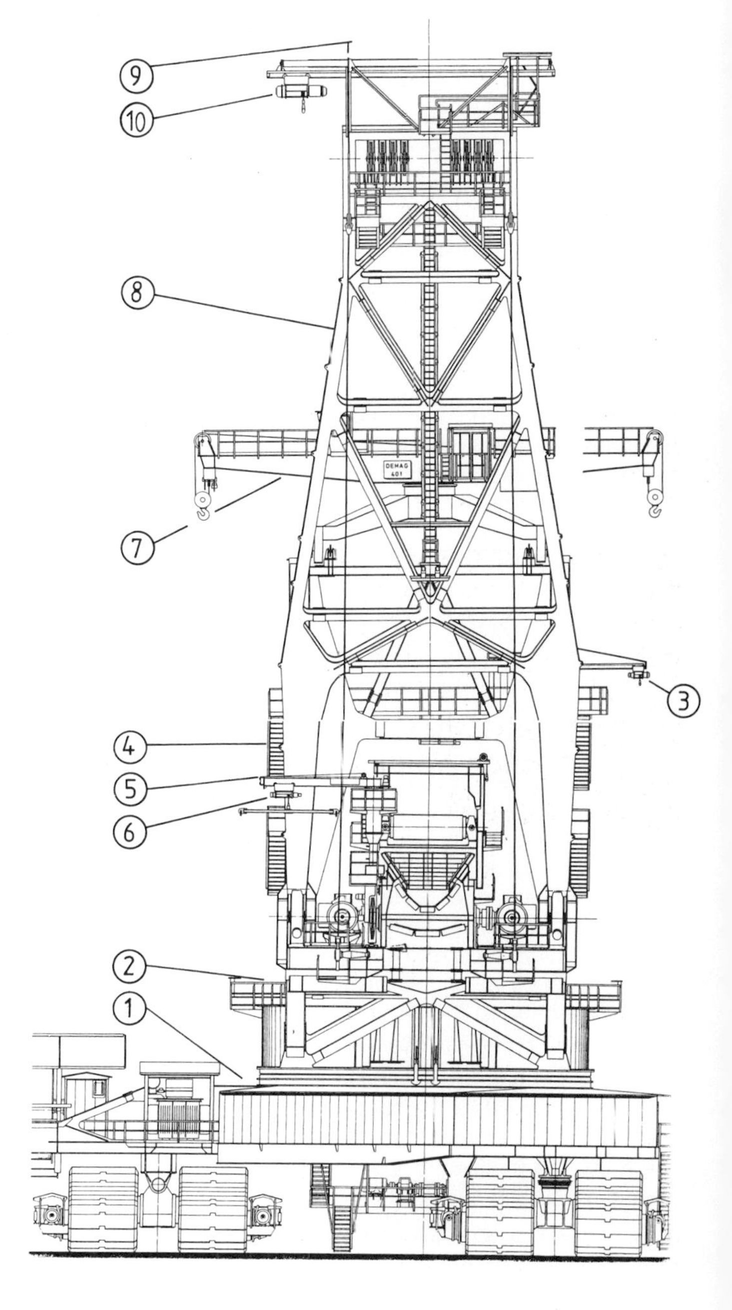

Fig. 3

For example, the following auxiliary devices are incorporated (and Fig. 3 shows their arrangement on the machine):

1. Lifting device for the workshop at the substructure
2. Lifting device for maintenance of the lubrication system at the substructure
3. Lifting device for maintenance of the lubrication system at the superstructure
4. Dismantling carriage and loading station for replacing the belt catenaries in the bridge conveyor
5. Dismantling carriage and loading station for replacing the belt catenaries in the intermediate conveyor
6. Dismantling carriage for dismantling the belt catenaries in the discharge conveyor
7. Crane for replacing the conveyor gearboxes and drive pulleys in slewing centre
8. Crane for maintenance of the hoisting system and electric equipment on the counterweight boom
9. Lifting device for replacing the rope sheaves at the rope mast counterweight boom
10. Lifting device for replacing the rope sheaves on the rope mast discharge boom.

An additional precondition for the installation of such expensive auxiliary devices is a careful and detailed investigation into the type of joint to be adopted for these components. For example this point was decisive for the selection of flanged joints at the drive and belt conveyor drives.

Non-scheduled shutdowns by defective catenary idlers are largely avoided by the incorporation of quick release devices, detachable during running operations (see Fig. 4).

Fig. 4

In the normal working position a pin prevents lowering of the catenary, and so one single action is required in case of a defect (see Fig. 5) to take the damaged idlers out of operation. A complete replacement is then made during the scheduled repair and maintenance times.

Detailed planning of spare parts storage in connection with "on-board" equipment guarantees short shutdown times in case of any non-scheduled outages due to component failure.

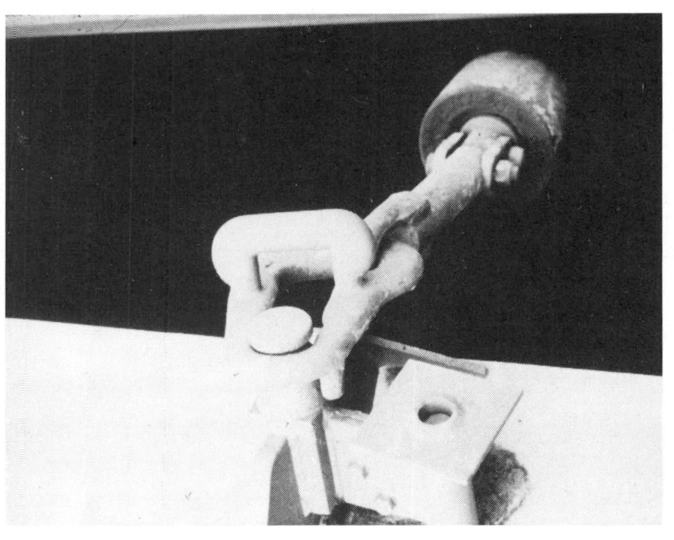

Fig. 5

3.3.2 Influence of design on the system availability

The availability of the total system, related to the mechanical and electrical operating condition, is calculated with the following formula:

$$f_{AT} = \frac{1}{1/f_{AN} - (n - 1)}$$

where f_{AN} indicates the availability of the individual machines. This calculation does not, however, include any time factor, produced by the change-over work during advance of mining operations. For example, a repeated belt conveyor shifting would reduce the total availability to a considerable extent. For the common operation of the dump bench conveyor system and the spreader this means that the number of interruptions required for shifting work decreases with increasing dumping volume.

The dumping volume that can be obtained depends not only — as already stated — on the discharge boom length, but also on the length of the receiving boom, irrespective of the admissible values derived from the dump mechanics. The receiving boom increases not only the general outreach or working range, but also serves to bridge over differences in level between the belt conveyor line and the spreader. Two different design types are being operated. Firstly, the freely suspended version, Fig. 6, and secondly, the boom-version supported on an additional crawler assembly, Fig. 7.

Fig. 6

Fig. 7

This alternative offers several advantages and will be dominant in future applications. The third spreader of this type has recently been fabricated, erected and put into operation by Demag Lauchhammer.

4. Selection of the Most Adequate Design Types from the Point of View of Construction

4.1 Presentation of the Different Design Types

The following description shall only cover the arrangement of the belt conveyor systems on the machine types, neglecting any structural characteristics on e.g. design versions adopted for the boom. These subjects are influenced by other criteria which are without any importance for the operational sequence in opencast mines.

One has to proceed from the requirement that the used components can withstand the loads imposed, have a minimum weight, are simple to fabricate and competitive with respect to price. These basic requirements are achieved by the manufacturers by means of detailed static calculations, vibration tests, wind — tunnel tests and adequate quality control during fabrication.

4.1.1 Spreader (Stacker) with two conveyor systems, Fig. 8

This spreader type incorporates one receiving conveyor and one discharge conveyor. Considering its simple design and the relatively low number of joints, this type seems to meet all preconditions for a good availability. A careful study of this type, however, revealed some disadvantages which must be considered:

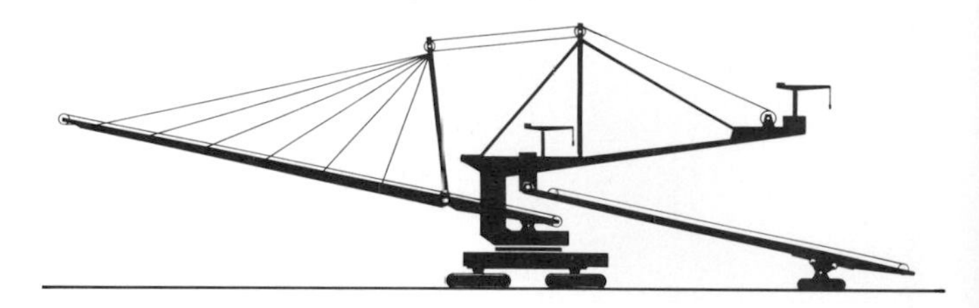

Fig. 8

1. The location of the boom pivot point close to the slewing center of the C-frame results in a framework for the cross member of the C-frame. Transversal forces have to be absorbed, too.
2. The angle of inclination at the receiving end of the discharge boom conveyor changes considerably during the different operational positions.
3. Since the horizontal angle in the feed zone is constantly modified by the slewing motion during the stacking operation, the material flow is subject to permanent change.

Factor 1. leads to high design weights, while factors 2. and 3. require a complicated central feeding chute, since the tendencies mentioned involve the possibility of spillage and blockage at the chutes. This contributes further to the danger of damage to belts and mechanical components.

Should the conception adopted call for an acceleration of the material on the belts — this latter running at high speed — more problems are created. Discharge boom belts, running at high speed, have an important influence on the total design. Since belt loads can be reduced — maintaining identical throughput rates — this in turn allows the boom weight and the counterweight to be reduced, thus keeping the total mass of the substructure as low as possible.

4.1.2 Spreader (Stacker) with three conveyor systems, Fig. 9

For this version the material flow develops via an additional short intermediate conveyor onto the discharge boom. This arrangement results in four important advantages:

1. The boom pivot point is located in the front section of the supporting structure. Thus the framework for the forward cross-member can be deleted and a continuous latticework structure be provided.
2. The angle of inclination between the receiving boom conveyor and intermediate conveyor remains constant.
3. The belt speeds of the two conveyors (same as under 2.) are identical, resulting only in a directional change in material flow at this transfer point.
4. Material acceleration is done at the receiving end of the boom conveyor, however, without any change in direction of material flow.

Comparisons between spreaders with 2 and 3 conveyors reveal that the 3-conveyor spreader can be built 10% lighter in weight and should be preferred for reasons of cost calculation derived from this lower weight.

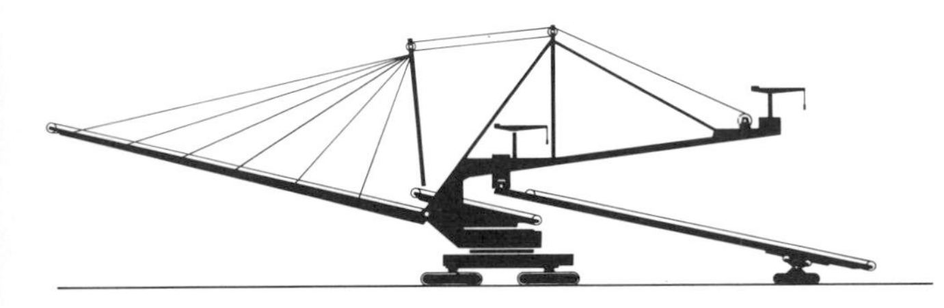

Fig. 9

4.1.3 Spreader with suspended receiving conveyor, Fig. 10

At first sight this variant seems to be a simpler solution, since it incorporates only one central travelling assembly, an advantage when comparing the manoeuverability of the different systems. However, the simple adaptation of the receiving chute to the tripper car can only be regarded to a limited extent as an advantage, since normally the tripper car already has a mobile receiving boom.

A clear disadvantage can be found during stability analysis. Strongly differing working positions — angular positions of the booms to each other — and uneven material flow over the total length cause heavy gravity displacement which can only be compensated by high design weights.

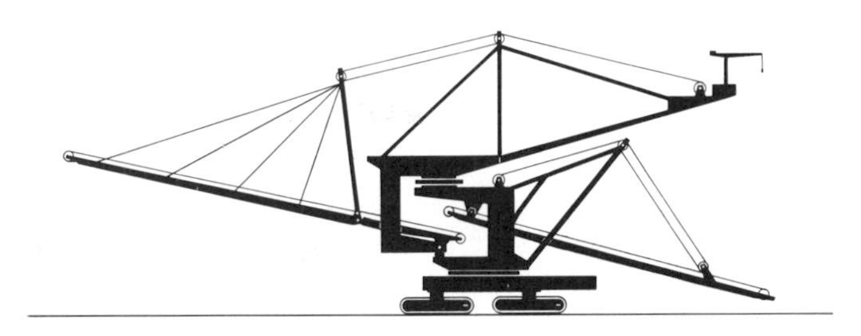

Fig. 10

4.1.4 Spreader with receiving conveyor supported on crawler assemblies, Fig. 11

An additional crawler assembly supports the receiving end of the boom and leads to a stable total system. In order to compensate the disadvantage of the additional travelling system, all structural components are identical to those of the main travelling system. Further advantages are the low total weight, combined with the simplified control of material transfer from the tripper car onto the boom of the spreader.

Fig. 11

5. Summary

The present report on development trends in spreader construction cannot naturally be valid for any specific application. External influences often involve the need to make decisions contrary to the statements made earlier. These influences might be of a geological or climatic type and create specific problems which cannot be touched on in an article of a general nature.

For an opencast mining operation with important production figures and average influencing parameters, the recommendation can be given to plan spreaders for the dumping side having the following basic design:

> Spreaders on crawler assemblies with receiving boom also crawler supported, one intermediate conveyor and one discharge belt conveyor system running at high speed, this equipment completed by an adequate number of auxiliary devices.

Development Trends in the Construction of Spreaders for Opencast Mine Dumps

Part 2
Cross-Pit-Spreaders

Aby Weiss, Germany

1. Introduction

Direct haulage of overburden over the shortest distance across the open pit to the mine area where the useful seams were already exploited represents the oldest and most economic method of exposing deposits by application of opencast mining techniques. The working range of the discontinuously operating Draglines is restricted to 30—40 m due to their relatively short outreach, in spite of the constant increase in size of these units. An extension of the haulage distance by introducing scrapers or trucks, or through rehandling of the material by Draglines, leads to a considerable increase in operating costs.

For increased digging heights, as long as 60 years ago the method of continuous overburden transport by means of conveying bridges was introduced. These systems require an additional travel path at the dumping side and their design needs to be exactly tailored to the geometric dimensions in a specific mine. They are only applicable for higher material flow volumes, large and flatly embedded deposits, as well as for overburden characteristics which permit the construction of stable dumps. Based on today's state-of-the-art, the economic limit of conveyor bridges is reached at a digging height of 60 m.

For exposal of deeper seams in large opencast mines, extensive application of the "around-the-pit" method is made, facing no limits as far as both size and depth of a mine are concerned. The overburden, mined by bucket wheel excavators in several steps, is transported by belt conveyors around the pit for final dumping by spreaders. These systems, comprising several units and very long relocatable belt conveyor systems, cause both high investment and operating costs that are only economic for large mineable deposits.

Efforts were made to find other solutions for small and medium-size operations, i.e., different combinations of continuously and discontinuously operating systems [1]. A most promising combination consists in a direct coupling of the digging unit to a long discharge conveyor travelling on the excavator side and extended in a free-supported manner across the open pit. These so-called "direct-dumping combinations" are remarkable for their excellent maneuvering capability, since they need no rails and floor-mounted belt conveyor systems are not required. The reduced number of machines operating in this combination provides a high availability of the total system.

Mr. A. Weiss is Project Engineer with Mannesmann Demag Baumaschinen, Division Lauchhammer, Düsseldorf, West Germany.

2. Hitherto Available Designs of Direct-Dumping-Combinations

One method practised until now for direct dumping of overburden consists in the application of a combination comprising one bucket wheel excavator and one slewing spreader, whilst the latter is identical to the type used for the continuous opencast mining technique, but with longer discharge boom. This direct-dumping combination, practised in Eastern Europe, is mostly operated at the deepest bench of a multi-step opencast mine [2], but also above an overburden cut made by a Dragline [3].

Cross-pit-spreaders of this conventional design were built with boom lengths up to 165 m. In order to extend the range of application of these spreaders, several companies proposed spreaders of identical design, moving on crawlers or walking pads, but with longer discharge booms. Considering, however, the overproportionate increase in dead weight and the associated disadvantages, these projects have not been realized until now.

Other proposed combinations of excavators and two successively operating spreaders or excavators, intermediate bridge conveyor and spreader, where one spreader should disadvantageously travel at the dumping side, did bring no change for the better, since disadvantages were incurred by their increased number of transfer points and low availability factors.

Another stage in the development of direct-dumping combinations is represented by the direct-dumping excavators used in US bituminous coal mines [4]. These excavators with long discharge boom are preferably operated according to the tandem method, with another excavator on the same bench or the next higher bench. Due to design aspects, the discharge boom length of these versions is restricted, too.

3. New Cross-Pit-Spreader

In recent years, coal deposits were explored and partly exposed in several countries, with the coal seams embedded in several layers, sometimes deeper than 30 m, i.e., beyond the optimum working range of a Dragline. Since the costs inherent in the movement of overburden by Draglines are lower than those of all other mining systems, attempts are naturally made to use these systems for deeper deposits, too.

A reasonable solution for this type of opencast mine is to combine the Dragline operation with a direct-dumping unit. Exposure of the deeper seams is left to the Draglines which form a preliminary dump. The top layers are mined by bucket wheel excavators. The overburden is directly transferred to the cross-pit spreader for deposition onto the material previously dumped by the Dragline (Fig. 1). If an overburden thickness of 30 m is allocated to the Dragline, the upper 20—30 m require a cross-pit spreader having a boom length of about 200—230 m, based on normal geological conditions.

If spreaders of this boom length were built according to commonly practiced construction types, they would be very complicated, cost-intensive and difficult to erect. This is especially due to the fact that the hitherto used spreaders have as special feature a superstructure that can be slewed relative to the undercarriage. The presence of a ball slew bearing results in a considerable increase in dead weight of the spreader, since many components are subjected to bending load and must be built as fully-web girders.

The overburden is normally transported to the rotating centre of the machine via a bridge conveyor supported on an additional travelling system. The requirement for maximum rotating capability of the superstructure, a prerequisite for the normal continuous opencast mining technique, leads in this case also to a complicated superstructure. The long counterweight boom, carrying the weight for compensation of the discharge boom mass, is mounted at a ligh level to permit slewing over the receiving conveyor. The gravity centre of this spreader version is relatively high.

In order to find a simplified version for the cross-pit spreader, the mining method with direct-dumping-combinations has been developed further in such a manner that slewing of the discharge boom becomes superfluous.

One condition for the stability of the high dumps in deep pits is to form a preliminary dump prior to banking up the final dump. This can be achieved by slewing of the discharge boom, but also by dumping at different boom lengths. In this case, the preliminary dump is banked up by the Dragline and the final dump by the cross-pit spreader.

The lack of slewing capability allows full use to be made of the discharge boom length, so that a higher overburden thickness can be transported, compared to a slewing boom of identical length. This solution requires, however, levelling of the ripple-shaped dump by means of auxiliary equipment.

The problem of accomodating the overburden masses, especially at the bench end, normally can be fully managed by the slewing capability of the receiving boom.

Another prerequisite for a perfect operation is that the equipment operated at both the same and different levels, including the machines for mining of the pay mineral, can operate independently of each other. The cross-pit-spreader has to guarantee that the Dragline moving at a lower level can continue its operation underneath the boom without any restrictions. When Dragline and cross-pit-spreader approach each other at the same level, time and production losses shall be reduced to a minimum. This requirement is solved by the Dragline leaving its travel path in the direction of the overburden highwall and moving to a position adjacent to the cross-pit spreader to enable this latter to pass by. Subsequently, the dragline returns to its operating place. Passing of the two systems can be realized on a relatively small level, since the spreader receiving boom is slewed by 90°. This eliminates the need for slewing of the discharge boom for this by-pass operation.

Based on the considerations given above, a new design version of cross-pit-spreaders has been developed (Fig. 2). Noteworthy features are low dead weight, compared to the conveying length required, ease of erection and the possibility of dismantling and relocation to another mine, when necessary, at low expenditure.

Compared to former versions, the slewing capability of the discharge boom was removed. This caused a considerable simplification of the structure. In order to maintain the flexibility of the system, the spreader can be equipped with one or two luffing and slewable receiving booms, so that all necessary work, such as new block cuts, ramp cuts etc. can be made to the extent as required.

In addition this spreader can travel both perpendicular to its discharge boom and at a definite angle to it. This permits a certain compensation of the masses excavated at the dumping side.

Simple design is a characteristic feature of this new cross-pit-spreader. The relationships between all components of the supporting structure are arranged in such a manner as to form statically favorable triangles of relatively large angles. This type of supporting structure, only subjected to tensile and compressive loads of the webs, results in low dead weight with optimum strength.

The rope-suspended boom can be built with a relatively narrow cross-section, since the suspensions are widely sup-

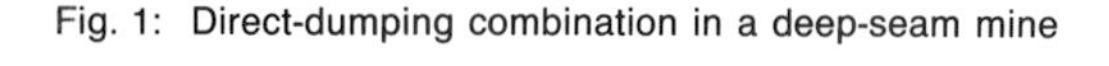
Fig. 1: Direct-dumping combination in a deep-seam mine

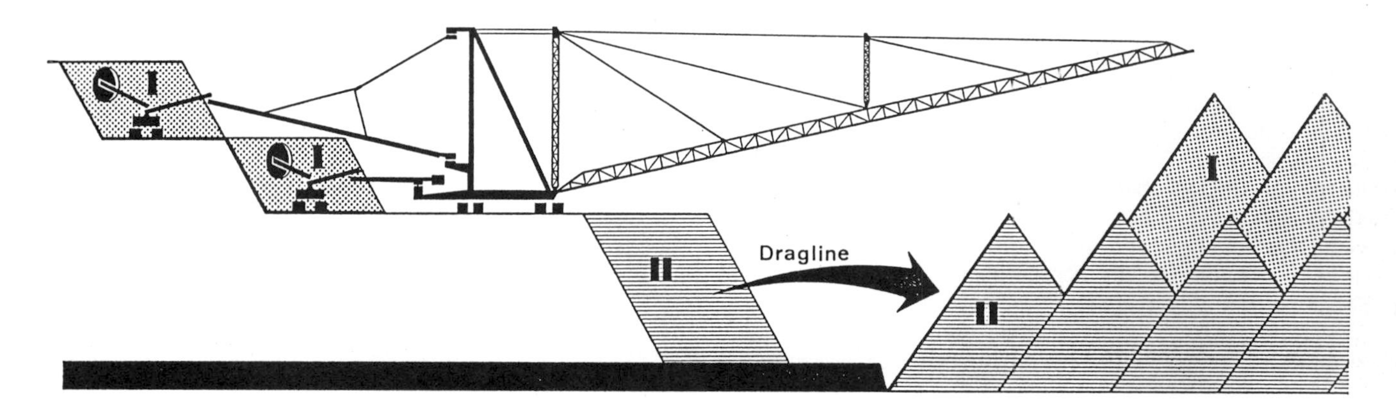

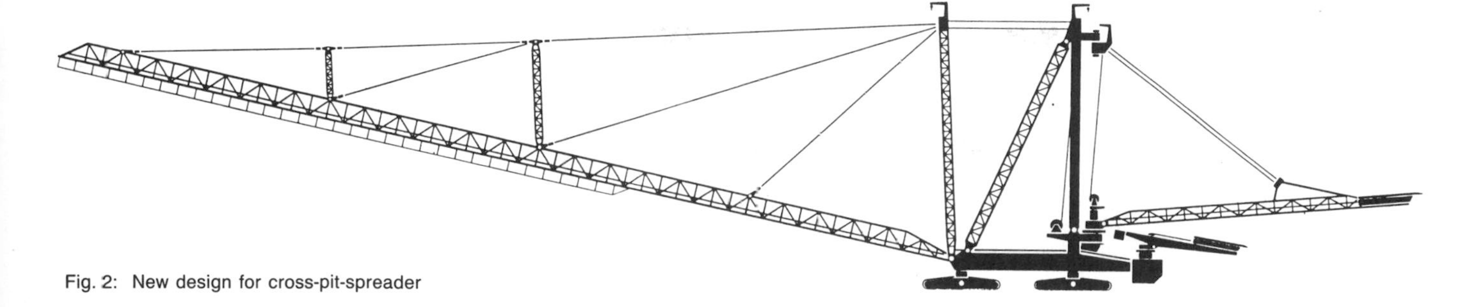

Fig. 2: New design for cross-pit-spreader

ported in the central structure. This is a point of special importance, considering the cross wind forces applied to the boom. The favorable arrangement of the horizontal suspensions avoids any modification in length of the suspended ropes during boom luffing motions.

The machine is supported in a statically determined manner by two travelling systems at the deep slope edge and one travelling system at the high slope edge. The supporting platform of the boom is seated on a triangular horizontal structure directly above the travelling system. A compressional stay in the axis of symmetry of this supporting structure takes the horizontal loads of the long boom. This strong beam is extended in the opposite direction to serve as a counterweight boom. Most of the heavy mechanical parts and the total electrical equipment are mounted on and in this beam. The deep position of these components, including the counterweight, garanties a relatively low position of the centre of gravity. This results in a high stability with a positive influence on the possible discharge boom length.

The belt conveyor system of the cross-pit-spreader consists of the receiving conveyor, accelerating conveyor and the discharge conveyor. The accelerating conveyor is rigidly supported on the central stay beam and serves also to even out the material flow. The material is then transferred onto the discharge boom without change in flow direction.

Different design versions can be adopted for the receiving section of the belt conveyor system, to suit the loading machines. If loading is done by a giant bucket wheel excavator or one single feeder, one slewing receiving boom will be supported on the counterweight boom. When using two compact bucket wheel excavators, two independently slewing receiving booms are arranged at the counterweight boom. One of these booms can be of the rocker type, whereas the second one is suspended from the vertical web of the supporting structure. For all possible combinations, a minimum slewing angle of 90° has to be maintained for the receiving booms, to permit by-pass of spreader and Dragline. A further possibility consists of direct loading of the accelerating conveyor via a chute mounted at the counterweight boom, e.g. by means of a belt wagon or a hopper spreader.

Fig. 3: Perspective view of a open pit mine with two bucket wheel excavators and cross-pit-spreader combined with dragline operation

4. Conclusions

In order to maintain the Dragline operation in deeper mines, long cross-pit-spreaders are required for transport of the upper layers. The new development presented shows that machines of this type can be economically designed if both their layout and the mining method used are matched to each other.

References

[1] Steinmetz, R., "Die Direktförderung von Abraum — eine rationelle Technologie zur Mineralfreilegung in Tagebauen", Berg- und Hüttenmännische Monatshefte 2, 1982, pp. 33—39

[2] Ökrös, M., Koos, G., "Die Direktversturzkombination im technologischen System des Tagebaues Thorez", Neue Bergbautechnik, Vol. 9 (1979), No. 3, pp. 140—148

[3] Prokopenko, V. J., Wendler, H., "Untersuchungen zur Vergrößerung des Anwendungsbereiches der Technologie des Direktversturzes in der URSS", Bergbautechnik, Vol. 20 (1970), No. 1, pp. 3—8

[4] Gaertner, K., "Tagebau-Aufschluß im Stripmining-Verfahren", Braunkohle, Wärme und Energie, 1, 1966, pp. 1—6

Overburden Stacker for Loy Yang Open-Cast Mine

W. Quaas, Germany

Abraum-Absetzer im Loy Yang Tagebau
Empileur de recouvrement pour la mine à ciel ouvert de Loy Yang
Apiladora de escombros para la mina a cielo abierto Loy Yang

ロイ・ヤング露天掘り炭坑用オーバーデンスタッカー

罗伊扬露天矿装设的过载堆积机

وحدة التكديس ذات التحميل المفرط الخاصة بالمنجم المفتوح لوي يانج

Summary

This case study describes the new overburden stacker in the most recent expansion project of the S.E.C. of Victoria. Design details and technical data are given.

1. Introduction

Following Morwell and Yallourn, the Loy Yang open-cast mine is the most recent expansion project of the State Electricity Commission of Victoria in the Latrobe Valley coalfield, approximately 120 km to the east of Melbourne. The Latrobe Valley extends over a length of 30 km at a width of roughly 12 km. A thin overburden of but 20—30 m covers 180 m of almost pure lignite with a mean calorific value of 1,900 Kcal/kg. This is equivalent to an overburden to lignite ratio of 1:6. In Europe's best open-cast mines the overburden to lignite ratio is 4:1. The large-scale open-cast mine currently being developed at the Hambacher Forst between Cologne and Aachen in West Germany will even have an overburden to lignite ratio of 6:1. Even though the overburden in the Latrobe Valley is not very thick, substantial problems are being encountered in connection with its disposal because of its extremely adverse physical and chemical properties and difficulty of handling it. The overburden predominantly consists of adhesive clay and tends to form lumps in sizes of up to 1 m diameter at a weight of approximately 1 t. During the dry season the clay becomes as hard as concrete and entrained sand makes it extremely abrasive.

2. Overburden Stacker System

In 1978 a contract was placed with a joint venture consisting of Johns Perry Ltd. and Maschinenfabrik Augsburg-Nürnberg AG under the technical leadership of M.A.N. for a stacker system to pile or dump the overburden.

The stacker system consists of a stacker mounted on 10 crawlers, a connecting bridge and a belt tripper with built-in conveyor drive system for the dump conveyor. The equipment weighs over 2,100 tonnes and is scheduled to go into operation within the next few days. The main technical data are given in Table 1.

Table 1: Main technical data

Overall length of stacker system	168.0 m
Overall height	35.0 m
Overall total installed capacity	3800 kW
Outreach of discharge boom	50.0 m
Belt width of discharge conveyor	2.6 m
Number of crawlers under stacker	2x3 = 6
Number of crawlers under belt tripper	2x4 = 8
Maximum soil pressure of stacker	0.9 kg/cm²
Maximum soil pressure of belt tripper	1.2 kg/cm²
Width of crawler pads	3.8 m
Supporting width of connecting bridge	55.0 m + 2.5 m
Transport speed of conveyor system	3.5 to 4.5 m/s
Travel speed of stacker system	3.0 to 10 m/min
Slew speed of boom	30 m/min

The stacker system is designed to handle 17,000 t of overburden per hour. The theoretical handling rate is 200,000 m³ per day. A stacker comparable in size and handling rate has so far not been used in Australia (Fig. 1).

As early as in the tendering stage meetings were held with the responsible engineers for maintenance, repair and operation of the machine to discuss all major aspects and the specific overburden conditions in order to warrant a high availability of the equipment. Models were made to study and facilitate the removal of parts subject to wear and tear.

3. Revolving Superstructure

The structural steelwork of the stacker and belt tripper is predominantly of full web and box girder construction — this reflecting the latest technological progress. This method of construction has also gained acceptance for open-cast mining because it offers the best resistance to corrosion and ingress of dirt (Fig. 2).

Even though the investment is higher than for lattice construction, subsequent maintenance is much easier. Corners do not corrode and patching the paintwork is much cheaper.

Oberingenieur Werner Quaas, M.A.N. Maschinenfabrik Augsburg-Nürnberg AG, P.O. Box 440100, D-8500 Nürnberg 44, Fed. Rep. of Germany

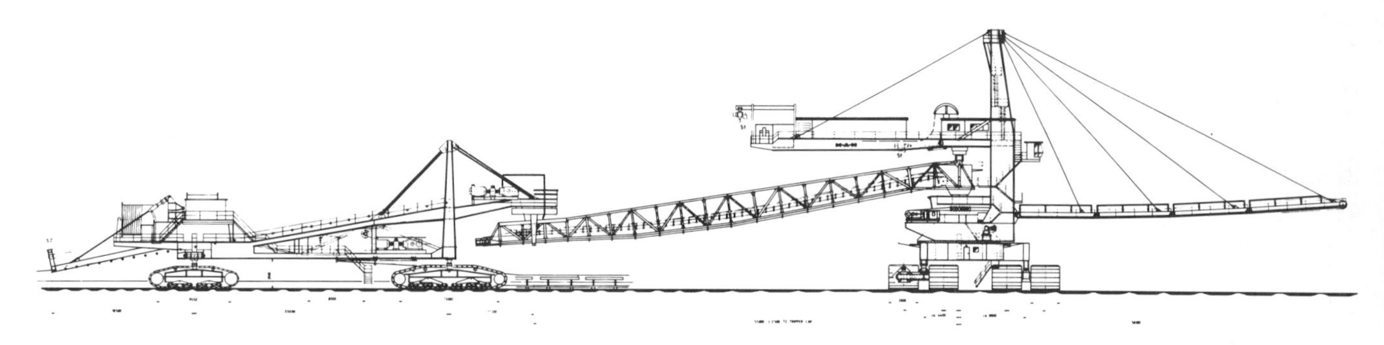

Fig. 1: General arrangement drawing of stacker system

Apart from that, a full web/box girder construction is easier to keep clean.

The connection between the revolving superstructure and the stacker gantry takes the form of a 2-row large-diameter ball slew bearing. The contact faces of the gantry and revolving superstructure are machined to warrant a positive seat. The slew bearing is secured by friction-grip bolts to the internal ring of the gantry.

The ball slew bearing is lubricated by means of an automatic centralized lubrication system. The ring gear is sprayed with grease. The slew bearing is protected by an additional metal cover with rubber skirt.

The two spur/planetary gearboxes of the slew motion are self-contained units and driven by squirrel-cage motors. The power is transmitted by overhung pinions to the ring gear located separately on the gantry.

For similar machines — also for reclaimers — the self-contained slew system takes the form of an M.A.N. slew drive unit with pin gear which is used when customers prefer the ease of interchanging the pins. The use of a pin gear necessitates a subdivision of the large slew drive units into several small slew units. On wearing and during repairs the individual pins are easy to remove without having to dismantle the ring gear. Besides that, the new pins are easy to make in the owner's workshops (Fig. 3).

When several small slew drive units are used it is a quick and easy matter to swing one slew drive unit out of position and continue operation without further disturbance or at a reduced handling rate.

The self-contained slew drive units can be replaced in a minimum of time. Repairs on the machine are therefore unnecessary.

Fig. 2: Stacker system during erection

Fig. 3: Self-contained slew drive unit

After swinging out the slew drive unit it can be inspected on the machine without the need to interrupt operation.

4. Conveyor System

For the sake of standardisation and spare parts inventory all conveyor drives are equipped with 610 kW squirrel-cage motors — one 610 kW motor for the boom conveyor, one 610 kW motor for the bridge conveyor and three 610 kW motors for the dump conveyor.

Oil-hydraulic couplings are incorporated as a starting aid for the boom conveyor and the bridge conveyor. The conveyors are reversible in order to counteract chute blockage. Three water-cooled hydraulic couplings are provided for the three drive units of the dump conveyor. These serve as a starting aid and for infinitely variable control of the transport speed from 3.7 to 5.3 m/s.

Service cars are provided for removal and re-installation of the upper and lower conveyor idlers.

The boom, the connecting bridge and the main girder of the belt tripper are designed such that the return run of the rubber conveyor belt is located underneath the supporting structure, thereby enabling dribblings and dirt to fall down freely.

5. Crawler Mechanism

While piling the overburden the stacker system will have to be moved many times. Special attention has therefore been given to the design of the crawler mechanisms (Fig. 4).

The crawler mechanisms are driven by slipring motors and planetary gearboxes. The crawlers of the belt tripper are mechanically linked in pairs, a squirrel-cage steering motor driving the guarded steering spindles via propeller shafting and reduction gears. The steering drives are amply dimensioned to enable the crawlers to be turned in completely while the actual stacker is at a standstill.

The stacker is equipped with two sets of steering crawlers; the third set of crawlers is stationary.

Removable guards on the crawler beams protect the drive unit, the rollers, the take-up system and the lubricating pipes from dirt.

6. Auxiliary Equipment

For the operating personnel there are washrooms with toilets, a crew room with kitchenette and air-conditioned operator's cabs. A workshop is provided for minor repairs. From four lubrication centres all lubricating points are automatically supplied with grease and oil via centralised lubrication systems. Lifting beams, cranes and electric lifting tackle with capacities of 0.5 to 10 t are provided for maintenance

Fig. 4: Crawler beam

and as handling aids. There is also a splicing unit for repairs on rubber belts.

The stacker system is equipped with a central fire fighting system which is connected by hose pipes to the central water supply system. Two standby tanks of 5,000 litres capacity each are available for emergencies. Additional manual fire extinguishers are arranged at all important points, e.g., at the drive units and in the electrical equipment houses.

Thorough and detailed coordination between the owner and the manufacturer and the application of modern design and manufacturing techniques have made it possible to match open-cast mining machinery to specific geological and mining requirements and to build such machinery within a reasonable period of time.

 Volume 5, Number 5, October 1985

Overburden Spreader for the Mae Moh Power Station, Thailand

J. Goller, Australia

As is the case in other countries worldwide, the kingdom of Thailand is, due to its dependence on oil, also developing its coal reserves and power stations. As part of this expansion programme, the power station at Mae Moh in the northern part of Thailand, near the Golden Triangle, is presently undergoing a major investment programme.

In relation to this programme, a contract for the removal of 90 Mm^3 overburden was awarded to the Thai company BME/Sahakol in 1983. The overburden is to be removed over a period of six years, keeping in line with the overall mine development. After fierce international competition and exhausting negotiations, Eglo Engineering, Sydney, was successful tenderer for the conveyors and the spreader, and it in turn awarded the spreader contract to PHB Weserhutte (PWH) Pty. Ltd., Sydney. The concept of the spreader was developed by PWH, Sydney, with part assistance from PWH Weserhütte AG, Cologne, Federal Republic of Germany.

The spreader consists of three key elements: the rail mounted tripper car, the connecting bridge between the tripper car and the spreader, and the spreader itself (Fig. 1).

The tripper car, with an aerial gauge of 3.5 m, travels along a shiftable conveyor with a belt width of 1,800 mm. The maximum feed capacity of the system is 6,500 m^3/h, which is equivalent to 10,000 t/h of overburden. The tripper car is only equipped with long travel drives, whereas the conveyor drives are at the front and the rear station of the shiftable conveyor. In addition, a small carriage travels on the shiftable conveyor's tails, supporting the rear end of the connecting bridge, with the head end being suspended underneath the counterweight boom of the spreader. In order to ensure reliable feed to the connecting bridge conveyor, this bridge was designed 45 m long, resulting in an inclination of 10°. The 1,800 mm wide belt travels at 5 m/s on 45° fixed mounted trough idlers. High-voltage motors with fluid couplings and bevel helical shaft mounted gear boxes are employed to drive the conveyor.

Mr. Joachim Goller, General Manager (Sales), PHB-Weserhutte Pty. Ltd., P.O. Box 730, Crows Nest, N.S.W. 2065, Australia

The spreader itself is mounted on two fixed mounted, fully balanced crawlers, which maintain an average ground pressure of 0.9 kg/cm^2 (Fig. 2). The two-crawler concept was selected as it proved, within the parameters of this plant, to be very economical, flexible and of extreme manoeuvrability. The requirement of travelling, for example, around the head station of the shiftable conveyor could easily be fulfilled with the two-crawler concept as the machine can be turned into one crawler.

The spreader boom is 50 m long and employs two conveyor drives. As was the aim of the connecting bridge, the loading point on the spreader boom was reduced to an inclination of 11° to ensure proper loading of the conveyor under dry and wet conditions. This led to a banana-shaped boom design, resulting in an 18° inclination at the boom tip in order to maintain the discharge height required.

The entire crawler undercarriage was designed and supplied in an extremely short time from PHB Weserhütte AG. The upper slew platform was placed by PHB Weserhutte, Sydney, to a company in Hong Kong, whereas the entire remaining structural fabrication was carried out in Thailand to the benefit of local employment as well as gaining of local experience.

Coordination of this contract, as well as supply of mechanical components such as idlers, pulleys and bogies, was carried out in Australia by PWH's Sydney office. The task was to complete the contract within fourteen months, the contract being awarded to PHB Weserhutte in September 1983 with final commissioning to be carried out in December 1984. By now, the plant has already shifted some 3 Mm^3 and is well on target to reduce the amount of time initially envisaged to remove the 90 Mm^3.

Execution of this contract has shown the flexibility of PHB Weserhütte as a multinational or, better, a multi-country company. Also, benefits could be drawn from the fact that similar machines have been built elsewhere.

Australia has realised over the last few years that the South Pacific area is an area of potential growth in the years to come. The region can be well served by Australian companies, and, as such, it is the aim of PWH, Sydney, to expand into this market.

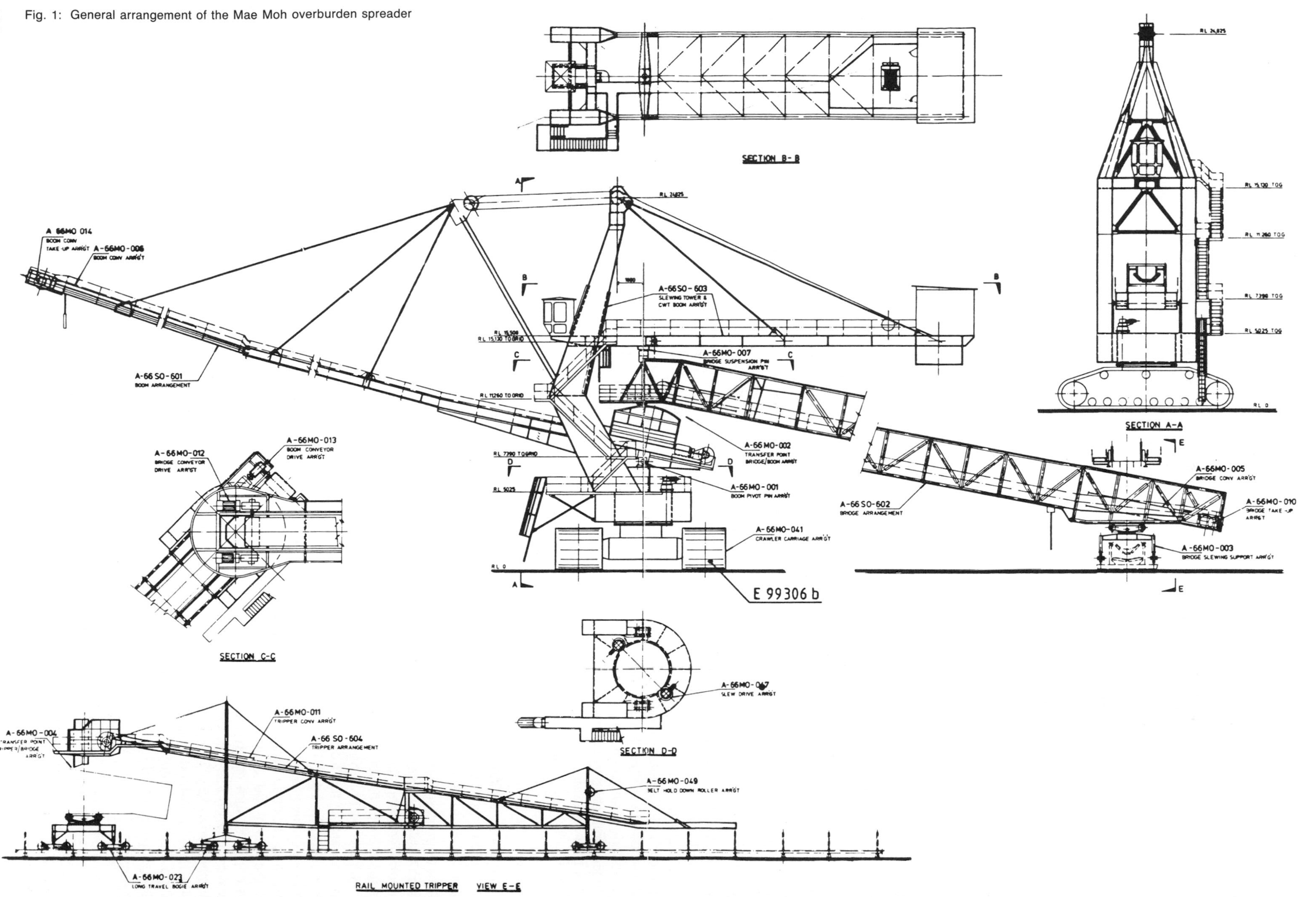

Fig. 1: General arrangement of the Mae Moh overburden spreader

Fig. 2: Mae Moh overburden spreader mounted on two crawlers

The Best of bulk solids handling:

Pneumatic Conveying of Bulk & Powder

Summer 1986
250 pages, US$ 26.00
ISBN 0-87849-063-9

This publication comprises over 30 technical papers written by internationally known experts on R & D studies and application possibilities for pneumatic conveying and transporting.

An important book for everyone engaged in the design and operation of pneumatic pipelines.

Contents

Order Form

Please send me the book: **Pneumatic Conveying of Bulk & Powder (D/86).**
I enclose cheque drawn to a U.S. bank in the amount of US$ 26.00.

Name: ______________________ Title: ______________________

Company/Institution: ______________________

Address: ______________________

City: ______________________ Country: ______________________

Cheque No.: ______________________ Signature: ______________________

Pro-forma invoice upon request.

Please mail this Order Form together with your cheque to:

TRANS TECH PUBLICATIONS
P. O. Box 266 · D-3392 Clausthal-Zellerfeld · West Germany
Tel. (0 53 23) 4 00 77 · Telex 9 53 713 ttp d

Volume 2, Number 1, March 1982

Continuous Surface Mine Materials Handling Systems

T.M. Brady, J.M. Goris, T.W. Martin and E.M. Frizzell, USA

Summary

This paper presents a review of three ongoing research projects dealing with continuous materials handling systems in open-pit mines in the USA. The systems include mine-run-rock conveyors, high-angle conveyors, and movable in-pit crushers. These material handling systems are being evaluated and applied to mining systems primarily to reduce the dependency on diesel fuel and to reduce mining costs.

1. Introduction

Open-pit mining operations in the United States handle about $2.3 \cdot 10^9$ metric tons* of ore and waste each year excluding coal, sand, and gravel. Most of this material is handled by trucks. In recent years, truck haulage costs have increased to where they account for more than half of all pit operation costs. It is expected that this figure will continue to rise, primarily because of increasing fuel and labor costs.

During the past 20 years, several large mining operations have been very successful in developing high capacity conveyor systems. Some examples are: the brown coal industry in Germany, iron mining in the USSR, copper mining in Zambia, and tar sand operations in northwest Canada. One of the largest mines in the world is at Fortuna, Germany, where over 14,000 t/h are handled by conveyor belts. In the USA, several high-capacity conveying systems have been developed for applications other than mining. Examples are the Oroville Dam project and a system which handles 18,100 t/h loading iron ore barges on the Great Lakes. The outstanding feature regarding these installations is their high continuous capacity, which results in substantially lower costs.

Most of the systems mentioned handle alluvial-type material consisting mainly of fines, and have the following equipment in common: (1) wheel excavators or reclaimers, (2) shiftable conveyors, (3) crawler-mounted stackers, and (4) steel cables core belting. According to Dennehy [1], a comparative recent assessment made in the U.S. and Europe revealed that there are some 80 different types of conveyors, 10 types of elevators, and 50 types of feeders. Conveyor technology in the lignite deposits of West Germany is perhaps the most developed and is now spreading to the tar sand and coal projects of western North America.

Many surface mines in the United States today move comparably large volumes of material; however, they must handle rock, not fines or alluvial material. As compared to alluvial material, rock normally must be put through a primary crushing unit and reduced in size prior to conveying. Rock handling systems also have higher capital and operating costs per ton than those handling fine material, because a more rugged installation is required and more belt wear and damage will occur.

For some years now, the merits of employing belt conveyors in the U.S. mines for primary open-pit haulage have been discussed and argued. Few hard rock operators have been influenced by the advantages of belt conveyors, preferring instead the more flexible and initially less costly truck haulage. There are two significant rock moving operations in North America now using conveyors for haulage in open-pit mines. The first is at Twin Buttes in Arizona, where conveyors have been used since 1965. In 1979, this mine moved 45,268,494 tons of ore and waste. The second operation is at the Sierrita open-pit copper mine, also in Arizona, which moved 58,065,598 tons of ore and waste in 1979. Because of this limited use of conveyor haulage systems in U.S. surface mines, it is difficult to analyze and compare their productivity with present truck haulage.

Both Twin Buttes and Sierrita require trucking of ore to the in-pit crushers. Recent developments in large front-end loaders, where bucket sizes have now surpassed the 19 m^3 mark, suggest that a loader-conveyor combination with load haul dump units feeding directly to movable conveyors could be more efficient. However, it appears that further development of alternative rock handling systems to reduce trucks depends on either (1) the availability of large movable crushing systems feeding conventional or high angle conveyors or (2) the development of conveying systems to handle run-of-mine material with no crushing necessary.

Both approaches have advantages; the first alternative is the more technically feasible from the standpoint of the crusher, since manufacturers presently have such machines on the market and the capability of developing improved models.

* all tons (t) in this paper are metric tons

T.M. Brady, General Engineer; J.M. Goris, Supervisory Mining Engineer; T.W. Martin, Supervisory Mining Engineer and E.M. Frizzell, General Engineer, U.S. Department of the Interior, Bureau of Mines, Spokane Research Center, East 315 Montgomery Avenue, Spokane, WA 99207, USA

However, a high angle conveyor will require more research and development. From an economic assessment, the second alternative, by eliminating the crusher and trucks from the pit, could be the lowest cost system. Technically, again, the development of conveying systems to handle plus 50 cm material is still in the research stage.

Based on this analysis, the U.S. Bureau of Mines is conducting three research projects involving alternative haulage systems. They are:

1. A mine-run-rock conveyor system capable of handling rocks up to 150 cm in size.
2. A high angle conveyor capable of moving material at angles up to 45 degress.
3. A large movable crusher with feed through capacities of 1,800 to 3,600 t/h.

2. Mine-Run-Rock Conveyor

The main factor that restricts lump size for conventional belt conveyors is collision impact between the rock on the moving belt and the fixed idlers, as such impact damages the conveyor belt idlers. Major improvements have recently been made in this area. In addition to steeper troughing angles, the flexible roller base, known as the Garland* system, has been developed. Its design principle is the interlocking of several rollers with chain links, offering a limited amount of roller flexibility. The Garland can be deformed to change its trough angle, dependent on the material passing over it. This results in elimination of part of the kinetic energy of the impact load. However, even using a Garland conveying system, it would be necessary in most U.S. metal and nonmetal surface mines to provide primary crushing ahead of the belt system.

This section of the paper describes the results of work completed to date by R.A. Hanson Company, Inc., under contract to the U.S. Bureau of Mines. This research and development effort covers the design, fabrication, and testing of two prototype mine-run-rock (MRR) or large rock conveyors.

3. Current Applicable Conveyor Technology

A review of the literature, visits to mining and manufacturing operations, and discussions with researchers led to the investigation of five conveyor concepts as possible candidates for handling run-of-mine material in U.S. mines. The five systems are:

1. Idler Supported Endless Belt (ISEB).
2. Car and Rail Supported Endless Belt (CRSEB).
3. Cable and Pulley Supported Endless Belt (CPSEB).
4. Car Train.
5. Belt Train.

3.1 Idler Supported Endless Belt (ISEB)

Idler supported endless belts carrying large rocks are basically conventional conveyors with stronger components and closer idler spacing. They are characterized by flat belts with side structures to keep material on the belt. Many crushers are fed with a short belt of this type from the loading pockets. Two firms, Bando Chemical Industries and Réalisation d'Equipments Industriels (REI), have developed and manufactured ISEB conveyors capable of handling material to 1.5 m in size. The Bando equipment has been used in a municipal land reclamation project near Osaka, Japan; and REI of Paris, France, has designed and manufactured rockbelt loaders for over 10 years with installations in Spain, Greece, France, Cuba, Great Britain, and Italy.

3.2 Car and Rail Support Endless Belt (CRSEB)

The CRSEB consists of a belt supported by moving cars. Each car is a troughed crossmember with wheels that roll on upper carrying and lower return rails. These cars are connected by roller chains, and pulled by the force of friction between the loaded conveyor belt and the car. Patents on this concept have been filed in the Soviet Union.

The first prototype was installed in 1970 at the Karatau mining operation in Hazakhstan, USSR, to transfer blasted phosphate ore from the open pit to the crushing plant. As shown in Fig. 1, material up to 1.5 m in size was handled at rates of

Fig. 1: Large rock on USSR CRSEB conveyor (Photo courtesy of Licensintorg, USSR)

1,500 t/h up an inclination of 20 degrees. This 52 m long prototype conveyor used a 1.2 m wide belt of fabric core construction driven by a single drive pulley. At the loading area, several rows of truck tires supported the belt and provided shock absorption.

An operational rock conveyor that is 1,250 m long in three flights is now being built at the Karatau mine (Fig. 2). This conveyor carries a 1.6 m belt and incorporates several improved design changes. The cars are curved rather than troughed and use smaller diameter wheels.

Joy Manufacturing has experimented with a car and rail conveyor system with a flexible belt that is hard-bolted to small cars that run on rails. This equipment is designed to convey 15 cm minus material around horizontal curves.

3.3 Cable and Pulley Supported Endless Belt (CPSEB)

This method uses wire rope as the power transmitting medium. The belt is made of rubber vulcanized around a fabric envelope and stiffened laterally with transverse

* Reference to specific equipment, trade names, or manufacturers does not imply endorsement by the Bureau of Mines.

Fig. 2: First flight of 1,250 m CRSEB conveyor at Karatau mine (Photo courtesy of Licensintorg, USSR)

flexible steel rods and is carried on the wire ropes supported by pulleys. These pulleys are mounted on line stands located at intervals along the conveyor route.

Cable Belt Ltd. of England is the designer and sole manufacturer of this conveyor method. A paper by Ian M. Thomson [2] outlines the design evolution of the method and discusses in detail the major components of the system. This conveyor was designed to transport run-of-mine coal or crushed rock at high speeds over long distances. It is used extensively in Great Britain, with several installations in the United States. One of these is a 9,910 m length conveyor at the Anamax Company's copper mining operations near Sahuarita, Arizona.

3.4 Car Train

The quasi-continuous car train system is comprised of a series of mine cars traveling on railroad-type trucks. Electrically-driven, rubber-tired drive wheels act against continuous steel flanges on the sides of the cars, moving the train forward. Several trains travel on a single loop of track and are propelled by multiple-driven stations positioned along the tracks.

The car train was developed in France for use in a nickel mine of the Societé le Nickel (SLN) of Paris. The method is referred to as SECCAM. Two SECCAM systems, 19.3 and 2.4 km long, were built on the island of New Caledonia to transport nickel ore from a mine in the mountains to a port facility. Ultimately, both installations were abandoned in favor of conventional conveyor systems.

3.5 Belt Train

Two systems have been developed that use a belt riding on, or suspended from, a rail-mounted car; one in West Germany, and one in the USSR.

The ASBZ system developed in West Germany is an experimental quasi-continuous, high-speed facility for transporting bulk raw materials over long distances. The trains are driven by linear motors on an elevated, column-supported, two-rail track. The prototype system handled 19,000 m³/h of crushed rock at speeds of 1,200 to 1,300 m/min over a 4,900 m track. This installation has now been dismantled. Applications for the ASBZ system are limited to high volumes and long distances due to the extensive and costly loading and unloading installations required.

The Soviets have built a belt train which is another quasi-continuous system composed of a series of connected cars which travel on rails. On horizontal or low-incline tracks, the motive power is furnished by rubber tires rotating in a horizontal plane reacting against the structure of the cars. On high-incline tracks, a linear induction motor is used. This Soviet system, consisting of ten 122 m long trains, was used at the Sarbai Quarry. It handled material up to 1.0 m in size at 1,500 m³/h over a distance of 6.4 km. The system is still under development.

4. Crib and Cable Supported Endless Belt

Following their investigation, the R.A. Hanson Company designed another conveyor system to handle large rock. The crib and cable supported endless belt (CCSEB) is shown in Fig. 3 and integrates design features of both the CRSEB and CPSEB methods. It uses a steel core belt supported on steel, troughed cribs attached by mechanical clamps to a wire rope located on each side of the crib. The wire rope is supported by pulleys mounted on line stands along the length of the conveyor. The belt is driven by a conventional drive system and the cribs are pulled along by friction of the loaded belt.

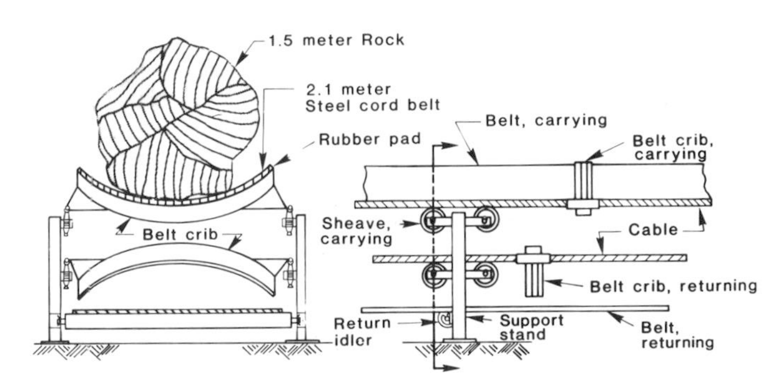

Fig. 3: Crib and cable supported endless belt

Preliminary design for three of these systems, the ISEB, CRSEB, and CCSEB, was completed to the extent that budgetary acquisition costs and operating costs for production models could be established. Acquisition costs for a 360 m section are shown in Fig. 4.

	ISEB	CRSEB	CCSEB
Labor ($1000)	$ 553	$ 553	$ 518
Materials ($1000)	1,921	1,619	1,483
SGA and Profit ($1000)	1,423	1,249	1,151
Total Cost ($1000)	$3,897	$3,421	$3,152
Cost per Foot	$3,330	$2,924	$2,694

Fig. 4: Estimated production model costs of MRR conveyors

The operating costs of the three different systems were also calculated as part of the investigation. Based on both the acquisition and operating costs, it was concluded that the CRSEB and CCSEB systems could transport run-of-mine material for the least cost per ton kilometer — $ 0.102 and $ 0.115, respectively. The estimate for the ISEB system was $ 0.155 per ton kilometer.

5. Prototype Testing Program and Facility

Based on this investigation and analysis, two concepts have been selected for testing, the Crib and Cable Supported Endless Belt and a modified Car and Rail Supported Endless Belt. A test facility, illustrated in Fig. 5, is being fabricated. The facility will be approximately 46 m in length, 12 m wide, and stand 14 m high. Large rocks, 1.5 m in diameter, will be loaded with a boom crane into the vertical chute at one end of the facility. From the bottom of these chutes at the end of the facility, the material will be carried by 2.2 m wide belts up to a 20° incline and dropped into the chute at the other end, thereby recycling the rock in a continuous conveyor-chute-conveyor-chute system. Both conveyor belts will be powered by an electric motor through a chain and sprocket drive.

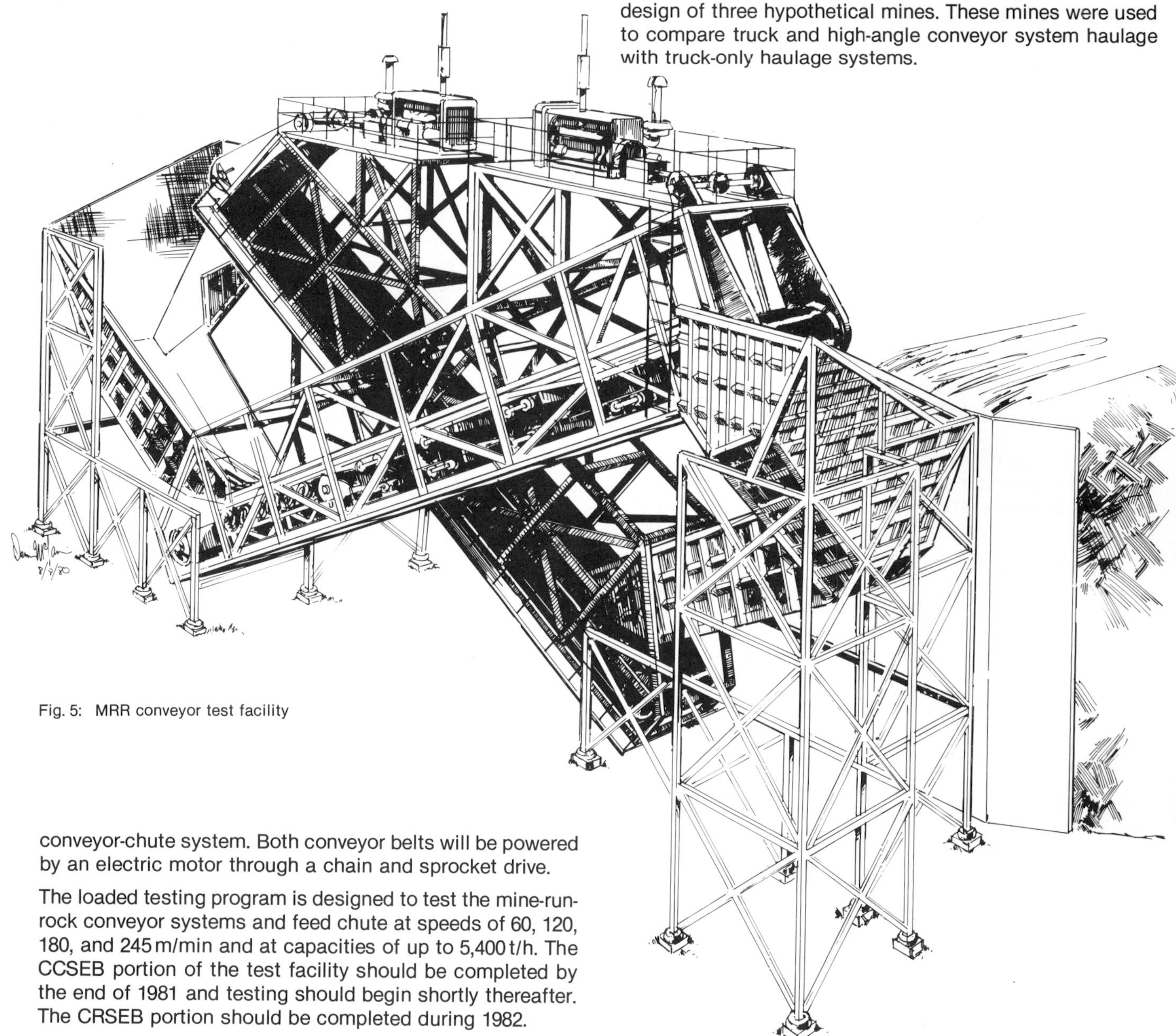

Fig. 5: MRR conveyor test facility

The loaded testing program is designed to test the mine-run-rock conveyor systems and feed chute at speeds of 60, 120, 180, and 245 m/min and at capacities of up to 5,400 t/h. The CCSEB portion of the test facility should be completed by the end of 1981 and testing should begin shortly thereafter. The CRSEB portion should be completed during 1982.

6. High-Angle Conveyors

One of the difficulties of conveyor haulage is the inherent slope restriction of approximately 18° for handling material. Unfortunately, this does not coincide with the typical slope of open-pit walls, which generally range between 38 and 45 degrees. Conventional conveyors must, therefore, either be placed in a notch to reduce this angle or run out of the pit in a switchback configuration. Both methods pose serious problems to the day-to-day operation of the mine. A possible solution is a conveyor with a lift capability of approximately 45° capable of handling high tonnages and high lifts.

Dravo Corporation of Pittsburgh, Pennsylvania, under contract to the U.S. Bureau of Mines, conducted a study of the technical and economic feasibility of using a high-angle conveyor system in open-pit mines in the U.S. as an alternative to truck haulage. First, a survey of the metal and nonmetal surface mining industry in the U.S. was conducted to determine the performance requirements for high-angle conveyors. The mines included copper, taconite, phosphate, molybdenum, and gold. The data collected and the input from mine planners and operators formed the basis for the requirements of a high-angle conveyor system and the design of three hypothetical mines. These mines were used to compare truck and high-angle conveyor system haulage with truck-only haulage systems.

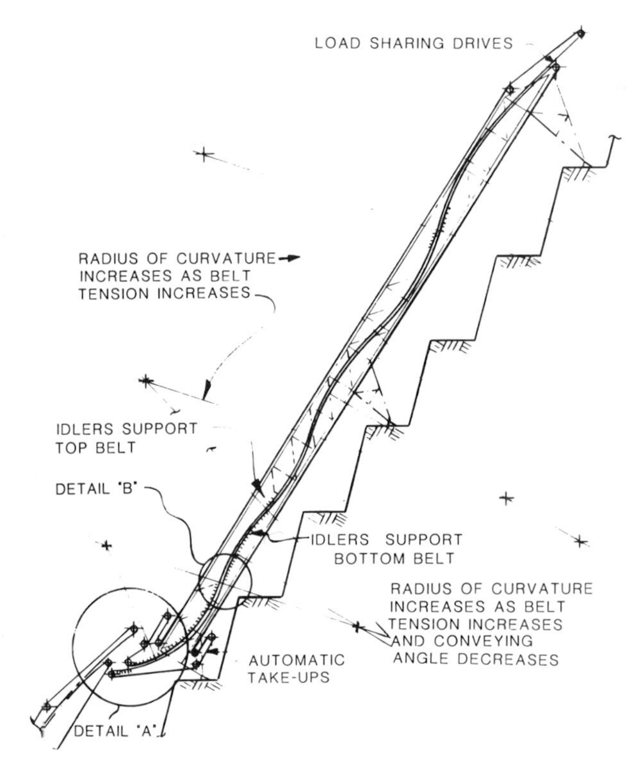

Fig. 6: Snake sandwich conveyor

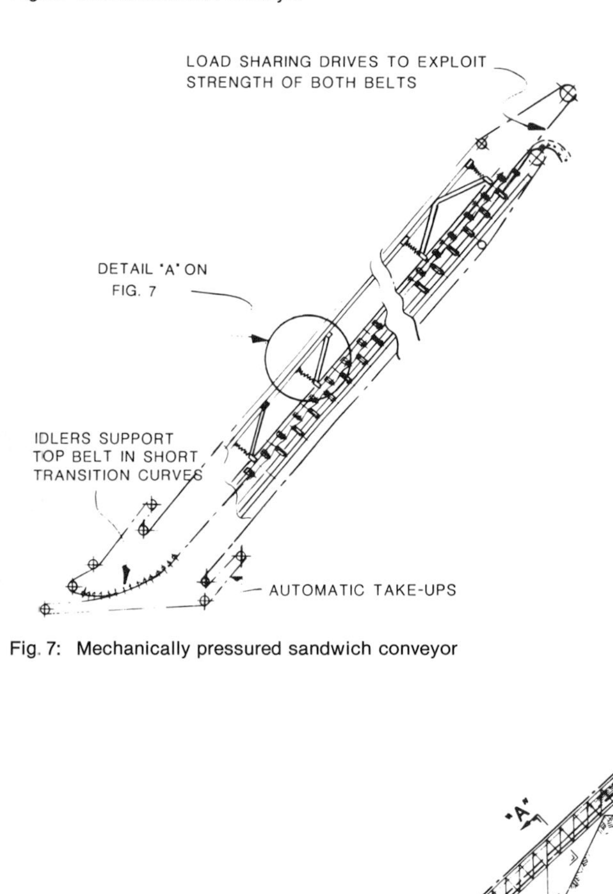

Fig. 7: Mechanically pressured sandwich conveyor

6.1 Minimum Performance Requirements for a High-Angle Conveyor

After reviewing all the information collected, the following average performance requirements were developed:

1. **Material Characteristics:** angular, hard, slightly abrasive, 38° of repose.
2. **Lump Size:** 203 mm maximum lump size.
3. **Capacity:** 3,200 t/h
4. **Lift:** maximum possible based on equipment limitations.
5. **Mobility:** move at least once every 2 years.

Next, a state-of-the-art study of high-angle conveyors was conducted and involved such conveyors as bucket ladders, belts with partitions, and sandwich belts. After a thorough analysis of the collected data, an evaluation was performed in order to determine which conveying concept and system for a high-angle conveying unit best satisfies the needs of the mine operators.

Sixteen methods were considered including a linearly accelerated bucket column. These included high-capacity bucket ladders to be carried by chains, cables, or belts; pocket belts; pipe belts; sandwich belts; slurry conveyors; and screw conveyors. An evaluation of these methods showed that the sandwich belts would best meet the performance requirements for the surface mines surveyed. Three sandwich belts were then investigated:

1. The snake sandwich belt (Fig. 6).
2. The mechanically-pressured sandwich belt (Fig. 7).
3. The pneumatically-pressured sandwich belt (Fig. 8).

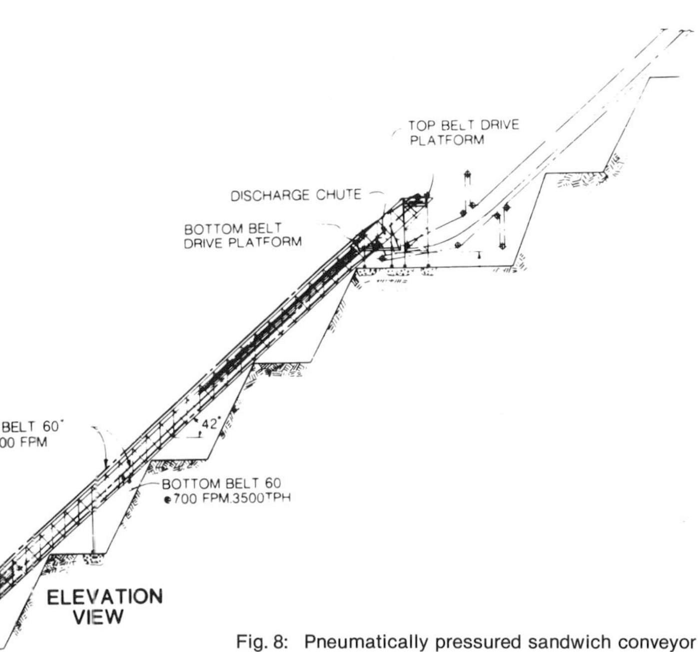

Fig. 8: Pneumatically pressured sandwich conveyor

Of these, the pneumatically-pressured belt was selected as the preferred method; while the mechanical and snake belts ranked a close second and third, respectively. A pneumatically-pressed sandwich conveyor module of 96 m of lift to convey at 3,200 t/h was then developed in more detail. Erected price, operating, maintenance, and mobility costs were determined. Although the conveyor design was based on 3,200 t/h, the pneumatically-pressed sandwich belt is capable of conveying tonnages in excess of 5,000 t/h.

7. Movable In-Pit Crusher

The introduction of mobile crushers occurred in 1954 when Krupp first built a movable primary crusher on crawlers. Since then, others have been built and mounted on walkers, tires, skids, and rails. Their principal use today is in limestone pits, where lower tonnage rates are generally required (under 1,000 t/h), and where the operation is generally not around the clock, thus providing time for maintaining the crusher system.

The use of movable crushers in large open-pit mines is presently being studied by Gard, Inc. under contract to the U.S. Bureau of Mines. The objective is to determine the need, applicability, and economic limitations of movable in-pit crushing/conveying systems. The areas of investigation involve a survey of a number of large open-pit mines to determine their needs and requirements for in-pit crushing, an examination of the state-of-the-art of movable crusher technology, and an integration of the results of these two into a unified concept for in-pit crushing/conveying.

Mine operators surveyed were all interested in movable crushers and had an almost universal preference for gyratory-type primary crushers, citing their inherent reliability and durability and their relative freedom from clogging. All operators preferred large crushers with capabilities of 2,700 to 3,600 t/h to minimize the number of conveyors needed. One operator indicated the possible need to move a crusher every 9 months to 1 year; most agreed a move every 2 years would be sufficient, and some would tolerate a system at one location for 5 years.

It was determined that large primary crushers can be moved from the main grinding circuits out into the pit by setting the crusher on a large frame and using tracks, wheels, or walking mechanisms. This portability promotes minimum haulage distances between the shovels and the crusher by allowing the crusher to be placed at a location central to the shovels. The expense of the large supporting structures and moving mechanisms is justified by their contribution to reduced fuel consumption and fewer haulage vehicles and, consequently fewer operators and maintenance personnel. In addition, the ease of crusher movement contributes to increased production levels.

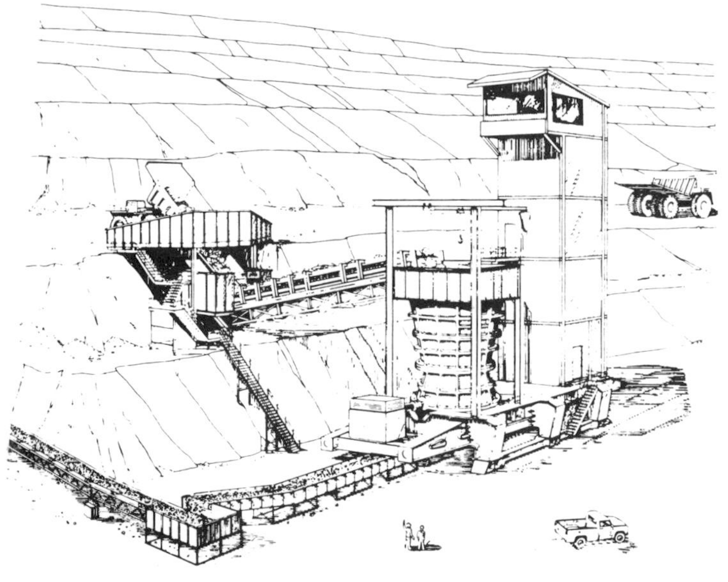

Fig. 9: Movable crusher

Although a fully-movable large gyratory crusher is a large piece of equipment to move, the technology for moving such large structures is available. Walkers are used in the mining industry already on large draglines and are compatible with this type of equipment movement. When the need to move the crusher arises, commerically available transporters can be attached to the heavy structural frame and removed when the move has been completed. The transporter should serve the moving needs of all crushers in the mine complex and be available between moves for other plant requirements.

A movable crusher could also be recessed in a pit bench and a feed hopper installed above the crusher to facilitate the unloading of the trucks, or the crusher station could be positioned parallel to the toe of the bench with rock being fed by a rock belt or apron feeder from the next higher bench. Fig. 9 represents an artist's concept of the crusher at the toe of the bench.

Removal of the crushed product can be by conventional idler, steel cable belt conveyors set in the mine at the normal 18°, or high-angle sandwich-belt conveyors. Fig. 10 shows another artist's concept of a movable crusher with a high-angle conveyor.

8. Comparison of Haulage Systems

The pneumatically-pressed sandwich belt was compared to truck haulage using three hypothetical mines representing the copper, taconite, and phosphate industries. Included in the comparative analysis was a movable in-pit crusher. Based on the need expressed by mine personnel for optimum flexibility within the pit, it was determined that in all three cases trucks would be used to haul the blasted material from the mine face to the in-pit crushers. The crushers then fed the high-angle conveyors, which elevated the ore and waste material to the top of the pit. However, the out-of-pit haulage systems differ for the three different hypothetical mines.

The possibility of truck haulage outside the pit was considered in all three high-angle conveyor applications. Such a system proved to be economically favorable in the hypothetical taconite mine when compared to the conventional haulage system and examined within the specified return on investment (ROI) of 10 to 15 percent. The high-angle conveyor system for the phosphate mine proved to be economical in only part of this range. Because of the small amount of material that is actually hauled on the surface from the high-angle conveyor in this mine plan, the large capital expense required for the installation of a stacker/spreader system could not be justified. It was apparent that surface haulage by trucks was not logical in the copper mine where the out-of-pit truck haulage system would constitute a large portion of the total haulage cost. The high-angle conveying system in the copper mine was complemented by a level, conventional conveying system to the plant and waste dumps, and shiftable conveyors with spreader/stackers at the dump sites. Such a system proved economically favorable. It was also found that a level, conventional conveying system could be used in the taconite mine to further lower the already favorable costs of the system described.

For conventional truck systems, haulage profiles were analyzed to determine the required number of trucks at different stages of the mine life and thus reflect the cost increases in haulage as a function of increasing depth. The number of trucks was also used as the basis for selection of support and maintenance equipment, the cost analysis of the operation being based on 1980 dollars. Manpower costs represented the average requirements throughout the mine life. Capital costs for all major equipment were obtained on budgetary price quotes from various manufacturers during January 1980. The costs of power to operate the facilities are based on the rates charged in a state where the mine might be located.

The dollars per ton of ore figures that were generated represent the price at which the ore must sell to get the specified return in investment. With a 10 percent return on investment after taxes, the selling prices per ton of ore for the truck systems versus the high angle conveyor/movable crusher systems in the copper, taconite, and phosphate mines are $1.53 versus $1.47, $1.26 versus $1.22, and $2.99 versus $2.88 respectively.

9. Conclusions

The rising costs of truck haulage combined with the threat of diesel fuel shortages suggest that the next major advance for U.S. open-pit mining will be increased use of conveyors for both ore and waste.

As mines grow larger and deeper, the amount of savings due to less truck usage becomes greater. Both mine distances and lifts are expected to increase at a rate that will adversely affect even the largest haul truck. Studies have shown that the cost-per-ton per truck goes up exponentially as the distance uphill increases. This is due to lower production-per-hour per truck because of increased cycle time (lower speeds during hauling at high grades over long distances) and additional operating costs per truck (increased fuel consumption and maintenance costs).

At this point, Bureau investigation and analysis have determined that movable crushers, high-angle conveyors, and mine-run-rock conveyors are economically competitive with truck haulage. Feasible concepts have been developed for both the high-angle and mine-run-rock conveyor and the major task ahead is the technical development and testing of these concepts. Presently, ongoing Bureau research will do this testing and development for two mine-run-rock conveyor concepts. In the case of mobile crushers, the technology remains to be demonstrated in operating mines. In 1980, a semi-mobile jaw-type crushing plant was installed in the Meirama open-pit lignite mine in northwestern Spain. Repositioning of this unit will be necessary every 6 months to 2 years, and the feasibility of its use should be confirmed over the next 5 years. As mentioned previously, a gyratory crusher is presently being erected at the Sishen Iron mine in South Africa. It is scheduled to commence operating at the end of 1981.

There are additional advantages not taken into account in an economic analysis of conveyor systems. A conveyor system is far more efficient in terms of energy usage, less labor intensive than a trucking system, and less sensitive to inflation over the years, as replacement and maintenance parts are considerably less. The disadvantages are lack of flexibility, the need to crush large rocks (for conventional conveyors), and the potential consequences of shutdown.

In summary, the benefits to the mining industry from the development and use of these alternative haulage systems are:

1. Elimination of the dependency on diesel fuel.
2. Reduction of mining cost.
3. Reduction of labor force.
4. Reduction of maintenance.
5. Reduction of noise and air pollution.
6. Reduction of haul road cost.

Bibliography

[1] Dennehy, J.T., "Belt Conveyor Transport System", The Ars. I.M.M. Conference, Illawarra, May 1976.

[2] Thomson, I.M., "Operational Experiences and the Development of Cable Belt Conveyors", CIM Bulletin, v. 68, no. 763, November 1975.

[3] Martin, T.W., J.M. Goris, and T.J. Crocker, "Large Rock Conveying Systems and Their Application in Open-Pit Mines", SME-AIME Annual Meeting, Chicago, IL, February 1981.

[4] Frizzell, E.M., R.N. Johnson, and E.A. Mevissen, "Conveying at High Angles in Open Pits", Proc. SME-AIME Annual Meeting, Chicago, IL, February 1981.

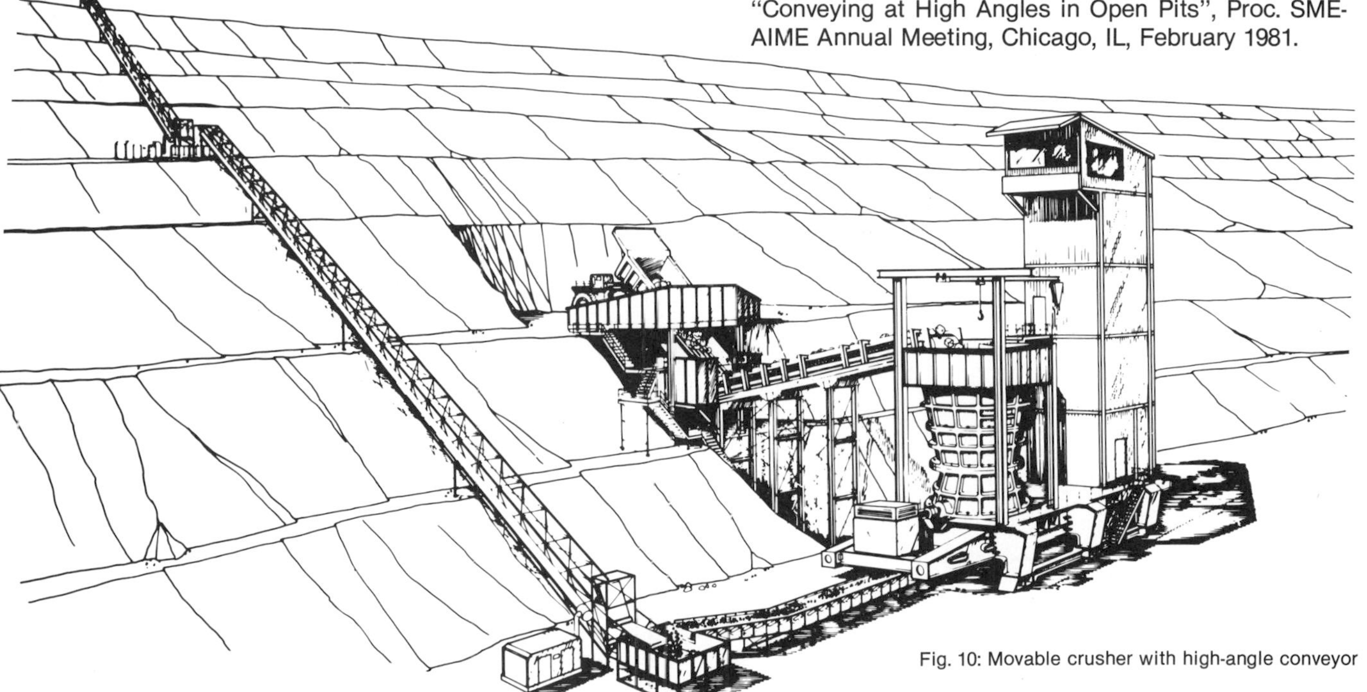

Fig. 10: Movable crusher with high-angle conveyor

 Volume 4, Number 1, March 1984

Sandwich Belt High Angle Conveyors

Applications in Open Pit Mining

Joseph A. Dos Santos, USA

Summary

High angle conveyor (HAC) systems in open pit mining offer many advantages over the traditional, truck only, haulage systems, including: (1) superior energy efficiency, (2) less dependency on petroleum products, (3) less sensitivity to inflation, (4) less labor, (5) less excavation for the amount of ore recovered, and (6) less ramp construction and maintenance costs. Previous studies have revealed the technical and economic feasibility of high angle conveyors. The present paper introduces the Continental Conveyor sandwich belt HAC as the vital link to optimization of in-pit conveying systems. A test and demonstration unit, featuring 1,524 mm (60 in) belts to lift materials up to 19.5 m (64 ft) at angles to 60° with conveying rate to 2,721 mt/h (3,000 st/h), has been built and demonstrated. Discussion continues with the HAC systems to suit the various open pit mining operations and the corresponding characteristics of the system components, particularly the high angle conveyor. The range of HAC systems from the low capacity [up to 1,814 mt/h (2,000 st/h)]*, low lift [10.7 to 36.6 m (35 to 120 ft)], highly mobile to the moderate capacity [1,814 to 3,629 mt/h (2,000 to 4,000 st/h)], moderate lift [36.6 to 76.2 m (120 to 250 ft)], intermittently mobile to the high capacity [greater than 3,629 mt/h (4,000 st/h)], high lift [greater than 76.2 m (250 ft)], semi-mobile, are discussed along with some possible combinations of these.

1. Introduction

Open pit mines are the principle source of minerals, soft coal and lignite throughout the world. In efforts to achieve energy independence and to be reliable energy suppliers to world markets, the U.S. and Canada are seeking to exploit their abundant deposits of coal, oil shale, tar sands, and diatomaceous earth to produce synthetic fuels (synfuels) that will reduce dependence on imported oil. Open pit mining methods are presently used in Canada to recover vast deposits of tar sands and promise to be the economical mining method for many future synfuels projects.

* mt = metric tons
st = short tons

Joseph A. Dos Santos is Manager/Advanced Systems, Continental Conveyor & Equipment Company, Inc., Winfield, Alabama, USA

Extended version of a paper presented at the American Mining Congress Mining Convention, San Francisco, September 11—14, 1983, and Society Mining Engineers — American Institute Mechanical Engineers Fall Meeting, Salt Lake City, October 19—21, 1983

To date, trucks have been the primary haulage means in open pit mines. It has been acknowledged generally that conveyor systems are more economical for transporting and dumping mined materials at very high volumetric rates. Such systems have been and are used, in conjunction with continuous mining machines such as bucket wheel and bucket chain excavators, in the lignite fields of Germany and in other parts of the world. These have been generally limited to application in soft or unconsolidated materials which will not consistently result in large boulders deposited on the conveyor belts. Mining of hard rock and consolidated materials throughout the world typically involves blasting of the bank, loading by discontinuous means, such as shovels, backhoes, front end loaders, etc., and haulage to the plant and waste dumps by large off-highway trucks in the 73 to 154 mt (80 to 170 st) payload category. The many large boulders which result after blasting make continuous belt haulage impractical without further reduction of the material.

Truck systems offer flexibility in the mining operation because each truck can be rerouted on a hourly basis. Truck haulage costs have, however, increased steadily over the last two decades and dramatically in the late 1970s through 1980.

This steady trend is expected to continue into the future. Unit capital and operating costs have increased due to inflation, the rapid increase in crude oil related commodities, such as fuel, lubricants and tires, and due to increased pit depths which result in lowered truck efficiency and availability [1]. The truck fleet sizes are continually increasing due to increasing waste to ore ratios, decreased ore grades, increased haulage lengths and lifts as mines expand and deepen. Large support facilities, equipment, and maintenance crews are required to service the truck fleet and to construct and maintain large haulage ramps. Truck haulage costs now comprise as much as 50% of the total mine operating costs in some mines [2].

There are serious doubts about the future economic viability of trucks as the prime means of material transport. In response to increasing truck haulage costs and in the face of increasing tonnages to be moved, two large U.S. copper mines [181,400—226,800 mt per day (200,000—250,000 st/d), 229—305 m (750—1,000 ft) deep] [1, 3], responded in the late 1960s and early 1970s by installing high capacity belt conveyors with large, permanent type, gyratory, in-pit crushers. The in-pit crushers serve to reduce large boulders to a conveyable size. These systems did not seek to eliminate all trucks, but rather to displace them, by belt conveying, in elevating the ore and waste from within the pit and to the plant and waste dumps. In each case, the flexibility of a trucking system was maintained within the pit from the mine face to the in-pit crusher.

These systems proved more economical than the truck only systems [1, 3], and the loss in overall mine flexibility was not initially a problem because of the local in-pit truck haulage. As the mines continued to expand and deepen, however, truck haulage costs once again began to climb due to the increased haulage lengths and lifts to the in-pit crushers. Permanent crusher locations were also found to obstruct the recovery of newly discovered ore and of material which had been reclassified as ore due to lowering of the cut off grade. It was evident that the next generation of large in-pit crushers and conveying systems must provide for economical relocation of the system components to follow and not dictate nor obstruct the direction of mining.

Mobile crushers with theoretical rates of 300 to 1,000 mt/h (331 to 1,102 st/h) have been used to reduce material to conveyable size for over twenty-five years [4]. Only a few larger units have outputs of 1,000 to 1,600 mt/h (1,102 to 1,764 st/h), and application has been predominantly in limestone.

High capacity, mobile or movable crushers to reduce hard rock materials have only recently been developed. The U.S. Bureau of Mines (BOM) conducted a study from August of 1979 to November of 1980 which ultimately led to the design of a movable crushing system employing a large 1.52×2.26 m (60×89 in) gyratory crusher with a throughput capacity of 3,629 mt/h (4,000 st/h) [5]. The crushing system which consists of separate, self-contained, crusher unit, apron feeder, and truck loading hopper can be moved once every year to once every two years by independent, crawler type, transporter units. Industry too has introduced several developments of similar movable crushing systems to feed movable and shiftable high capacity belt conveyors [6, 7]. Such systems retain the local flexibility of truck haulage from the mine face to the loading hopper. The remaining material haulage is by elevating conveyors to the pit perimeter and overland conveyors to the ore stockpiles and waste dumps, with waste spreading by shiftable conveyors and crawler mounted mobile spreaders. Overall mining flexibility is also retained because all conveyors are either shiftable or movable.

Duval Corporation began installation of the first, movable, in-pit crushing and conveying system of this kind in 1981 at its Sierrita Copper and Molybdenum Mine near Tucson, Arizona. Operation began December, 1982, and tentative plans are already underway for another two, duplicate systems at the same property [7]. Operation of another similar system also began in 1982 at Sishen's North Iron Ore Mine in South Africa [8]. Many other companies are now conducting studies or making plans to install similar systems.

Completely self-contained, self-propelled, mobile, gyratory crushers, highly mobile connecting conveyors, shiftable conveyors, and spreaders were also recently installed in a South African coal mine to reduce and transport shovel loaded overburden at 3,000 mt/h (3,307 st/h) design rate to the out-of-pit waste dumps. This system has completely eliminated truck haulage [9]. Highly mobile, high capacity, crushing and conveying systems for coal are also possible with high capacity, crawler mounted, feeder breakers [10] and roll crushers.

Since the installation of the first, large, in-pit, crushing and conveying systems, the trend has moved slowly towards more mobile and, therefore, more flexible crushing and conveying systems. Movable, shiftable and mobile conveyors have significantly improved overall system flexibility. The vital link which permits optimization of any in-pit conveying system, however, has until now been missing.

2. High Angle Conveyor (HAC) In-Pit Conveying Systems

That vital link is a high capacity high angle conveyor (HAC) which meets the performance requirements of open pit mines. The resulting high angle conveyor (HAC) system differs from the latest mobile or movable, in-pit, crushing and conveying systems only in the use of the HAC as the continuous elevating means. The result is, however, a more compact, mobile and, therefore, more flexible system. HAC systems offer many advantages over traditional truck haulage systems including:

1. **Superior Energy Efficiency**
   Trucks must transport their own dead load in addition to the payload. Dead load can range up to 45% of the total gross hauled weight when the truck is loaded to capacity. In a high angle conveyor, the energy goes towards elevating material with only a small amount lost to idler and pulley friction, material acceleration, etc.
2. **Less Dependency on Petroleum Products**
   Electrical power for a high angle conveyor system is increasingly generated by coal, hydro, and nuclear power. The dependency on imported oil and the prospect of diesel fuel rationing need not concern the mine planner.
3. **Less Sensitive to Inflation**
   A high angle conveyor will have a longer life and much lower maintenance costs when compared to trucks which are replaced every six to eight years.
4. **Less Labor Cost**
   A HAC system requires less operators, maintenance personnel and facilities. Trucks require one driver per shift each, and large maintenance crews must provide continuous upkeep and repair to keep the truck fleet operating.
5. **Less Total Excavation**
   High angle conveyors may be supported along any stable slope. Total excavation is, therefore, determined by geotechnical stability considerations and not by the 8% to 10% maximum slope requirements of the truck haulage ramps. This also makes a high angle conveyor superior to a conventional elevating conveyor with maximum slope of about 15°.
6. **Less Haulage Road Maintenance**
   A high angle conveyor system requires only small access roads for maintenance vehicles. This greatly reduces the large equipment fleets and thus costs required for large truck haulage ramps.

In a coordinated effort, the U.S. Bureau of Mines (BOM) conducted a "High Angle Conveyor Study" [2] from August, 1979 to December, 1980, simultaneously with the aforementioned mobile crusher study [5], with interfacing of the two, otherwise independent, studies to agree on compatibility of the performance requirements as revealed during surveys of the metal/non-metal open pit mines of the U.S. The BOM study also revealed all, then, state-of-the-art high angle conveying methods and found none adequate for the operational requirements of large U.S. open pit mines. A development criterion was then formulated and used in an engineering evaluation of all HAC methods. That evaluation found the sandwich belt high angle conveying approach the most operationally appropriate and economical solution. Significant speculatory advances conforming to the criterion were then made in the sandwich belt conveyor technology.

In similar studies over the past four years, Continental Conveyor & Equipment Company, Inc. reached a similar conclusion. The advancement of sandwich belt high angle conveyor technology, however, has been taken beyond speculation to the building and successful demonstration of a high capacity high angle conveyor test and demonstration unit at the company headquarters in Winfield, Alabama.

3. Continental Conveyor HACs

Continental Conveyor HACs are based on the sandwich belt concept. Before discussing the test and demonstration unit and the advantages of this system, it is appropriate at this point to provide insight into the theory of sandwich belts and to relate this to the Continental Conveyor sandwich belt method.

3.1 The Sandwich Belt Principle

The sandwich belt approach employs two ordinary rubber belts which sandwich the conveyed material. Additional force on the belts provides hugging pressure to the conveyed material in order to develop sufficient friction at the material to belts, and material to material interface so that sliding back will not occur at the design conveying angle. Fig. 1 is a simplified illustration of the interaction of forces. If the cover belt is not driven, then the lineal force N to provide the required hugging pressure at conveying angle α is given by the following equation:

$$N \geqslant W_m \left(\frac{\sin \alpha}{\mu} - \cos \alpha \right)$$

Where: $\mu = \mu_m$ or μ_b whichever is smaller

A more realistic model is shown in Fig. 2. An ample belt edge distance assures a sealed material package during operation even when belt misalignment occurs. A more comprehensive treatment of force interaction for a complex model along with the implications of driving both belts is not within the present scope and can be found in Dos Santos and Frizzell [11].

The Continental HACs consist of a carrying conveyor belt which is supported on closely spaced troughing idlers and a floating cover belt which is softly pressed onto the conveyed material by closely spaced, fully equalized, pressing rolls. The required material-hugging pressure varies according to the conveying angle, material characteristics and the dynamics of the system. The hugging pressure device is, therefore, designed for the specific requirements of the application with provision for field adjustment.

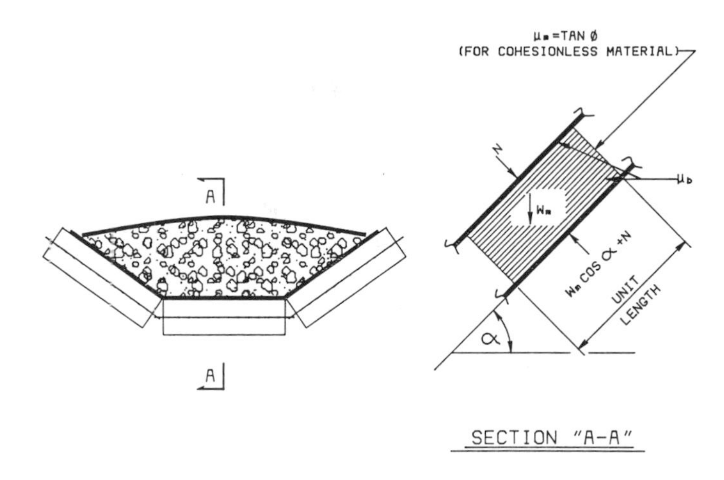

α = CONVEYING ANGLE

Ø = INTERNAL FRICTION ANGLE OF BULK MATERIAL

μ_m = COEFFICIENT OF FRICTION FOR BULK MATERIAL ON BULK MATERIAL

μ_b = COEFFICIENT OF FRICTION FOR BULK MATERIAL ON CONVEYOR BELT

N = NORMAL LINEAL HUGGING LOAD EXERTED BY THE TOP SURFACE

W_m = LINEAL WEIGHT OF BULK MATERIAL

Fig. 1: Idealized model #1 of Sandwich Belt Conveyor

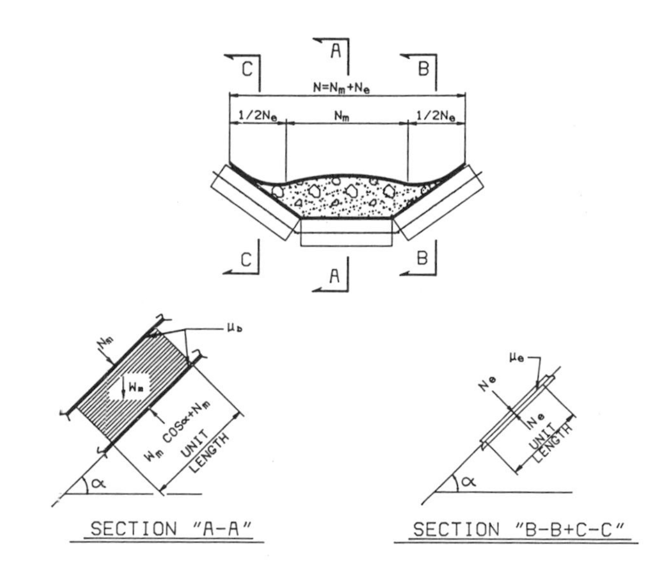

α, μ_m, μ_b, N, W_m ARE AS DEFINED IN MODEL #1

N_m = THAT PORTION OF N WHICH BEARS DIRECTLY ON THE CONVEYED MATERIAL

N_e = THAT PORTION OF N WHICH BEARS DIRECTLY ON THE EDGES OF THE BOTTOM BELT

μ_e = COEFFICIENT OF FRICTION AT THE INTERFACE OF THE COVER AND CARRYING SURFACES

Fig. 2: Idealized Model #2 of Sandwich Belt Conveyor

3.2 The HAC Test and Demonstration Unit

The HAC prototype unit, illustrated on Figs. 14—20, features 35 m (115 ft) overall length with 1,524 mm (60 in) belts to convey a wide variety of materials at incline angles from 30° to 60° and at conveying rates to 2,721 mt/h (3,000 st/h) when

at the 60° conveying angle. Main technical features are given in the Appendix.

Study of the sandwich belt concept had been ongoing for quite some time at Continental Conveyor. In June of 1982, however, conceptual engineering intensified and by September of that year the sandwich belt HAC had been conceived. Final engineering of the test and demonstration unit began in October with first equipment orders and start of fabrication by early December. Ground breaking was by the first week of January, 1983, and foundations were completed by end of February. Erection followed immediately, and the HAC prototype conveyed its first material, run of mine coal from an Alabama open pit. at 33° conveying angle, on June 13, 1983. The conveyor was raised the following day to transport the same material at 45°, and several days later at 60°. To date (12-9-83) the HAC prototype has sucessfully demonstrated conveying of Alabama coal (rom and sized), Texas lignite and coarse copper ore at angles to 60 degrees. Coal and lignite have been conveyed at rates above 1,814 mt/h (2,000 st/h) while coarse copper ore has been conveyed at up to 2,540 mt/h (2,800 st/h).

Why a HAC test and demonstration prototype? Though troubled by ever increasing haulage costs, the mining industry must rely on proven equipment and techniques. Installation of a complete HAC system requires too high a capital expenditure to risk on engineering speculation alone. The HAC test unit affords the opportunity to test performance with a wide variety of materials and permits demonstration on a large enough scale to convince the mining and materials handling industries.

3.3 Advantages of the Sandwich Belt HAC

The Continental Conveyor HACs can take on various forms as illustrated in Figs. 3—7 and offer many advantages over other systems, including:

1. **Simplicity of Approach**

   The use of all conventional conveyor hardware means interchangeability of components, fast delivery of replacement parts. Operating experience with conventional conveyors leads us to expect high availability and low maintenance costs.

2. **Virtually Unlimited in Capacity**

   The use of conventional conveyor components permits high conveying speeds. Available belts and hardware to 3 m (120 in) wide make possible capacities well in excess of 9,072 mt/h (10,000 st/h).

3. **High Lifts and High Conveying Angles**

   Lifts to 107 m (350 ft) are possible with standard fabric belts and single run lifts greater than 305 m (1,000 ft) are possible with steel cord belts. High angles in excess of 45° are possible without excessive wear because of the soft, floating, fully equalized, hugging pressure device.

4. **Flexibility in Planning and in Operation** (See Figs. 3—7)

   The sandwich belt lends itself to a multi-module conveying system using self-contained units as well as to a single run system using an externally anchored, high angle conveyor. In either case, the conveyor unit may be shortened or lengthened or the conveying angle may be altered according to the requirements of a new location. High angle conveying modules may be mounted on rails, rubber tires or crawler type transporters or may be equipped with walking feet for optimal mobility.

5. **Easy Cleaning and Quick Repair**

   Smooth surfaced belts allow continuous cleaning by belt scrapers or plows. This is especially important in handling wet and sticky material. Smooth surfaced belts present no obstruction to quick repair of a damaged belt by hot or cold vulcanizing. Quick repair means less costly downtime.

6. **Dust-Free Operation**

   During operation, the material is sealed between the carrying and cover belts. Well centered loading and ample belt edge distance result in no spillage along the conveyor length. Dust control efforts are only needed at transfer points.

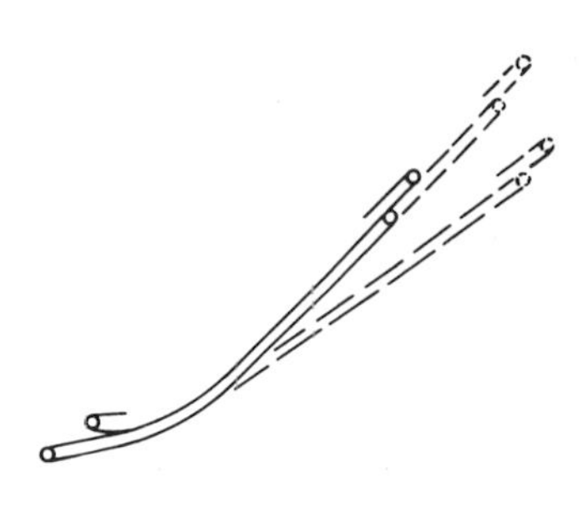

Fig. 3: HAC may be lengthened or shortened and the angle may be altered as required

Fig. 4: HAC modular approach in deep open pit mines

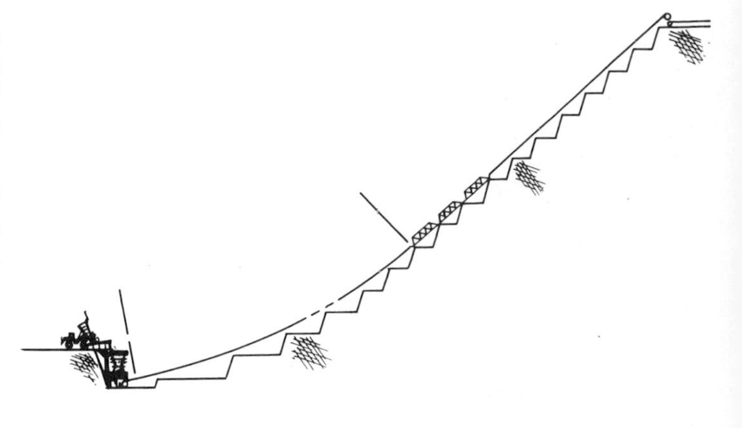

Fig. 5: Very high, single lifts are possible with steel cord belts

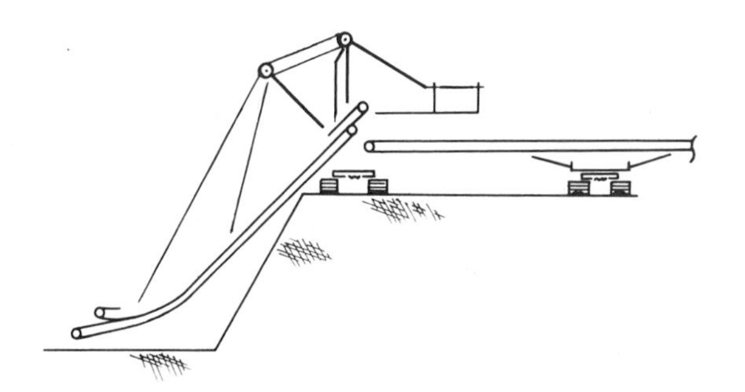

Fig. 6: HAC with single point suspension is combined with beltwagon for optimal mobility in high wall applications

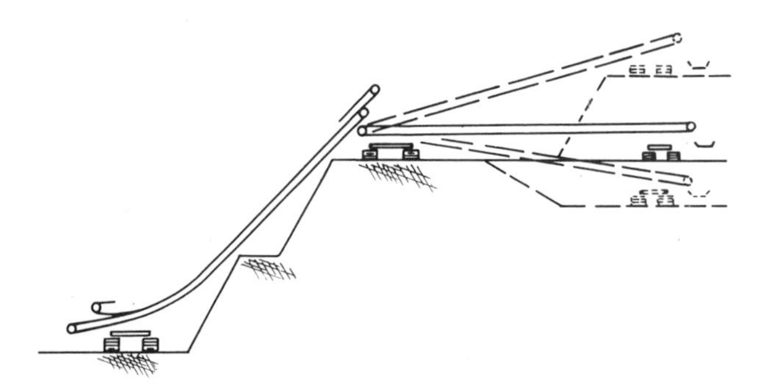

Fig. 7: HAC with two point suspension is equipped with conveyor bridge for intermittent mobility in end wall applications

4. HAC Systems — Applications in Open Pit Mining

Each form of the HAC unit, as illustrated by Figs. 4—7, is especially suited for the mobility, flexibility, lift and tonnage requirements of particular types of open pit mining operations.

The resulting HAC systems may take on many forms. As a minimum, any HAC system consists of: (1) an in-pit crushing system to reduce the material to a conveyable size, (2) the HAC to elevate the material out of the pit, (3) an out-of-pit conveying system to take the material to the plant or to the waste dumps. The following describes some of the many possible HAC systems.

4.1 Deep Open Pit Mines

Multiple HAC units may be arranged in series (see Figs. 4 and 8), with the lower HAC module discharging onto the upper HAC to achieve high lifts from deep open pit, copper, taconite, or western phosphate mines. The system in Fig. 8 consists of truck haulage from the mine face to an in-pit, movable, crushing system, the movable crusher, one or more HAC units, and an out-of-pit conveying system to the plant and waste dumps. Waste and overburden disposal at each dump site is by means of a shiftable conveyor, traveling tripper and a crawler mounted waste spreader. Multiple systems or subsystems may be used in any deep open pit mine.

The modular approach offers great flexibility in overall mine planning. The self-contained modules may each have up to 106.7 m (350 ft) of lift. Such modules are pontoon-mounted at two or more support points with provision for transport by rubber tired or crawler mounted, independent, transporter units, or by attachable/detachable walking feet. Whichever transport means is chosen, the same will be used at the movable crushing system, the HAC units and the drive and tail stations of the movable and shiftable conveyors. Relocation of the movable crusher and extension or relocation of the HAC modules may be required every one to two years [2, 5], and this does not warrant permanently attached transport means. Conveying angles are determined by stable mine slopes which may vary from approximately 38° to 45°. Design rate for the movable crushing system, the HACs and the out-of-pit conveyors may be from 1,814 mt/h (2,000 st/h) in western U.S. phosphate mine applications to greater than 6,350 mt/h (7,000 st/h) in deep, open pit, copper mines. Even greater throughputs may be required in future oil shale and tar sands projects.

The movable crushing system will receive blasted material with a maximum 1.52 m (60 in) boulder size and reduce it to a discharged product of 152 to 254 mm (6 to 10 in) maximum lump size, depending on the design rate, belt width, etc. Except in the smaller phosphate mines, this requires a large, gyratory type crusher in the 1.52×2.26 m (60 x 89 in) size range with additional pre-screening to achieve the very high throughputs.

A single run HAC with steel cord belts may be used in lieu of the modular approach to elevate materials more than 1,000 ft from within the deep open pit. The system of Figs. 5 and 9 does not differ in any other way from the HAC system just discussed. System performance is, however, significantly affected. Very high lifts produce extremely high belt tensions at the conveyor drive pulleys. Such high tensions are beyond the working strength of fabric ply belts, but not beyond that of steel cord belts. Such tensions make a self-contained HAC unit impractical, and large anchorage foundations are required at the head (drive) and tail ends to relieve the intermediate support spans of the belt tension forces. A long transition curve from the HAC loading area to the ultimate conveying angle is also required with a troughed steel cord belt, because of the high axial stiffness of the steel cords. The result is a less mobile HAC system with increased imposition into the pit because of the long transition curve. A single run HAC will be less expensive, however, and the elimination of material transfer points will improve the system's availability.

4.2 Thin Seam Strip Mines

The highly mobile HAC unit of Figs. 6 and 10 eliminates the need for trucks and haulage ramps traditionally required to haul coal out of single or multiple, thin seam, coal or lignite, strip mines. The HAC system is used to convey the coal from pit to plant with primary overburden stripping by dragline in direct cast or extended bench mode. Prebenching and recontouring may be by other means, such as bucket wheel excavators with end around conveyors and spreaders, by scrapers, or by shovels and trucks. Coal loading into the mobile crusher may be by shovel, backhoe, or front end loader. Continuous excavating machines, such as bucket wheel excavators or Easy Miners may also be used to load the coal directly onto the HAC without the need for additional crushing.

The HAC system consists of a mobile, crawler mounted, high capacity, feeder breaker or roll crusher to reduce the lump size, a mobile, crawler mounted, HAC unit to elevate the coal from within the pit, a crawler mounted beltwagon, a shiftable conveyor, and an overland conveyor to take the coal to the plant (Overland conveyor is not shown on Fig. 10).

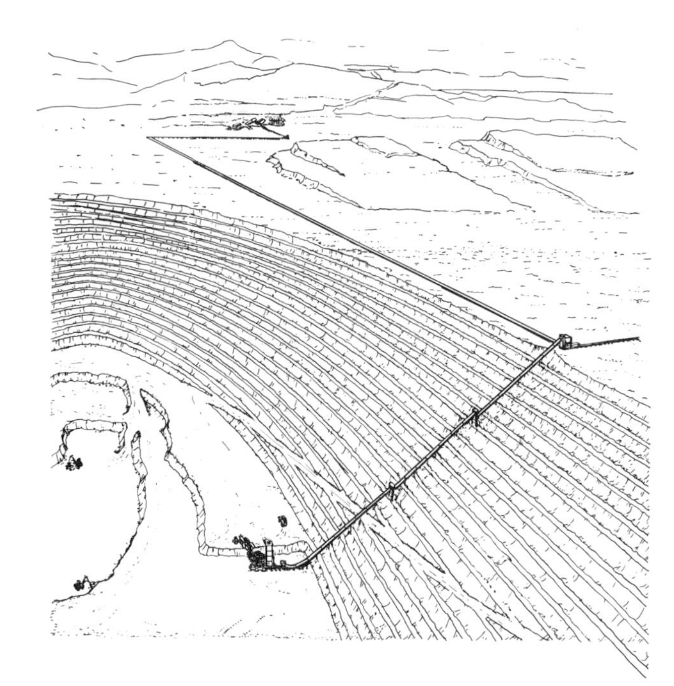

Fig. 8: Modular HAC System in deep open pit mine

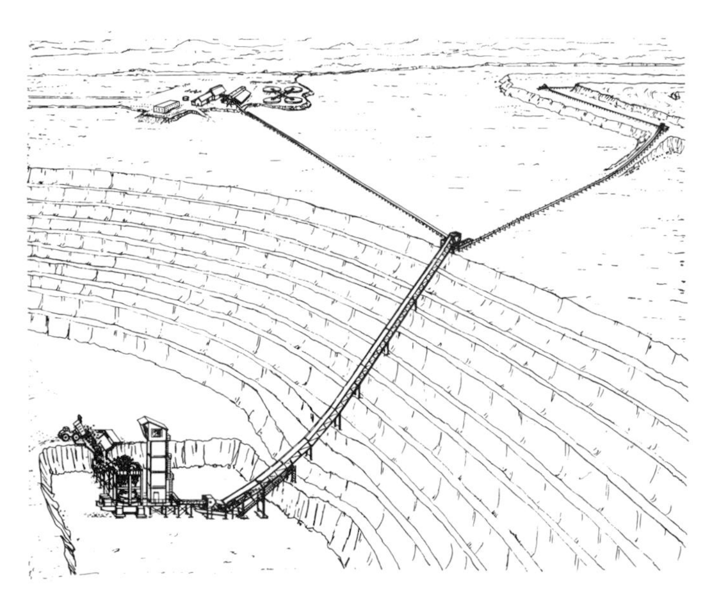

Fig. 9: Single lift HAC system in deep open pit mine

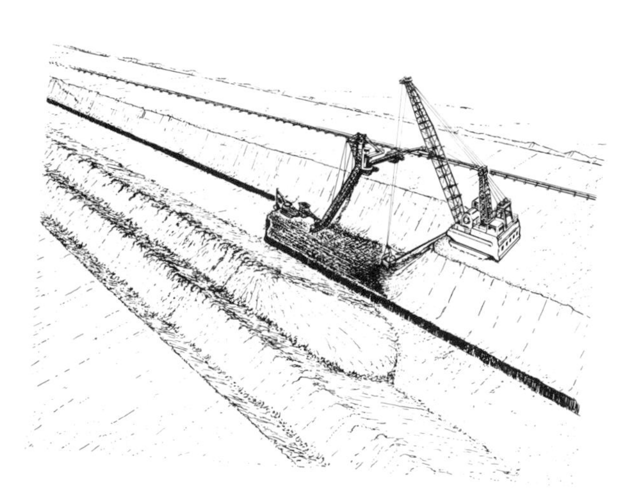

Fig. 10: High Angle Conveyor in strip mine

The present HAC unit is ideally suited to work with large, stripping draglines because of its high mobility and the single bench suspension. This makes possible continuous travel along an irregular dragline bench and around the dragline cut when the HAC and dragline must cross each other's path.

Depending on the overburden thickness and the size of the stripping dragline, the HAC unit may have a lift of 18.3 to 36.6 m (60 to 120 ft). Depending on the bank stability, the conveying angle may be 45° to 60°. Design rate for the mobile crusher, HAC, and the other conveyors is in the range of 907 to 1,814 mt/h (1,000 to 2,000 st/h). Mobility must be continuous for the mobile crusher, HAC, and connecting beltwagon since they follow the continuous advance of the dragline cut. Frequency of shiftable conveyor advance depends on the advance distance, pit length, overburden thickness and production rate of the dragline. This may be from once every several weeks to once every several months. The overland conveyor is lengthened at the tail end in 152 to 305 m (500 to 1,000 ft) increments or once every several months.

The mobile feeder breaker or crusher must reduce the coal to a 152 to 254 mm (6 to 10 in) maximum lump size depending on the design rate, belt width, etc. Such a crusher must be equipped with a luffing, swivel, discharge boom of ample length to permit easy loading of the HAC even when the dragline bench height varies slightly along the pit length.

4.3 Thick Seam Mines

Very thick seams, up to 42.7 m (140 ft) of sub-bituminous coal are presently mined in the western U.S. by shovel truck open pit methods. Overburden thickness, initially 12.2 to 18.3 m (40 to 60 ft) will increase to approximately 91.4 m (300 ft) before end of mine life in some mines. This makes necessary multiple benches in the overburden with one to three thick benches in the coal seam. Such requirements make the HAC configuration of Figs. 7 and 11 most appropriate. The HAC system is again used to convey the coal from pit to plant. Overburden removal and spoiling may be by bucket wheel excavators with end around conveyors and spreaders, by scrapers, or by shovels and trucks. Coal loading into the mobile crusher may be by shovel, backhoe, or front end loader.

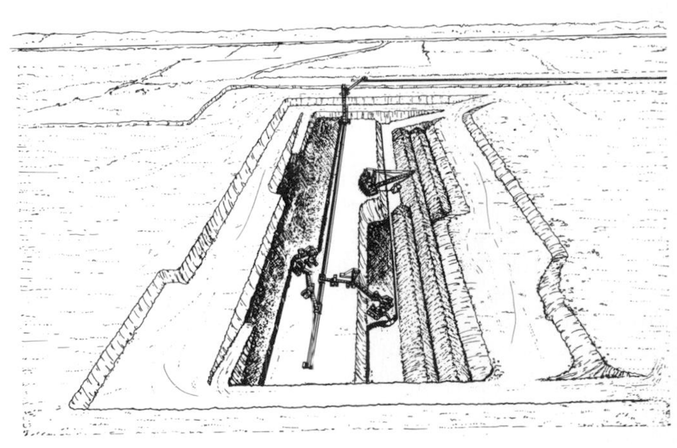

Fig. 11: Thick seam coal mine — with High Angle Conveyor at end wall

The HAC system consists of one or more mobile, crawler mounted, high capacity, feeder breakers or roll type crushers, each with crawler mounted beltwagon, an in-pit shiftable conveyor with hopper cars as required, a HAC with connecting conveyor bridge and loading hopper car and an overland conveyor to the plant. In a three bench operation, the single in-pit shiftable conveyor must be located on top of the middle bench with long beltwagons or bench type high angle conveyors feeding the coal from the lower and upper benches. In-pit shiftable conveyors may be employed at two or more coal benches for system redundancy and to minimize conveying between benches. This, of course, requires as many HAC units as there are shiftable conveyors.

A multiple bench operation makes impossible a single HAC lift out of the pit at the advancing high wall. Such a single lift can only be at the end wall where the multiple benches come together as two or three, very high, non-advancing benches. The HAC unit consists of the HAC, a conveyor bridge and a hopper car. The HAC has lower support within the pit and an upper support at a strategic elevation with regard to the ultimate thickness and variation of the coal seam and overburden. The conventional conveyor bridge spans between the HAC and overland conveyor, changing its angle to lower or elevate the material as required by the varying total elevating height. A rail mounted hopper car supports the discharge end of the conveyor bridge and directs the coal onto the overland conveyor. The HAC with support at two bench elevations spans an intermediate bench which can be used for pit access, truck haulage or conveying the overburden to the spoil side of the pit.

HAC lift requirements may be from 36.6 to 76.2 m (120 to 250 ft) with combined HAC and conveyor bridge unit from 45.7 to 91.4 m (150 to 300 ft). The HAC conveying angle is approximately 45°, while the conveyor bridge is limited to ±15°. Design rate for the mobile crushers and beltwagons will match the maximum production of a 30.6 m³ (40 yd) shovel at 2,722 mt/h (3,000 st/h). Each shiftable conveyor and HAC unit are designed for 2,722 mt/h (3,000 st/h), 3,629 mt/h (4,000 st/h), or 5,443 mt/h (6,000 st/h) depending on whether these are carrying the production of one, two or three coal loading shovels. The overland conveyor is designed for 3,629 mt/h (4,000 st/h) or 5,443 mt/h (6,000 st/h). Mobility must be continuous for the mobile crushers and beltwagons since they follow the advance of the bench cut along the entire pit length. These are, therefore, permanently mounted on self-propelled crawlers. The frequency of advance for the shiftable conveyor and the HAC unit along the endwall depends on the advance distance, pit length, coal thickness and production rate. This may be from once a month to once every several months, and this does not warrant permanently attached transport means. The HAC unit is, therefore, mounted on pontoons at the lower and upper ends with provision for transport by rubber tired or crawler mounted, independent transporter units, or by attachable/detachable walking feet. Whichever transport means are chosen will also be used at the drive and tail stations of the shiftable and overland conveyors. The tail end of the connecting conveyor bridge is supported on the HAC structure while the discharge end is supported on a rail mounted hopper car which is not self-propelled and must be towed during the advance of the HAC. Extension of the overland conveyor may be in 152 to 305 m (500 to 1,000 ft) increments or once every several months.

The mobile feeder breaker or crusher for this application must be equipped with luffing, swivel discharge boom as in the thin coal seam trip mine applications.

Fig. 12 illustrates another form of the HAC system in thick seam coal mining. The approach is similar to the thin coal seam system of Fig. 10 except the shiftable conveyor trails on a prepared spoil bench rather than advancing on the prepared dragline bench. The mobility requirements are similar to the thin coal seam system, while the design rates for the highly mobile HAC, the connecting beltwagon, shiftable conveyor and overland conveyor are much higher at 2,722 to 3,629 mt/h (3,000 to 4,000 st/h). The HAC lift is approximately 30.5 to 45.7 m (100 to 150 ft), and the conveying angle may be 30° to 40° depending on the stability of the spoil material. In this scheme, much of the stripped overburden is spoiled in the mined out pit to form a prepared bench for the HAC and shiftable conveyor. This requires much spreading by dozers after dumping. The remaining overburden is spoiled behind the shiftable conveyor and can, therefore, be contoured and seeded according to reclamation requirements.

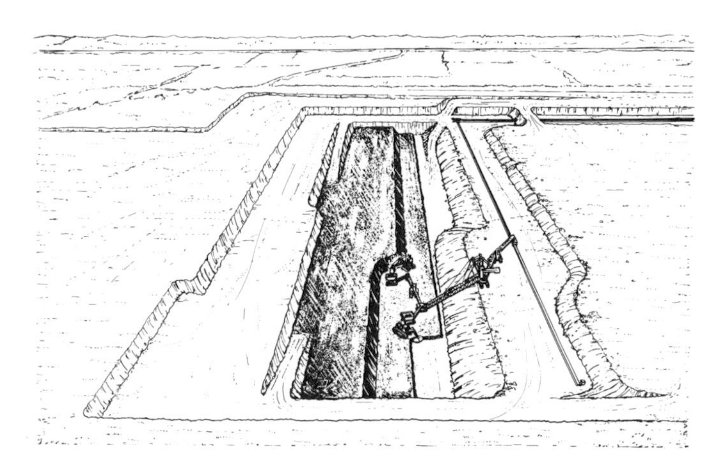

Fig. 12: Thick seam coal mine with High Angle Conveyor at spoil

4.4 Multiple Thin Seam Mine with Thick Parting

The HAC system of Fig. 13 illustrates the use of both a highly mobile HAC at the high wall and a movable HAC unit with connecting bridge and hopper car at the end wall. The characteristics of the mobile crushers, beltwagons and high wall HAC are as described in the thin coal seam system of Fig. 10 while the shiftable conveyor, end wall HAC unit and overland conveyor are as described in the first thick coal seam system of Fig. 11.

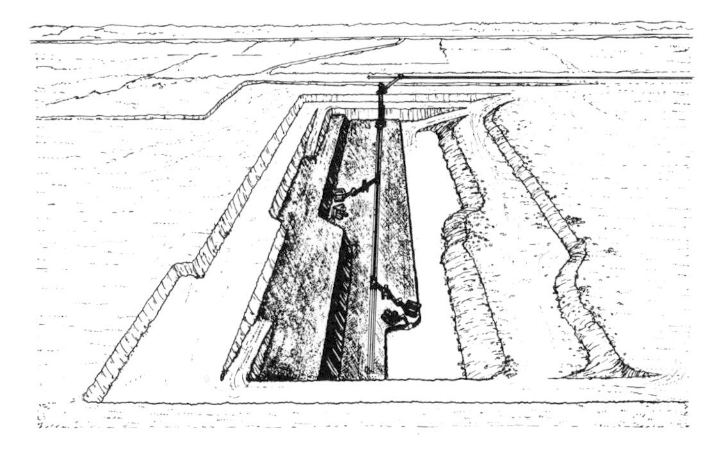

Fig. 13: Multiple thin seam coal mine — with High Angle Conveyors at highwall and at end wall

4.5 HAC Bench Conveyors

Pontoon, crawler, or rubber tire mounted, HAC, bench conveyors for 10.7 to 15.2 m (35 to 50 ft) bench heights may be used with beltwagons, crawler mounted, self-aligning face conveyors, and/or cascading conveyor modules to produce a highly mobile, highly flexible, haulage system which can be used locally to advance one or several overburden benches to uncover a desired amount of ore. The system can then be removed and relocated for operation in another area of the mine or at another nearby mine. Such a system requires that all components be highly mobile or of a size and weight that makes them easily towed by other mine equipment or lifted by mobile crane. With a projected operating weight in the 18.1 to 36.3 mt (20 to 40 st) range, the HAC bench conveyor qualifies for the latter.

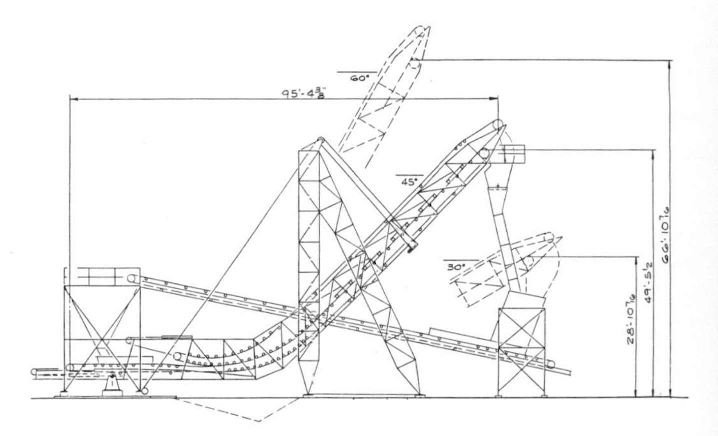

Fig. 14: HAC prototype — schematic

5. Summary and Conclusions

There are serious doubts about the future viability of trucks as the primary means of material transport in open pit mines. Two large U.S. copper mines realized large cost savings, initially by installing permanent type, large, gyratory crushers and conveyors to displace trucks in elevating material from within the pit. These had the disadvantage of reducing mining flexibility and limiting the amount of ore recovered. The introduction of large, movable, gyratory type, in-pit crushing systems and movable and shiftable, elevating and overland conveyors have improved the overall flexibility of the mining operation. The remaining disadvantage is that the conventional elevating conveyors must be routed along approximately 15° conveying angles. This does not permit optimal mine slope angles, and additional excavation is required to accommodate the low angle conveyors. The high angle conveyor (HAC) is, therefore, the link to system optimization when used as the continuous elevating means.

Studies have revealed the economic advantages of HAC systems in open pit mining. Past state-of-the-art searches, however, found no high angle conveying method to meet the requirements of large open pit mines. This need has been filled by the development of Continental Conveyor HACs and HAC systems. Various forms of these have been discussed, and several applications in open pit mining have been illustrated. The building, testing and demonstration of the HAC prototype has advanced large capacity, high angle conveying technology beyond mere speculation. The HAC prototype and Continental Conveyor's commitment to supply HACs to the industries' requirements give validity and substance to the discussion of HAC systems-applications in open pit mining.

Appendix

1. HAC Prototype — Description and Features

The HAC prototype is shown schematically in Fig. 14 and at 60° conveying position in Fig. 15.

The HAC prototype was designed for two basic purposes.

a) To conduct extensive testing on a wide variety of materials conveyed at various angles from 30° to 60°, at various conveying speeds, with various degrees of applied hugging pressure to (i) determine these interrelations, (ii) explore the limits of the concept, and (iii) formulate a reliable HAC design model which includes accurate determination of the drive power requirements.
b) To serve as a convincing demonstration unit for customers from a variety of industries. The HAC is, therefore, designed to handle a wide variety of materials at high tonnage rates.

The HAC prototype consists of (a) the high angle conveyor featuring two, 1,524 mm (60 in) belts, each with its own drive and automatic take-up means, (b) a hoist and tower with wire rope reeving to permit positioning of the HAC at various conveying angles from 30° to 60°, (c) a 1,219 mm (48 in) conventional return conveyor, (d) a loading hopper and connecting chutework, and (e) a control building to house all electrical control and monitoring equipment.

The high angle conveyor, conventional return conveyor, loading hopper and connecting chutework form a closed recirculating system to permit conveying of the same material for as long as desired. A feed gate beneath the HAC loading hopper controls the material flow to the HAC. A belt scale located at the return conveyor with remote readout at the feed gate permits monitoring and adjustment of the flow rate. A control/display panel at the control house monitors HAC performance, such as conveying angle, belt speed, tonnage rate, horsepower and load sharing at the HAC drives, differential HAC belt speed, accumulated operating hours. A six pen graph plotter provides a permanent record of conveyor performance.

2. HAC Prototype Design Parameters

2.1 Materials to be Handled

— Lignite, coal, copper ore and waste rock, crushed taconite, iron ore fines and pellets, sand, gravel, crushed stone, various grains, wood chips
— 165 mm ($6^1/_2$ in) is the design maximum lump size for predominantly lumpy material.
— The conveyed loose density may be up to 2.4 mt/m³ (150 PCF) for the design cross-section conveyed at a 60° incline. Higher densities are possible at lower conveying angles and/or lower cross-sectional filling.
— Materials handled will vary from the extremely soft and friable to the extremely hard and abrasive.

2.2 Design Conveying Rate at Maximum (60°) Angle

(Higher capacities are possible with the medium to high density materials when conveyed at angles lower than 60°.)

Fig. 15: HAC at 60°

— 0 to 1,955 mt/h (0 to 2,155 st/h) for materials of 0.8 mt/m³ (50 PCF) density
— 0 to 2,468 mt/h (0 to 2,721 st/h) for materials of 1.6 mt/m³ (100 PCF) density
— 0 to 2,684 mt/h (0 to 2,959 st/h) for materials of 2.4 mt/m³ (150 PCF) density

2.3 Conveying Angle

— Infinitely variable from 30° to 60°

3. Main Technical Features

(The following is not a comprehensive list of all features of the HAC prototype, but rather a limited list of the more important features.)

3.1 HAC Conveyor

Overall Length	— 35 m (115 ft)
Vertical Lift	— Variable, 7.9 to 19.5 m (26 to 64 ft)
Belt Speed	— Infinitely variable from 0 to 6.1 m/s (0 to 1,200 FPM)
HAC Belts	— two endless belts, 1,524 mm (60 in) belt width, nylon fabric reinforced
Hugging Pressure Device	— Fully equalized standard impact rolls on spring loaded pressing modules
HAC Conveyor Drives	— 74.6 kW (100 HP) top belt drive — 111.8 kW (150 HP) bottom belt drive — Each drive is shaft mounted onto each head pulley shaft extension. Each consists of an infinitely variable speed DC motor, 0 to 1,800 RPM with overspeed capability to 2,300 RPM, mounted onto a right angle hollow shaft reducer with transmission from motor to reducer input shaft by drive sheave to driven sheave thru multiple V-belts. — The nominal reducer ratio is 10.0 : 1. Three sheave sets for each drive permit overall drive reduction ratios of 12.0 : 1, 17.0 : 1, and 22.5 : 1.
HAC Take-ups	— Automatic by constant pressure hydraulic cylinder — 101.6 mm (4 in) diameter cylinder at top belt — 127 mm (5 in) diameter cylinder at bottom belt — Each take-up consists of the tail pulley on a traveling take-up carriage, a ten-

sioning hydraulic cylinder and wire rope reeving to connect the two.

— A single, variable setting, hydraulic pump unit provides a constant predetermined pressure to both cylinders.

HAC Head, Drive Pulleys	— 610 mm (24 in) O.D. with 13 mm (1/2 in) Chevron lagging
HAC Tail, Take-Up Pulleys	— 457 mm (18 in) O.D., mounted on traveling take-up carriage

3.2 HAC Luff Hoist and Reeving

Hoisting Rope	— 19 mm (3/4 in) diameter
Hoist	— Single drum, multiple wrap driven by 7.5 kW (10 HP), 1,200 RPM motor thru drive chain and integral worm gear reducer
Reeving	— Six part reeving between hoisting tower and HAC truss with anchor and hoist at HAC pivot foundation.

Section "A—A"

α = conveying angle

ϕ = internal friction angle of bulk material

μ_m = coefficient of friction for bulk material on bulk material

μ_b = coefficient of friction for bulk material on conveyor belt

N = normal lineal hugging load exerted by the top surface

W_m = lineal weight of bulk material

Section "A—A"   **Section "B—B + C—C"**

α, μ_m, μ_b, N, W_m are as defined in model #1

N_m = that portion of N which bears directly on the conveyed material

N_e = that portion of N which bears directly on the edges of the bottom belt

Fig. 17: At HAC entrance the loaded bottom belt joins the top belt to sandwich the material between.

Fig. 18: Fully equalized standard impact rolls on spring loaded pressing modules provide a floating blanket of evenly distributed hugging pressure.

Fig. 16: Each automatic take-up, at bottom (shown) and top belts of HAC, consists of the tail pulley on a traveling carriage, a tensioning hydraulic cylinder, and wire rope reeving to connect them.

Fig. 19: At HAC entrance the loaded bottom belt joins the top belt to sandwich the material between. The two belts do not separate until the discharge area where the material is released. Tail pulley on traveling take-up carriage is shown at upper left.

Fig. 20: HAC head end showing top and bottom belt drives. Each drive is shaft mounted at the corresponding drive pulley. A DC motor drives a hollow shaft right angle reducer thru multiple V-belts.

References

[1] Coile, J. J., "In Pit Crushing and Conveying Vs. Truck Haulage", Mining Congress Journal. January, 1974, pp. 23—27

[2] Mevissen, E. A., Siminerio, A. C. and Dos Santos, J. A., "High Angle Conveyor Study", by Dravo Corporation for Bureau of Mines, U.S. Department of the Interior under BuMines Contract No. J0295002, 1981, Volume I, 291 pages, Volume II, 276 pages

[3] Anon., "Pit Crushers and Conveyors Move Sierrita Ore and Waste", Engineering and Mining Journal. June, 1977, pp. 102—103

[4] Kok, H. G., "Use of Mobile Crushers in the Minerals Industry", Mining Engineering. November, 1982, pp. 1584—1588

[5] Johnson, R. N., "In Pit Crushing and Conveying with Movable Crushers", by GARD, Inc. for Bureau of Mines, U.S. Department of the Interior under BuMines Contract No. J0295004, 1980, 211 pages

[6] Almond, R. M. and Schwalm, R. J., "In-Pit Movable Crushing/Conveying Systems", Mountain States Mineral Enterprises, Inc., presented at the AMC International Mining Show, Las Vegas, Nevada, October 11—14, 1982

[7] Iles, C. D., "Costs of In-Pit Crushing", Mining Engineering, April, 1983, pp. 319—320

[8] Anon., "Iscor's Sishen Grows and Grows", Engineering and Mining Journal, November, 1982, pp. 122—125

[9] Alberts, B. C. and Dippenaur, A. P., "An In-Pit Crusher Overburden Stripping System for Grootegeluk Coal Mine", South African Institution of Mechanical Engineers, One Day Symposium on Materials Handling In Opencast Mining, September 18, 1980, 21 pages

[10] Anon., "Feeder Breakers", Mining Magazine, February, 1982, pp. 149—155

[11] Dos Santos, J. A. and Frizzell, E. M., "Evolution of Sandwich Belt High-Angle Conveyors", CIM Bulletin. Volume 576, Issue 855, July, 1983, pp. 51—66

 Volume 5, Number 6, December 1985

Mobile Elevator Conveyors

H. von Blomberg, G.L. James and I. Ozolins, Australia

Summary

Recent publications have shown several concepts of mobile elevator conveyors (MECs). These machines feature a conveying system capable of elevating material at a steep angle (65° to 70°), mounted on crawlers. They are being developed for open cuts where trucks are the primary cost factor in transporting bulk material from the working face to the surface. The MEC concept permits the use of cost-effective conveyors in place of the trucks without interrupting the mining operation.

To match the operation of most open cuts, the following design criteria have been used:

1. The MEC must be self-propelled and capable of following the operation of the excavator or shovel.
2. It must be easily movable around the mine.
3. It must provide uninterrupted flow of material when fed by front-end loaders or a shovel.
4. It must elevate material over bench heights of up to 55 m and possibly greater.
5. It must incorporate good access to all mechanical and structural components for inspection and maintenance.

This paper describes two MEC arrangements presently being developed, for which provisional patent applications have been lodged.

1. Mobile Elevator Conveyor Using a Segmented, Hinged Retractable Boom (by Minenco)

1.1 Description

This machine is supported on crawlers running on the surface at the top of the high wall. A segmented boom reaches down to the lower level (Fig. 1).

The tail end of the segmented boom carries a loading boot. The boom itself supports the steep-angle elevating conveyor, which discharges onto a horizontal conveyor. The latter delivers the bulk material to the surface conveyor system. The segmented boom and its conveyor can be retracted inside the structure supporting the horizontal conveyor.

The entire machine is constructed from standard tried and proven components, and the concept is feasible for elevating material up to and beyond existing bench heights.

1.2 Boom Structure

The boom consists of a number of segments hinged at intervals along the bottom chords. The location of the hinge allows the segments to follow a curved guide system when the boom is being extended or retracted. When extended, the segments form a stable structure due to:

— the position of the hinge
— the moment from the wire rope support
— the mass distribution
— the system geometry.

The boom structure supports the boom conveyor, the loading boot and any services that may be required on the boom (e.g., lights).

The above boom structure is supported by the wire rope support clear of the pit wall. An alternative arrangement is one that rests on the wall. The segments would then be supported on wheels, skids or by similar means. The boom is free to be moved up and down the wall, as required. This system requires a stable and relatively flat wall surface. The self-locking effect of the hinge and segment arrangement is not required, and also the boom wire rope support is not necessary.

This alternative system is not as flexible as that with the preferred boom structure, i.e., the machine can only extend the boom where the wall is stable and relatively flat. Also, as the machine must retract its boom prior to travelling, the mining operation would stop during machine relocation.

However, this system could prove successful and economic in some mines.

1.3 Segment Guide System

The curved guide system connects to straight guide sections extending the length of the machine (see Figs. 2 and 3).

This guide system allows the boom to be completely retracted inside the machine to provide a compact arrangement for travelling.

Mr. H. von Blomberg, Chief Design Engineer, Krupp (Australia) Pty. Ltd., Mr. G.L. James, Senior Mechanical Engineer, and Mr. I. Ozolins, Chief Mechanical Engineer, both with Minenco Pty. Ltd., P.O. Box 5202 AA, Melbourne, Vic. 3001, Australia

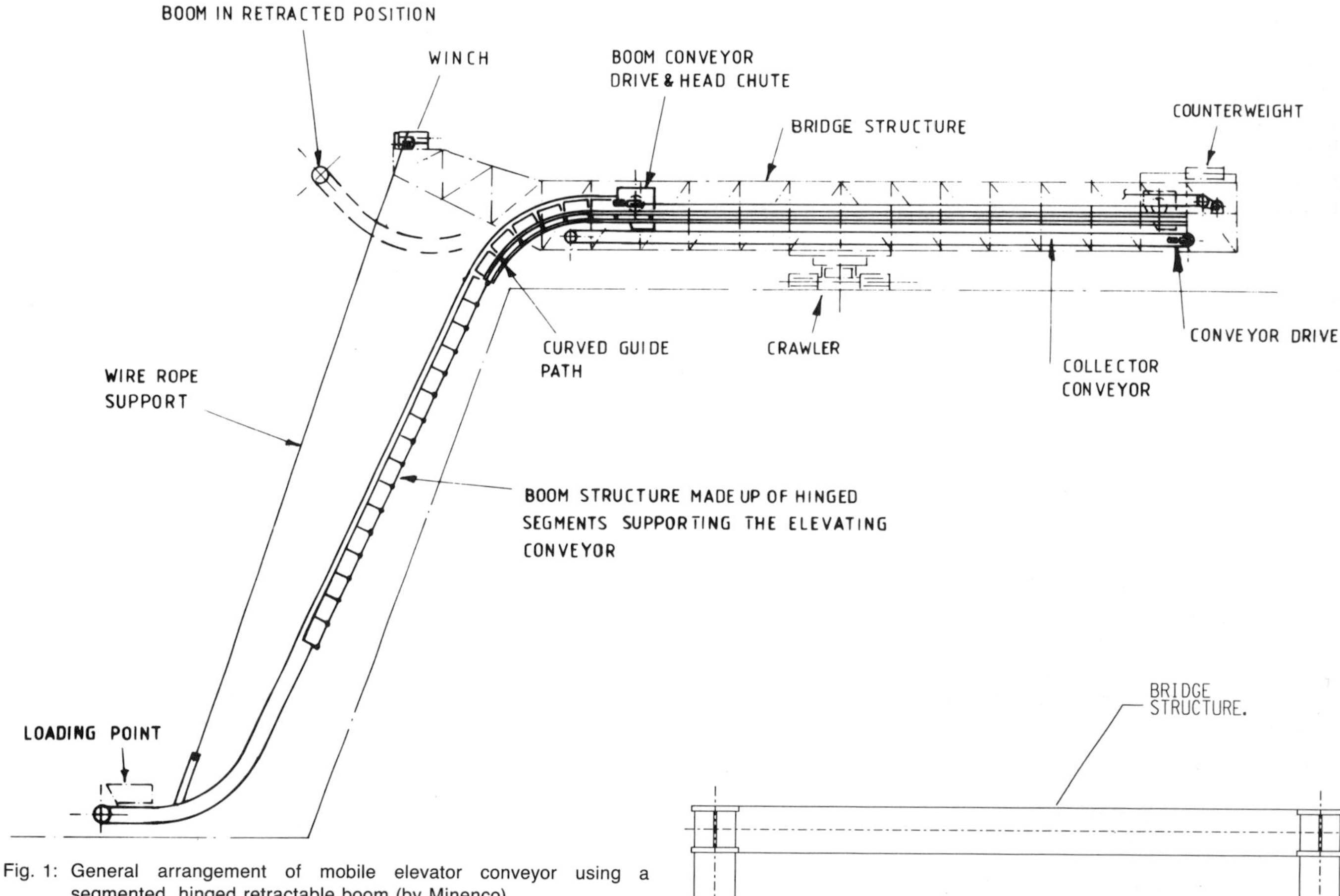

Fig. 1: General arrangement of mobile elevator conveyor using a segmented, hinged retractable boom (by Minenco)

The guide system also allows the boom to be extended to any length within the maximum and minimum constraints of the machine. This allows the machine to operate on varying bench heights.

The guide system also resists the force couple due to the boom conveyor drive.

The guide system consists of wheels located at the segment hinge points on both bottom chords and two guides (adjacent to each segment bottom chord) comprising an upper and a lower rail. The wheels are flanged, and they are restrained by the upper and lower rails.

Alternatives to this guide system could be metal or synthetic slides running on a fabricated guide.

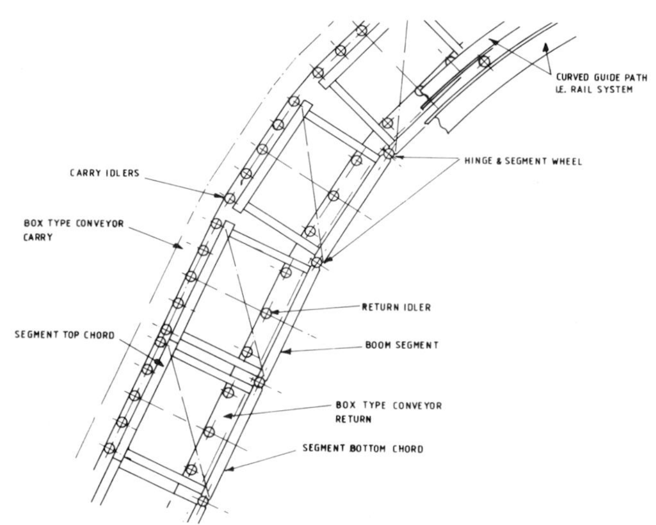

Fig. 3: Boom segments

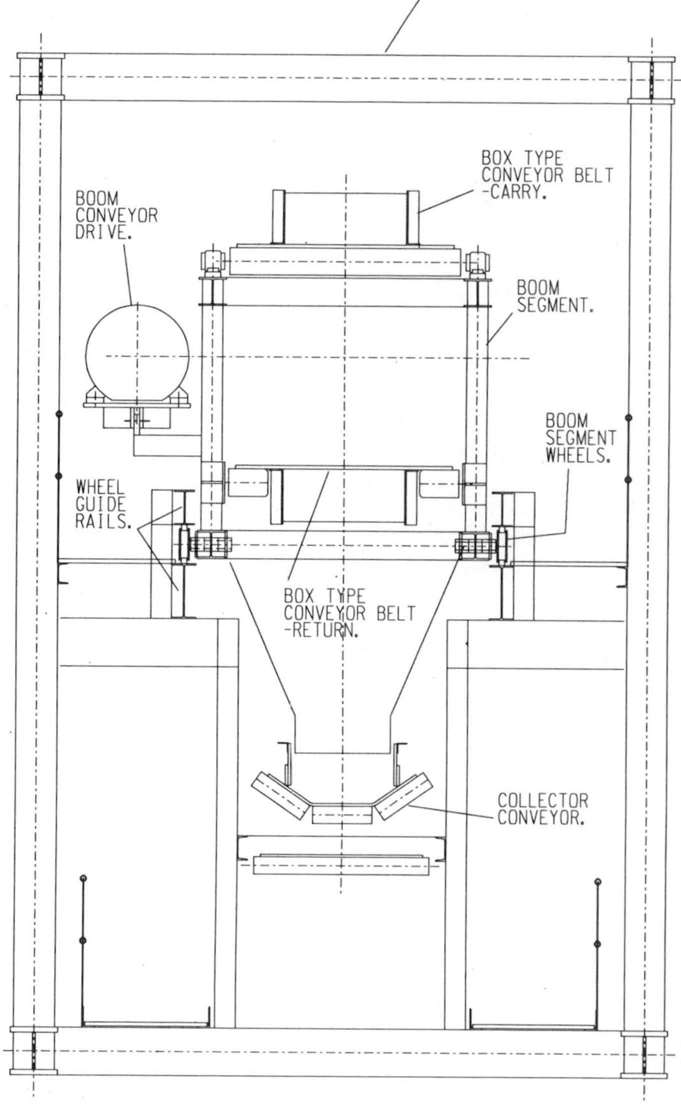

Fig. 2: Cross-section of bridge

1.4 Discharge Point and Collector Conveyor

The material elevated from the pit floor or lower bench is discharged from the boom conveyor onto a conventional troughed belt horizontal collector conveyor. Extending the length of the machine, this conveyor allows the boom conveyor to discharge at any point along it. Hence, the segmented boom can be extended and lowered to any depth, as required.

The collector conveyor also provides the additional benefit of collecting the carry-over spillage which falls from the return side of the boom conveyor. Cleaning devices are used to assist this spillage to fall onto the collector conveyor.

1.5 Benefits over Existing Concepts

1. This machine is much lighter than existing proposed single-support mobile elevator conveyors. The expected savings could be as much as 50 %, depending on the particular machine application.
2. The machine height has been greatly reduced from that of existing proposals for single-support mobile elevator conveyors. The benefits of this are reduced windloads, greater stability and faster relocation.
3. In the travel mode, with the boom completely retracted, the machine is compact, allowing rapid and easy relocation in difficult mine arrangements. Existing proposals are cumbersome, making relocation slow and difficult.
4. The boom length can be varied to suit the bench height.
5. The boom angle can be varied to suit the wall angle.
6. The machine is built up from standard items all of which have been tried and tested and accepted as common and reliable by the industry.
7. The boom can be readily serviced when it is fully retracted. Hence, the boom does not require heavy walkways and stairs.
8. The machine is feasible for bench depths exceeding present limits, i.e., 55 m and above.

2. Mobile Elevator Conveyor Using a Hinged Bridge at Each End (by Krupp)

2.1 Description

This machine consists of a hinged bridge simply supported by undercarriages at each end (see Fig. 4(a)). The hinges at the upper and lower end parts enable the machine to accommodate various high wall angles and bench depths (see Fig. 4(b)). Existing proposals for bridge-type machines do not include these hinges. Hence, they could not accommodate varying bench depths and angles. Also, the hinges allow the machine to form a straight bridge, allowing movement around the mine (see Fig. 4(c)). The machine is constructed from standard items, and the concept is feasible beyond existing bench depths.

2.2 Bridge Structure

The whole supporting structure is a truss bridge braced in horizontal and vertical plane, resulting in a torsional stiff design. The intermediate truss bridge is supported by hinged connections at the upper and lower ends. Inspection during operation and maintenance is possible by means of a self-adjustable stair, which makes the accessibility of the bridge independent from the actual inclination.

The discharge truss bridge on the upper bench accommodates the drive unit, which is designed for the power requirements when operating at the maximum high wall angle. The cleaning device behind the drive pulley is equipped with a beater mechanism to ensure that no material remains in the belt-boxes on the return way. The spill material from the cleaning device is collected on a spillage conveyor underneath the main belt and transported into the discharge chute. This protects the undercarriage from falling material and keeps the ground level clean of material.

The loading end on the lower bench features a storage bin, which feeds the material on the belt by means of vibrating feeders. During the loading cycle of the front-end loader or the shovel, the storage bin ensures a continuous flow of

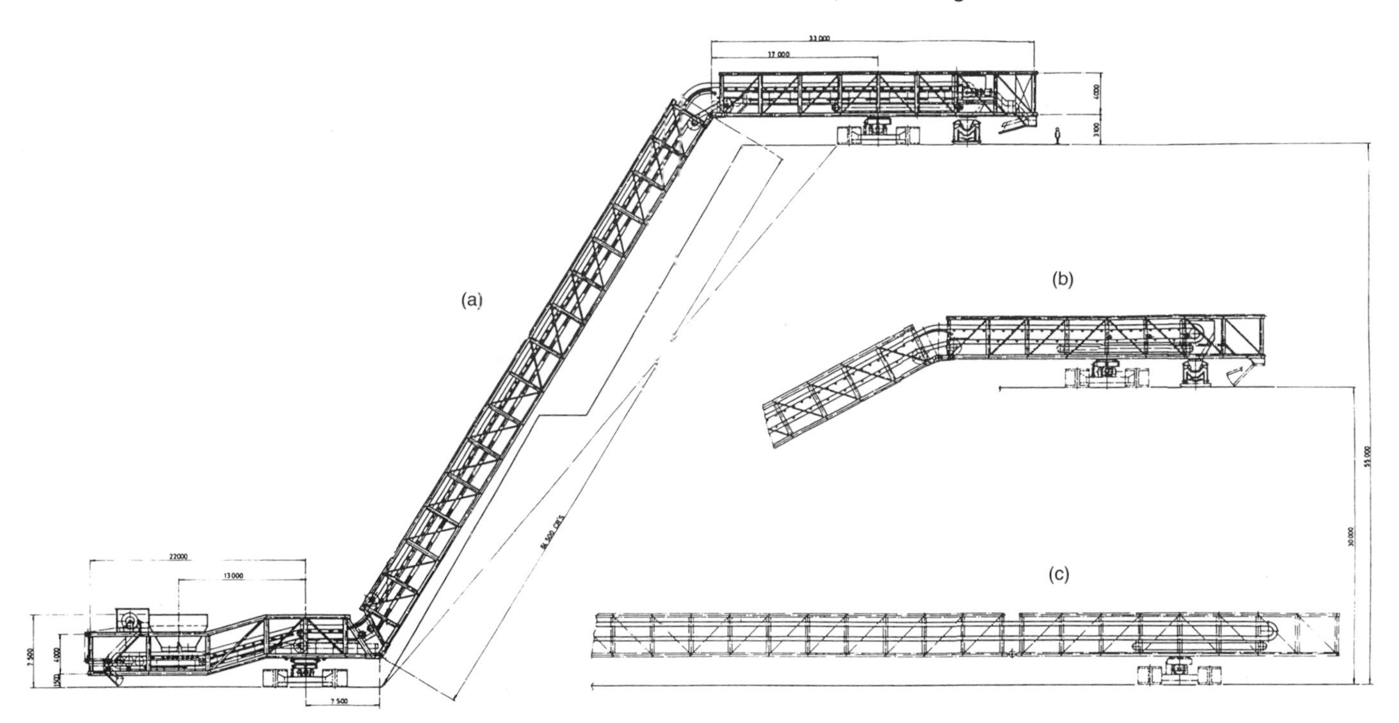

Fig. 4: General arrangement of mobile elevator conveyor using a hinged bridge at each end (by Krupp)

material to the upper bench without interruption. The vibrating feeders discharge the material from the bin closely above the belt cleats to protect them from impact load. The tail pulley is supported on a movable carriage and tensions the belt by means of a take-up winch. Depending on the inclination of the intermediate truss bridge, the relative belt length between the loading and the discharge point differs, requiring adjustment by the winch.

All mechanical items of the machine are easily accessible, and wide platforms give enough space for maintenance.

2.3 Undercarriages

The machine can be equipped with self-propelled undercarriages, which may be of wheel- or crawler-type. They are levelled on both horizontal planes by means of hydraulic cylinders and fully slewable around their vertical axes. The bridge truss on the upper bench is fixed to the undercarriage, whilst the lower one is shiftable in longitudinal direction to allow for horizontal movement. This system equalises unequal travel of both undercarriages as the combined slew and longitudinal motion gives flexibility to the machine without causing damage to the truss bridge structure.

Crawler undercarriages can be equipped with various crawler widths to suit the allowable ground pressure of the pit.

2.4 Movement around the Mine

Any movement within the mine can be carried out. This may be necessary when the machine is relocated to another bench or when a dragline is to be passed. In such a case, the lower bench undercarriage has to travel to the upper bench on a prepared road. At the same time, when it is travelling upwards, the upper bench undercarriage moves in diagonal direction to allow for the stretching of the intermediate truss section. As the undercarriages are levelled in the horizontal plane, the inclinations do not have any effect on the truss bridge, which always remains in the horizontal plane.

2.5 Benefits of the System

1. The machine has all equipment on board to transport material upwards.
2. The machine can operate at pit walls of various depths and angles.
3. Only one to two operators are needed to operate the machine.
4. Two undercarriages give a sound stability to the machine.
5. The machine can move around the pit.
6. The machine is built up from standard components and equipment with high reliability.
7. The machine can be readily serviced without any repositioning.
8. The machine is feasible for bench depths exceeding present limits, i.e., 55 m and above.

3. Equipment Common to Both Machines

3.1 Elevating Conveyor

The proposed conveyor is based on the "Flexowell" from Conrad Scholtz AG, Federal Republic of Germany, or similar belt conveyors. This type of conveyor belt consists of a special cross-stabilised flat base belt with flexible corrugated side walls fitted to each side. The flexible side walls allow the belt to run over head and tail pulleys and, most important, the curved section formed by the boom or bridge segments.

The "Flexowell"-type belt conveyor has cross-cleats spaced at regular intervals, which lift the conveyed material. If required, this type of conveyor can lift material vertically.

Another manufacturer of this type of belting is, amongst others, Hartmann Band GmbH, Federal Republic of Germany.

Other conveying methods that could be used successfully on these machines include:

— the sandwich belt by Continental Conveyor and Equipment Co. Inc. or similar
— individual metal buckets attached to the conveyor belt, wire rope or chains or similar.

3.2 Loading Point

This can have several arrangements, depending on the individual mine requirements:

1. A simple skirted section on the boom or bridge, fed by a shovel, hopper, crusher and feed conveyor/bandwagon system. This system could include a truck, depending on the distance between the material collection point and the elevating machine. An alternative to the shovel could be a front-end loader (FEL) or a similar machine.
2. If the material lump size is acceptable, the crusher could be deleted from 1. above. The shovel of the FEL could dump directly into a mobile hopper. A feeder (belt, apron or vibratory) at the bottom of the hopper would control the feed onto the loading point.
3. A continuous miner such as a "Voest-Alpine" or a "Dosco"-type road header, a bucketwheel reclaimer or another continuous miner feeding material directly into the boom loading boot.
4. The elevating machine forms a link in a long conveyor system, i.e., the elevating machine is fed by a long conveyor and discharges onto a long conveyor. Several elevating conveyors could occur throughout the length of this type of system.

3.3 Transport System

The preferred method to transport these machines is by one or several (diesel, electric or other) crawler units.

These offer low profile, low ground pressure and the ability to slew using the crawlers. Also, they can include a slewing table that can be used to rotate, level, raise and lower the machine. The crawler units can be integral or detachable. If detachable, the units could be used to move other equipment (e.g., mobile crushers) or be shared among several mobile elevating machines.

Alternatives to the crawler units are multiple pneumatic tyre transporters and walking legs. If the machine is not large, it could be mounted on unpowered tyres or crawlers. The machine could then be relocated using a towing vehicle.

(Note: Walking legs are not preferred.)

 Volume 4, Number 1, March 1984

Steep Angle In-Pit Conveying

Walter H. Diebold, USA

Summary

In-Pit Steep Angle Conveyor Systems are expected to start replacing truck haulage in the foreseeable future for reasons of energy conservation, reduction in maintenance and operating costs, less sensitivity to inflation and, in general, greater economy and efficiency in bulk materials handling. The sidewall type steep angle conveyor has been used in many heavy duty applications the world over and can be adapted to the specific requirements of in-pit steep angle conveying, where mobility or semi-mobility is required. An example of such an application is discussed and explained.

1. Introduction

High-angle in-pit conveying is an idea that is being seriously studied by a number of important mining companies in North America. Its application in coal, iron ore, and copper mining is expected to reduce mine operating costs considerably compared with the conventional discontinuous methods, such as truck haulage, which are fuel, labor, and maintenance intensive. Therefore, a mine introducing high-angle in-pit conveyors will gain a competitive advantage over mines which continue to use conventional haulage.

While the capital outlay for in-pit conveyor systems obviously varies depending on pit depth, the angle of incline, and the capacity required, it generally will not exceed that of truck haulage when all factors are taken into consideration. These factors include the cost of excavation, as measured in relation to the volume of minerals mined, the cost of building truck ramps and roads, and the capital outlay for the trucks and maintenance facilities. In many cases, in fact, the capital outlay will be less than for truck haulage. This cost advantage will best be realized by new mines, whereas in existing producing mines the switch from truck haulage may involve the abandonment of a substantial capital asset.

It is in operating cost that in-pit conveyors are clearly superior to trucks. Consider the following:

- A conveyor system is much less labor intensive in that it can be operated with a minimum of supervision. There is no need, as on trucks, to have one operator for each 120 or 150 ton unit of haulage capacity. There is less impact from regular wage increases.
- In general, a conveyor system can be powered by electricity. Its power consumption costs are, therefore, more predictable, less subject to inflation than diesel oil, and less dependent on petroleum-based fuels and their availability.
- As a bonus, conveyor systems are quieter and produce less air pollution than trucks.

Mr. Walter H. Diebold is President of Scholtz-EFS Corporation, Annandale, VA 22003-3278, USA

2. The Scholtz-EFS Flexowall® Type Belt

Over the last 10—15 years, different types of steep angle conveying systems have been developed, such as pneumatic systems, sandwich belt systems, and sidewall belt systems.

Scholtz has elected to develop and bring to the market its sidewall belt system which it based on its Flexowall® type belt. When our engineers evaluated the various approaches to steep angle conveying, pneumatics was discarded as too energy intensive and too limited in capacity. Further, in many cases pneumatic systems lead to unacceptable product degradation.

The sandwich belt system using conventional belts at first seemed a natural for Scholtz, which is a large manufacturer of fabric and steelcord belts. After studying the operational problems of sandwich belt systems using conventional belts, Scholtz decided not to pursue this principle, except in very special applications in which two Flexowall® belts are married, one with cleats and one without. This type of application, however, is not of interest for in-pit conveying. The sandwich systems using conventional belts suffer from the difficulty of keeping the two belts, i. e., the carrying belt and the cover belt, closely united to prevent material roll back or spillage on the sides, or both. In general, a substantial number of mechanical components are needed to achieve this aim and, of course, as in any machine, the more components there are, the more maintenance, repairs, and shutdowns.

As a result of these considerations, Scholtz researched the development and application of flexible sidewall belts which offered the opportunity to design simpler systems with fewer components and the possibility of increased capacities. Further, the pocket type design of Scholtz's Flexowall® belt prevents product degradation and spillage since the material rests securely in the pockets. This characteristic also accounts for the fact that Flexowall belts have a longer life than

 Volume 4, Number 1, March 1984 

conventional belts since abrasion is curtailed by the material's restricted movement on the belt.

Over the last five years or so, Scholtz systems have been installed in North America and overseas with capacities up to 2,500 STPH* of coal, or 5,000 STPH of iron ore. An expansion of Scholtz's belt making capability is presently underway and will lead to belt capacities approximately 50% greater than that.

3. Proposed Installation

To illustrate the use of Scholtz-EFS steep angle conveyors in an in-pit application, an example of a proposed installation is described here. A conceptual drawing is attached to this article, Fig. 1.

The application is for a capacity of 1,000 STPH R.O.M.** coal and a vertical lift up to 61 m (200 ft), at an angle of 36 degrees. The system is mobile. Movement ranges from 1.5 m/h (5 ft/h) to 7.9 m/h (26 ft/h).

The major components of the system are:

1. A 1,000 STPH mobile feeder breaker with boom discharge conveyor
2. A 146.3 m (480 ft) self-propelled high angle linear stacker (conveyor)
3. A track-mounted load-out unit consisting of a surge bin and a load-out conveyor. The load-out unit also serves to support the head station of the linear stacker.

The system receives R.O.M. coal from a 992C front end loader, sizes, conveys, and ultimately loads the material on 120 t belly dumps at the mine crest. The system is priority based on truck availability. The system is controlled by an operator at the surge bin. He batches material into the 120 ton belly dumps as it becomes available. The material flow is controlled to ensure a constant material flow to the load-out conveyor.

The system is completely self-contained and ambulatory. The propell system is controlled at the load-out unit and a servo system maintains the appropriate attitude of the high angle conveyor tail section. The feeder breaker is independently driven and is paced by the tail section.

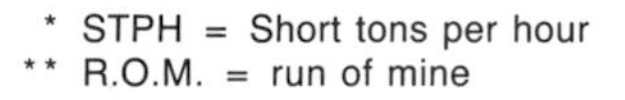

* STPH = Short tons per hour
** R.O.M. = run of mine

4. Functional Description

4.1 Track Mounted Feeder Breaker

The track mounted feeder breaker receives R.O.M. material (maximum lump size 40 inches) from a 992C front end loader, sizes material to nominally 6 inches, and meters the sized material onto a high angle conveyor.

The feeder breaker has a 50 ft luffing and slewing boom conveyor. The entire machine is track mounted. The control is remote and interfaced at the truck load-out unit.

4.2 High Angle Linear Stacker

The high angle stacker receives material from the feeder breaker and transports it to the crest of the pit at an angle of approximately 36 degrees. The conveyor is a 55 inch wide Flexowall® type with support stands located at 15 m intervals within a 23 ft deep gallery truss. The gallery truss is completely self-supported and is trunnion mounted to the load-out unit at its head. The tail is supported by a floating trunnion mounted on a track carrier. The conveyor varies in angle to accommodate up to a 50 ft variation in lift height.

4.3 Self-Propelled Load-Out Unit

The load-out unit receives sized material from the high angle conveyor and batches it into 120 ton belly dump trucks. This is accomplished by utilizing devices that meter the material uniformly onto the load-out conveyor. The unit is carried on tracks which are steerable and driven. The unit also supports the head trunnion of the high angle conveyor. The weight of coal is continuously monitored, and flow is controlled by a process controller and load cell units incorporated in each bin.

5. Conclusions

While this example is based on an actual potential application, it should not be thought of as indicating in any way the present limits to capacities and lift heights of Scholtz-EFS Steep Angle Conveyors. There is no doubt that the flexible sidewall belt technology, which has proven itself in thousands of installations all over the world, will be found in quite a few open pit mines.

In addition to mobile applications, stationary conveyors of this type offer extremely simple means of elevating material. They only need to be supported typically at 15 m (49 ft) intervals which reduces the requirement for mechanical components far below even a conventional troughed belt conveyor.

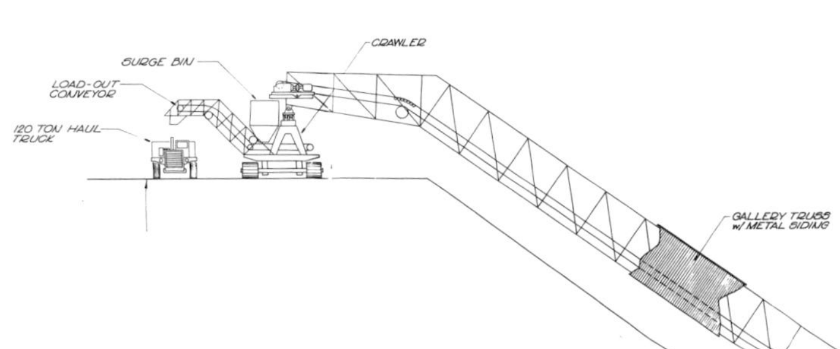

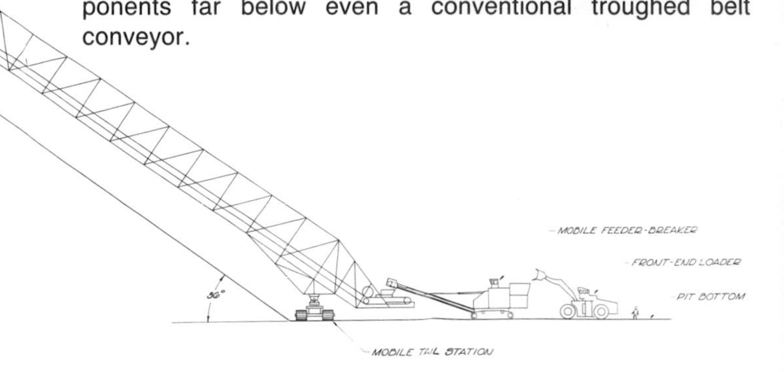

Fig. 1: Proposed steep-angle mobile in-pit conveyor system

Technical Data

Material	Coal
Capacity	1,000 t/h
Density	55 lbs/ft³
Lump size	6 inches
Belt speed	450 ft/min
Belt length	950 ft
Belt width	55 inches
Effective carrying width	23 5/8 inches
Sidewall height	15 3/4 inches
Cleat height	11 3/4 inches

Moveable Suspended Belt Conveyors for Overburden Transport in Open Pit Mines

J. Kogan, Israel

Summary

The author describes the use of moveable suspended belt conveyors for the transport of overburden in across-the-pit operation. Substantial savings in cost and energy are concluded to be possible compared to other systems of comparable capacity.

1. Introduction

In the development of open pit mining, difficulties arise in the stripping and removing of the overburden. The cost of this process limits the depth of modern open pits.

Large capacity belt conveyors present the most efficient means for transportation of overburden from the face to the dump. If the dump is inside the open pit, utilization of moveable cross conveyors gives the simplest solution ("Abraumförderbrücke") — transferring the overburden directly from excavator to dump. Before World War II, many such conveyors were built in Germany (see, for example, Fig. 1).

With increase in the overburden layer and relevant increase of conveyor length, weight and price also increase. According to data of existing conveyors this weight is proportional to the length to a factor of 2.5. Therefore, after the war in Germany such equipment was inappropriate and a new system of round-the-pit conveyors with the use of stackers for distribution of the overburden on dump (Fig. 2) was developed [3].

This system was very expensive because of the great length and additional expense of the stacker, but for a time it was the most efficient one.

In the USA attempts were made to revert to direct transportation of overburden in the dump with the use of long stackers (∿200 m). But such stackers were very large and their ground pressure high.

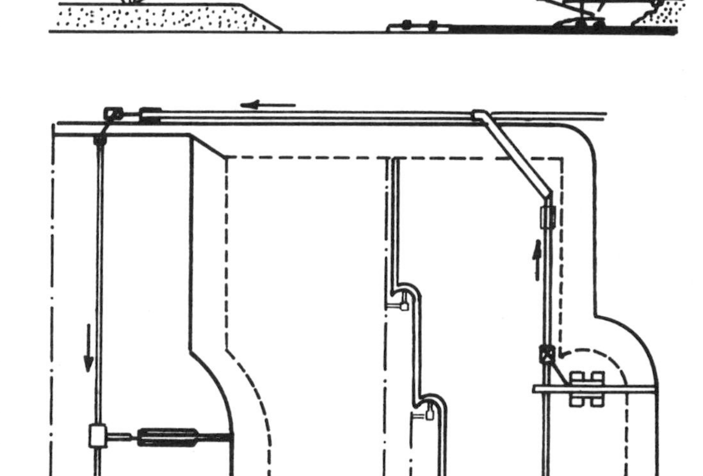

Fig. 2: Round-the-pit conveyor system

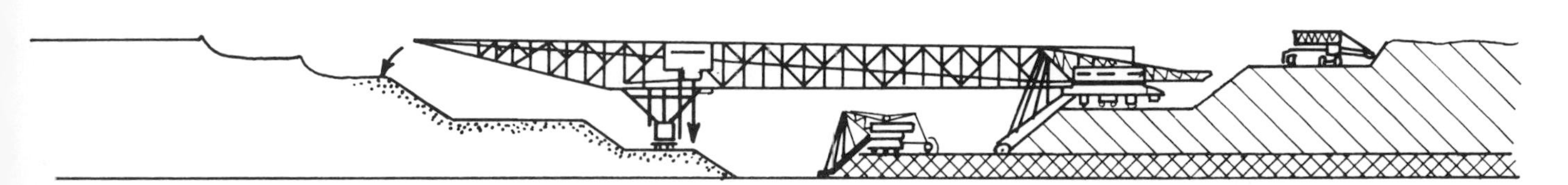

Fig. 1: Moveable cross conveyor (Abraumförderbrücke) in an open pit

It is possible to improve conveyors for direct transportation from excavator to dump by means of replacing the rigid structure of the conveyor bridge by ropes. This idea was first proposed by Schultze-Manitus for transportation of concrete [2].

Dr.-Ing. J. Kogan is Professor at the Faculty of Mechanical Engineering, Technion — Israel Institute of Technology, at Haifa, Israel.

2. Moveable Suspended Conveyors

The author of this paper has also carried out research in the field of open pit mining in connection with conditions in an open clay pit [1].

Initially an experimental conveyor of 100 m length was built for checking the stability of belt movement. It had one span with a moveable two-pulley tripper. The idlers for a belt of 600 mm width were installed on two parallel ropes of dia. 21.5 mm. The end points of the ropes were permanently fastened. At the time of testing, the ropes swung in the transverse direction, but these oscillations did not affect the operation of the belt. We now know that bending of the belt in the transverse direction over a large radius is completely acceptable.

On the basis of research results of this experimental model, the first industrial prototype for the Kaolin mine in Donbass (USSR) was constructed. It moves on rails which are situated along the mine (Fig. 3). Specifications are listed in Table 1.

non-qualified staff, when precision of laying equals ± 0.5 m, the stability of support is sufficient.

At first it was not clear what happens to rails moving under the action of horizontal forces. As in the experiment, dowel-pins fastened to sleepers from below were planned, but it turned out that displacement takes place only at the beginning, after the initial laying of the railways. Later, when the sleepers press into the ground, displacement is not possible. Therefore, the dowel-pins were removed and for protection against the initial displacement a pile of length 1 m (through every tenth sleeper) was driven in, in front of the sleeper.

When moving a conveyor along rails, skewing is inevitable and it is necessary to align and balance the ropes, otherwise considerable lateral inclination of the conveyor and a sudden change of its axis will appear, which will disturb the action of the conveyor. Experience has shown that where possible, this disturbance will be avoided if on one side an equalizing traverse is installed. Two end guides are also needed to turn the ropes when skewing the conveyor.

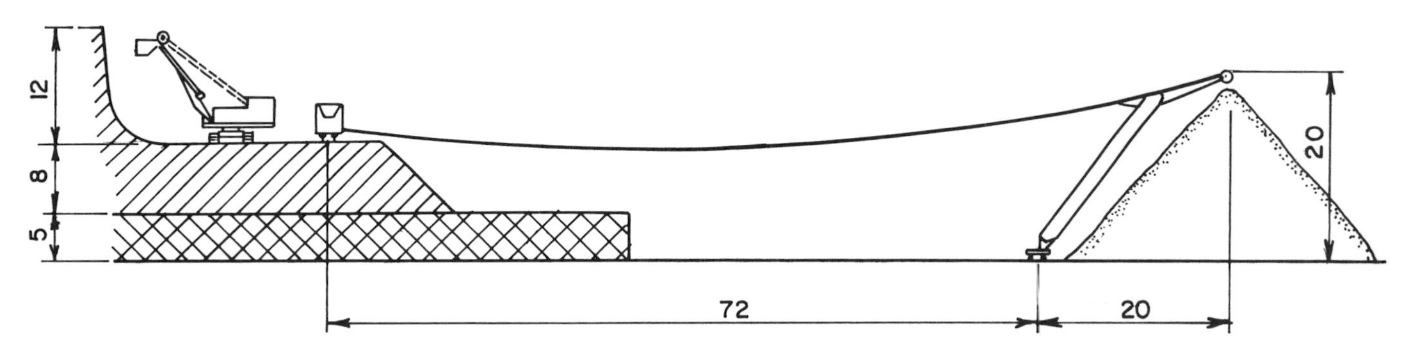

Fig. 3: Moveable suspended cross-pit conveyor

Table 1: Characteristics of moveable suspended belt conveyor in Kaolin mine

Total length	100 m
Distance between railways	77 ± 0.5 m
Belt width	600 mm
Belt speed	2.5 m/sec
Capacity	250 t/h
Rope diameter	20 mm
Rope sag	5 m
Rope tension	2 x 8 t
Total weight of conveyor	56.5 t
Weight without counterweight	38 t

To load the conveyor a shovel with a 2 m³ bucket is used. The load passes to the conveyor through a hopper equipped with a crushing device and an apron feeder. The work of the conveyor in the mine is continuous all year round, in the summer and winter, from temperatures of 30 °C, down to —30 °C, and with winds of up to 25 m/sec. This gives an indication that the swing in the transverse direction does not affect the operation of the belt. It was discovered, however, that a wind gust blowing from below can throw a belt from the upper idlers. For protection against this, precautionary arches were installed.

It is known that it is impossible to lay rails exactly, but experience has shown that even when the rails are laid by

The conveyor drive described was installed on a face support, see Fig. 3. This is not convenient, as the load is carried along the belt with little tension. Installation of the drive above, as usual, is better. Such installation on a large span with high capacity can cause difficulties if sway support is used on the dump side. In such cases it is probably better to install a dump support directly on the dump itself which is divided into two ledges (Fig. 4, A).

Large suspended conveyors must obviously move on crawler or walking mechanisms. It is very desirable to use the weight of the face excavator itself as the counterweight of the conveyor. For this, the conveyor ropes must be fastened to the chassis of the excavator.

For modern open pit mines the possibility of skewing the conveyor in its plane to an angle of ± 10° is also necessary. In consequence the length of the conveyor must be changed so that the swinging of the dump support will be sufficient and an automatic change in the length of rope at constant tension will be necessary. For example, a span length of 290 m should be able to reach

$$290\left(\frac{1}{\cos 10^\circ}\right) \approx 6\,\mathrm{m}.$$

This is easy to achieve by means of rope tackles or hydro-cylinders.

For dump stability, it is important to fill the dump from the bottom to the top, or at least at the beginning to make a rest

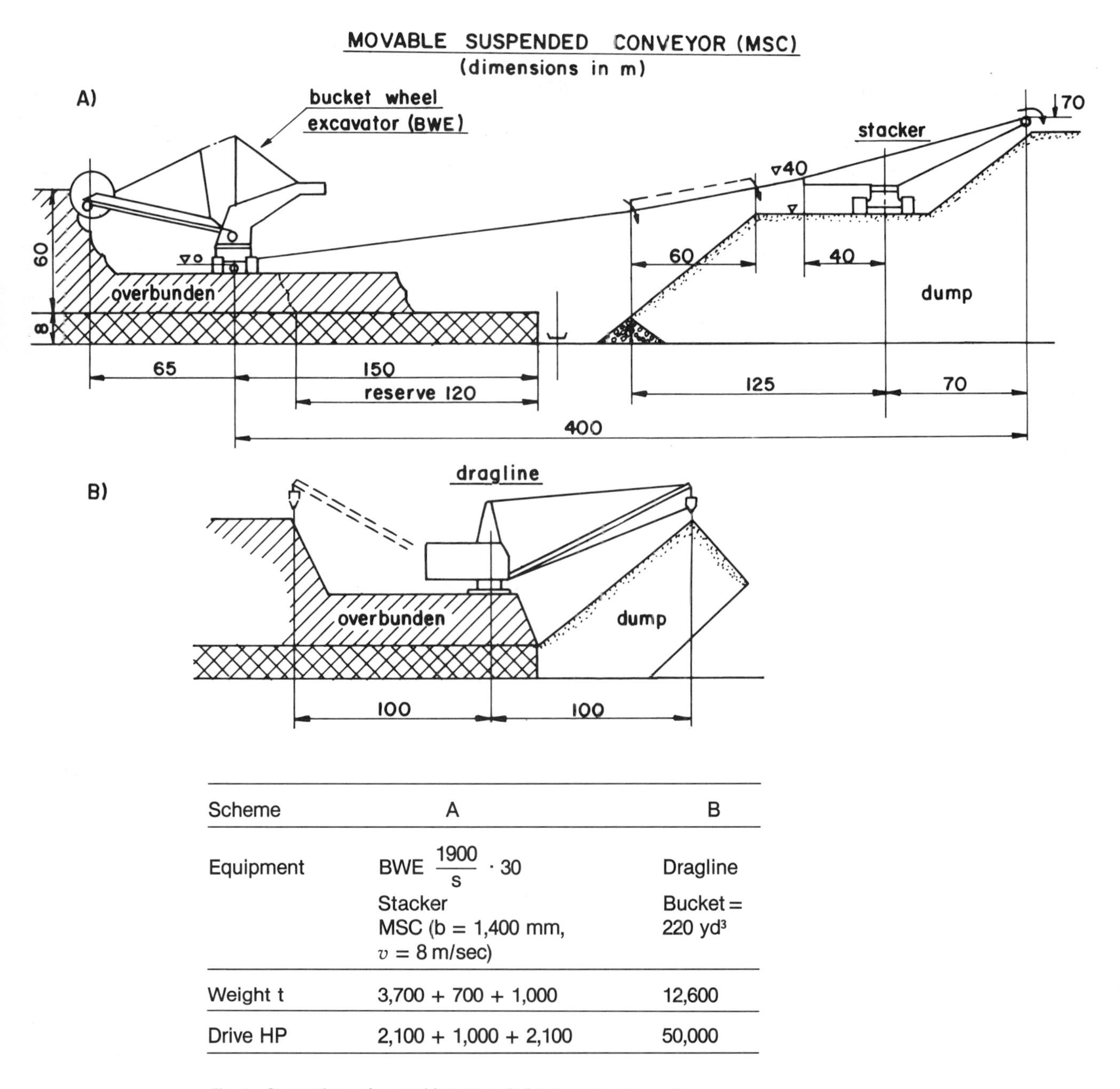

Scheme	A	B
Equipment	BWE $\frac{1900}{s} \cdot 30$ Stacker MSC (b = 1,400 mm, v = 8 m/sec)	Dragline Bucket = 220 yd³
Weight t	3,700 + 700 + 1,000	12,600
Drive HP	2,100 + 1,000 + 2,100	50,000

Fig. 4: Comparison of moveable suspended conveyor system with conventional dragline for a capacity of 5,600 m³/h

prism at the bottom. For this purpose in moveable cross conveyors with rigid structure an intermediate discharging device was used (see Fig. 1). This device is not desirable for use with suspended conveyors. A better arrangement is an intermediate discharging device from the belt. A solution for this for a belt which moves with large velocity has not yet been found. Detailed research of the possibility of intermediary discharge is thus necessary. The cost of this research would be justified by the end savings by using suspended conveyors.

3. An Economic Comparison

Fig. 4 gives a preliminary comparison of a moveable suspended conveyor with a walking dragline for a capacity of 5,600 m³. (This comparison is evidently possible in mine conditions where a bucket wheel excavator could work). From this comparison it is clear that weight and consequently price of the equipment for this conveyor version is lower by more than half, and the installed power lower by nearly nine tenths. At the same time, conveyors, as is well known, are very convenient in operation; it is possible to advance the overburden transport more than the mineral recovery by 2—3 months, i.e., it is possible to stop overburden work in the winter and use this time to check and repair equipment.

References

[1] Kogan, J., "Podvesnie lentoschnie transporteri", Mashgiz, 1949, (Russian)

[2] Schultze-Manitus, H., "Neue Kabelkrane für Gießbetonarbeiten", Fördertechnik No. 15/16, (1935)

[3] Editorial, Mining Congress Journal, Sept. 1981

Continuous Conveying in Open Pit Mines

W. Glanz and J. Weißflog, Germany

Summary

Various open pit mining systems with hard rock and unconsolidated overburden are analyzed and the main differences between discontinuous, semi-continuous and fully continuous mining and transportation methods are outlined.

It is demonstrated that future open pit mining systems will tend to use continuous or semi-continuous haulage and mining operations in which detailed, long-term economic and technical planning and evaluation with the help of computer programs will play a major role.

1. Introduction

When analyzing the various opencast mines it can be established that there exist only two basic types of mining systems, i.e., the discontinuous strip mining system as used in the USA (Fig. 1) and the continuous mining system as is commonly applied in Europe (Fig. 2). The reason why these two mining methods have been developed parallel to and independently of each other to such a high technical standard results from the varying geological conditions of the mineral deposits.

Whereas pay minerals such as coal, phosphate, bauxite, iron ore and copper ore are mined in the USA from relatively thick and horizontal seams with low to moderate overburden depths, the mineral deposits in Europe lie at much greater depths and usually not in homogeneous seams. Due to the very favourable overburden to coal ratio of 2:1 in the USA, mining and haulage are possible by means of shovel excavators and draglines and heavy trucks, whereas the considerably higher mass movement in European opencast mines, with an overburden to coal ratio of up to 10:1, can only be mastered by continuous haulage systems. In the meantime the conditions in the USA have also undergone a change to such an extend that it is no longer possible to uncover usable coal seams without removing increasing depths of overburden, while coal quality often deteriorates. Therefore several opencast mines in the USA are becoming deeper and the volume of material that has to be moved increases accordingly. Higher production rates are also a consequence of the substitution process of the expensive oil by coal.

During recent years the opencast mining scene in the USA has altered and the first steps towards continuous haulage have been made. The authors are of the opinion that this development will continue in the future due to the fact that governments, not only the American government, have introduced environmental laws and regulations which require very high standards of reclamation of mined-out areas.

Dipl.-Ing. W. Glanz and Dipl.-Ing. J. Weißflog, PHB Weserhütte AG, D-4970 Bad Oeynhausen 1, Federal Republic of Germany

Based on a paper given at the "Conference & Expo VI" at Louisville, KY, USA, October 27—29, 1981

2. Truck Haulage System

Considering the overburden and intermediate coal seams that have to be exploited and the various good and medium quality coals, which to a certain extent have to be mined selectively, the classical truck haulage system shows an ever increasing uneconomical tendency as the price for diesel fuel, labour costs and the necessary costs incurred for land reclamation as well as the overburden depths increase.

Increased haulage distances will result in an expansion of the truck fleet even for constant production rates. With the aim of decreasing production costs in this important indus trial field, national and international institutions have conducted investigations to minimize operating costs. These investigations do not concentrate only on individual pieces of mining equipment but stress an analysis of various opencast mining operations and their planning and control by means of computer programs.

The necessary preliminary investigations were based mainly on individual systems in view of the costs for haulage per ton kilometer. These economic diagrams can only be used in a qualitative sense and in most cases results cannot be generalized since the conditions in each opencast mine are quite unique, i.e., parameters such as length of mine, its width and depth, the volume of material to be moved, reclamation requirement, lifetime of the pit, type of man power available etc.

It can be established that within the last ten years the operating costs as well as the haulage diagram for individual systems as a function of haulage distance show a detrimental result for the truck/shovel or truck/dragline systems especially for opencast mines with unconsolidated overburden (Fig. 3).

Fig. 1: Typical strip mining operation with dragline

Fig. 2: Mining operation with bucket wheel excavator, conveyor belt system and stacker

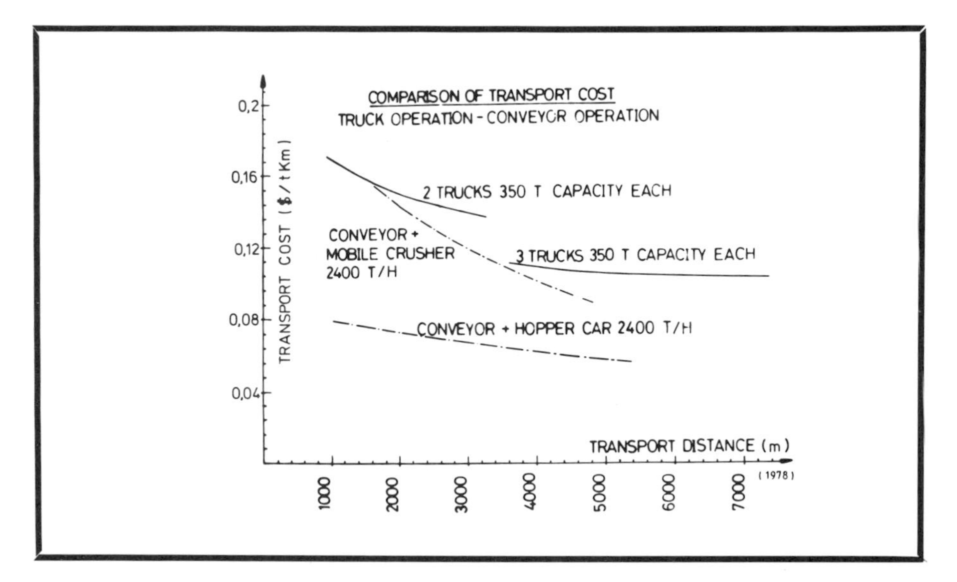

Fig. 3: Comparison of transport costs: truck vs conveyor operation

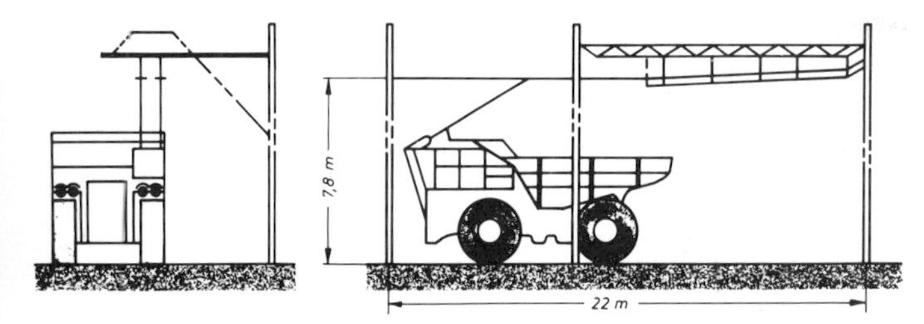

Fig. 4: Trolley-operated truck system

The classical mining method for hard rock open pit mines is the truck/dragline or truck/shovel system which nowadays calls for an increased number of graders and dozers due to the reclamation regulations.

In order to keep down the operating costs for such systems as far as possible, larger trucks are nowadays built with up to 350 t capacity and, for instance, trolley operated (Fig. 4). In this manner a decrease in operating costs of up to 35 % is possible. Much work has also been done in order to increase the capacity of draglines and shovels and to lower their operating costs, for instance by increasing dragline working radius and bucket capacity.

However, when handling more than 1 million bank m³/year over a haulage distance in excess of 1,500 m from a pit depth of over 80 m, this system is uneconomical when compared to semi — or fully continuous mining operations. Particularly detrimental to the extensive use of the truck system is the necessity to provide and maintain the necessary access roads. It should also be mentioned that long distances to overburden dumping areas have a negative operating cost effect even if the haulage distance for the mined coal should be short.

If reclamation of the spoil area is required it becomes apparent that costs involved for levelling by means of draglines, dozers and graders are considerably higher than if these operations were carried out by stackers in the form of a precultivated dump.

Of course open pit mines exist, for instance Phalabora in South Africa, where due to the inherent conditions only a truck haulage system can be used, and a change to a combined mining system would no longer be economic, but in other instances, as for example the SWICC, a change of mining system proved to be necessary.

3. Combined Mining Systems

The combined mining system was introduced in order to cope with the deteriorating overburden to coal ratio and with the reclamation regulations. The conveying system was introduced for handling the overburden whereas exploitation of the pay mineral, i.e., coal, is handled by a shovel/dragline and truck system. Conveyor belts can be introduced according to the so-called "around-the-pit" or "cross-the-pit" method.

With the "around-the-pit" system a shiftable conveyor is fed with overburden via a hopper car. The hopper car can travel on the sleeper-supported rails of the conveyor system or alternatively as an independent crawler-mounted unit. The hopper car can be fed directly by means of a shovel or a dragline but also by means of a bucket wheel excavator (Fig. 5).

If the overburden and interbedded material which has to be removed requires blasting, a mobile crushing unit should be incorporated between the reclaiming unit and the conveyor system in order to reduce the size of the material to enable conveyor haulage thus also reducing costs for additional blasting.

The advantage of truck haulage lies in the fact that oversize material can be transported without the necessity of primary crushing. However, this will lead to considerably higher costs for reclamation or the dumps will have to be built up selectively.

The outgoing pit conveyor transfers the material onto a mainly stationary belt conveyor which in turn passes the material to the dump conveyor and then onto the travelling stacker which evenly distributes the material so that a homo-

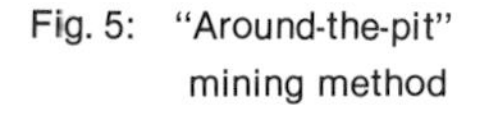

Fig. 5: "Around-the-pit" mining method

geneous dump with a precultivated surface is built up. In American open pit mines the conveyor system usually takes a U-shaped form while the dump generally travels parallel to the exploitation. In European opencast mines the belt conveyor systems can take any required form as the dump can be built up at any required position. However, in the planning stage it is made sure that the turning point is built so that the outgoing conveyors coordinate in a circuit with the recovery and convey the material to a stationary conveyor system that is connected with the shiftable dump conveyor (Fig. 6).

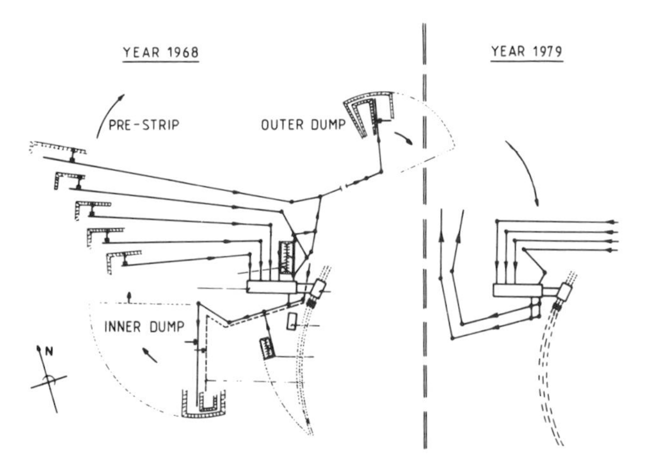

Fig. 6: Open pit conveyor system at Ptolemais, Greece

The previously mentioned "around-the-pit" system is practiced by SWICC and is also planned for NCPC.

The "cross-the-pit" system was applied in an extremely compact form by Peabody, for example, or in a very flexible form by Blue Circle, the latter being influenced by the German lignite opencast mining systems with conveyor bridges and bucket wheel excavators. These systems were introduced in the 1920s. With the cross-pit mining system the material is mined by means of shovels, draglines or bucket wheel excavators. These machines are followed by a mobile, crawler-mounted bridge conveyor which is supported on the lower pit bench and the dump bench (Fig. 7).

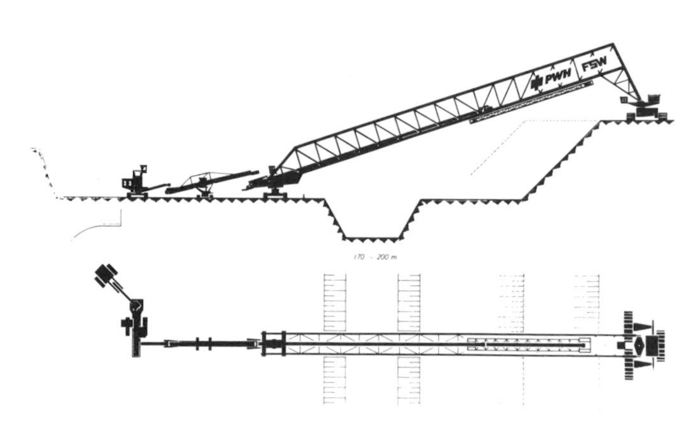

Fig. 7: "Cross-pit" mining at Blue Circle open pit mine

An essential criterion for this mining system is the distance between the face front and the subsequent bench. A new cross-pit system built by PWH was commissioned in December 1981 in England. The conveying capacity is 2,800 t/h with a conveyor bridge length of 145 m and an elevation of 26 m. Similar systems were built for irrigation projects in Iraq and are also successfully used in South Africa and Canada. A precultivated dump is achieved by the around-the-pit system due to the shuttle conveyor, which is located before the crawler unit on the dump bench, or as a slewing boom behind the travel assembly.

In combined systems for mining solid material, in-pit-crushing has meanwhile been introduced, for instance by Foskor in South Africa for handling phosphate or in Grooteluk for handling overburden with capacities of up to 3,000 t/h (Fig. 8 & 9). It remains to be seen if and when in-pit-crushing will be accepted in continuous open pit mining transportation systems. A novelty in this respect is a bucket wheel designed for a European country in which crushing of the mined material is incorporated directly adjacent to the bucket wheel of the wheel excavator. The continuous mining system transports the overburden and pay minerals continuously to their final destination and thus replaces the dragline, shovel and truck operations in the case of mining unconsolidated materials. In hard rock mining, blasting must take place and the material must be transferred by means of a dragline or shovel into the crusher from where it is fed onto the belt conveyor.

4. Shiftable Conveyor Systems

The shiftable conveyor systems in use today present no problems and achieve a utilization index of 98 %. The lateral movement of the conveyor line is executed by pipe-laying bulldozers whereby shifting is effected at rates of up to 2,000 m²/h and more. The drive and return stations of the conveyor system are moved by dozers, by crawler mechanisms or by hydraulic walking pads, and in the past they were also shifted into position on appropriate rail paths.

Today the elongation of the longitudinal axis of a shiftable belt conveyor no longer presents problems. Three different principles are in use whereby the rear sections of the conveyor line are taken out and placed into the advancing front section of the line.

Without the necessity of splitting the rubber belt the drive and return stations are moved in the direction of advance in the transverse position of the shiftable belt conveyor. Subsequently the additionally required idler sections are fitted in.

A further technique is to use a very long reserve belt loop in the lower belt run which can be paid out in accordance with the advance. Here, however, the longitudinal movement is limited by the length of the reserve belt loop. The extension of a conveyor line by means of new conveyor frames and belts is the method usually used.

As conveyor systems do not require a specially prepared ground level, the expensive construction of truck haulage roads is not necessary. The belt conveyor system must be maintained and held clean with special equipment but ordinary maintenance vehicles fitted with additional equipment can also be used.

In semi-continuous mining systems the overburden and interbedded materials are transported by a conveyor system while coal is transported by trucks. In fully continuous systems the coal can also be transported by separate belt conveyors to its final distination. This, however, involves a

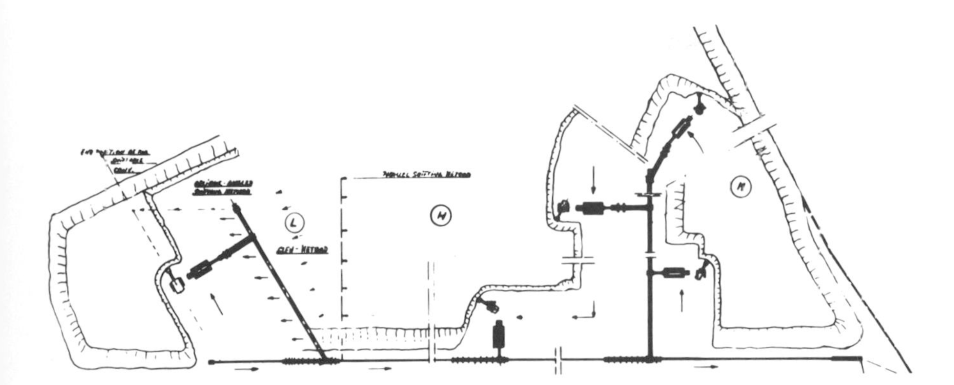
Fig. 8: Mining plan of Foskor Phalabora, South Africa

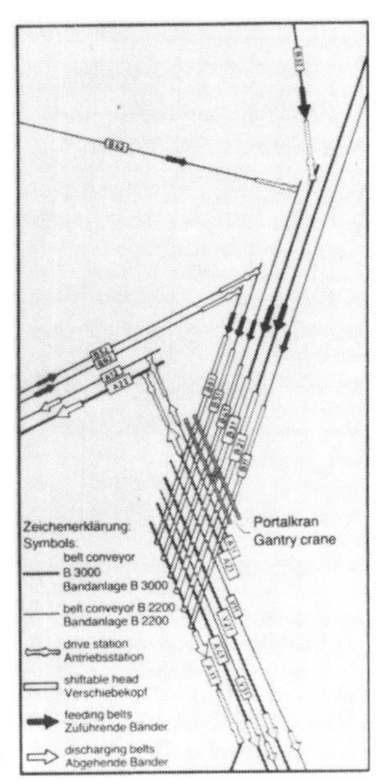

Fig. 10: Conveyor distribution point

Fig. 9: Blasting, in-pit crushing and conveyor system at Grooteluk South Africa

Fig. 11: Conveyor distribution point with extending conveyor heads

high investment and increased operating costs, but results in the highest degree of flexibility. In order to utilize the conveyor systems to an optimum extent and to reduce the investment cost as much as possible, conveyor systems with extensible conveyor heads are built into the system. This allows only one conveyor system to cope with two varying types of material (Figs. 10 & 11). The flow is directed by means of the extending conveyor heads at conveyor distribution points onto subsequent conveyors.

In the following, a few examples of systems designed and built by PWH will be discussed.

5. Captain Mine Conveyor System 1978

The Captain Mine consists of three sub-systems each with a total length of approximately 4.4 km (2.7 miles). Each system includes three shiftable conveyors: one face conveyor, one connecting conveyor and one dump conveyor (Fig. 12). Additionally, two elevating conveyors can be employed in the sub-systems. Each sub-system operates together with one bucket wheel excavator on the digging side and one spreader on the spoil side.

The unconsolidated overburden with depths of up to 12.8 m is mined by bucket wheel excavators. The top soil is transported via a crawler-mounted bandwagon (mobile bridge conveyor) to the hopper car feeding face conveyor A. From here the top soil moves via a conveyor bridge or directly to the connection conveyor C and further on to the spoil conveyor D. On the dump side the material leaves the conveyor via a rail-mounted tripper car with slewable discharge conveyor towards the stacker/spreader unit. The top soil is then positioned on top of the consolidated overburden and recultivation for agricultural purposes in then easily possible.

The consolidated overburden, blasted hard rock, is dumped directly across the pit by the main stripping machine, i.e., the shovel, from the virgin to the spoil side of the pit. The coal is excavated by shovels and transported by 130 ton capacity haulage trucks and by 150 ton coal carriers to Southwestern's 2,800 t/h preparation plant.

The main specifications of the conveyor system are:

Belt width	54 inch (1,400 mm)
Belt speed	950 ft/min (4.8 m/sec)
Design capacity	4,800 loose y^3/h (4,400 m^3/h)
Max. installed power per conveyor	4 × 400 HP (4 × 300 kW)
Total installed power approx.	12,000 HP (8.8 MW)
Number of conveyor flights	11
Total final length approx.	7.5 miles (12 km)
Troughing of upper belt	35 degrees
Troughing of lower belt	10 degrees
Type of carrying idlers	garland of 3 idlers
Type of return idlers	garland of 2 idlers
Type of impact idlers	garland of 5 idlers
Type of steel cord belt	St 1850—St 825
Cover plate thickness	$^3/_8$"—$^{15}/_{64}$" (9.5 mm—6 mm)

The main technical data of the conveyor bridge are:

Conveyor length	266 ft (81 m)
Elevation height	60 ft (18 m)
Conveying capacity	4,800 y^3/h (4,400 m^3/h)
Belt width	54 inch (1,400 mm)
Troughing angle	35 degrees
Belt speed	950 ft/min (4.8 m/sec)
Conveyor drive unit	2 × 300 HP (2 × 220 kW)

The availability of the complete system, from the bucket wheel excavator (BWE) on the digging side to the spreader on the spoil side, is 75 % on a monthly average for the first few months of operation and is bound to increase with the operating crew gaining experience and after the start-up delays are phased out.

The conveyor systems advance in steps of 60 m (200 ft) at an estimated frequency of four to five times per year depending on mining conditions. The shifting process takes 2—3 days at the Captain Mine, and again this time might be reduced with increased experience.

The complete system from the BWE to the spreaders is operated by a crew of 5 and a shift supervisor. Two beltmen control the conveyor system. Electricians and welders are provided from the mine pool when they are needed. There is one master mechanic on the day shift assigned to the complete system.

Daily maintenance of the conveyor system includes cleaning, checking belt cleaners and idlers and this is executed by the beltmen. The oil level in the conveyor drive gears is checked and the belt drive shaft is greased once a week.

Overall, the introduction of the new conveyor system was not completely free of difficulties. At the beginning some specific operational and soil problems were realized and had to be solved.

6. Opencast Mine Most, Czechoslovakia 1979

A new shiftable conveyor belt system with the following main technical data was installed in the lignite open pit mine at Most (Fig. 13).

Total length	8 km (5 miles)
Installed power	23 MW (31,000 HP)
Number of belt conveyors	9
Belt width	2,200 mm (84 inch)
Belt quality	St 2500
Rated conveying capacity	17,000—20,000 t/h
Belt speed	5.2 m/s (1,024 ft/min)

The steadily increasing depth of the overburden from initially 60—70 m (190—230 ft) to finally 200 m (660 ft) led to the installation of a conveyor system between the excavators and the spreaders for handling overburden, in 1978. The overburden above the coal is stripped and transported to an inner dump behind the coal seams. This system is now in operation since January 1981 and has fulfilled all expectations in respect of efficiency and utilization factor. Belt conveyors achieved utilization factors of 94 % of the available shift periods.

The computer-controlled central monitoring system switches all operating sequences when required. Defects and downtimes are automatically recorded, mathematically evaluated and compared with previously determined criteria so that written reports are not necessary any longer.

7. Goonyella Mine, Australia, 1980

One of the most important high-quality coking coal deposits in the world is located in the Bowin Basin in Queensland, Australia and is exploited by Utah Development Company (UDC) in the Goonyella Mine and by three other opencast mines.

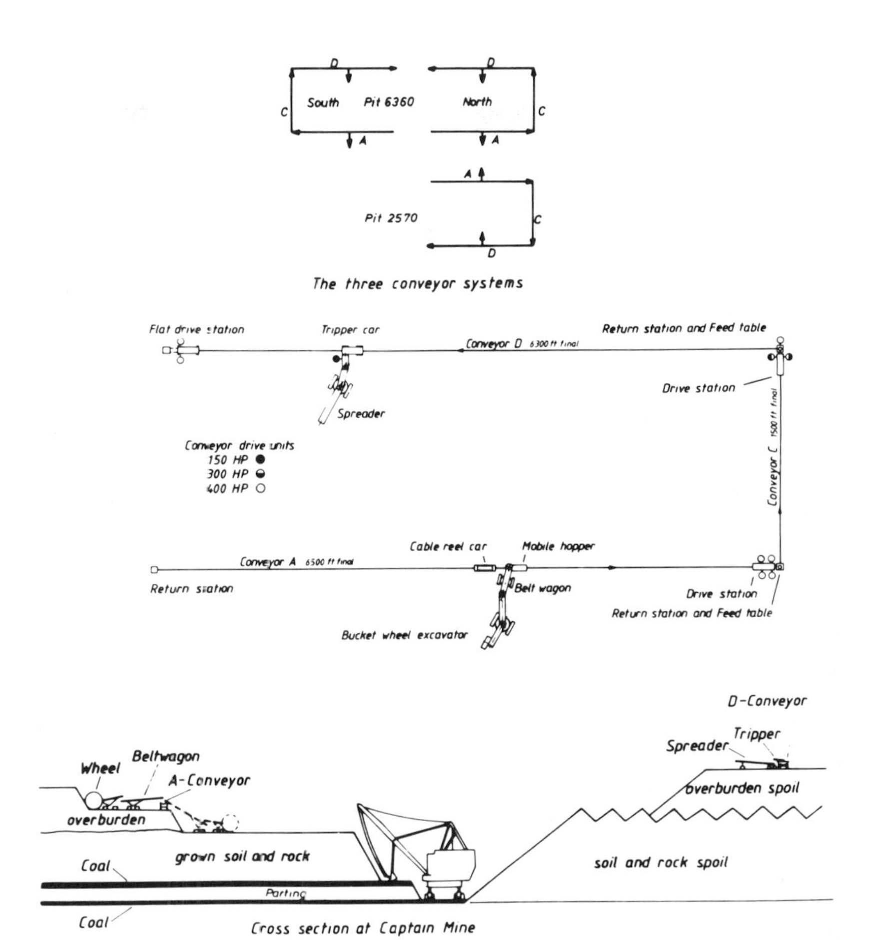

Fig. 12: Mining plan with conveyor system at Captain Mine, Illinois, USA

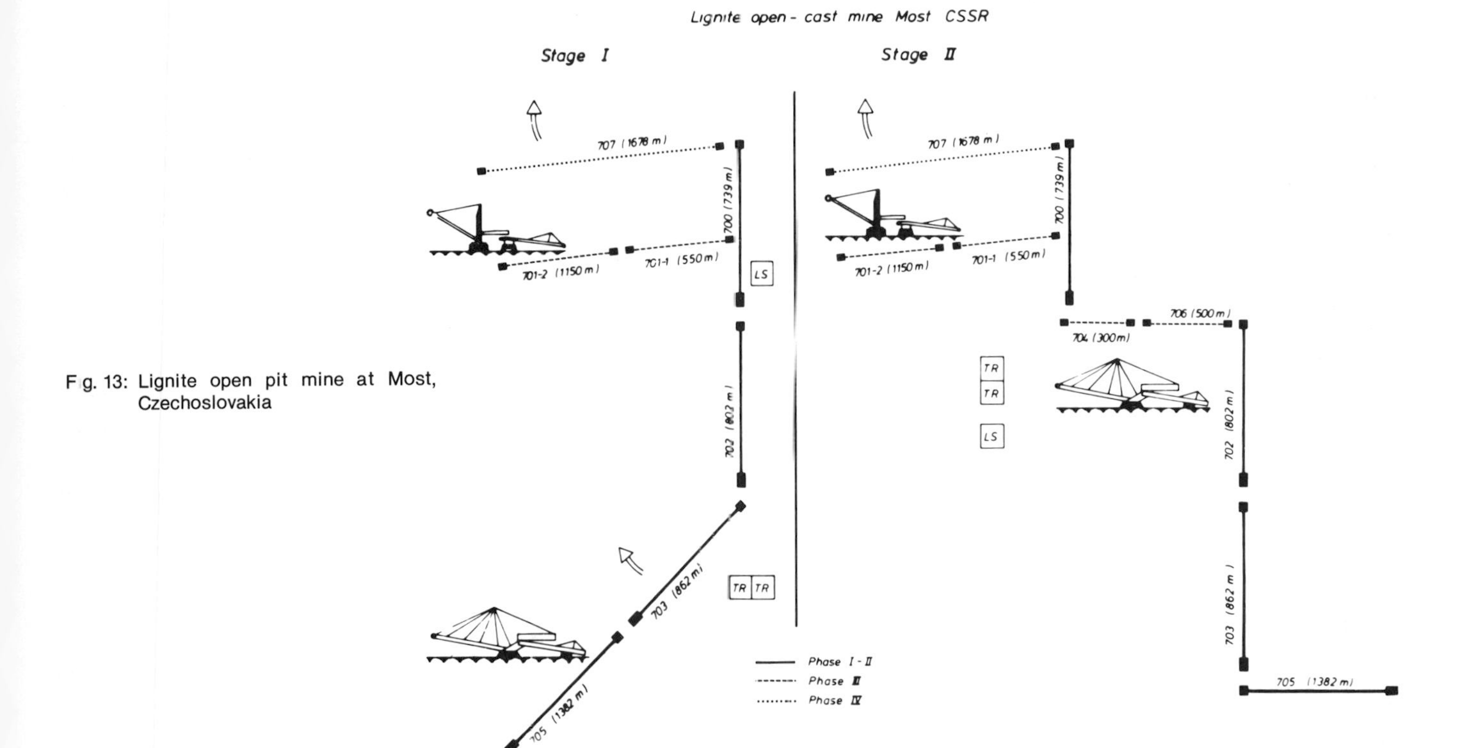

Fig. 13: Lignite open pit mine at Most, Czechoslovakia

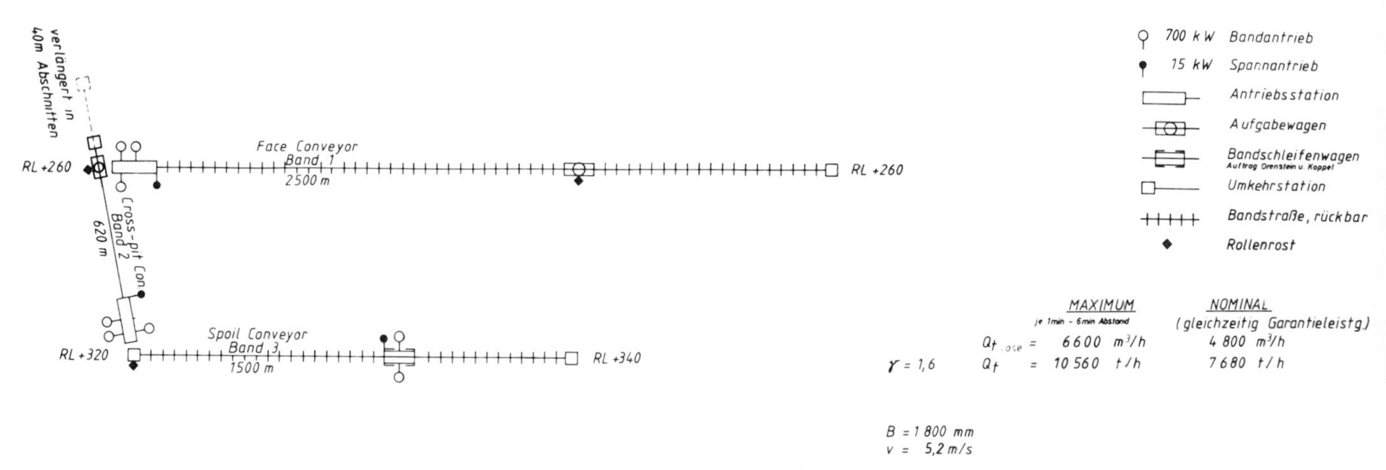

Fig. 14: Conveyor system at Goonyella Mine, Queensland, Australia

In the Goonyella Mine, the traditional American strip mining method with dragline operation for overburden removal and face shovels and truck haulage for coal extraction is used. Due to the increasing depth of the coal seam, however, the overburden cannot be handled by the dragline in a one-bench operation. Various operating alternatives for overburden removal were considered and evaluated econometrically as well as technically. The outcome of these investigations led to the decision to use a mining system with bucket wheel excavator, shiftable conveyors and spreader to mine the top bench of overburden with a thickness of up to 36 m (120 ft).

PWH supplied the conveyor system which must be capable of carrying blasted material with lumps of up to 500 mm (20 inch) diameter and a mass of 300 kg but also extremely moist and sticky material, (Fig. 14). This equipment consisted of:

2 hopper cars

1 bench conveyor; 2,400 m long; 1,800 mm wide; St 2200

1 transverse conveyor; initially 620 m long and in the final stage 1,200 m long; with 102 m lift; width 1,800 mm; St 4000

1 dump conveyor; 1,500 m long; 1,800 mm wide; St 2200.

The entire system is shiftable with hydraulic walking mechanisms for the 300 t drive head stations and a "shiftable head" installed on a crawler tractor to shift the conveyor. Transfer points are equipped with roller grates to protect the belt and impact idlers when blasted material is conveyed. The transverse conveyor has a reserve belt loop 80 m to permit simple extension.

The main technical data are:

Capacity:	7,680 t/h (average) and 10,560 t/h (maximum) with a belt speed of 5.2 m/sec 2,000 t/h with a belt speed of 2.6 m/sec creep speed for inspection purposes 1 m/sec
Drives:	700/350 kW (950/475HP) at 1,500/750 rpm
Idlers:	3-roll carrying garland idlers, 45 degree troughing 2-roll return garland idlers, 15 degree troughing

The design was based on the following data: Calculated expected lifetime: 5,000 hours/year for 20 years. The ambient temperature was assumed to be 45°C; a maximum wind velocity of 200 km/h (124 miles/h) was used. Several cold starts were carried out successfully and the system was taken into operation in late 1981.

8. Outlook for the Future

The above mentioned PWH installations illustrate the different structure and philosophy which are inherent to the continuous mining system. While previously short term planning was performed for the truck, shovel/dragline system with the basic thinking of "... this mine has coal deposits for x number of years at a certain output, let's mine it with a dragline or shovel and several trucks...", the planning periods for continuous mining systems continue to extend and modern electronic data processing equipment will be used for planning and evaluating the economic feasibility of various alternatives. These procedures will also be used in the future to continuously monitor and compare operating data of conveyor systems, bucket wheel excavators, crawler-mounted bridge conveyors, mobile inclined conveyors and spreaders etc, as well as compact mining machines with integrated crushers and such equipment which can be powered via trailing cables and by electric motors instead of diesel-mechanical or diesel-electric means.

Fig. 15 shows the open pit mine of the future. At the focus of this system is the continuous miner which is capable of mining the hardest coal layers with depths ranging from a few centimeters to up to 4 meters.

With this machine one has worked on the assumption that proven underground mining machines with capacities of up to 1,200 t/h can also be used in open pit mining, where the limited space of underground mining is of no consequence. Machines with capacities of 2,000 t/h and a deadweight of about 200 t are conceivable. After modification of the mining method, such a machine will prove successful once it has completed its trials. The machine is designed to be able to load trucks and conveyor belts and is capable of handling fine grained material (Fig. 16).

Summing up, one can observe that many open pit mines, especially in North America, the majority of which operate with truck/shovel or truck/dragline configurations, will for economic reasons switch over steadily to the combined conveyor-truck-shovel-dragline or conveyor-bucket wheel excavator — truck — dragline configuration to achieve an automated fully continuous mining operation with the cross-pit method in unconsolidated overburden formations.

In the field of open pit hard rock mining, a semi-continuous operation with conveyor systems, conveyor distributor stations and extending conveyor heads, continuous miners and trucks as well as shovel/dragline plus in-pit crushers

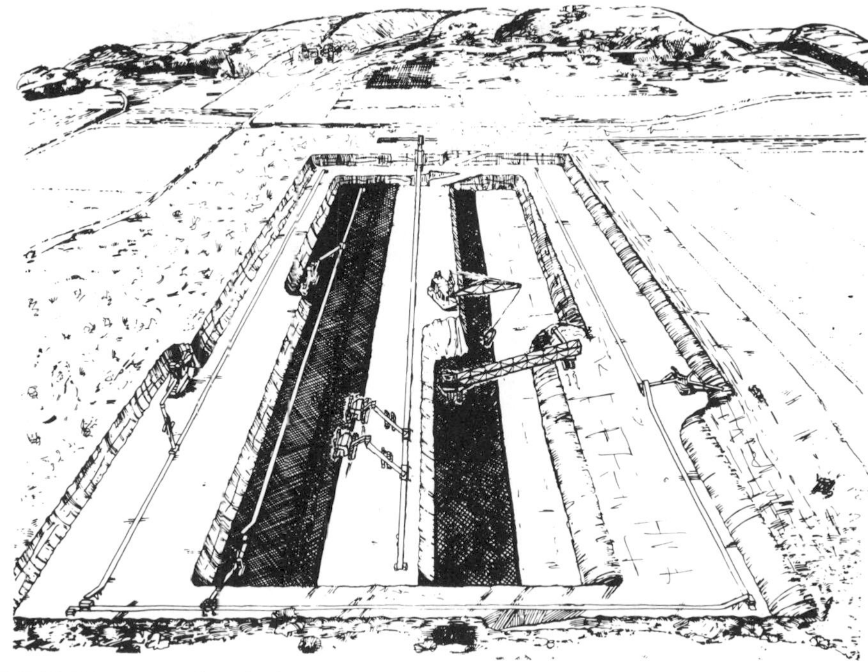

Fig. 15: Open pit mining system of the future

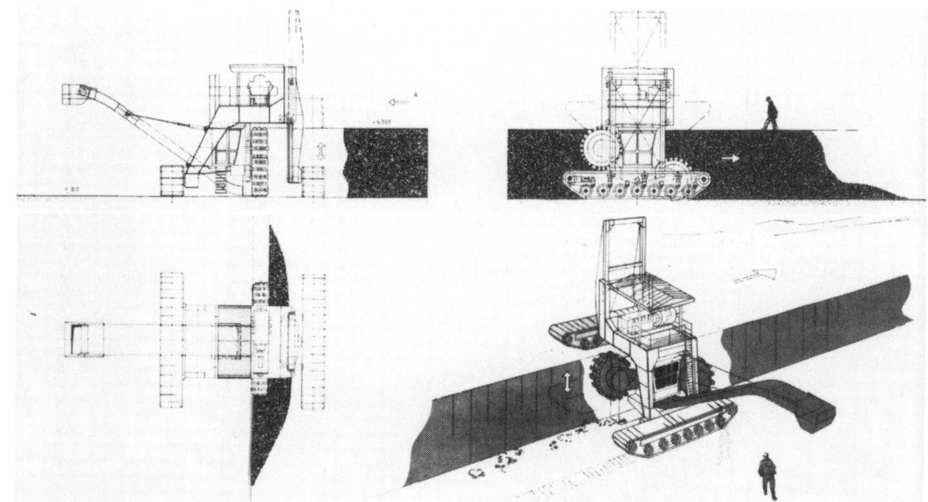

Fig. 16: Open pit continuous coal miner of the future

and mobile conveyor bridges/steep conveyor configurations will become an established practice.

For these types of open pit mining systems, project design and routine operation with respect to short, medium and long-term planning, control, monitoring and the acquisition of working stock-data as well as machine assignment schedules will be taken over to an increasing degree by main computer centers.

References

[1] Morgan, B.V., "Off-Highway Haulage", The Quarry Manager's J., Dec. 1968

[2] Lake, D.M. and Brzezniak, W., "Truck Haulage Using Overhead Electrical Power to Conserve Diesel Fuel and Improve Haulage Economics"

[3] Korak, J., "Technisch wirtschaftliche Untersuchung unter der Berücksichtigung der Transportmittel Kombination im Festgesteinstagebaubereich", Dissertation TH Aachen 1978.

[4] Learmont, T. and Chare, H.B., "Area Stripping Productivity Solutions in Deep Overburden", American Mining Congress 1980

[5] Bandopadhyay, S., Ramani, R.V. and Manula, G.B., "A Computer Simulation Model for a Surface Coal Mine", bulk solids handling No. 1 Vol. 1 (1981) pp. 129—135

[6] Schlemme, U.H., Schriefer, D. and Weißflog, J., "A Shiftable Conveyor System for the Captain Mine in Percey, Illinois, USA", PWH 1981

[7] Weißflog, J., "Großbandanlagen mit Bandschleifenwagen im nordbömischen Braunkohlenrevier Most", Braunkohle, April 1979

[8] Alberts, B.C. and Dippenaar, A.P., "An In-Pit Crusher Overburden Stripping System for Grooteluk Coal Mine", ISCOR-Pretoria 1980

[9] Jackson, D. and Smith, L. "New Horizon Looms for Lignite", Coal Age, May 1981

[10] Weißflog, J. and Glanz, W., "Rückbare Bandanlagen", Bergbau, March 1981

[11] Gupta, B.K., "The Arch Mineral Captain Mine — Reclamation Innovation for the 80s", Mining Magazine, May 1980

New Developments in Continuous Mining and Materials Handling

Werner Rixen, Germany

Summary

Bulk mining and handling technologies have been affected in two ways by the soaring cost of oil and oil products in recent years.

Firstly, the substitution of oil by coal, oil sands and oil shale demands the development of high-capacity mining and handling equipment, and secondly, the rising fuel costs favour a trend away from cost-intensive heavy truck transport.

To cope with the growing volumes of material in mining and handling, larger equipment units have to be designed. This tendency becomes apparent from the examples of the Hambach open pit mine in Germany.

A combination of the flexibility of heavy truck transport and the cost advantages of belt conveyors is achieved by the system of semi-mobile crushing installations described in this paper.

1. Causes of New Development Trends in Mining and Handling Technology

Bulk mining and handling technologies have been affected in two ways by the growing scarcity of low-cost energy resources in recent years.

In the first instance, the increasing cost of oil and natural gas led to intensive efforts throughout the world to develop fossil fuel resources for energy generation.

To open up new coal, oil shale and oil sand deposits, adequate equipment has to be created for mining and handling such materials.

Since especially the more accessible coal and lignite deposits are exhausted, it is necessary to turn to deposits with less favourable overburden/coal ratio and less easy stripping conditions: higher overburden/coal ratios in turn mean handling of greater overburden volumes for production of a given quantity of coal.

The increasing share of coal in world energy supply thus leads to an over-proportional increase in mining and handling volumes which can only be attained by mining and handling equipment designed for higher capacity. This development is illustrated below by some examples.

A second aspect influencing the present development in mining and handling technology is the enormous rise in cost of gasoline and diesel fuel, causing a change from expensive heavy truck transport in the pits to cheaper continuous flow of materials by means of belt conveyors.

While the continuous-flow system of bucket wheel excavators, shiftable conveyors, and boom stackers for spoil disposal, has already been firmly established in open pits with non-compact overburden, it is only lately that a system has been developed for hard rock open pits by means of heavy truck transport in the pit, a semi-mobile crusher, and subsequent conveyor haulage; this combines the flexibility of truck transport for the mining advance with the cost advantages of conveyor transport. The concept, too, is described by an example as follows.

Dr.-Ing. Werner Rixen, Chief Project Engineer, Excavator Department, Krupp Industrie- und Stahlbau, P.O. Box 14 1960, D-4100 Duisburg 14 Rheinhausen, Federal Republic of Germany

2. Open Pit System with a Daily Output of 240,000 bank m³

2.1 Reasons for Development of the System

The lignite district of the Rhineland extends over an area of about 2,500 km² west of the Rhine, between the cities of Cologne, Bonn, Aachen and Mönchengladbach. With a total lignite reserve of 55,000 million tons, of which 35,000 million tons are profitably mineable by today's standards of technology, this is the largest developed lignite deposit in Europe [1,2]. About one quarter of the West German electricity supply is based on lignite from this district in the Rhineland.

Industrial utilization of these lignite deposits began around the turn of the last century, mostly in the southern part of the district where seams of up to 100 m thick were found under a shallow overburden cover. Since 1955, these extremely favourable reserves have been exhausted. In order to meet the growing power demand nevertheless by sufficient coal production, it was necessary to open up deep-going lignite pits, with higher overburden/coal ratios.

The latest example of this new development is the Hambach mine. Rheinische Braunkohlewerke AG, which also owns and runs the opencast mine in Hambach, is the sole mining company operating in the Rhenish lignite field. Initially, operations will begin in the Hambach I coalfield where reserves amount to 2,400 million tons.

In its final stage around the middle of the 1990s the Hambach I open pit is scheduled for a coal plus overburden volume of $350 \cdot 10^6$ m³ per year. If this output rate had to be achieved with equipment of the maximum dimensions existing until now, then so many equipment groups, each comprising a bucket wheel excavator, conveyor system and spoil stacker, would be required that difficulties in operation management would have arisen. Therefore a new equipment generation was developed for Hambach with an output capacity of 240,000 m³ per day.

2.2 Description of the 240,000 m³/day Equipment

2.2.1 Bucket wheel excavators

Bucket wheel excavators with a daily capacity of 240,000 bank m³ are the largest earthmoving units existing in the world today (Fig. 1).

Their design is characterized by the tripartite arrangement of the overall plant, — the actual excavator, a connecting conveyor bridge, and separate mobile loading unit — which has been introduced for all high-capacity bucket wheel excavators. The connecting bridge is a telescopic structure, its ends resting on the excavator and loading unit. The arrangement represents a considerably improved serviceability of the excavator plant; for instance, the face conveyor and loading unit may be placed 14 m higher or 16 m lower than the excavator part; set-up movements of the excavator in partial block digging may proceed without having to move the loading unit. From one bench level, in high and deep cut and intermediate high and deep steps the bucket wheel can then dig a total face height of 98 m.

Among the technical data of the bucket wheel excavator plant, particularly impressive figures are the total length of about 200 m, a height of 87 m, bucket wheel diameter of 21.6 m, and bucket wheel drive rating of 3360 kW, as well as the total weight of the plant in operating order which is 13,400 tons [3].

2.2.2 Conveyors

All belt conveyors in the open pits are shiftable, so that they can follow the advance of mining without major loss of time.

At the Hambach mine, the belt width in the shiftable conveyors is 2.8 m and the belt speed is 7.5 m/sec. A conveyor may be equipped with up to six 2000 kW drives at the head and tail terminal, and its handling rate may be up to 37,500 t/h. [4].

Drive terminals weigh approximately 700 tons; return terminals weigh from 80 tons up to 220 tons depending on the installed equipment (Fig. 2).

Fig. 1: Bucket-wheel excavator with a capacity of 240,000 bank m³/day

Head and tail terminals, as well as the transition trusses of the drive terminals, can be shifted or transported over longer distances by transport crawler units.

The transport crawler system was originally developed by Krupp Industrie- und Stahlbau for conveyor systems of 3 m belt width, but is meanwhile being used also for smaller conveyors and semi-mobile crushers.

2.2.3 Boom stacker

The stacker of the 240,000 m³/day equipment generation is a three-conveyor design. It has a feeding conveyor whose one end rests on a crawler-mounted supporting carriage while the other end is mounted in the slew center of the stacker, followed by an acceleration belt and the actual discharge conveyor (Fig. 3). Overburden is transferred from the dump conveyor to the stacker feeding conveyor by a crawler-mounted tripper car with integrated slewing belt.

Among the technical data of the boom stacker, the most noteworthy are a discharge boom length of 100 m, feeding boom length of 80 m, and more than 60 m total height of the machine. Belts of the feeding and accelerating conveyor are 3.2 m wide and run at a speed of 5.2 m/sec, while the discharge belt is 3 m wide and has a speed of 7.8 m/sec. The working weight of the stacker is about 5,700 tons [3].

3. Semi-Mobile Crushing Plant

3.1 Development of the Semi-Mobile Crusher System

In today's open pit hard rock workings, overburden and pay minerals are blasted from the mining face and loaded onto dump trucks by power shovels or wheel loaders. The trucks take the overburden to the dump and deliver the mineral to the processing plant.

Truck haulage in open pits provides great flexibility for the mining operations. Trucks can collect different grades of mineral from various points of the mine and deliver them to the processing plant in such a way that the mixture is uniform at all times.

On the other hand, truck transport is much more costly than conveyor transport, and the comparison between the two systems has recently shifted even more in favour of conveyors due to the soaring fuel costs. Operating costs of a truck fleet tend to increase especially if steep gradients have to be negotiated on the exit route from the pit.

One possibility of combining the flexibility advantage of dump truck transport with the low cost of conveyor systems is the use of semi-mobile crushers followed by conveyors for onward haulage [5, 6].

In this system, too, overburden and pay mineral are blasted from the face. The material is then loaded onto trucks by

Fig. 2: A 4 x 2,000 kW conveyor drive unit being moved by a crawler-mounte

Fig. 3: Boom stacker with a capacity of 240,000 bank m³/day

shovels or wheel loader. The trucks take the material to the semi-mobile crusher set up in a central location within the pit, so that no great gradients occur on the truck route.

In the crusher, the material is sized down sufficiently for conveyor transport, and delivery to the conveyor system for onward haulage to the dump sites outside the pit, or to the processing facilities.

The semi-mobile crushers are relocated from time to time to follow the advance of mining; the new site is chosen so as to provide optimum truck access.

This concept maintains the flexibility of truck transport; pay minerals can be collected from the different mining points and taken to the processing plant in the grades momentarily required for the beneficiation process. It also makes use of the cost advantages of conveyor transport, especially where major differences in level have to be overcome to take the material out of the pit to surface.

Depending on the pit layout and mining advance, the crusher has to be relocated only every six months to two years. These time intervals make it expedient to use a semi-mobile crushing plant which is relocated by a transport crawler.

While in operation, this crushing plant is set on two or three pontoon bases spaced with sufficient clearance between them for entry of the transport crawler. For relocation of the crusher, the transport crawler lifts the crushing plant by means of its built-in jacking system, and carries it to the new site.

3.2 Application Example: An Open Pit Lignite Mine in Spain

The Meirama lignite deposit is located in a narrow valley near La Coruna in north-western Spain. Its overburden is partly interspersed with granite boulders and shale horizons; in addition, solid granite strata occur at the edges of the deposit, some of which has to be removed along with the mining operation.

The ordinary overburden is stripped by bucket wheel excavators, and taken to the dump site by conveyors. The granite and shale are broken up by blasting, the lumps are loaded onto trucks by power shovels, and delivered to the semi-mobile crushing plant. The sized-down overburden is discharged from the crusher onto the overburden conveyor system and taken to the dump site together with the ordinary overburden.

The core of the crushing plant is a Krupp jaw-type gyratory crusher No. 135—190, Type Esch, with lateral enlarged feed opening for a throughput rate of 600 t/h; maximum feed size is 1,800 mm x 1,250 mm x 1,000 mm, product size is 0 to

Fig. 4: Transfer of a semi-mobile crushing plant by a crawler-mounted transporter

250 mm. Material is fed to the crusher by a reciprocating plate feeder from a bin into which the heavy trucks dump the material.

The weight of the crushing plant in ready operating order is 500 t, it is 17.6 m high, 8 m wide and 36.9 m long. The crushing plant and the ramp structure for heavy truck access to the bin are moved by a transport crawler, Type R 500.

Fig. 4 shows the crushing plant being moved by the transport crawler to its new destination. According to present plans the crushing plant is scheduled to be moved every one to three years.

References

[1] Leuschner, H.-J. "Planungskriterien für den Aufschluß des Braunkohlentagebaues Hambach", Braunkohle, Nr. 24, S. 41 (1972)

[2] Leuschner, H.-J. "Der Braunkohletagebau Hambach — eine Synthese von Rohstoffabbau und Landschaftsgestaltung", Braunkohle, Nr. 28, S. 111 (1976)

[3] Krumrey, A., "Schaufelradbagger und Absetzer für 240000 m³ Tagesleistung", Braunkohle, Nr. 28, S. 90 (1976)

[4] Sartor, W., "Die Entwicklung der 3-m-Bandanlage", Braunkohle, Nr. 31, S. 267 (1979)

[5] Korak, J., "Vergleich der Transportmöglichkeiten Schwerlastkraftwagen mit Kombination fahrbare Brechanlage und Gurtbandanlage für den Haufwerktransport im engeren Festgestein-Tagebaubereich", Braunkohle, Nr. 32, S. 229 (1980)

[6] Rixen, W. and Benecke, K.J., "Energy saving ideas for open pit mining", World Mining 34—6, P. 84 (1981)

All articles published in this book have appeared originally in the bi-monthly journal **bulk solids handling** — The International Journal of Storing and Handling Bulk Materials, the leading technical journal in this field throughout the world. If you are interested in receiving a free sample copy or information on subscription rates, please write to:

Trans Tech Publications, P.O. Box 1254, D-3392 Clausthal-Zellerfeld, Federal Republic of Germany.
Telex: 9 53713 ttp d

The Best of bulk solids handling:

Mechanical Conveying, Transporting & Feeding

Summer 1986
250 pages, US$ 26.00
ISBN 0-87849-066-3

This publication comprises over 30 technical papers written by internationally known experts on various aspects of mechanical conveying and transporting techniques.

An important book for mechanical and process engineers who have to move bulk solids.

Contents

H.H. Oehmen, *Univ. of Hannover, Germany:*
Theory of Vibrating Conveyors

A. Musschoot, *General Kinematics Corp., USA:*
Vibrating Feeders for Stockpiling and Reclaim

P.E. Henriksen, *Skako A/S, Denmark:*
Feeder Unit with Efficient Activation of Bulk Solids in the Storage Bottom

Ch. Fuchs, *Joest GmbH, Germany:*
Unbalance Motors — The Reliable Drive for Modern Vibratory Systems

G.D. Dumbaugh, *Kinergy Corp., USA:*
The Evolution of the First "Universal" Vibratory Drive System for Moving and Processing Bulk Solid Materials

D.K. Reynolds, *International Combustion Australia Ltd., Australia:*
Evaluation and Selection of Feeding Equipment for Bulk Materials

K.W. Lonie, *MINENCO Pty. Ltd., Australia:*
A Practical Model for Sub-Resonant Vibratory Feeder

G.D. Dumbaugh, *Kinergy Corp., USA:*
The "Induced Vertical Flow" of Bulk Solids from Storage

R. Wahl, *Vibra Screw Inc., USA:*
The Bin Activator

L. Rombach, *Maschinenbau Louise GmbH, Germany:*
Rotary Discharge Machines for Reclaiming Bulk Materials from Stockpile

B.R. Gabbett, *Redler Conveyors Ltd., U.K.:*
New Concept in Coal Feeders

K.H. Koster, *Univ. Essen, Germany:*
Centrifugal Discharge of Bucket Elevators

G.J. Beverly, A.W. Roberts et al., *Univ. of Newcastle, Australia:*
Mechanics of High Speed Elevator Discharge

W.B. Anderson, *Rexnord Inc., USA:*
A Maintenance-Free Splice for Belt Bucket Elevators

P.E. Wingrove, *Bethlehem Steel Corp., USA:*
Steep Angle Conveying of Coal Refuse

K. Shibata, *Shinko Kiko Co., Ltd., Japan:*
Universal Tray Lifter

S.P. Paul, *Cement Corp. of India Ltd., India:*
The Latest Concepts of Mined Rock Handling and Transport Systems in the Indian Cement Industry

D. Hill, *Strachan & Henshaw, U.K.:*
Developments in Wagon Tipplers

C.M. Rader, *Heyl & Patterson, USA:*
Prototype Railcar Dumping System

W.H. Kerr, *Bateman Materials Handling Ltd., South Africa:*
Rapid Train Loading Practices at South African Export Coal Mines

B.W. Hall, *State Rail Authority of N.S.W., Australia:*
Transport of Coal by Rail in the Hunter Valley, N.S.W.

BHP Engineering, Australia:
Train Loading at the Saxonvale Coal Mine

T.S. Cooke, Jr., *Ortner Freight Car Co., USA:*
Railroading of Phosphate

P.K. Chakravarty, *Tractel-Tirfor, India:*
Mechanised Composting of City Refuse

D. Bullivant, *BRECO, U.K.:*
Modern Aerial Ropeways and the Environment

V. Spyer, *Pohlig Heckel do Brasil S.A., Brazil:*
Aerial Cableways as a Transport Mode in Brazil

D. Werner, *Univ. of Karlsruhe, Germany:*
Investigations of a Model for a Circular Bag-Filling Machine for Cement

J.C. Thorne, *Reed Medway Packaging Systems Ltd., U.K.:*
The Filling, Closing and Handling of Sacks

J.D. Bapat et al., *Cement Research Institute of India, India:*
Appropriate System Design for Bulk Packaging of Cement

W. Binzen, *Dravo Wellman Co., USA:*
The Integrated Approach to Dust Collection

R.N. Slater, *Telemotive U.K. Ltd., U.K.:*
Radio Control in Bulk Solids Handling Applications

D. Wünsche, *Raco International, Inc., USA:*
Electric Cylinders in Bulk Material Handling

Order Form

Please send me the book: **Mechanical Conveying, Transporting & Feeding (G/86).**
I enclose cheque drawn to a U.S. bank in the amount of US$ 26.00.

Name: ______________________ Title: ______________________

Company/Institution: ______________________

Address: ______________________

City: ______________________ Country: ______________________

Cheque No.: ______________________ Signature: ______________________

Pro-forma invoice upon request.

Please mail this Order Form together with your cheque to:

TRANS TECH PUBLICATIONS
P. O. Box 266 · D-3392 Clausthal-Zellerfeld · West Germany
Tel. (0 53 23) 4 00 77 · Telex 9 53 713 ttp d

 Volume 2, Number 1, March 1982

Combined Mining Systems for Open Pit Mines

R. Franke, Germany

Summary

Production rates in opencast mines will have to be increased in the future, especially in the energy sector. Reasons for this development are an increased demand for coal, oil-sand and oil-shale, the steadily growing share of opencast mining operations as opposed to a decreasing share for underground mining, and the increase of the overburden to pay mineral ratio.

Mining systems can be classified into discontinuous operations (for small to medium mass movements) and continuous operations (for medium to very high mass flows). Due to the cost advantages of continuous systems, a tendency towards the application of combined mining systems becomes obvious, in which loosening and loading is executed by discontinuous methods whereas transport and dumping of the spoil is effected by continuously operating equipment.

The various possible combined opencast mining systems are presented and explained by practical examples.

1. Introduction

The energy sector occupies with 49 % a dominant position in todays total mass movement (including overburden) of world mining operations which can be estimated to be around $27 \cdot 10^9$ t. Other important fields are sand, gravel and aggregates (31 %), iron-ore mining (9.5 %), and copper mining (5 %), whereas the rest such as phosphate, bauxite, uranium, heavy minerals etc., only amounts to 5.5 %.

Concerning the growth rates which can be expected for the years to come, data from various sources indicate that there will be hardly any growth in the field of sand and gravel. All estimates made during the last years and decades for the iron ore industry have proven to be too high. The development in copper mining cannot be predicted with any certainty. The absolute volume of mass movement for phosphate, uranium and bauxite will surely increase, but considering their low share, this increase will not create a real challenge for the future.

Dr.-Ing. R. Franke, Managing Director, Mannesmann Demag Construction Machinery, P.O. Box 180362, D-4000 Düsseldorf 13, Federal Republic of Germany

Based on a paper delivered at the Conference on Conveying Technology TRANSMATIC 81, September 30 — October 2, 1981, organised by the Department of Conveying Technology (Institut für Fördertechnik), University of Karlsruhe, Fed. Rep. of Germany

A completely different situation can be found in the energy sector. Based on the world energy consumption of the year 1980, i.e., $7.9 \cdot 10^9$ t bituminous coal units (conversion of the consumption of all primary energy carriers to the equivalent consumption of bituminous coal), the actual estimates predict an increase of 2.5 % p.a. (i.e., about 50 % of the consumption increase between 1965 to 1973). This figure considers both a considerably lower growth rate of the gross national product and also the recognizable tendency that an increase in GNP will, in the future, go in hand with a considerably lower increase in energy consumption.

According to estimates the total consumption for the year 2000 will be around $19 \cdot 10^9$ t bituminous coal units. The share of oil will decrease while natural gas and water power will remain constant. The share of nuclear power is estimated to rise from 2 % to 10 %. The share of coal will increase slightly from 26 % to 28 %, and the share for synthetic energy, i.e., oil-sand and oil-shale, will rise from about 0 % in 1980 to 4 % in 2000.

In order to permit a calculation of the masses which must be moved in open pit mines of the energy sector in the year 2000, the following assumptions were made:

Share of open pit bituminous coal mining:

1980 = 38 % 2000 = 49 %

Overburden to coal ratio for bituminous coal:

1980 = 10:1 2000 = 16:1

Overburden to coal ratio for lignite:

1980 = 3:1 2000 = 5:1

Share of open pit synthetic energy mining:

1980 = 100 % 2000 = 60 %

Overburden to oil shale (oil sand) ratio for synthetic energy materials:

1980 = 1:2 2000 = 1:1

Based on these assumptions an increase in exploitation of pay minerals in the energy sector from $1.86 \cdot 10^9$ t in 1980 to $6.09 \cdot 10^9$ t in the year 2000 can be forecast (growth factor 3.2 = 6 % p.a.) The increase for overburden removal will be from $11.1 \cdot 10^9$ t in 1980 to $42.25 \cdot 10^9$ t in 2000 (growth factor 3.8 = 7 % p.a.). These figures prove the importance of opencast mining operations for the future.

In the following the main mining methods, i.e., discontinuous and continuous systems will be discussed and the advantages of combined systems will be explained by analyzing several planned or already operating open pit mining systems.

2. Discontinuous Mining Systems

The individual mining operations are executed by the following equipment (Fig. 1):

Loosening	Drilling rig blasting
Loading	Hydraulic excavator rope shovel wheel loader
Haulage	Off-highway trucks
Dumping	Off-highway trucks
or alternatively by	
Loosening	Drilling rig blasting
Loading, Transporting, Dumping	Dragline

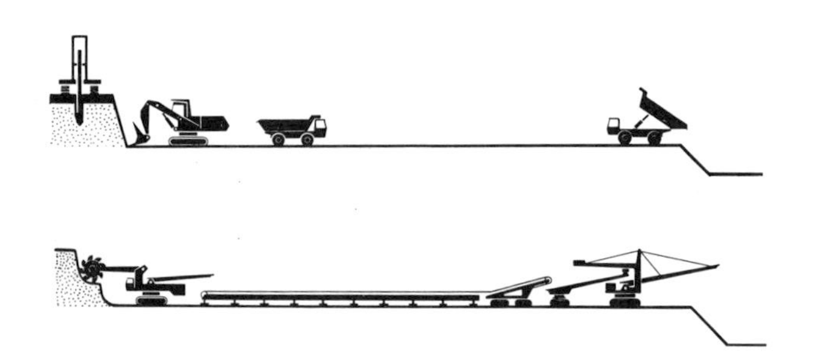

Fig. 1: Discontinuous (top) and continuous mining system

Typical applications for this system are:

- Mines with small to medium mass movements (100 — 7,000 t/h per haulage chain)
- Minerals to be mined: Lignite, bituminous coal, iron-ore, copper-ore, phosphate etc.
- Type of overburden: Unconsolidated, inhomogeneous formations, semi-consolidated and hard rock (cemented conglomerates, shale, clay, sandstone, limestone etc.)

3. Continuous Mining Systems

The individual mining operations are executed by the following equipment (Fig. 1, lower part):

Loosening, loading	Bucket wheel excavator bucket chain excavator
Transporting	Belt conveyor system (shiftable and stationary) conveyor bridge
Dumping	Spreader (stacker)

Typical applications for this system are:

- Mines with medium to very high mass movements 1,500 — 35,000 t/h per haulage chain
- Minerals to be mined: Lignite, bituminous coal (soft), chalk (soft), gypsum (soft)
- Type of overburden: Soft soil and unconsolidated formations (sand, gravel, loam, loess, clay, slightly cemented conglomerates).

4. Combined Mining Systems

Worldwide, discontinuously operating mining systems predominate. The extremely high costs of transportation and haulage, which have increased during the last 10 years from about 45 % to 60 % of total investment and operating costs for discontinuous systems, has led to the evaluation and introduction of combined mining systems. In these systems the operations "loosening" and "loading" are executed discontinuously, whereas the operations "transportation and haulage" and "dumping" are effected by continuously working equipment.

This tendency is also a consequence of an increase in mass movement due to higher production rates also in hard rock and semi-consolidated rock mines, an increase of mining depths (the present limiting depth is about 300 m, planned for the future are 600 m). Of major importance are also the increasing thicknesses of overburden and especially the disproportionate rise of the price of diesel fuel and tires.

In the following the various combined mining systems are discussed in detail:

4.1 Combined Mining Systems with Intermediate Truck Haulage

4.1.1 Direct feeding onto conveyor (Figs. 2 & 3):

Excavator — heavy truck — storage trench — reclaimer — conveyor

In order to be able to use conveyor belt transportation in a certain mine area of a Canadian oil-sand mine of Suncor, in which oil-sand is mined with rope shovels, Mannesmann Demag supplied, in 1979, a trench reclaimer with subsequent belt conveyor system.

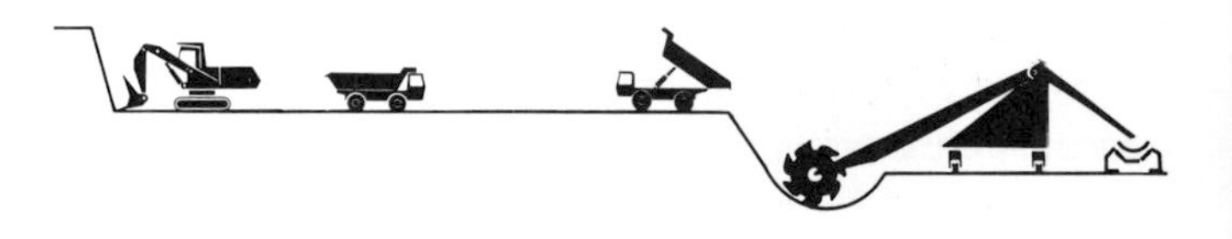

Fig. 2: Combined mining system with intermediate truck haulage and storage trench reclaimer

The oil-sand has an approximate oil content of 14 % and is interbedded with individual large boulders. This material has to be blasted during the winter months due to the prevailing permafrost conditions with a freezing depth of several meters. In the winter time, oil-sand has to be transported in large, sharp-edged lumps and in the summer time the oil-sand has the consistency of softened asphalt.

Fig. 3: Storage trench reclaim system at Suncor, Canada

The intermediate transport to the 1,100 m³ storage trench is accomplished using 120 t trucks. The trench, serving mainly as an intermediate storage for equalizing the material flow, is supported by sheet piles on the dumping side only and was dug by the trench reclaimer itself. The bucket wheel reclaimer with a guaranteed capacity of 4,000 t/h is of the luffing boom type with which the oil-sand in the trench can be reclaimed in 4 cuts. Operation of the machine can be automatic or manual. Special design features can be found in the following areas:

— Design of the bucket wheel and the buckets. Boulders with an edge length of up to 1 m have to be reclaimed and oversize material must be rejected.
— Design of chutes, belt conveyor system and transfer points. The chutes can be sprayed with Kerosene as well as heated. The lumpy material required special impact idlers at transfer points.
— Easy erection and dismantling must be possible as the machine has to operate at different positions in the mine.

This combined mining system works successfully and might serve as a model for other operations involving direct feeding onto belt conveyors.

4.1.2 Grading prior to feeding onto conveyor (Fig. 4)

Excavator — heavy truck — grizzly — conveyor

An installation supplied by Kawasaki for a mine of the Kobe Development Board (Myodani, Japan) represents a good example of a system including truck transport, grading and feeding onto a belt conveyor. This plant has a capacity of 2,500 t/h and is fed by trucks of relatively low payload, dumping at short cycling times at 6 dumping points located side by side, in order to obtain the desired equalizing effect for feeding the conveyors. The material slides over a grizzly where the minus 200 mm material passes directly onto the belt conveyor. Oversize material (approximately 16 %) slides onto a second conveyor after having been slowed down by a chain curtain. This second conveyor discharges into a crusher with a throughput of 400 t/h, with subsequent feeding onto the main conveyor. The height difference between dumping level and lower belt is 12 m.

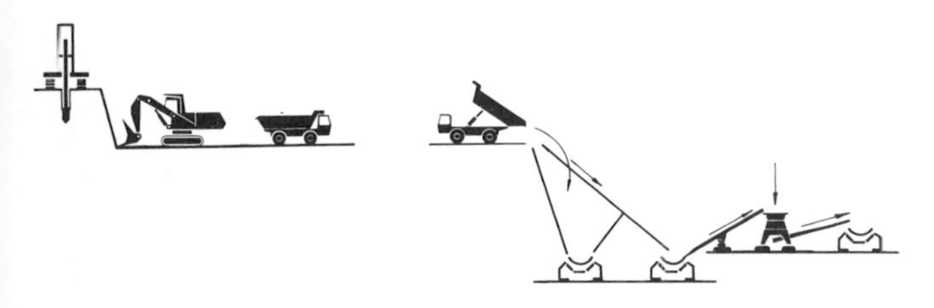

Fig. 4: Combined mining system with intermediate truck haulage and grading of material prior to loading onto belt conveyor

This considerable height difference and the lumpy material require special design features for the discharge conveyor, which was supplied by Bando under the name Rock Belt. The belt width is 2,000 mm, the thickness of the steelcord belt is 40 mm, with 20 mm top cover and 10 mm bottom cover. The special strengthening corresponds to a class St 3000 belt. This belt is said to be extremely resistant to impact. The non-troughed belt is supported on intermediate frames which carry the spring supported impact idlers of 250 mm diameter with 350 mm spacing, and which are themselves supported by the main structure via spiral springs. The belt speed is 0.5 m/sec.

Experience so far allows an estimate of useful life of approximately 3 years. The belt was turned after about 1.5 years, since it was worn out on one side due to the irregular impact of the material.

4.1.3 Crushing prior to feeding onto conveyor (Figs. 5 & 6)

Excavator — heavy truck — mobile crusher — conveyor

A suitable model for the operation of a semi-mobile (in-pit) crusher fed by heavy trucks and connecting to a continuous

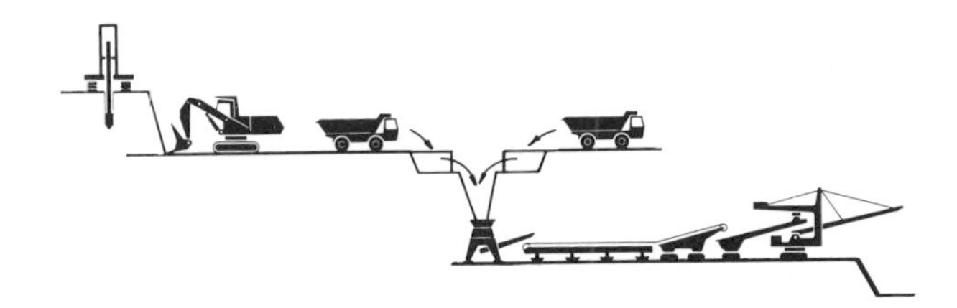

Fig. 5: Combined mining system with intermediate truck haulage und mobile crusher

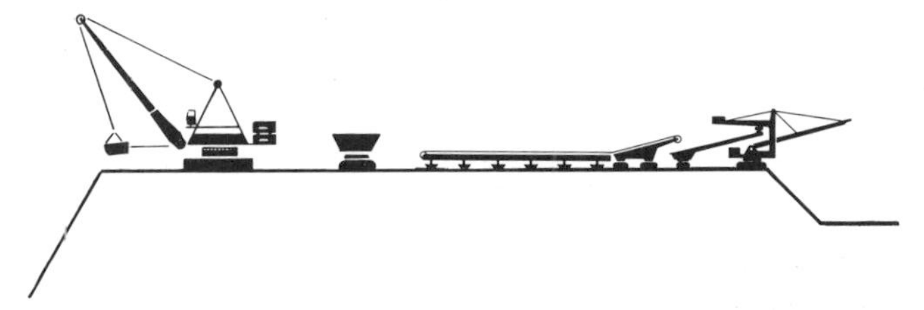

Fig. 6: Discontinuously working dragline feeding material via hopper car onto belt conveyor

transport and dumping system is the overburden mining system at Sishen, an iron-ore mine of ISCOR (South African Iron and Steel Corp. Ltd.).

This installation was built by Krupp Industrie- und Stahlbau in early 1980 and represents the biggest and most modern plant ever as far as throughput is concerned. According to the Tender Specifications it comprises three major sections:

1. Heavy trucks feeding a crusher and crusher discharge
2. Belt conveyor system for transport out of the pit
3. Belt conveyor system on the spoil pile and spreader.

A gyratory crusher with a throughput of 6,000 t/h, positioned in a trench between two retaining walls, is fed from two sides by heavy trucks of 155 t payload dumping material into the 170 m³ feed hoppers.

A Nordberg crusher which can accept lumps of 1.2 m diameter and "fishes" of 2.4 m length discharges into an outlet hopper of 150 m³ content. A discharge conveyor transfers the material onto the 1,800 mm belt conveyors running underneath the crusher and effecting the transport out of the pit. These conveyors are positioned in two sections of 500 m length each. They can be considered as quasi-stationary. The belt conveyor line can be extended on the crusher side when the crusher is relocated at intervals of several years, depending on the advance of the mining operations.

The dumping installation on the spoil consists of two 1,800 mm belt conveyor systems located at right angles to each other (one shiftable and one extendable section), one tripper car and one crawler-mounted spreader with a boom length of 56 m connected with the belt conveyor system via a 30 m long conveyor bridge.

A transport crawler with 1,000 t carrying capacity is an important component of this system and serves to relocate the crusher with supporting structure designed as a 3-point supported ring girder. Other parts, such as feed hoppers, service cranes etc. have to be dismantled and transported separately. Retaining walls and foundations have to be removed.

This installation represents a completely new concept as far as design and size are concerned.

4.2 Combined Mining Systems Without Intermediate Transport

4.2.1 Direct feeding onto conveyor (Fig. 7)

Dragline — hopper car — conveyor

For the combined opencast mining systems without intermediate transport by trucks one example is given here of an installation with direct material feeding onto conveyors, without grading or crushing.

The Fortuna opencast mine of the Rheinische Braunkohlenwerke, Germany, uses the continuous mining method with bucket wheel, belt conveyor and spreader. However, a dragline operation can be found on the deepest level of this mine where the coal formation is influenced by several faults. On this level a bucket wheel excavator has been working for several years, with an output of 1,040 bank m³/h in a 20 m high cut, as well as a Ruston Bucyrus RB-480-W dragline with an effective capacity of 600 bank m³/h. The dragline with a 66 m boom length is equipped with a 15 m³ bucket and has a digging depth of 40 m. With a deadweight of 775 t this machine is lighter and cheaper than a bucket chain excavator of identical capacity and digging depth.

Fig. 7: Semi-mobile crusher at ISCOR, South Africa

The dragline transfers the material — and this is the critical operation — into the 100 m³ crawler-mounted hopper car with a hopper opening of 7.5 m by 6.4 m. Despite these dimensions it is sometimes difficult to position the bucket above the hopper. Discharge from the hopper is by means of an apron feeder running at 0.2 m/sec. The service weight of the hopper car is around 400 t. It became obvious that the belt conveyor system should have a width of 1,600 mm due to the lumpiness of the material which is naturally considerably larger than if mined by a bucket wheel excavator. The belt speed can be relatively high, i.e., 5.2 m/sec if catenary idlers are used.

4.2.2 Grading prior to feeding onto conveyor (Figs. 8 & 9)

Excavator — grizzly — hopper belt wagon — conveyor

A real simple solution for combining discontinuous and continuous transport systems was presented by Mannesmann Demag at the Mining Exhibition Bergbau '81 in Düsseldorf, Germany. This system comprises two hydraulic excavators Type H 241 and one so-called hopper spreader HS 5000-56, i.e., a hopper belt wagon with a very long feeding boom, which can operate as a spreader and dump material directly into the mined out pit areas, similar to the operation of a dragline.

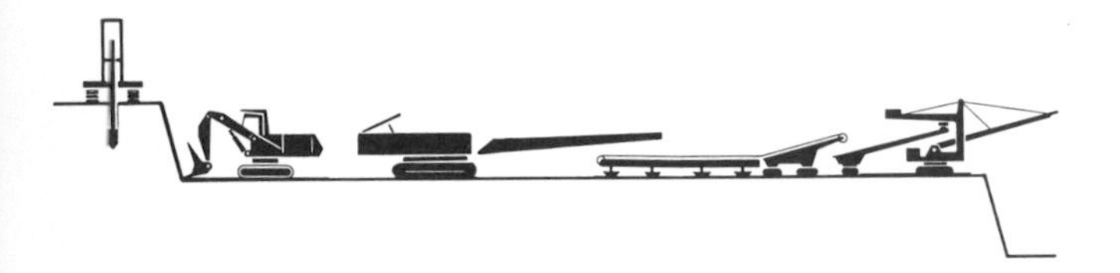

Fig. 8: Grading by hopper belt wagon prior to feeding onto conveyor

In another and considerably more interesting alternative the feeding boom establishes the connection to a shiftable belt conveyor system which in this special case is fed via a small hopper car running on the shifting rails of the conveyor system.

The hydraulic excavator H 241, especially developed for mining operations, has already been in operation for two years in several installations with very good results. These excavators were supplied as diesel-hydraulic and electro-hydraulic versions with a drive power rating of 1,000 kW. Their service weight is 270 t. The excavators can be equipped with buckets of 10 m³, 14 m³ (for overburden) and 21 m³ (for coal). It is envisaged that two excavators will alternately feed into the hopper belt wagon (at approximately 30 sec intervals) via a hydraulically collapsible grizzly with a 400 mm mesh width. The hopper can take the contents of two buckets.

A discharge conveyor with an infinitely variable discharge speed of maximum 0.5 m/sec controls the loading of the 1,400 mm wide boom belt which runs at a constant speed of 4.5 m/sec.

The maximum outreach of the 56 m boom (rotating center to center discharge pulley) guarantees sufficiently long time intervals between shifting of the bench conveyor line.

Lifting and lowering of the boom is effected by two hydraulic cylinders; the boom itself is a latticed tubular construction. The crawler assembly with 7.5 m gauge and 7.5 m centre to centre tumbler length permits a low ground pressure of 8 N/cm² with a service weight of the machine of 300 t. Drive motors of 700 kW as well as hydraulic motors and pumps are, as far as possible, part of the hydraulic excavator program.

In order to avoid off-center belt run during unfavourable ground conditions, the total superstructure is tiltable around the longitudinal axis (boom direction) by means of hydraulic cylinders.

The equipment combination of H 241 and HS 5000-56 has found considerable interest since its presentation at Bergbau '81.

Fig. 9: Model of Mannesmann Demag "hopper spreader" HS 5000-56

4.2.3 Crushing prior to feeding onto conveyor (Fig. 10 & 11)
Excavator — mobile crusher — conveyor

The presentation of application examples will be completed with an installation supplied by PWH (PHB Weserhütte), Germany to the Grooteluk coal mine in South Africa. Here a mobile crusher with intermediate bridge and belt conveyor system is used for the transportation of the overburden. The total mass movement is around 32 million t/year with an overburden share of 8 million t. The overburden thickness varies between 15 to 20 m, it consists of 85 % shale and 15 % sandy soil and is blasted, excavated by P + H 2300 rope shovels and discharged into a mobile crusher on walking pads. The mobile crusher with a weight of about 1,000 t follows the rope shovel at 25 m intervals.

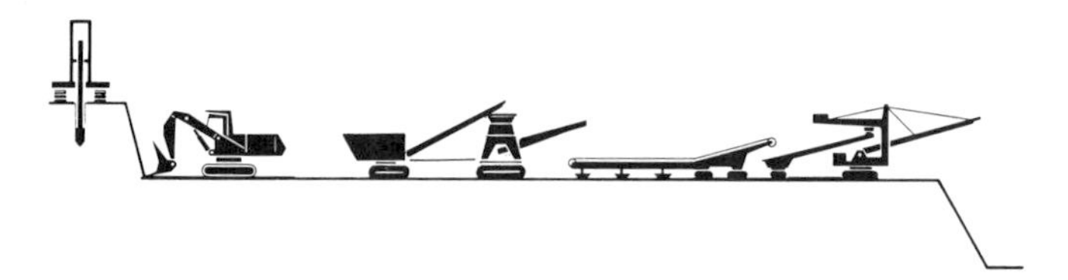

Fig. 10: Crushing by in-pit crusher prior to feeding onto conveyor

The total material passes through a gyratory crusher and is transferred via a 23 m long boom onto a crawler-mounted 110 m long conveyor bridge transporting the material to the bench conveyor system. The subsequent belt conveyor system consists of 4 sections with a total length of 3,600 m. The 1,400 mm wide belt runs at a relatively low speed of 2.7 m/sec. Considering the characteristics of the crusher, the belt system with max. 3,000 t/h is oversized compared to the required nominal rating of 1,800 t/h.

Crushing of the material is required due to the fact that the overburden has to be mixed in a subsequent stockpile arrangement with waste material from the coal processing plant, which still contains about 25 % carboniferous material, in order to avoid self-ignition.

Several alternatives to this total system were analyzed and compared with respect to investment and operating costs. Since the individual cost figures in Rand/ton are not significant, the comparison was related to a pure shovel — truck system (100 % value). The following data were worked out:

1. Heavy trucks + stationary crusher outside the pit + conveyor system — 75 %
2. Heavy trucks + mobile crusher outside the pit + conveyor — 75 %
3. Heavy trucks + mobile in-pit crusher + conveyor — 66 %
4. Selected solution — 42 %

Fig. 11: Mobile crusher with intermediate bridge at Grooteluk coal mine, South Africa

Volume 4, Number 3, September 1984

Planning Aspects for the Application of Continuous Transport Systems in Hard Rock Open Pit Mines

S. Kutschera, Germany

1. Introduction

Base metal prices are currently running at their lowest level for decades and the future does not seem to hold much promise of significant price increases.

Efficient ore removal can make an existing mine more profitable, or can make a marginal new deposit worth developing.

Currently, the majority of hard rock open pit mines use large haul trucks with the advantages of flexibility they offer as the life of the mine progresses.

However, recent experience has proved that, at certain depths, in-pit crushing and belt conveying can show significant cost savings. These cost savings are maximized if the mine plan takes into account both short-term and long-term needs by proper mine design.

Fig. 1 shows a comparison between typical truck haulage and in-pit crushing with belt conveying for capital costs in relation to pit depth.

Fig. 2 shows a comparison between typical truck haulage and in-pit crushing with belt conveying for operating costs in relation to pit depth.

The relative cost figures are based on typical installations and the break-even point would vary from mine to mine. However, as a general rule, the deeper the pit and the larger the daily tonnage mined, the more beneficial the belt conveying system will be.

The costs for the in-pit crushing and belt conveying system include:

- **Trucks**
  For transporting the run of mine material from the shovels to the in-pit crusher
- **In-Pit Crushing**
  Includes all equipment that is required over and above the crusher, which is common to both systems
- **Belt Conveyor System**
  To move the material from the in-pit crusher to the pit perimeter.

Dipl.-Ing. S. Kutschera is employed with Lurgi GmbH, Frankfurt/Main, Federal Republic of Germany

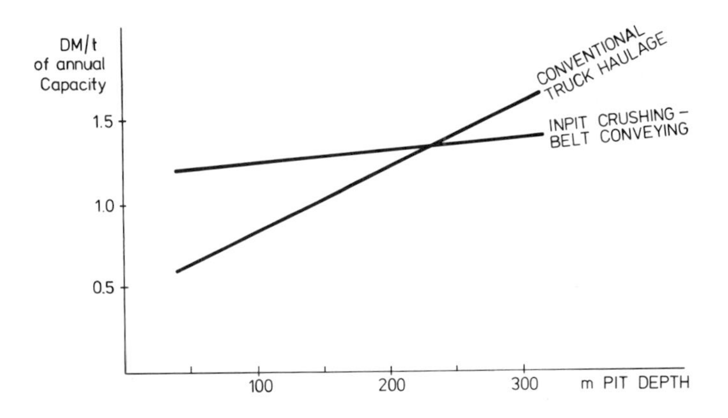

Fig. 1: Comparison of the capital costs of typical truck haulage with in-pit crushing using belt conveying, in relation to pit depth

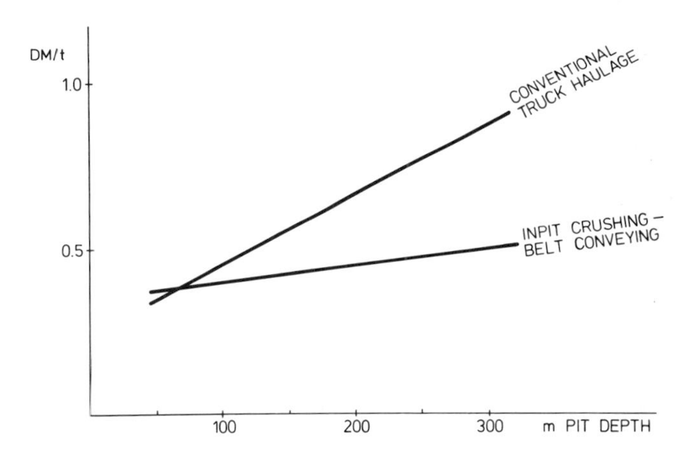

Fig. 2: Comparison of the operating costs of typical truck haulage with in-pit crushing using belt conveying, in relation to pit depth

As can be seen, operating costs are less for all but the shallowest pits, whereas the capital cost break-even point takes place at increasing depths.

However, if electric power generation is provided by some fuel other than imported oil, significant foreign currency savings may accrue. Additional savings may be made in devel-

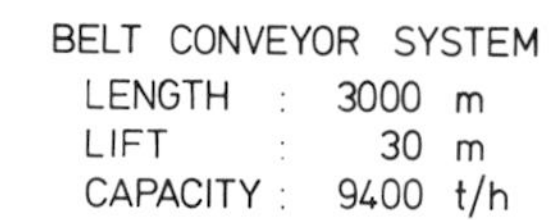

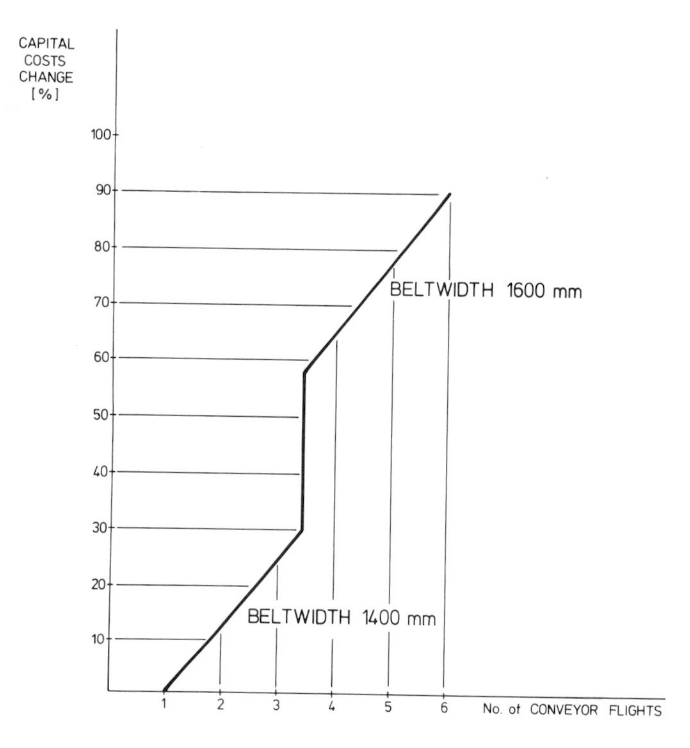

Fig. 3: Higher capital costs resulting from increased number of belt sections

oping countries if there is the ability to manufacture a portion of the conveyor systems locally, rather than importing the large haul trucks.

In spite of these advantages, the belt conveying system can be at a disadvantage if recognition of its unique requirements is not taken into account during mine planning.

The two most significant problems are:

a) Flexibility to react to deposit variations and changing operating requirements, and
b) System availability.

2. In-Pit Belt Conveying Systems

A wide range of publications provide information on the "state of the art" of belt conveying systems, including operational characteristics and technical data. Until just a few years ago, the use of belt conveying systems was limited to soft rock open pits, quarries and material handling. Now, however, the use of these systems for hard rock in pit transport is becoming more and more widespread.

What requirements with regard to material quality, operating size and haulage distances must be met to make the use of belt conveying systems economic?

In the vast majority of cases, it will be necessary to reduce the run of mine material from hard rock open pits to a size which reduces belt and idler wear and tear. Belt conveyors are highly efficient equipment for long-distance transport, and therefore should be used to convey large quantities of material. For small quantities and the transport of run of mine material over short distances, conventional truck haulage has been proven to be more flexible.

The mine plant design must consider the needs of the entire deposit for the life of the mine. This has a significant impact on the selection of proper excavating, crushing and conveying systems.

3. Mine Planning

A mine plant represents a dynamic system which is subject to constant change, depending upon the geology of the deposit as well as varying production requirements. The conveying system selected must be capable of adapting to the changing requirements. With conventional truck operation it is possible to react rapidly to changes relating to the deposit and operating conditions. However, (unless initially planned for) this can only be achieved with delays and additional expenses in the case of in-pit crushing and belt conveying systems.

A more detailed design is required for a belt conveying system than a conventional truck haulage plan, to ensure that the maximum cost advantage offered by the belt conveying systems is obtained. Belt conveyors represent long-term investments, which must be designed for use over the entire life of the pit.

The following example can be used to demonstrate the extent to which proper planning can influence not only the capital costs, but also have a significant impact on future operating costs of belt conveying systems. An open pit mine, which has been operating exclusively with truck haulage, is to be converted to an in-pit crushing and belt conveying system. For this purpose, a 500 m long belt conveyor is required, partially arranged on a ramp.

According to the progress of the mine, the belt conveyor is to be extended gradually to an ultimate length of 3,000 m. If, in this case, the final conveying requirements are taken into account during the initial planning, a straight conveyor belt course can be achieved for the entire life of the operation.

It is then theoretically possible to use just one belt conveying section. If this is not possible, or if future operating conditions cannot be determined accurately enough initially, installation of additional sections will be required in subsequent operations. If this is the case, one must accept the resulting higher capital costs stemming from the increased number of belt sections (see Fig. 3).

As each belt section requires its own drive head station and tail station, the shorter the belt system, the greater the influence these units have on the cost per meter of the system. Therefore, capital cost will increase with the number of conveyor sections.

This increase is caused not only by the high cost per meter described above, but also takes into account the necessity of increasing the system capacity in order to compensate for the decrease in overall system availability, since system availability is contingent upon the individual number of components that comprise the entire system.

In Fig. 4, the availability of the entire system is demonstrated as a function of the availability of the individual components. Because of the relatively low number of individual components, it will not be possible in practice to assume that failures will occur simultaneously. The down times of the

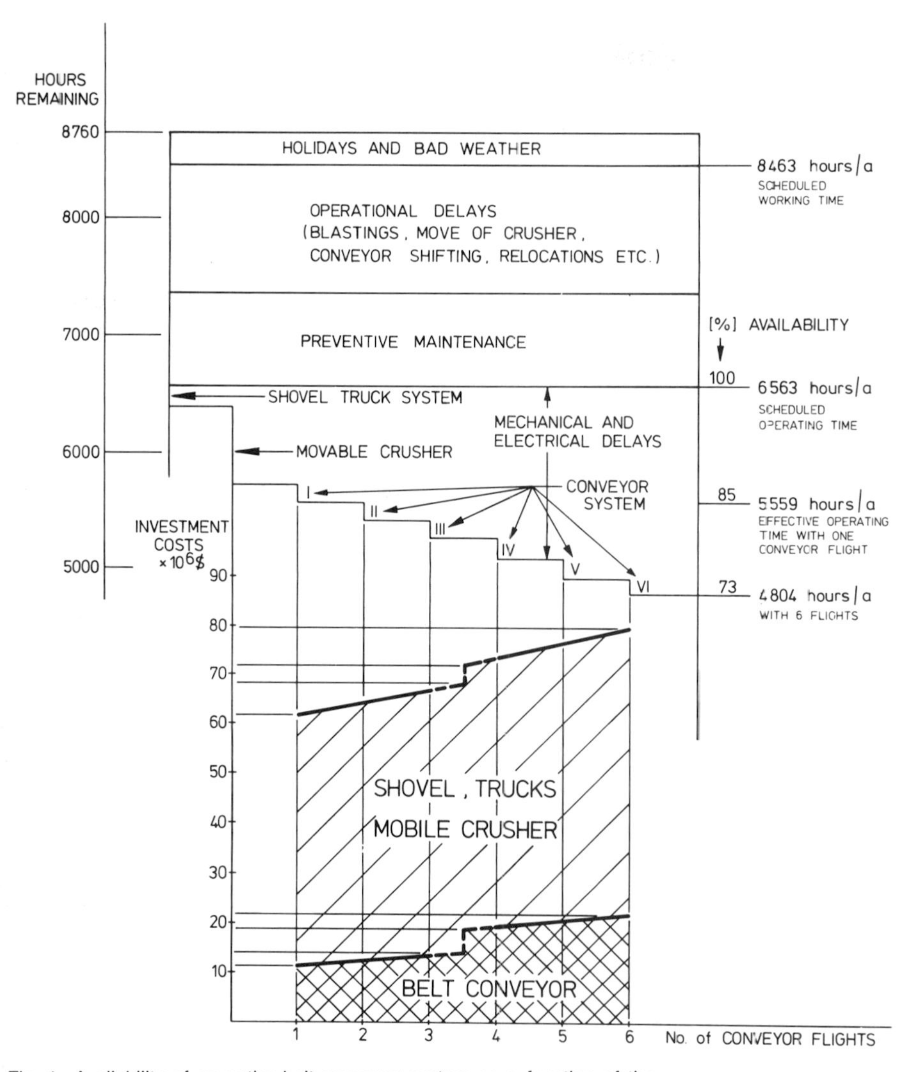

Fig. 4: Availability of an entire belt conveyor system as a function of the availability of the individual components

Fig. 5: Effect of an increase in the number of sections of a belt conveyor system on operating costs

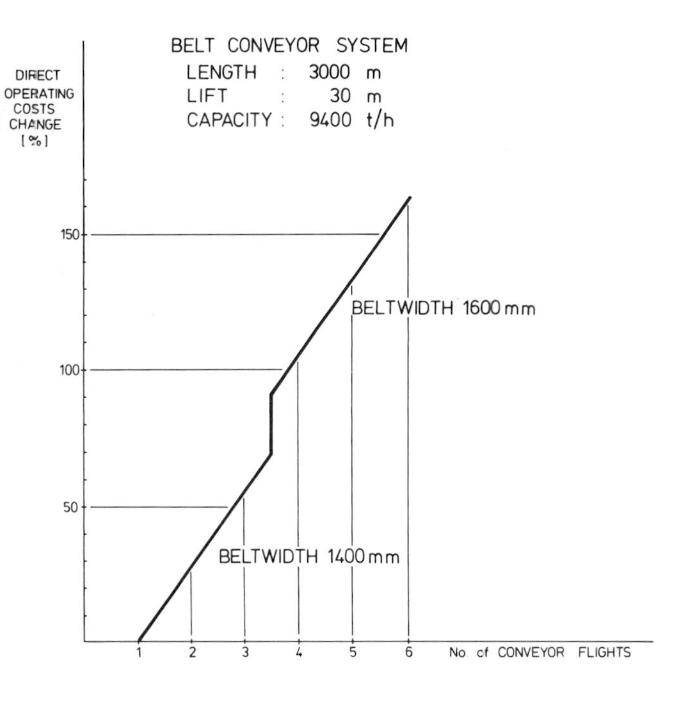

components are additive, and therefore have an impact on the effective operating time of the entire system each time an additional component is used.

This means that not only the haulage capacity of the belt conveyor system, but also the loading capacity of the excavators, the haulage capacity of the trucks for the horizontal intermediate transport, and the throughput capacity of the in-pit crushers, have to be increased accordingly. Fig. 3 demonstrates the influence of a capacity increase on the capital costs of the entire system. What is the result, for example, if four separate belt sections are needed instead of two? The availability of the entire system is consequently reduced from 82.4 to 77.8%, that is to say by 4.6%. The capital costs for the entire system, however, will increase by approximately $ US 10 million, or 16 %.

If one considers the progression of operating costs for the belt conveyor system (Fig. 5), it becomes clear that these increase more rapidly with a greater number of sections than do capital costs. This can be explained by the higher personnel requirements — each transfer point normally requires one man — higher energy consumption and, above all, belting costs.

Since wear and tear to the belt occurs mainly at the transfer points, and feeding onto a shorter belt increases belt wear, operating costs increase significantly with the number of sections. Fig. 6 shows the operating costs of the belt conveyor system, broken down according to different types of costs.

Volume 4, Number 3, September 1984

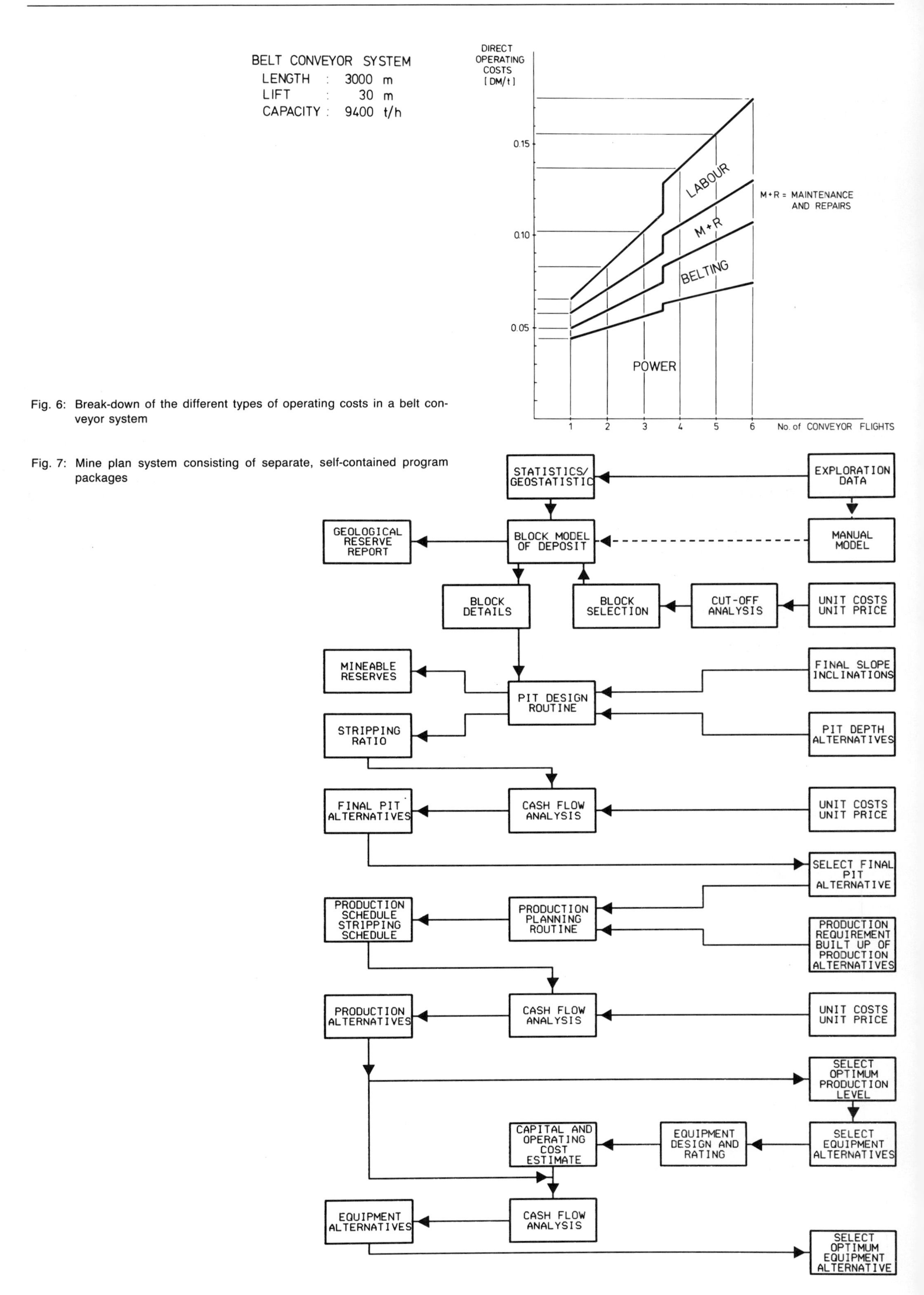

Fig. 6: Break-down of the different types of operating costs in a belt conveyor system

Fig. 7: Mine plan system consisting of separate, self-contained program packages

The fact that the cost of ore and overburden transport can now represent as much as 50 % of total mine operating costs clearly shows that proper planning can increase the profitability of a mining project. Proper planning becomes even more significant if we are to develop the low grade deposits in future with attendant unfavorable stripping ratios.

How can we meet the challenge of developing these low grade deposits while maintaining profitability? Lurgi believes the answer may be found with in-pit crushing and conveying systems. Whatever system is chosen, a detailed mine plan is a must.

The relatively high degree of sensitivity of this conveying system to changes in operating conditions requires extensive planning. In view of the various interdependencies and numerous variables, Lurgi have developed modern software to assist in identifying the optimum mine plant design concepts.

The prime component of our software is the computer simulated block model of the deposit. The model which exhibits the smallest degree of deviation from the exploration result and thus comes closest to actual conditions, will serve as a foundation for the block model.

The Reserve Report is established from the block model using various "cut-off" grades. The next step is to investigate the various mining approaches that can be applied in order to establish an optimum production plan. This mining concept must be capable of providing:

- Minimum overburden handling
- Proposed production rate
- Proposed production quality
- Optimized capital investment
- Low operating cost.

This calls for the development of a realistic mine plan using state-of-the-art computer programs. We find it helpful for the mining engineer to be able to influence all stages of the mine planning. It has been shown that a system design consisting of separate, self-contained program packages is easier to control than a fully integrated system. This method allows the mining engineer to analyze and adjust the results of each program prior to continuing with the next step.

A concept such as this (which is presented in Fig. 7), will enable the mining engineer to dispense with time-consuming manual calculations. The in-put by the engineer, as well as the experience and conclusions to be drawn from the results, can be channelled directly into the computer.

We believe that the "tools" to ensure optimized planning of belt conveying systems in hard rock mining are now available. However, one must appreciate the large amount of design work required (which practically amounts to "tailor-making" the conveying system).

In this connection, hard rock miners can learn from the experience acquired in soft rock mining. That is to say, they can use the experiences acquired in large bucket wheel excavator-belt conveyor-spreader systems in lignite mining. Planning in this area has always been far more extensive and detailed than in conventional truck haulage operations. The utilization of such large machines, which in dimensions and capacity have to meet the requirements of the deposit and the production plan over the entire life of the pit, call for in-depth planning to an extent that was previously unnecessary in hard rock mining. As shown, the use of high-efficiency belt conveying systems in hard rock mining calls for a similar degree of work and effort.

All articles published in this book have appeared originally in the bi-monthly journal **bulk solids handling** — The International Journal of Storing and Handling Bulk Materials, the leading technical journal in this field throughout the world. If you are interested in receiving a free sample copy or information on subscription rates, please write to:

Trans Tech Publications, P.O. Box 1254, D-3392 Clausthal-Zellerfeld, Federal Republic of Germany.
Telex: 9 53713 ttp d

The Best of bulk solids handling:

Stacking, Blending & Reclaiming of Bulk Materials

Summer 1986
250 pages, US$ 26.00
ISBN 0-87849-061-2

This publication comprises some 40 technical papers written by internationally known experts on the various methods of stacking, blending and reclaiming of bulk materials.

An important book for everyone engaged in dry bulk handling.

Contents

Order Form

Please send me the book: **Stacking, Blending & Reclaiming of Bulk Materials (B/86).**
I enclose cheque drawn to a U.S. bank in the amount of US$ 26.00.

Name: ____________________ Title: ____________________

Company/Institution: ________________________________

Address: ________________________________

City: ____________________ Country: ____________________

Cheque No.: ____________________ Signature: ____________________

Pro-forma invoice upon request.

Please mail this Order Form together with your cheque to:

TRANS TECH PUBLICATIONS
P. O. Box 266 · D-3392 Clausthal-Zellerfeld · West Germany
Tel. (05323) 40077 · Telex 953713 ttp d

 Volume 4, Number 4, December 1984

An Appraisal of the Correct Transport System for a Surface Mine

David Bullivant, U.K.

1. Introduction

In general, the objective of any mining operation is to extract the ore from the ground and transport it to the process plant at the minimum cost. It can be argued that the winning of the ore does add value to the product, but certainly the transportation aspect is only a pure cost factor and therefore must be achieved in the most economic way. This is particularly relevant to surface mining, where the usual operational steps, discounting the primary means of extraction, are loading by shovels, truck transport, discharging the material into the primary crusher for ore and conveying waste material onto the dump.

Off-road trucks represent a very flexible means of handling material within a surface mine and this feature has made them the most widely-used method of transportation. This came about during the period when the supply of diesel fuel was abundant and cheap; however, today the position is very different. The ten-fold increase in oil prices over the past decade means that transportation costs have to be examined and analysed more carefully by engineers throughout the world. This analysis has to be applied generally to the overall costs of operation with particular regard to the energy consumption.

2. Analysis of Energy Consumption

With a truck the relation of pay load to dead weight is about 1.5 to 1.0. The import of this is that 60% of the energy requirements, when hauling material, are needed to move the pay load. When the complete round trip is considered, the energy utilised in moving the pay load drops even further to about 40%. This means that a great deal of energy and therefore money, is wasted in just moving the truck around the area of the surface mine.

In general, a truck will consume 2 l of diesel fuel to lift a pay load of 100 t through 10 m, and 8 l to transport the same pay load 1,000 m on the level. These figures have to be generalised, as many factors affect the fuel consumption, not least the condition of the haul road surface, or rolling resistance. For example, an increase in rolling resistance, a measure of the force required to roll a wheel over the ground and hence dependent upon load and surface conditions, from 3% to 11% can result in the fuel consumed by one truck increasing by almost 50% and a doubling of the fleet fuel consumption to haul a given quantity of material, taking into account the reduced truck productivity and the larger fleet required.

Mr. David Bullivant is Manager — Surface Mining Operations, Babcock-Moxey Ltd., Gloucester, U.K.

In a study of a surface coal mine designed to produce 2,700,000 t/a, a fleet of eight 90-t haul trucks, together with graders, supporting vehicles and water wagons, was considered suitable. For the overburden removal, with a stripping ratio (bench m^3/t coal) of about 9:6, a loader/truck system was selected with a fleet of seventeen 77-t trucks and support vehicles. The analysis of fuel consumption over a year's operation produced the following figures:

Overburden handling	8,700,000 l
Fuel haulage	2,700,000 l
Total	11,400,000 l

With diesel fuel having about 3,800 kJ/l, the energy expended is $433 \cdot 10^9$ kJ or 160,000 kJ/t of coal mined. With a sub-bituminous coal of 4,500 kcal/kg, this means that the ratio of Energy Expended to Gross Energy Recovered is 1:117.

This in itself would appear wasteful and this is especially so when the cost of fuel per ton of coal is considered. Taking Diesel fuel at US$ 0.35/l, the fuel cost alone per ton of coal mined is US$ 1.48, with an annual fuel bill of US$ 3,990,000.

What is the alternative?

With the modern design of belt conveyor, the relation of pay load to dead weight, dependent upon belt size, ranges from 2 to 1 up to 4 to 1, an improvement over trucks from 33 to 260%. Including the return of the dead weight, some 75% of the total energy consumed is utilised in moving the material, which represents a considerable improvement in energy utilisation. As a result of this, a belt conveyor requires about 3 kWh to lift a pay load of 100 t through 10 m and 11 kWh to transport the same pay load over 1,000 m on the level. The comparison with truck costs is shown in Fig. 1 and this indicates that the margin of cost saving increases in favour of conveyors as both the depth of the mine and the transport distance increase, factors which are to be expected as the mine is developed.

Overall, the energy requirements for a conveyor installation can be expected to be about 20—25% of the comparable

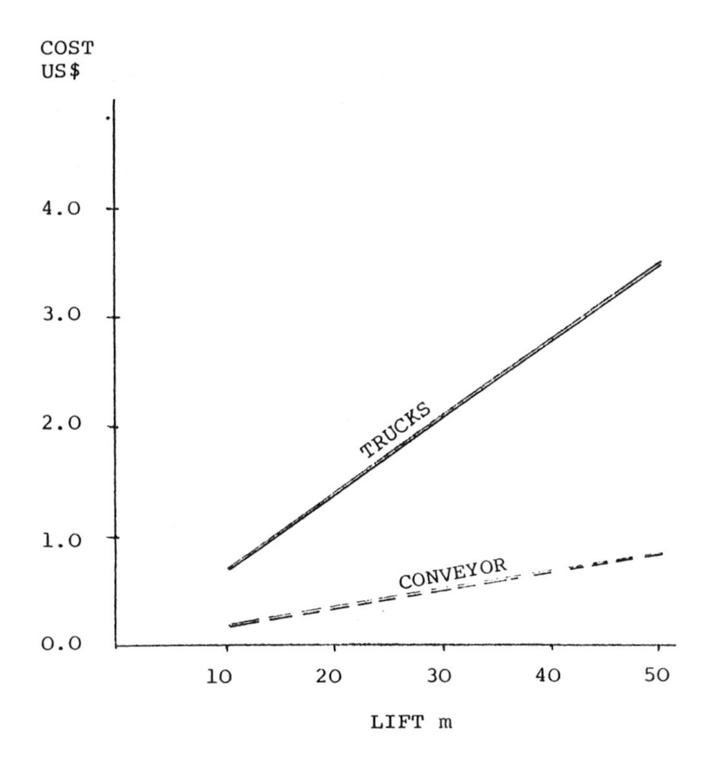

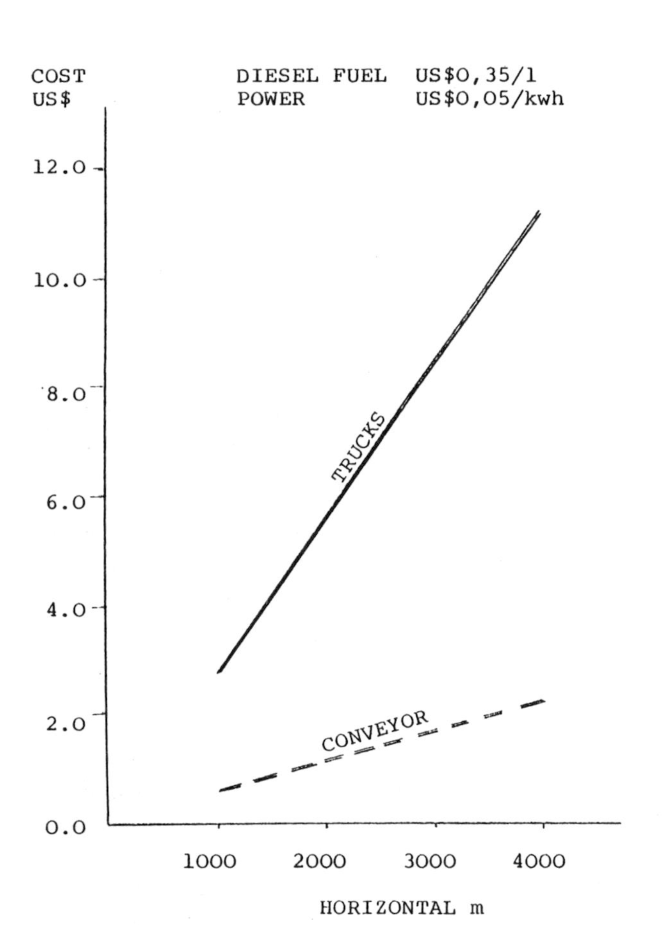

Fig. 1: Comparable costs to move a 100 t pay load

truck system. Using the previous figures, the energy expended is therefore about 35,000 kJ/t of coal and the ratio of Energy Expended to Gross Energy Recovered is 1:540, a considerable improvement on the truck situation.

It can be argued that energy in the form of either diesel fuel or coal has to be expended to generate the electricity used for the conveyor system. This is, of course, correct, but is it significant?

The direct savings in fuel costs between the two systems are about US$ 2,700,000 per year. Taking into account a captive diesel generator set, the net annual fuel savings for the conveyor system are about US$ 2,500,000 or US$ 0.93/t coal mined, which is very significant.

3. Conveyor System

When using a conveyor system it is essential that the material be reduced in size so that it can be handled on the belt. With belt sizes greater than 1,000 mm, the practical maximum material size is about 250—350 mm. In some cases, therefore, a crusher is required in front of the conveyor. It may be that the crusher will replace the primary crusher at the process or preparation plant, in which case the crusher will be sized accordingly. Otherwise, the function of the crusher is solely to produce the correct size for the belt, with process sizing being carried out at the plant.

The effect of the crusher on energy consumption is very slight as, dependent upon the required reduction ratio and material being handled, the power requirements will be of the order of 1.0 to 2.0 kWh/t.

Every surface mine has to be examined individually to determine the best mining plan and the arrangement of the crushing and conveying systems. It may be that the mine layout, particularly if more than one face is being worked at the same time, will be best served by having a limited truck haulage operation, transporting the material from the face to a semi-mobile crusher, which is moved infrequently as the face or faces advance. This arrangement maintains a certain flexibility of operation with reduced operating costs. Alternatively, a fully mobile crusher can be used to keep pace with the face movement. This aspect, together with the selection of the appropriate combination of the special conveyor designs available today (Fig. 2), has to be examined to produce the most effective and lowest operating and ownership cost system.

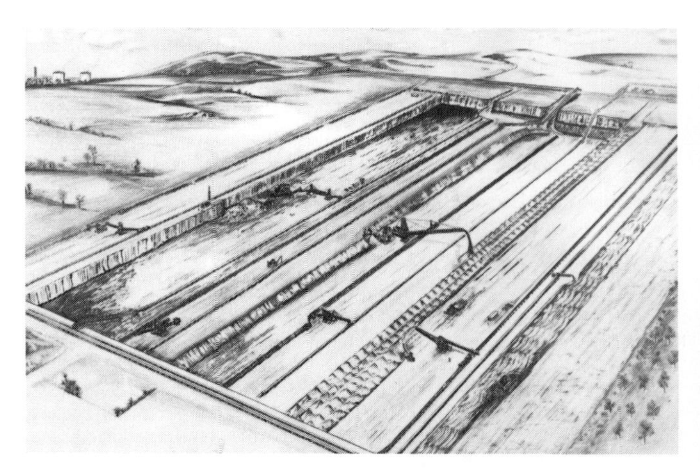

Fig. 2: Open-pit mine using various belt conveyors

It is clear from the foregoing discussion that the conveyor based system is the most energy effective method of material transport in a surface mine. It is also clear from an analysis of capital costs and operating costs that, in many instances, the conveying system is also the most economic (Fig. 3).

Further analysis can be used to demonstrate the effect on total cost of capacity and operating depth (Fig. 4).

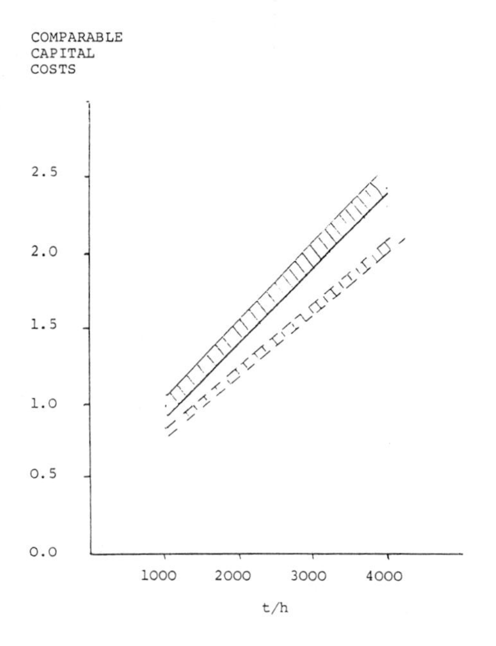

Fig. 3: Comparison of transport systems
Trucks — loading shovel and dump trucks to crusher
Conveyors - - - face shovel direct to crusher and conveyor system

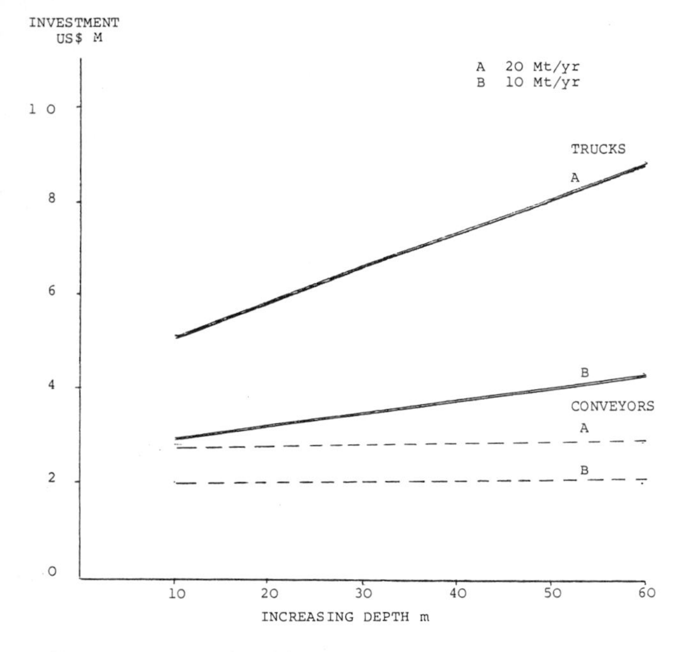

Fig. 4: Effect on investment cost of mine depth and capacity

4. Availability

The supporters of trucking in surface mining often argue that trucks provide a more reliable system of transportation than conveyors. This argument is based upon the premise that a truck can be held in reserve for use in the event of a breakdown, or one can be hired from a local contractor. This can be expensive, as it involves additional capital investment or inflated hire charges.

It is often assumed that if the average mechanical availability of a truck is, say, 70%, then there is a good chance that 7 will be available out of a fleet of 10 at any one time. This is not totally correct, as the number of trucks available at one time out of a fleet of a given size should be calculated assuming the probability that the truck is operational is equal to the overall availability and using the binominal distribution. The results of the calculation are shown in Fig. 5 for a fleet of 10 trucks.

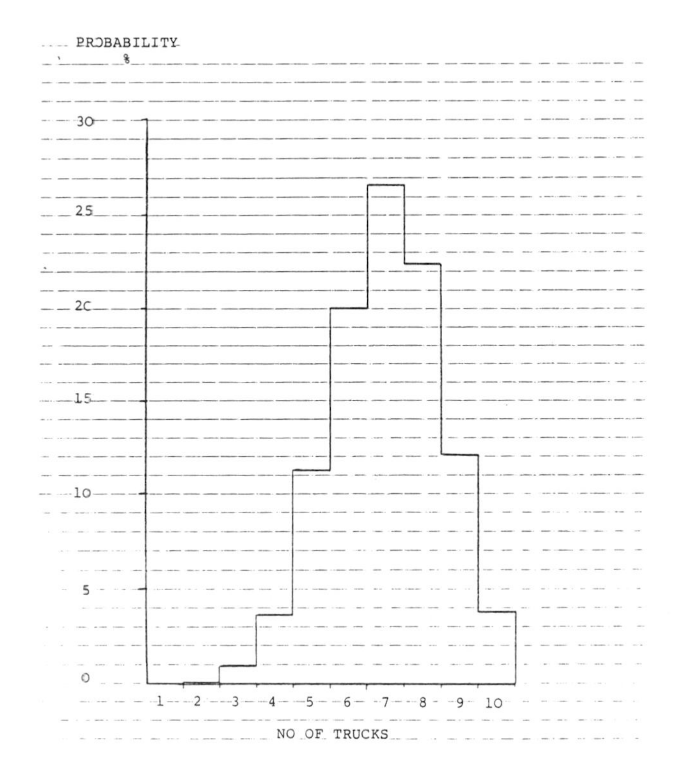

Fig. 5: Truck availability (probability that a particular number of trucks will be available in a fleet of 10 trucks)

Modern conveyor design, using the latest techniques and components, has demonstrated that the reliability of a system is very good. When allowance is made for the additional capital costs incurred in providing standby trucks, including the sophisticated maintenance facilities and the effectiveness of conveyors, the results support the claims of the conveyor system.

5. Advantages of an In-Pit Conveyor System

In addition to the various cost analyses, which in any particular situation will demonstrate the cost effectiveness of an in-pit conveying system when compared with conventional trucking, there are a number of other intangible advantages, which will be of benefit to the mine operator.

- Reduced environmental problems — noise, dust
- Improved long-term planning, due to the reduced influence of escalation on operating costs
- Reduced manpower requirements, not only direct labour, but also indirect labour, required for manning the maintenance workshops required for trucking
- Reduction of satellite cost consuming operations — haul road construction and maintenance — thereby permitting the primary excavator to operate more effectively

- The ability with over- and/or interburden handling of blending to prevent spontaneous combustion
- The more effective utilisation of downstream process plants as a result of the uniform flow of material.

6. Conclusion

The arguments presented in the foregoing demonstrate that the use of conveying systems in surface mining is not only the most energy effective method of transportation, but also, in the cases demonstrated, the most economic in terms of overall operating and ownership cost. The use of conveyors coupled with mobile crushing plants is not claimed to be the panacea for all surface mines. However, it has been demonstrated that savings can be made in operating costs by using conveyors instead of trucks. This technology is applicable not only to new mines, but also to existing operations, and it is recommended that the energy and cost conscious engineer should consider the situation carefully before deciding upon, or even staying with, a truck system and consult with the specialist equipment designers and suppliers to obtain an alternative conveyor system. A financial and objective analysis of the two systems will then permit the correct decision to be made on the transport system, taking into account the life of the mine.

Volume 4, Number 1, March 1984

A Comparison of Handling Systems for Overburden of Coal Seams

Nick Reisler and Conrad Huss, USA

1. Introduction

Many open-cast mines throughout the world are faced with unacceptable operating costs from their present haulage systems. This paper deals with handling burden for a hypothetical strip cut within the range of that typically encountered in the coal mining industry.

Three materials handling systems are applied to this hypothetical, typical strip:

System 1: Shovels/Trucks
System 2: Shovels/Trucks/Crushers/Conveyors/ Tripper-Spreader
System 3: Shovels/Crushers/Conveyors/Tripper-Spreader

Comparative capital and operating costs are developed in U.S. dollars. Drilling and blasting costs are not included.

2. Basis of the Study

The burden material is assumed to be of such a nature that it cannot be handled by a bucket wheel excavator. For all three systems coal will be hauled by truck.

The hypothetical open-cast mine is a simplification of several of the mines studied by Mountain States Engineers recently. As shown in Figs. 1, 2 and 3 there is basically a total of 40 m of overburden being removed in split level strips approximately 100 m wide by 2,400 m long. Coal is concentrated in one horizontal seam 40 m beneath the level ground surface. Some of the usual complicating factors which have been ignored include multiple seams, significant dip of the seams, and varying overburden depth. These complications ordinarily imply that the most economical haulage system for a specific pit would be a combination of the three systems described in this paper, probably in conjunction with a dragline operation. (In fact, this simplistic hypothetical mine considered in this paper could be handled quite nicely by draglines.)

For example, let us assume that instead of one seam a pit has two main coal seams, with the first seam being at 20 to 40 m below grade and the second seam being 30 m below the first. A likely solution to this mine would be to use shovels, trucks and conveyors for the overburden, and draglines for the interburden.

As a second example, let us assume that a mine has an uneven ground surface with multiple coal seams at a dip of 15 %. Such a system could require trucks for surface haulage of upper burden to excavate a flat working surface for two or three conveying systems to handle middle burden. The lower burden would again best be handled by trucks to avoid the frequent relocations that would be associated with a small lower pit working area.

The purpose of this study, then, is to determine relative costs for the three outlined materials handling systems. The mine is obviously simplistic and is used mainly as a basis for capital and operating costing.

Nick Reisler is Project Manager and Dr. Conrad Huss, P.E., is Vice President and Manager of Engineering, Mountain States Engineers, Tucson, AZ 85731, USA.

3. Description of the Mine

The mine has 45,650,000 bank m³ reserves with an associated overburden of 273,900,000 bank m³. Coal production is scheduled at 5,000,000 metric tonnes per year which implies an average annual overburden removal of 30,000,000 metric tonnes. The mine has an economical mine life of at least 21 years based on a 6 to 1 stripping ratio. More coal is available with a less desirable stripping ratio after year 21.

The coal seam and ground surface are virtually flat. The bank density of the overburden is 2.30 metric tonnes per bank m³, or 1.70 metric tonnes per loose m³.

4. Operating Parameters for the Haulage Systems

The mine is assumed to operate 340 days per year, 3 eight-hour-shifts per day. For all systems, equipment mechanical availability is assumed at 82 %, and personnel time efficiency is assumed at 83 %. Overall system utilization is thus at 68 %, or 16.3 hours per day, por 5,550 hours per year.

For a requirement of 30,000,000 metric tonnes per year, an average or flowsheet rate of 5400 mt/h is calculated. To permit temporary surges and recuperating capacity without overloading system components, a design flowrate of 6,000 mt/h is used.

Volume 4, Number 1, March 1984 

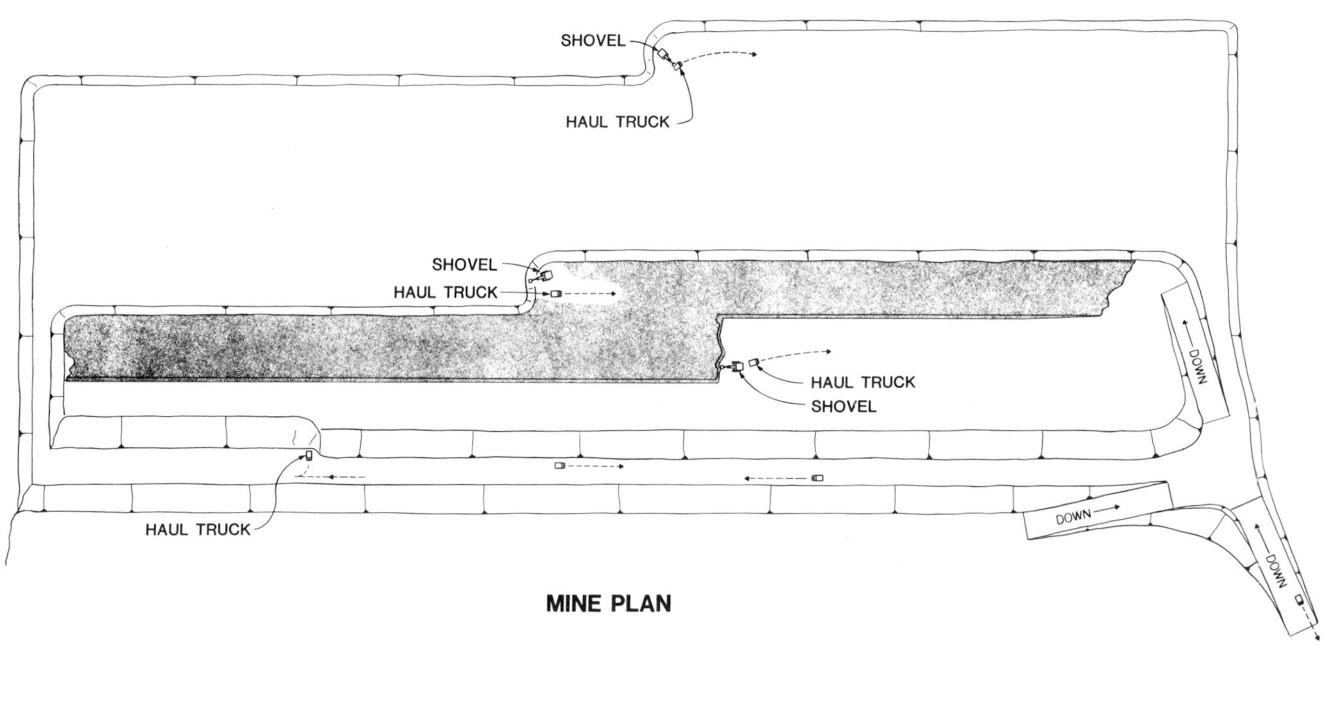

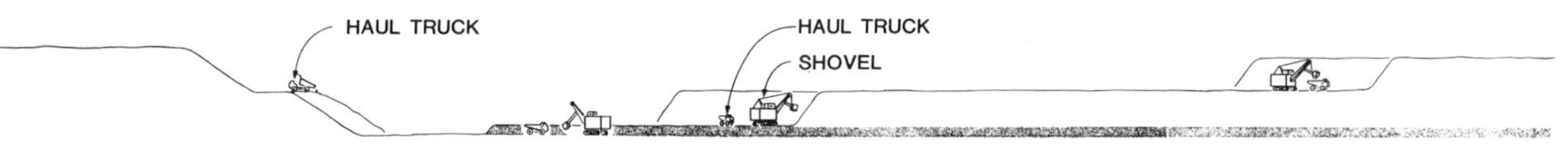

Fig. 1

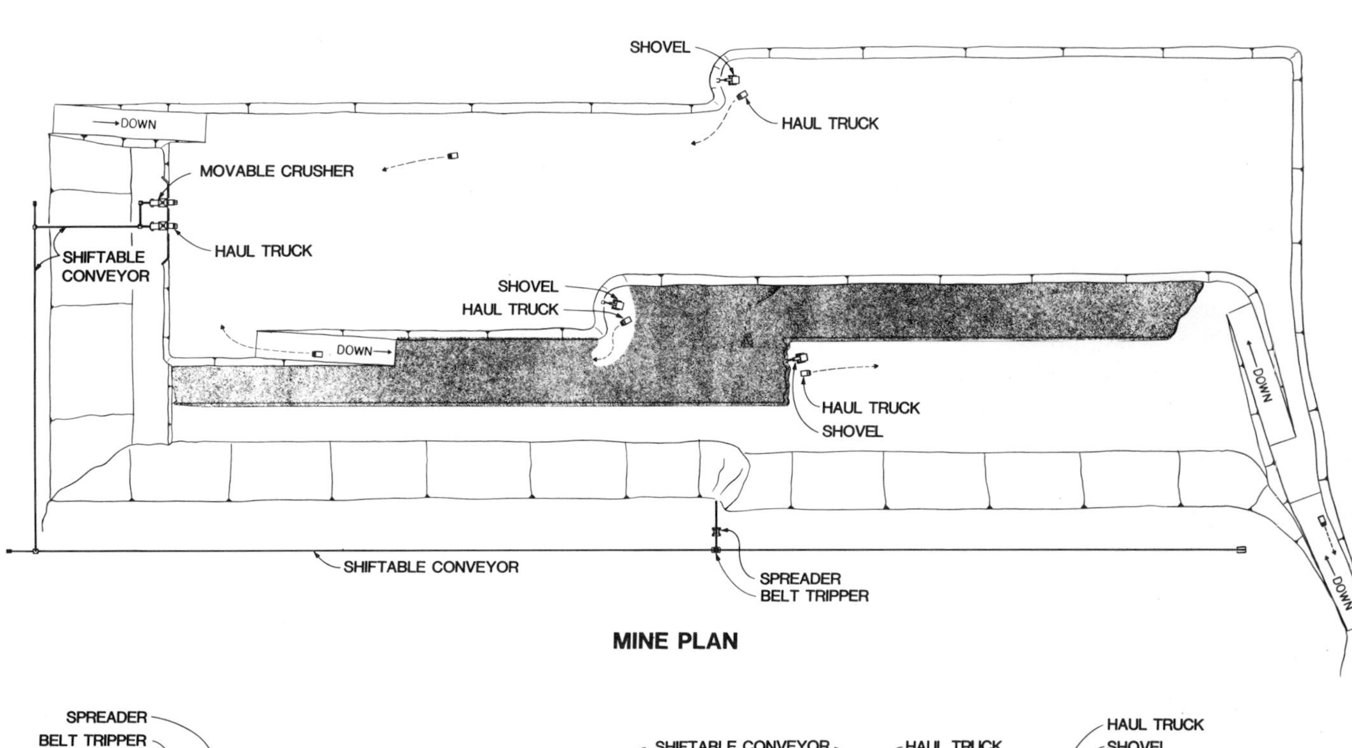

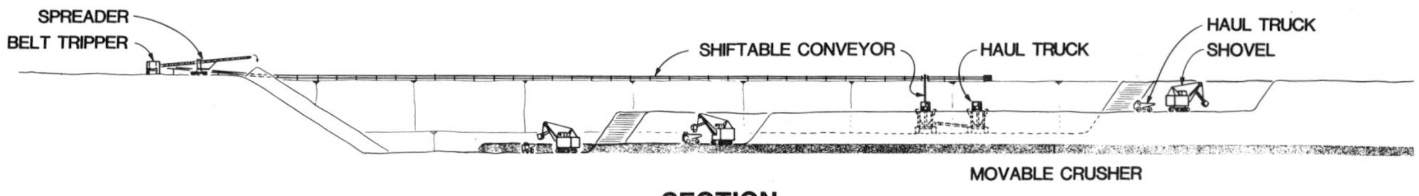

Fig. 2

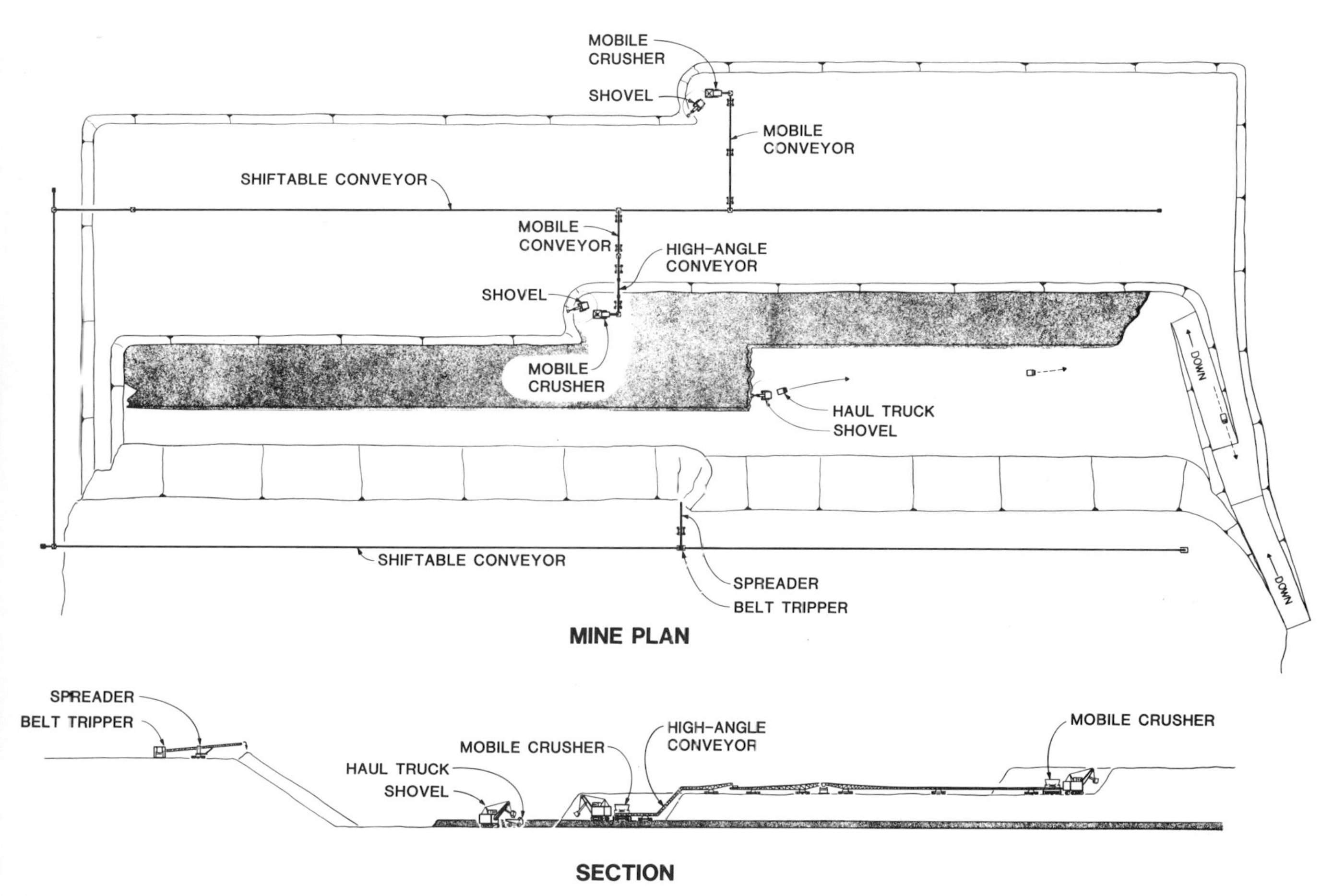

Fig. 3

5. System Description

In all systems, the basic strip encountered by each shovel is 100 m wide by 20 m deep by 2,400 m long, or 4,800,000 m³. Approximately 4 months will be required by one shovel to work a strip.

5.1 Shovels/Trucks (Fig. 1)

This system utilizes three shovels for overburden removal on two different 15 m benches. All haulage is by truck in a very accessible work area. The number of trucks required has been calculated by computer simulation.

Materials Handling Equipment List

3 Shovels, 19 m³ dipper each
15 Trucks, 155 metric tonne capacity each

5.2 Shovels/Trucks/Crushers/Conveyors/Tripper-Spreader (Fig. 2)

This system utilizes crushers and conveyors to decrease haul length and lift by the trucks. The movable crushers (each at 3,000 mt/h) are both located at one end of the strip cut to allow complete truck access at the other end. The number of trucks required has been calculated by computer simulation. Conveyors are 1.6 m wide and travel at a speed of 5 m/sec. Final material disposition is by a tripper-spreader combination.

Materials Handling Equipment List

2 Shovels, 19 m³ dipper capacity each
8 trucks, 155 metric tonne capacity each
2 Movable Crusher Stations, 3,000 mt/h capacity each
1 Shiftable Bridge Conveyor, 60 m long
1 Shiftable End Conveyor, 200 m long
1 Shiftable Side Conveyor, 2,400 m long
1 Belt Tripper
1 Spreader

5.3 Shovels/Crushers/Conveyors/Tripper-Spreader (Fig. 3)

This system eliminates trucks completely for burden haul by direct feed of the crushers with the shovels. A mobile crusher is located on each of the two benches. Mobile conveyors are 1.0 m wide and travel at 4 m/sec. Shiftable conveyors are 1.6 m wide and travel at a speed of 5 m/sec. Final material disposition is by a tripper-spreader combination.

Materials Handling Equipment List

2 Shovels, 19 m³ dipper capacity each
2 Mobile Crusher Stations, 3,000 m³ capacity each
1 High-Angle Conveyor, 20 m long
2 Mobile Conveyors, each 60 m long
1 Shiftable In-Pit Conveyor, 2,400 m long
1 Shiftable End Conveyor, 200 m long
1 Shiftable Side Conveyor, 2,400 m long
1 Belt Tripper
1 Spreader

6. Comparative Capital Cost Estimates

6.1 System 1: Shovels/Trucks

Item	Quantity	Cost
Shovels	3	9,000,000
Trucks	13	13,000,000
Maintenance/Warehouse	1	2,000,000
Fueling and Storage	1	2,000,000
Tire Shop	1	1,000,000
Total		27,000,000

6.2 System 2: Shovels/Trucks/Crushers/Conveyors/Tripper-Spreader

Item	Quantity	Cost
Shovels	3	9,000,000
Trucks	6	6,000,000
Movable Crusher Station	2	5,000,000
Maintenance/Warehouse	2	1,500,000
Fueling and Storage	1	2,000,000
Substation	1	500,000
Tire Shop	1	500,000
Shiftable Bridge Conveyor (60 m long)	1	500,000
Shiftable Overland Conveyors with Drives (2,600 m long total)	2	9,000,000
Belt Tripper	1	250,000
Spreader	1	3,000,000
Total		37,250,000

6.3 System 3: Shovels/Crushers/Conveyors/Tripper-Spreader

Item	Quantity	Cost
Shovels	2	6,000,000
Mobile Crusher Stations	2	7,000,000
Maintenance/Warehouse	1	500,000
Substation	1	500,000
High-Angle Conveyor (20 m long)	1	500,000
Mobile Conveyors (120 m long total)	2	1,000,000
Shiftable Overland Conveyors with Drives (5,000 m long total)	3	16,000,000
Belt Tripper	1	250,000
Spreader	1	3,000,000
Total		34,750,000

7. Comparative Operating Cost Estimate

Common labor costs are calculated based on $15.00/hour direct, plus 33 % fringes, or $20.00/hour without overhead. Electricity has been assumed at $0.035/kWh and diesel fuel at $1.00/gallon. Equipment life is 21 years for all equipment except trucks which are taken at 7 years. The costs are based on distances and lifts as depicted in the figures.

7.1 System 1: Shovels/Trucks

Item	Yearly Cost per Truck
Tire Replacement	80,000
Tire Repair	10,000
Mechanical Repair	200,000
Fuel	150,000
Preventive Maintenance	5,000
Drivers	175,000
Total	620,000

7.2 System 2: Shovels/Trucks/Crushers/Conveyors/Tripper-Spreader

Due to decreased lift requirements, yearly costs per truck will be slightly less than for System 1: $610,000.

Conveying and crushing system are assumed to run loaded 5,550 hours per year, empty for 2,410 hours per year, and shut down for 800 hours per year. Based on these operating conditions, the following annual energy consumptions are encountered:

Crushing	5,000,000 kWh
Conveying	8,500,000 kWh
Stacking	2,780,000 kWh
Total	16,280,000 kWh

At $0.035/kWh, the above consumption translates into an annual cost of $570,000. In addition, approximately 25 operating and maintenance personnel will be required at a cost of $41,600/year/man. Five percent of initial capital cost is allocated each year for operating supplies, parts and materials ($915,000).

7.3 System 3: Shovels/Crushers/Conveyors/Tripper-Spreader

Based on operating conditions stated for System 2, the following annual energy consumptions are encountered:

Crushing	5,000,000 kWh
Conveying	13,900,000 kWh
Stacking	2,780,000 kWh
Total	21,680,000 kWh

At $0.035/kWh, annual energy costs are $760,000. Based on a 30 man operating and maintenance crew, labor costs are $1,250,000 a year. Five percent of initial capital cost is allocated each year for supplies, parts and materials, ($1,400,000).

Year	SYSTEM 1				SYSTEM 2				SYSTEM 3			
	Capital	Operating	Combined	Cumulative	Capital	Operating	Combined	Cumulative	Capital	Operating	Combined	Cumulative
1	27,000	8,060	35,060	35,060	37,250	6,454	43,704	43,704	34,750	3,236	37,986	37,986
2		8,060	8,060	43,130		6,454	6,454	50,158		3,236	3,236	41,222
3		8,060	8,060	51,180		6,454	6,454	56,612		3,236	3,236	44,458
4		8,060	8,060	59,240		6,454	6,454	63,066		3,236	3,236	47,694
5		8,060	8,060	67,300	1,000	6,454	7,454	70,520	2,000	3,236	5,236	52,930
6		8,060	8,060	75,360		6,454	6,454	76,974		3,236	3,236	56,166
7		8,060	8,060	83,420		6,454	6,454	83,428		3,236	3,236	59,402
8	13,000	8,060	21,060	104,480	6,000	6,454	12,454	95,882		3,236	3,236	62,638
9		8,060	8,060	112,540		6,454	6,454	102,336		3,236	3,236	65,874
10		8,060	8,060	120,600	1,000	6,454	7,454	109,790	2,000	3,236	5,236	71,110
11		8,060	8,060	128,660		6,454	6,454	116,244		3,236	3,236	74,346
12		8,060	8,060	136,720		6,454	6,454	122,798		3,236	3,236	77,582
13		8,060	8,060	144,780		6,454	6,454	129,152		3,236	3,236	80,818
14		8,060	8,060	152,840		6,454	6,454	135,606		3,236	3,236	84,054
15		8,060	8,060	160,900	1,000	6,454	7,454	143,060	2,000	3,236	5,236	89,290
16	13,000	8,060	21,060	181,960	6,000	6,454	12,454	156,514		3,236	3,236	92,526
17		8,060	8,060	190,020		6,454	12,454	161,968		3,236	3,236	95,762
18		8,060	8,060	198,080		6,454	12,454	168,422		3,236	3,236	98,098
19		8,060	8,060	206,140		6,454	12,454	174,876		3,236	3,236	102,234
20		8,060	8,060	214,200		6,454	12,454	181,330		3,236	3,236	105,470
21		8,060	8,060	222,260		6,454	6,454	187,784		3,236	3,236	108,706

Table 1: Annual capital and operating costs in $1,000

8. Financial Summary

Table I summarizes capital and operating costs on a yearly basis. A Discounted Cash Flow analysis was run for this data. System 1 capital costs are the least and are taken as a base reference. Analyses are run for both Systems 2 and 3 based on the difference in their capital costs with those of System 1. Downstream cash flows are defined to include operating costs and post construction capital expenditures following the initial investment, all on a pretax basis.

For these stated assumptions, the DCF return on *additional* investment for System 2 is 18.1 % and the payout is 7 years after startup. The DCF return on additional investment for System 3 is 62.9 % and the payoff is 1.8 years after startup.

On a cumulative actual basis which does not take into account the present net value factor of money, per tonne capital and operating costs are as follows:

System 1: $0.35/tonne
System 2: $0.30/tonne
System 3: $0.17/tonne

Volume 5, Number 1, February 1985

Cost Reduction by In-Pit Crushing and Conveying

H. Althoff, Germany

Summary

The advantages and disadvantages of a mobile crushing plant with a belt conveyor system compared to a stationary crushing plant with haulage by heavy trucks are described. The capital expenditure and operational expenses are compared, and the individual cost factors are explained.

Every open-pit mine, however, is different and has to be assessed and analyzed individually in order to find the optimum and most economical solution. Due to high truck haulage costs, many open-pit mines tend to the more economical solution with mobile crushing plants and continuous belt conveyor operation.

1. Introduction

For material haulage within an open-pit mine, the heavy truck is the most flexible means of transportation. However, due to high fuel consumption, wear, and labour costs, it is also the most expensive one. Therefore, many open-pit mines are looking for alternative solutions for material transport.

A mobile belt conveyor system is most likely to meet the requirements because it is efficient and economical with regard to the power consumption. The material, however, has to be crushed at the mining face to lumps which are suitable for conveyor operation. For this purpose, a mobile crushing plant is required.

In some open-pit mines, the crusher is also used to crush the material to lumps of smaller particle sizes in order to save crushing costs in the subsequent preparation plant.

Although a cost comparison between a system with stationary crusher and truck haulage and a system with mobile crusher and belt conveyors favours in most cases the belt conveyor system, up till now only a few open-pit mines and quarries benefit world-wide from the advantages of a mobile crushing plant with continuous belt conveyor operation.

The reason may be that in most cases the initial investment costs for a mobile crusher with belt conveyor system are higher compared to a stationary crusher with truck haulage system.

However, the higher investment costs will soon be amortized by saving the high operational expenses of truck haulage. As a general rule, belt conveyor operation is more economical than truck haulage if the conveying distance exceeds 1,000 m.

Haulage by heavy trucks is expensive due to high capital expenditure, the short depreciation period of the trucks, and high operational expenses, such as Diesel fuel, wear of tyres, operational personnel, etc. These costs rise every year as a result of wage increases as well as longer distances to be covered when the mining face advances. For this reason, a cost comparison has to be prepared in each case for a period of several years.

2. Advantages and Disadvantages of a Belt Conveyor System

As costs are not the only criterion, all advantages and disadvantages of a continuously operating conveyor system have to be considered.

The disadvantages of a conveyor system compared to a truck haulage system are:

- less flexibility
- higher initial investment costs (in most cases)
- less suitability for big material lumps
- stoppage of the complete belt conveyor system if one belt fails

The advantages of a belt conveyor system compared to a truck haulage system are:

- lower capital expenditure due to a longer depreciation period
- lower energy costs
- lower personnel costs
- lower wear costs
- possibility of gradients of up to 18°
- possibility of bridges across roads, railroads, and rivers
- light-weight foundations, therefore no road and road maintenance costs
- possibility of automation

Stationary as well as movable belt conveyor systems have been improved and tested by advanced techniques and practical experience to such a degree that they will become increasingly important for open-pit mining in combination with mobile crushing plants and that they will be an economical alternative to truck operation.

Fig. 1 shows a mobile crushing plant with walking mechanism and a movable belt conveyor system as delivered

Mr. Heinz Althoff is employed with O & K Orenstein & Koppel, Zementanlagen und Aufbereitungstechnik, D-4722 Ennigerloh, Federal Republic of Germany

Fig. 1: Mobile crushing plant for Portland Zementwerke Heidelberg, Lengfurt, Federal Republic of Germany

in 1976 to Portland Zementwerke Heidelberg, Federal Republic of Germany. The throughput is 1,000 t/h limestone, the service weight is 920 t.

Since all open-pit mines are different regarding their situation and capacities, every open-pit mine has to be analyzed individually in order to find the optimum and most economical solution.

3. Cost Comparison

Table 1 shows a cost comparison between a stationary crushing plant with truck haulage and a mobile crushing plant with a belt conveyor system.

An exact cost calculation is different for every individual open-pit mine due to the many variables and has to be prepared for each year (e.g., year 1 to year 20) in tabular form according to capital expenditure and operational expenses. (Refer also to the dissertation of Dr. J. Korak in 1978.)

However, for a simplified cost calculation, the individual values can be obtained by assumptions and rough estimates. In the following, the items 1 to 7 of Table 1 are briefly explained.

The following basic data are required:

- material to be mined
- rated crushing capacity
- conveying capacity
- operating time
- production time
- yearly production
- lifetime of the plant components

Table 1: Cost comparison between a stationary crushing plant with truck haulage and a mobile crushing plant with a belt conveyor system

Costs of the crushing plant and costs of the transport system	System A Stationary crusher with truck haulage	System B Mobile crusher with belt conveyor
Capital expenditure/a		
1. Depreciation		
a) Crusher		
b) Transport system		
2. Interest		
a) Crusher		
b) Transport System		
3. Insurance and risk		
a) Crusher		
b) Transport system		
Operational expenses/a		
4. Wear costs		
a) Crusher		
b) Transport system		
5. Costs for spare parts		
a) Crusher		
b) Transport system		
6. Energy costs		
a) Crusher		
b) Transport system		
7. Personnel costs		
Total costs/a		
Costs/t	300—500 %	100 %

— initial investment costs
— costs of spare parts
— costs of wearing material
— energy costs
— number and capacity of trucks
— average distance in the open pit
— Diesel consumption
— rate of interest
— labour costs.

4. Notes and Assumptions Referring to the Cost Comparison

The investment prices of a mobile and a stationary crushing plant are known in most cases according to vendor quotations. For a mobile conveyor system, an average distance between the mining face and the preparation plant can be assumed, e.g., for the first four years. For truck haulage, the time per round trip and thereafter the number of trucks can be determined for the same average distance at an average driving speed (of, e.g., 20 km/h). One round trip includes loading, driving, and unloading time.

To 1. Depreciation:
The depreciation period for a crushing plant and a belt conveyor system is usually approx. 20 years. The depreciation period for a heavy truck is approx. 15,000 h or approx. four years, depending on its utilization.

To 2. Interest:
The interest is usually higher in the first years and then decreases. Therefore, these costs have to be determined according to the capital investment.

To 3. Insurance and risk:
This factor differs for every country, however, 0.3 to 0.5% of the investment per year can be presumed.

To 4. Wear costs:
The wear costs for liner plates, impact plates, hammers, impact bars, etc. vary considerably according to the material, crusher type, and throughput and have to be obtained from the equipment manufacturer. For truck haulage, one set of tyres, which costs about 4 to 6% of the value of a new truck, should be considered every year.

To 5. Costs for spare parts:
The costs for spare parts including maintenance and repair for a crusher and a belt conveyor system are approx. 1 to 2% of the initial investment per year. For truck haulage, this figure is higher, i.e., approx. 6% per year. For a belt conveyor system, however, a new belt should be included every ten years.

To 6. Energy costs:
- During crushing operation, the energy costs are: power consumption (kWh/t) x energy price ($/kWh) x throughput (t/a). The power consumption for the crusher is approx. 0.4 to 1.2 kWh/t, depending on the material.
- For travelling operation of the mobile crushing plant, the frequency and distance of crusher displacements have to be estimated. The energy costs result from: travelling power (kW) x travelling time (h/a) x energy price ($/kWh).
- The energy costs for the belt conveyor system result from: drive power (kW) x operation time (h/a) x energy price ($/kWh).
- The energy costs for truck haulage are determined by the fuel consumption (l/h) and the fuel price ($/l).

The fuel consumption for an empty or full truck is according to a rule-of-thumb Q = 0.07 l/kWh to 0.14 l/kWh. If one round trip is assumed with an average of 40% full load, then the fuel consumption depends on the drive power of trucks with 0.10 l/kWh, i.e., energy costs = number of trucks x fuel consumption (l/h) x driving time (h/a) x fuel price ($/l).

To 7. Personnel costs:
The personnel costs for a mobile crushing plant with a belt conveyor system are generally low. As the crusher and the belt conveyors are automatically controlled, two men can supervise and maintain the complete plant. For truck haulage, one driver is required for each truck per shift.

This rough estimate does not include the costs for shovels, front-end loaders, and bulldozers, as well as the infrastructure, such as buildings, workshops, road costs, etc. Also, additional operating facilities and preparation plants have not been included.

The cost comparison described above only refers to a system A "stationary crushing plant with truck haulage" and to a system B "mobile crushing plant with belt conveyor operation". The transport distance is always between the mining face and a point x, which may be, e.g., a preparation plant, a loading plant for a railway system, or a dump site, etc.

Most economical, however, would be a system where the material allows the use of a bucket wheel excavator at the mining face. The bucket wheel excavator already reduces the material ready for continuous belt conveyor operation.

5. General

Open-pit mines can generally be divided into three categories:

1. Production of up to 1 Mt/a:
Here, a stationary crusher with truck haulage is generally more economical than a mobile crushing plant.
2. Production of up to 8 to 10 Mt/a:
For these requirements, a mobile crushing plant with a belt conveyor system is more economical.
3. Production exceeding 8 to 10 Mt/a:
Here, a mobile or semi-mobile crushing plant with a belt conveyor system can be recommended, or a combined transport system with trucks and belt conveyors should be investigated.

These categories, however, are only guidelines which can have exceptions.

In addition to the price, other criteria for the choice of a mobile or a stationary crushing plant are of course the type of mining body (variations, depth, mixture, shape, and size) as well as transport distances, slopes, flexibility, availability, etc. The capital investment and operational expenses are, however, the most important factors when planning a new open-pit mine or when existing systems have to be extended or replaced.

Besides the initial investment costs and capital expenditure, operational expenses are of particular importance for the project analysis and the refinancing of the investment. In order to be able to evaluate this expenditure, a cost comparison between the two systems can be obtained despite the numerous variables and assumptions by a simplified cost calculation as described above.

bulk solids handling Volume 5, Number 1, February 1985

Modern Open-Pit Mining Technology

H. Ringleff, Germany

1. Introduction

The increasing demand of energy in the world as well as environmental legislation have brought forth methods of mining and overburden stripping which have been successfully implemented in opencast mining. The first system, the so-called "German opencast mining technology", which was introduced in 1955 with an output of 100,000 m³(bank)/d, revolutionized the open-pit mining industry worldwide with regard to continuous bulk material handling.

PHB Weserhütte (PWH) has earned a major share in the development of efficient opencast mining equipment, which provides its reliability worldwide under tough conditions in opencast mines.

The bucket wheel excavator has made it possible to mine deposits of multiseams at deeper levels or even in dipping seams. The development of new deposits, the majority of which are located at deeper levels, necessitates the removal of increasing masses of material on account of a more unfavourable overburden-to-coal ratio. Mining, transportation, handling, and dumping equipment must be adapted to these requirements.

2. Special Opencast Systems in Hard Rock

German technology does not only supply equipment for mining operations on cuttable material, where continuous excavating and conveying is applied. The intermittent mining method — especially in hard rock formations — needs other combinations of mining equipment. The most important are draglines, face shovels, and mobile or semimobile crushing units, all of which are covered by the PWH product range.

In this context, in-pit crushing and conveying systems have already proven a successfull technology in practical operations (Fig. 1). Shiftable and mobile crushing units play an important part in modern opencast mining techniques. Mobile crushers, wheel mounted or with crawler or walking mechanism, are able to follow the advance of the face in opencast mines and rock quarries. Semimobile crushing units can be

Fig. 1: Schedule of PWH opencast mining system

moved by versatile transport systems, such as wheel mounted piggyback transporters, transport crawlers, and detachable walking mechanisms, which are completely independent of the crushing units.

Since 1965, PHB Weserhütte has been supplying mobile crushers to the opencast mines of major mine operators. For instance, in 1980 Foskor Phalaborwa, South Africa, ordered an in-pit crushing and conveying system; in 1981, Iscor, South Africa, placed an order for an in-pit crushing and conveying system for overburden in their Grootegeluk coal mine (Fig. 2).

Fig. 2: Mobile walking-type crusher plant installed at Grootegeluk black coal opencast mine in South Africa; throughput 3,000 t/h

Dipl.-Ing. Horst Ringleff is Chief Manager (Mobile Crushers), PHB Weserhütte AG, D-5000 Köln, Division Bad Oeynhausen, Federal Republic of Germany

In South Africa and in opencast mining technology worldwide, both units represent a breakthrough from discontinuously operating shovel/truck systems to systems which combine discontinuously operating equipment, i.e., excavator/mobile crusher/conveyor system configurations.

3. Open-Pit Bucket Wheel System

The development of the open-pit phosphate mine of North Carolina Phosphate Corporation (NCPC), a company owned by Agrico Chemical Company, USA, has been reactivated with the start-up for the complete bucket wheel excavator system scheduled for mid 1986. The mining district is situated north of Aurora City at the mouth of the Pamlico River in North Carolina. The project has been delayed from an originally planned start-up in 1985 and will have a capacity of 3.7 million short tons/a calcined phosphate rock.

PHB Weserhütte equipment was chosen for the continuous around-the-pit mining system, which is also known as the "German" method, developed and used mainly for large opencast mines in Europe and exported as an idea around the world (Fig. 3). To move approx. 14 Mm³ waste and approx.

Fig. 4: Bucket wheel during workshop erection

Fig. 3: Artist's impression of NCPC's scheduled round-the-pit operation

Fig. 5: View to opencast mine with heavy-duty belt conveyor system and tripper car

7 Mm³ phosphate ore, the following equipment designed and supplied by PHB Weserhütte will be erected in 1985:

- 4 bucket wheel excavators type SR 630 (Fig. 4)
- 5 mobile transfer conveyors
- 2 combined hopper/cable reel cars
- 2 rail mounted tripper cars with slewing conveyors
- 4 km of shiftable conveyors with 1,400 mm and 1,800 mm wide belts
- 2 hydraulic walking mechanisms for slurry sumps.

The top layer of overburden will be stripped by two of the bucket wheel excavators at a rate of 4,500 t/h each, transported around the pit via mobile transfer conveyors, a shiftable conveyor system, and finally to the stacker unit, where the material is dumped back onto the exploited area. The auxiliary equipment such as the hopper/cable reel cars and the tripper cars is rail mounted and self-propelled (Fig. 5).

The distance between the two face conveyors and the dump side conveyor has been selected wide enough to allow removal of additional overburden, which is cast into the mined-out area, and removal of the phosphate ore by a dragline on a bench 18 m above the top seam of the ore. The excavated phosphate is cast into a stockpile on this dragline bench.

The ore will be reclaimed by the two reclaiming bucket wheel machines, which are feeding the ore via mobile transfer conveyors into the hoppers of the two self-propelled portable sumps. The ore will be slurried in the sumps and pumped in a pipeline to the beneficiation plant.

All mobile PHB Weserhütte equipment is mounted on extra wide crawlers to compensate for the extremely unfavourable ground conditions which are due to soil characteristics and the possibility of very high precipitation of more than 300 mm/d.

The equipment described will be operating parallel. After some years, it will be slewed around 180° so that it can operate in the same manner in the neighbouring field.

This conceptual design of a combination of continuous bucket wheel operation, dragline strip mining, reclaiming by means of bucket wheel, and pumping the phosphate slurry in pipelines is a unique solution for the USA and all of the world.

Transportation Equipment in Opencast Mines

K.-H. Althoff, Germany

1. Introduction

The price inflation in the raw material markets around the world and, consequently, the high investments necessitate, among other things, development of new mining and transport systems in opencast mines. With, for example, the coal-to-overburden ratio becoming more and more unfavorable, transport systems must follow the progress in mining flexibly and quickly.

Achieving this economically was only possible with the aid of suitable transport equipment which could be utilized for a wide range of loads.

For this purpose, PHB Weserhütte (PWH) developed a complete line of transport equipment which has, in the meantime, been put to use on a worldwide basis.

2. Transport Equipment

PWH transport equipment, utilized, for example, to move crushers and opencast mining machines, especially conveyor system assemblies, includes:

- detachable walking devices (Fig. 1)
- raft-type walking mechanisms (Figs. 2 and 3)
- transport crawlers (Fig. 4)
- wheel-mounted travel assemblies (Fig. 5).

Fig. 1: Conveyor drive station with two detachable walking devices

Mr. Karl-Heinz Althoff, PHB Weserhütte AG, P.O. Box 510 850, D-5000 Köln 51, Federal Republic of Germany

Fig. 2: Conveyor drive station with integrated walking device

Fig. 3: Mobile crusher unit with raft-type walking mechanism

Fig. 4: Conveyor drive station with transport crawler

Fig. 5: Mobile crusher unit with wheel-mounted travel assembly

Table 1 gives a breakdown of the transport equipment supplied by PHB Weserhütte.

Table 1: Breakdown of PWH transport equipment

Equipment	Europe	Overseas	Total	Total 1975
Detachable walking devices	80	18	98	34
Raft-type walking mechanisms	17	16	33	21
Transport crawlers	3	4	7	7
Wheel-mounted travel assemblies	1	1	2	2
Total	101	39	140	64

Rail-mounted travel assemblies are used as additional shifting systems. The movement possibilities of the load, however, are considerably reduced with this type of transport equipment because the rails which have been laid determine the shifting direction.

Lighter haulage loads mounted onto a pontoon can also be moved or pulled by dozers, respectively.

3. Selection Criteria

The decision as to which type of transport equipment is to be utilized in a specific opencast mine is dependent on a wide range of selection criteria. These includes, among others:

— ground conditions for the transport equipment
— possibilities within the mine to prepare the working level for the transport operation
— number of machines to be moved
— frequency of machine transport and transport routes
— influence of the load to be transported on the transporting equipment
— use of currently existing machines
— costs of the transport equipment.

The order in which these points have been itemized changes depending on the priorities set down in the different opencast mines, and the criteria can be supplemented with additional factors, if applicable.

This fact alone clearly shows that there is currently no transportation equipment on the market which is optimal under all applications and mining conditions. A thorough investigation of the local mining conditions is always required to insure proper selection of the appropriate transport equipment for the respective case of application.

4. Applications

4.1 Detachable Walking Devices

For 28 years, detachable walking devices have been employed to transport belt drive stations, return stations, extending conveyor heads etc., primarily in opencast lignite mines, around the world.

With the construction of detachable walking devices, it was for the first time made possible to transport larger loads without integrated transport equipment in any selected direction. A minimum of two — maximal four — walking pads is attached to the load which is to be transported. These walking pads are operated from a central control mechanism.

Fig. 6 shows a belt drive station in a Czechoslovakian mine. The station, which has a working weight of approx. 220 t, is moved with two walking pads, each of which has a lifting capacity of 130 t.

Fig. 6: Conveyor drive station in Czechoslovakian mine

The development of yet larger belt conveyor widths in mines and, thus, of heavier machines led to the construction of walking pads with a presently available lifting capacity of 250 t per walking pad.

Fig. 7 shows an extending conveyor head with four walking pads in an oilsand opencast mine in Canada. The walking pads have a lifting capacity of 200 t each.

Fig. 7: Extending conveyor head with four walking pads in oilsand opencast mine in Canada

The construction form of the loads which are to be transported is not influenced by the walking pads because they can be attached to the sides of the load's skid.

4.2 Raft-Type Walking Mechanisms

Raft-type walking mechanisms, on the other hand, are permanently attached to the respective loads to be moved.

They are a component of the main machine, as can be seen in Figs. 2 and 3.

As of the end of the 1960s, this type of walking mechanism construction supplemented the existing transport equipment such as fully integrated crawler travel assemblies or wheel-mounted travel assemblies, predominantly in hard stone opencast mines.

Since then, 58% of the mobile crushers supplied have been equipped with a raft-type walking mechanism. The remaining 42% are divided up between crawler-mounted travel assemblies (permanently installed crawlers and transport crawlers) — 28% — and wheel-mounted travel assemblies — 14%.

The raft-type walking mechanism affords the crusher's extensive motional flexibility within a small space and allows for great adaptability to the excavator. The feeding hopper can be respectively positioned in the most favorable excavator slewing range without having to change the discharge point between the boom and the conveyor system following the boom.

The maximum 2-m/min transport speed of mobile crusher units with raft-type walking mechanisms can only be evaluated under consideration of the specific conditions prevailing in a given mine, i.e., a hard stone opencast mine. These conditions depend on:

— feeding of the crusher
— throughput capacity of the crusher
— type of rock
— height of the wall
— design of downstream equipment.

These factors influence, first and foremost, the frequency of transport as well as the transport routes.

Based on an approx. 1,200-t working weight of the mobile crusher, as shown in Fig. 3, the weight portion of the raft-type walking mechanism amounts to approx. 80 t.

Due to the relatively large standing surfaces of walking devices as well as the sequence of movements during transportation, only minimal requirements are placed onto the working level in an opencast mine. Figs. 8 and 9 show that the transferring of loads is possible even when ground conditions are unfavorable.

4.3 Transport Crawlers

The initial research on transport crawlers to transport, for example, conveyor drive stations in opencast mines took place at the end of the 1950s, parallel with the development of walking devices.

Fig. 10 shows a sketch of a transport crawler from the year 1960.

The weights of the conveyor drive stations and their construction forms, with the skid over the full length of the station, gave walking devices their advantageousness as means of transport.

Fig. 8: Transferring of loads under unfavorable ground conditions

Fig. 9: Transferring of loads under unfavorable ground conditions

Fig. 10: Sketch of transport crawler (1960)

The first transport crawler was not introduced until the mid 1970s for the purpose of transporting conveyor drive stations in opencast mines.

Fig. 11 shows a transport crawler in the Public Power Corporation's (PPC) Southfield ("Suedfeld") mine in northern Greece.

The lignite field extends in southward direction, connecting to the previously opened up Kardia mine.

Fig. 11: Transport crawler in the Southfield mine, Greece

Table 2: Technical data of the PWH transport crawler for the Southfield mine, Greece

Length	9.0 m
Width	8.5 m
Height (with lifting cylinder retracted)	2.45 m
Length of stroke	0.76 m
Ability to climb with load, longitudinal	1:5
Permissible inclination	1:10
Travel speed, no load	0—30 m/min
Travel speed with load	0—15 m/min
Deadweight	120 t
Delivery weight of drive station	260 t
Average ground pressure with load	12.7 N/cm^3
Drive capacity, diesel motor	177 kW

Since August 1978, the transport crawler shown in Fig. 12 (Table 2) has moved all of the drive stations presently existing in the opencast mine with a working weight of 260 t each from the erection area to the area of operation. Besides this long-distance haulage, the transport crawler effected every change in position of the drive stations within the conveyor shifting in the opencast mine. Additionally, the transport crawler moved the tail end stations and transported other individual loads. All in all, this transport crawler has to date covered a total distance of approx. 2,500 km.

Fig. 12: Transport crawler moving conveyor drive station

Fig. 13: Transport of conveyor drive stations

In 1984, two additional transport crawlers of the same type were supplied to the expanded Southfield opencast mine.

After these machines had been erected, they were initially used to shift five drive stations approx. 18 km, from the erection site to the "Agios Dimitrios" lignite power plant (Fig. 13).

4.4 Wheel-Mounted Travel Assemblies

Wheel-mounted travel assemblies are predominantly employed in hard rock opencast mines due to the fact that the required working level with the allowable ground pressure which is necessary when using these travel assemblies can only be found here.

Very complicated designs, involving the use of several sets of wheels to reduce the ground pressure per wheel, did not prove to be successful.

Fig. 5 shows a mobile crusher unit with a permanently installed wheel-mounted travel assembly. The unit has a working weight of 420 t. The wheel-mounted travel assembly is relieved during the crushing operation, that is to say, the crusher rests on the hydraulic jacks.

5. Conclusions

The preceding descriptions illustrate once again that the development of suitable transport equipment for machines in opencast mines has contributed significantly to the application of shiftable belt conveyor systems worldwide.

 Volume 3, Number 1, March 1983

A Mobile In-Pit Crushing System for Overburden at Iscor's Grootegeluk Coal Mine

A.P. Dippenaar, C.O. Esterhuysen and N. de Wet, South Africa

Summary

In this paper the mobile in-pit crushing system for overburden, which was commissioned in July 1981, at Iscor's Grootegeluk Coal Mine, is described. Special reference is made to some engineering aspects of the system components, operating practice and an economic appraisal of the system.

1. Introduction

Grootegeluk Coal Mine, near Ellisras in the Northwestern Transvaal is located on the coal bearing strata of the Waterberg coal field and was officially opened on April 15, 1981 to produce blend coking coal for Iscor's blast furnaces at Vanderbijlkpark and Newcastle.

2. Geology

Numerous coal seams occur in the Waterberg coal field over a stratigraphic thickness of 129 m (Fig. 1). The lower 31 m is predominantly arenaceous with three well-developed seams (zones) of dull coal, and is considered equivalent to the Middle Ecca. The three zones, numbered 1 to 3 from bottom upwards, have average thickness of 1.35 m, 3.37 m and 7.82 m, respectively.

Zone 1 contains hardly any bright coal, but the bottom 1—2 m of zones 2 and 3 contain some bright coal and yield a low ash (5—6 %) fraction (RD 1.40) that could be suitable for form coke. Overlying the Middle Ecca is a 37 m layer shale, carbonaceous in the upper part, with two zones of dull coal: Zone 4A, 1.52 m thick and zone 4 on average 4.02 m thick. This predominantly shale succession is considered a transition between the underlying Middle Ecca and overlying 61 m of alternating layers of bright coal and shale, subdivided into seven zones (zones 5—11) and correlated with the Upper Ecca. The bright coal developed in zones 5 to 11 is the source of coking coal in the Waterberg coal field.

Drilling on the six Iscor farms proved in-situ reserves of 578 million tons of coking coal and a similar reserve of middlings suitable for steam generation for zones 5 to 11, as well as 1457 million tons of raw coal in zones 1,2,3,4A and 4.

3. Coal Products

In the first phase of development, the bright coal occurring in zone 5 to 11 is treated in a beneficiation plant to produce blend coking coal for use as a blast furnace feedstock at Iscor's steelworks. A middlings product is also obtained which in turn will be used as a power station feedstock for Escom's Matimba power station which is currently being constructed approximately 6 km away from the mine (Fig. 2). In this phase the production from the mine will be as follows:

Overburden stripping	8×10^6 t/a
Run-of-mine plant feed	15×10^6 t/a
Coking coal product	1.8×10^6 t/a
Middlings as power station feedstock	3.5×10^6 t/a
Plant waste	9.7×10^6 t/a

In the next phase of development it is planned to expand the mine to provide for an increase in the production of coking coal and the additional production of coal for the direct reduction processes. The middlings arising from these beneficiation plants plus the remaining raw coal will be used as power station feedstock for a total power station capacity of 3,600 MW. In this way optimum use is made of the coal reserves for metallurgical and energy generation purposes.

4. Mine Layout

The mine layout has been done for a 40 year period in order to cater for the demands of the power station over its expected life (Figs. 3, 4).

The Enkelbult pit, which is one of the coal sources, has been designed with an overburden bench height of 15 to 20 m, followed by four benches of approximately 17 m each in bright coal. Below bench No. 5, a separate selective mining operation has been planned for the extraction of raw coal from the lower four seams as a feedstock to the direct reduction coal plants.

Access to the various benches is by means of a double ramp system with a gradient of 1 in 12 and a road width of 50 m on the south side and 35 m on the north side. The ramps were designed to provide the shortest possible distance between the coal faces and the primary crushers.

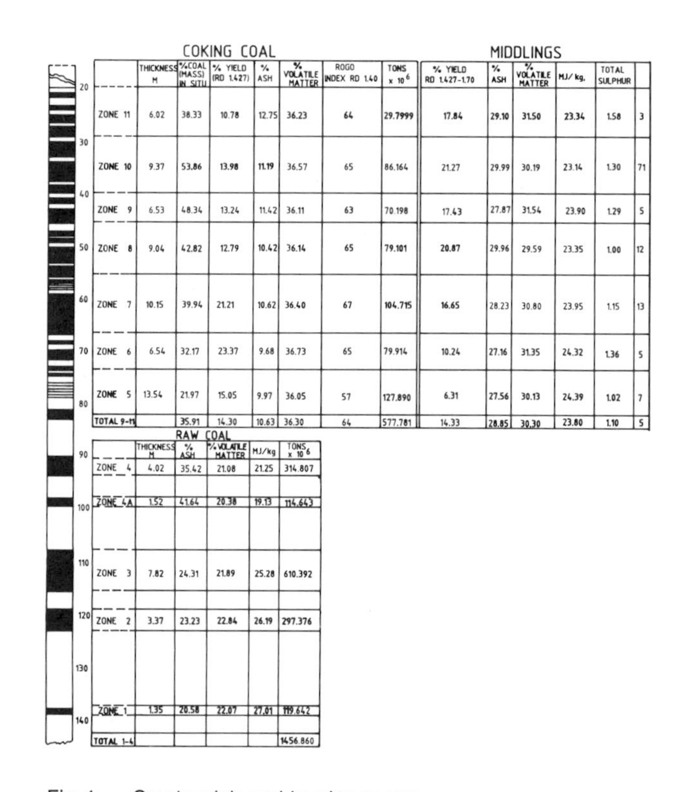

	COKING COAL							MIDDLINGS					
	THICKNESS M	%COAL (MASS) IN SITU	% YIELD (RD 1.427)	% ASH	% VOLATILE MATTER	ROGO INDEX RD 1.40	TONS x 10^6	% YIELD RD 1.427-1.70	% ASH	% VOLATILE MATTER	MJ/kg.	TOTAL SULPHUR	
ZONE 11	6.02	38.33	10.78	12.75	36.23	64	29.7999	17.84	29.10	31.50	23.34	1.58	3
ZONE 10	9.37	53.86	13.98	11.19	36.57	65	86.164	21.27	29.99	30.19	23.14	1.30	71
ZONE 9	6.53	48.34	13.24	11.42	36.11	63	70.198	17.43	27.87	31.54	23.90	1.29	5
ZONE 8	9.04	42.82	12.79	10.42	36.14	65	79.101	20.87	29.96	29.59	23.35	1.00	12
ZONE 7	10.15	39.94	21.21	10.62	36.40	67	104.715	16.65	28.23	30.80	23.95	1.15	13
ZONE 6	6.54	32.17	23.37	9.68	36.73	65	79.914	10.24	27.16	31.35	24.32	1.36	5
ZONE 5	13.54	21.97	15.05	9.97	36.05	57	127.890	6.31	27.56	30.13	24.39	1.02	7
TOTAL 9-11		35.91	14.30	10.63	36.30	64	577.781	14.33	28.85	30.30	23.80	1.10	5

RAW COAL

	THICKNESS M	% ASH	% VOLATILE MATTER	MJ/kg	TONS x 10^6
ZONE 4	4.02	35.42	21.08	21.25	314.807
ZONE 4A	1.52	41.64	20.38	19.13	114.643
ZONE 3	7.82	24.31	21.89	25.28	610.392
ZONE 2	3.37	23.23	22.84	26.19	297.376
ZONE 1	1.35	20.58	22.07	27.01	119.642
TOTAL 1-4					1456.860

Fig. 1: Grootegeluk coal bearing seams

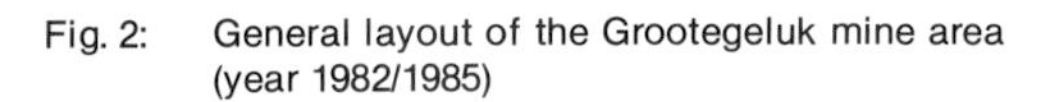

Fig. 2: General layout of the Grootegeluk mine area (year 1982/1985)

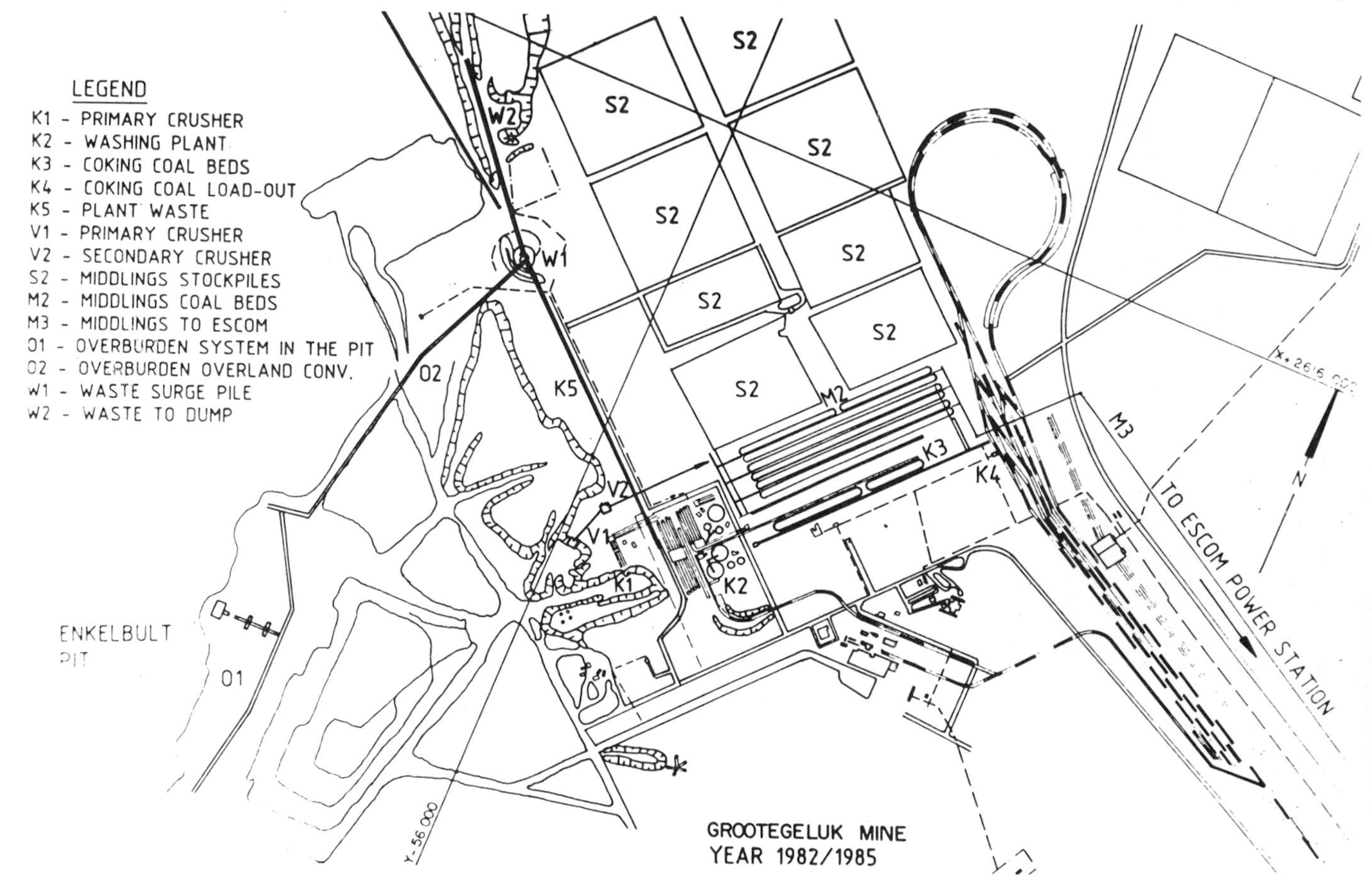

Since it is possible that the plant waste, which contains some 15% carbonaceous material, will burn when stockpiled, it was decided to blend the plant waste with the overburden from the pit in order to increase the ash content of the mixture, and thus prevent spontaneous combustion of the plant waste dumps.

Overburden and plant waste will initially be dumped east of the Daarby fault, outside of the possible open pit mining area. Back filling of the overburden and plant waste will commence as soon as sufficient pit floor has been exposed.

5. Mining Methods

5.1 Overburden Stripping

The original pit layout was based on the use of 150t diesel-electric rear dump trucks for the transportation of overburden.

A feature of the pit layout is long, straight mining faces — plus the crushing and mixing of the overburden with plant waste. The fuel costs involved in using dump trucks would have proved prohibitive and it was clear that alternative stripping methods had to be investigated.

The following alternative mining, crushing and transport schemes for in-put overburden stripping were studied with regard to practicability, capital and working costs:

- shovel and trucks (scheme A)
- shovel, trucks and fixed gyratory crusher (scheme B)
- shovel, trucks and fixed impact crusher (scheme C)
- shovel, trucks and semi-mobile gyratory crusher (scheme D)
- semi-mobile crusher with conveyors on coal surface (scheme E)
- fully mobile crusher with shiftable conveyors (scheme F)

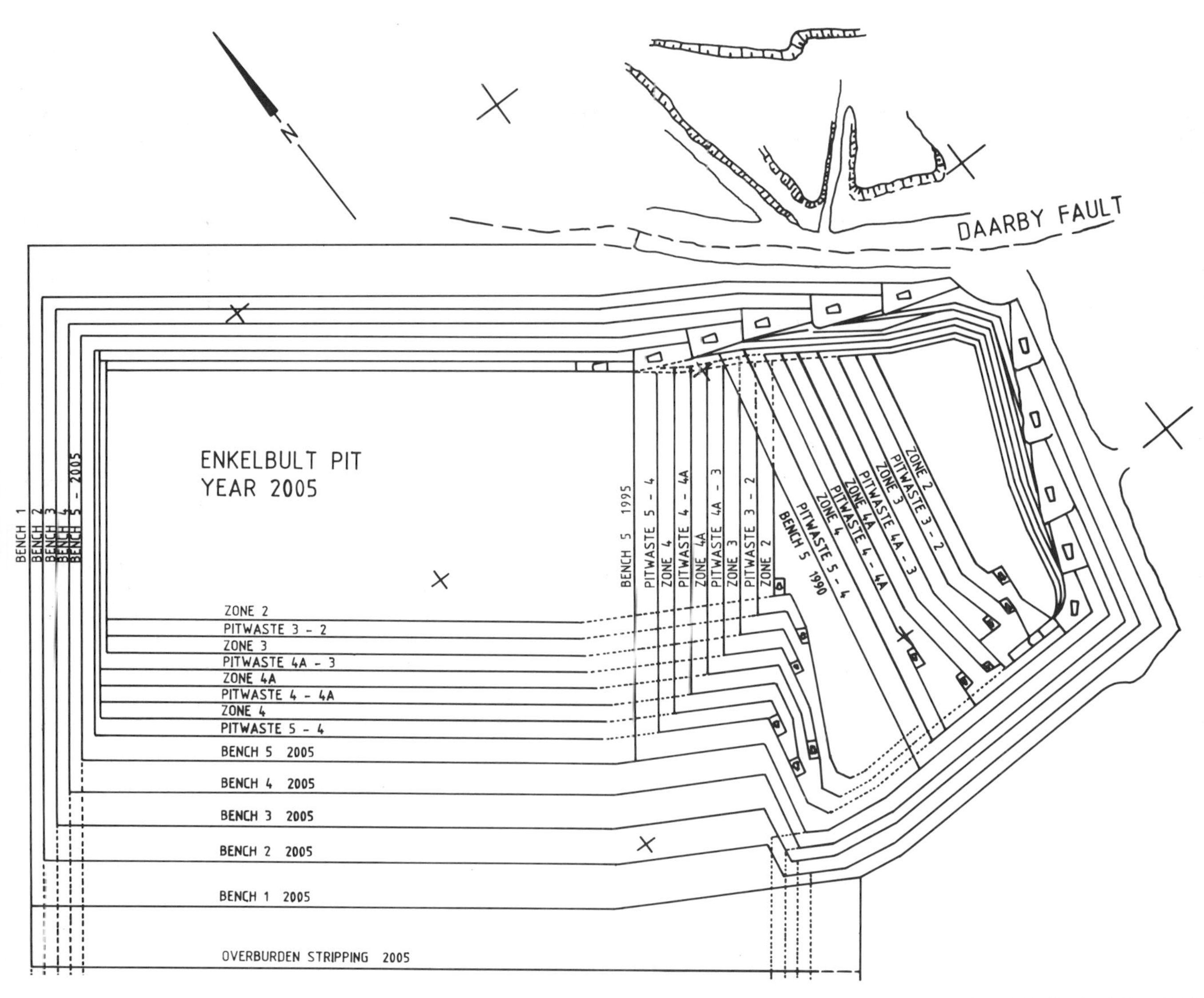

Fig. 3: Layout of the Enkelbult Pit (Year 2005)

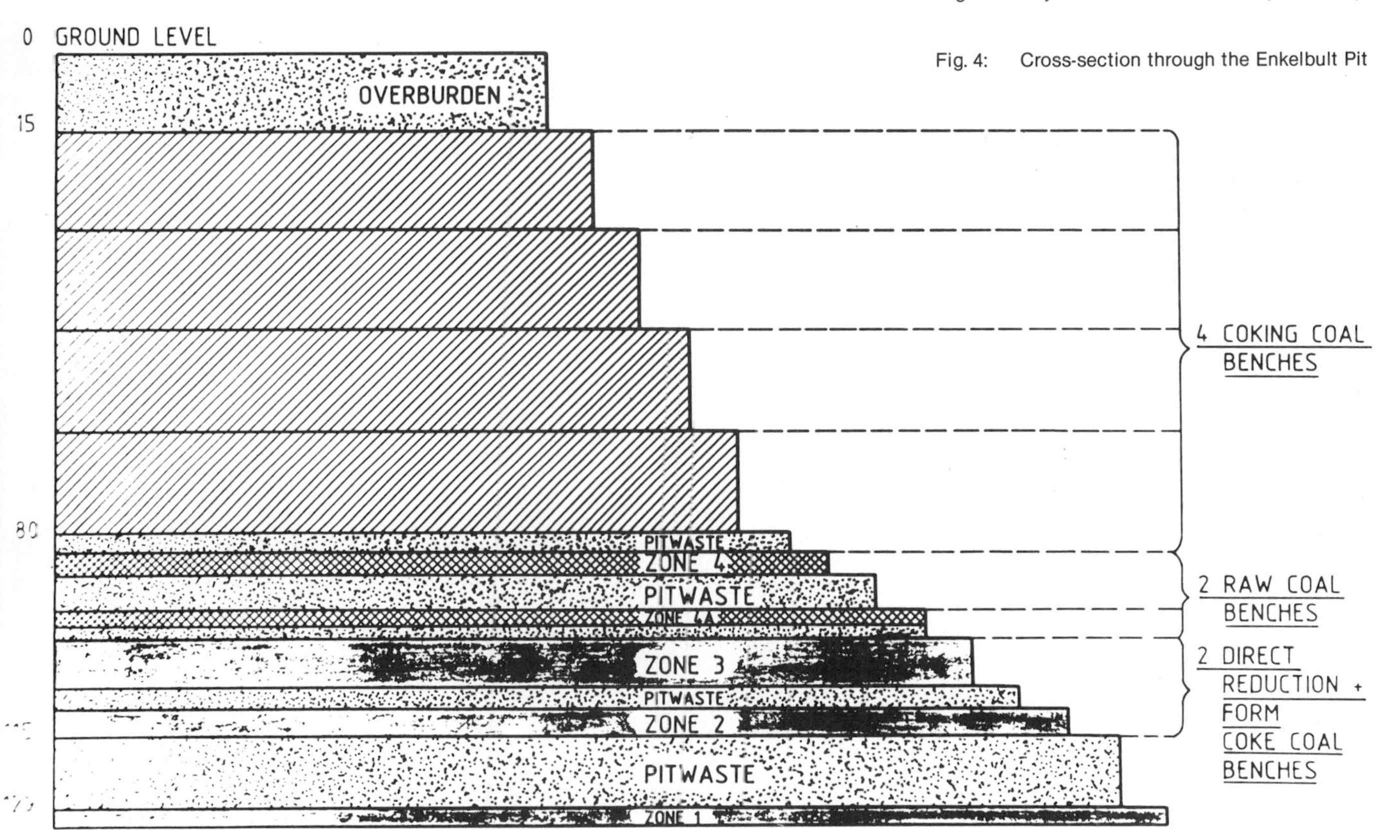

Fig. 4: Cross-section through the Enkelbult Pit

The straight-forward shovel-and-trucks scheme had the advantages of flexibility and low initial capital outlay. However, as the pit working face moved further away, haul distances would increase and more trucks would be required to cope with the required production rate. The main disadvantage — and the deciding factor, therefore — was the amount and cost of fuel required and the total operating cost.

Other schemes also had plus and minus factors and it was finally decided that fully mobile crushers and shiftable conveyors would be the most feasible and least costly.

The strategic and cost implications of diesel fuel, in addition to other technical aspects of the system, indicated that a healthy return on the capital investment would be realised when compared with a conventional shovel/truck mining system (Fig. 5).

After drilling and blasting, the broken rock is loaded by means of a 17 m³ electric rope shovel directly into the feed hopper of the mobile crusher (Fig. 6). A 110 m long crawler mounted bridge conveyor provides a safety distance between the blasting area and the shiftable conveyors and furthermore increase the effective operating block width and reduces the shifting frequency (Fig. 7). The rest of the transport system consists of 3 shiftable face conveyors, 1 extendable conveyor and 2 overland conveyors (Fig. 8). Mixing of overburden and plant waste and stacking by means of a crawler mounted slewing and luffing stacker is considered to form part of the plant waste stacking system. Figs. 9 and 10 show the relative movement of the main items of equipment.

SCHEME		ESTIMATED COSTS AFTER ESCALATION (COSTS ESCALATED FROM A BASE DATE OF JUNE, 1979					
		A	B	C	D	E	F
TOTAL CAPITAL	R/ton	0,089	0,214	0,196	0,214	0,222	0,289
	Present Value	2 702	6 527	5 967	6 495	6 777	8 808
OPERATING AND MAINTENANCE COST							
Fuel	R/ton	0,203	0,133	0,133	0,155	0,117	-
	Present Value	6 184	4 059	4 061	4 723	3 556	-
Tyres and Belting	R/ton	0,094	0,090	0,090	0,080	0,078	0,041
	Present Value	2 862	2 737	2 737	2 449	2 375	1 261
Other	R/ton	0,158	0,183	0,232	0,168	0,160	0,198
	Present Value	4 809	5 577	7 078	5 099	4 880	6 010
TOTAL OPERATING AND MAINTENANCE	R/ton	0,455	0,406	0,455	0,403	0,355	0,259
	Present Value	13 855	12 373	13 876	12 271	10 311	7 271
REPLACEMENT COSTS AND REMAINING VALUE	R/ton	-0,020	-0,093	-0,085	-0,098	-0,112	-0,149
	Present Value	- 588	-2 856	-2 597	-2 939	-3 424	-4 527
TOTAL COSTS	R/ton	0,524	0,527	0,566	0,519	0,463	0,379
	Present Value	15 969	16 044	17 246	15 827	14 164	11 552

Savings in Cost expressed as a D.C.F. Return (% p.a.) in respect of incremental capital expenditure.

21,4%
17,6%
44,8%
22%
28,2%

Fig. 5: Economic evaluation of various schemes for overburden stripping

Fig. 6: Mobile crusher/face shovel

Fig. 8: In-pit crusher belt conveyor system

Fig. 7: Mobile bridge conveyor

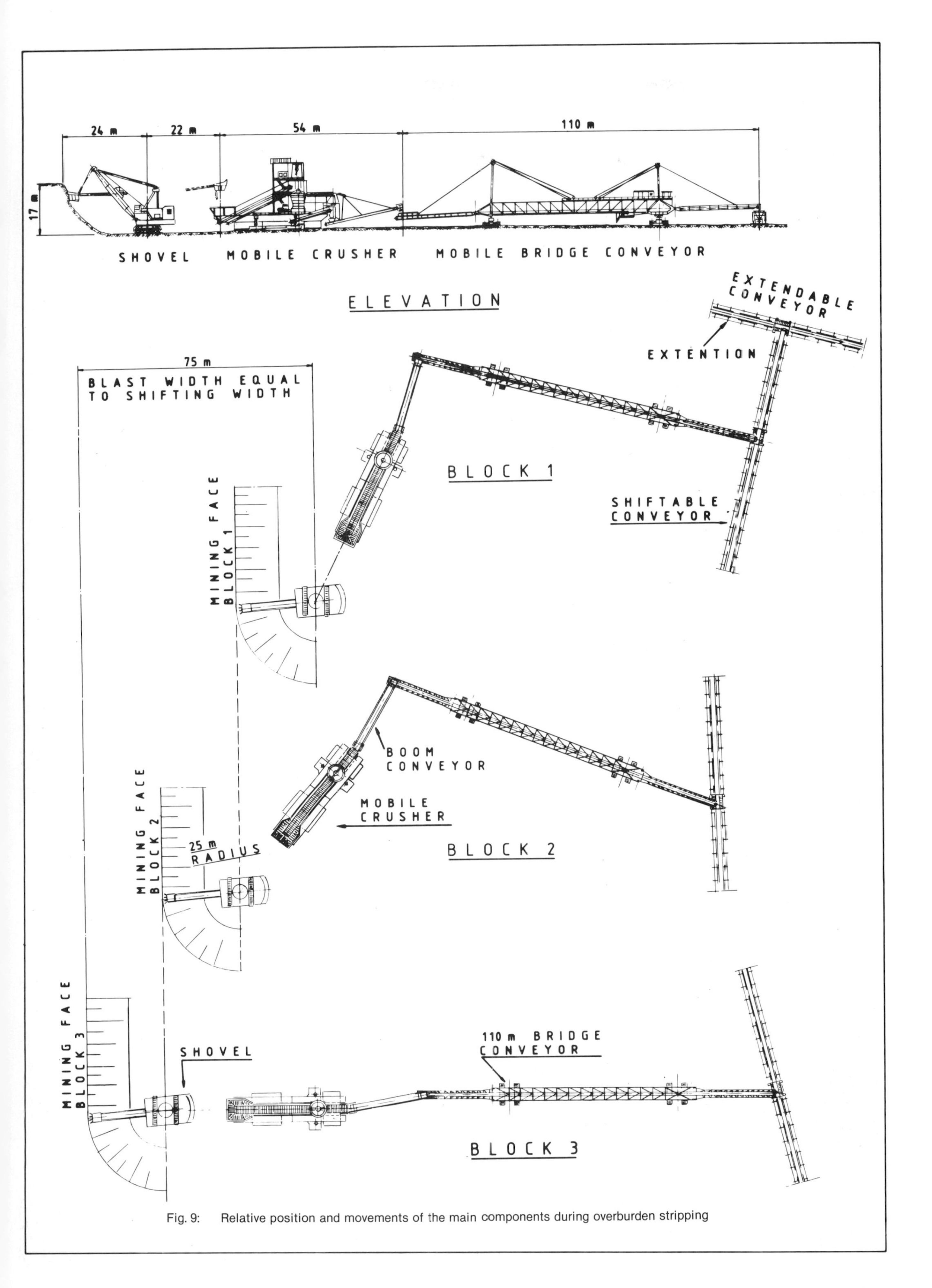

Fig. 9: Relative position and movements of the main components during overburden stripping

Fig. 10: Relative position of mobile crusher to working face

5.2 Bright Coal

Mining of the bright coal is carried out on four benches, approximately 17 m each, by means of drilling and blasting, loading with 22 m³ electric rope shovels and transporting to the beneficiation plant with 150 ton diesel electric rear dump trucks.

5.3 Dull Coal

The mining of the four seams of dull coal and their respective partings will be a selective mining operation with smaller equipment, for example 5 to 8 m³ shovels and 50 ton rear dump trucks.

6. Main Features of the In-Pit Crushing System

6.1 Shovel

The shovel is a P&H model 2300 LR fitted with a 19.8 m boom. The longer boom was chosen to maximise the amount of rock that could be loaded before the crusher has to be moved to a new position. The dipper size had to be reduced from the standard 22 m³ which is used in the coal to 17 m³ (struck). This machine is fitted with the swing drive of the next larger P&H shovel i.e., the 2800 to compensate for the longer boom. The hoist has also been increased in capacity with the result that the cycle times of this unit is similar to the standard unit.

Initial calculations indicated that the loading capacity would be in the order of 2,400 t/h based on the performance of the standard unit in combination with rear dump trucks. In practice it was found that an instantaneous loading rate of some 3,000 t/h can be achieved, mainly because no delays are experienced as is the case with spotting of trucks.

6.2 The Mobile Crusher Unit (Fig. 11)

The mobile crusher is a PHB-Weserhütte design with gyratory crusher type 54/74 and hydraulic walking mechanism fitted with 3 x 200 kW hydraulic pump drives.

Working weight:	1,100 t
Installed power:	approx. 1,300 kW
Material:	shale and sandstone, 0—1,800 mm 1.4 t/m³
Throughput:	3,000 t/h, 0—200/250 mm
Feed Hopper:	Capacity 60 m³
Apron Feeder:	2,400 mm wide, centre distance 24 m, incline 25°, 0.08 to 0.24 m/s variable speed drive, required power 2 x 200 kW
Spillage chain conveyor:	2,800 mm wide, centre distance 17.8 m, motor 2.2 kW
Crusher:	Allis-Chalmers gyratory crusher type 54/74, motor 400 kW, air cooling 75 kW
Crusher outlet Apron Conveyor:	2,600 mm wide, centre distance 12.1 m, motor 75 kW
Spillage Conveyor:	2,800 mm width, centre distance 9.1 m, motor 2.2 kW
Slewing Conveyor:	1,800 mm wide, centre distance 23.2 m, motors 2 x 45 kW
Mounting:	Raft — type coordinated walking mechanism speed up to 0.9 m/min., ground pressure when walking 1.5 kg/cm³, when working 1.0 kg/cm², motors 3 x 200 kW, 1 x 45 kW
Other features::	Maintenance crane, dust suppression.

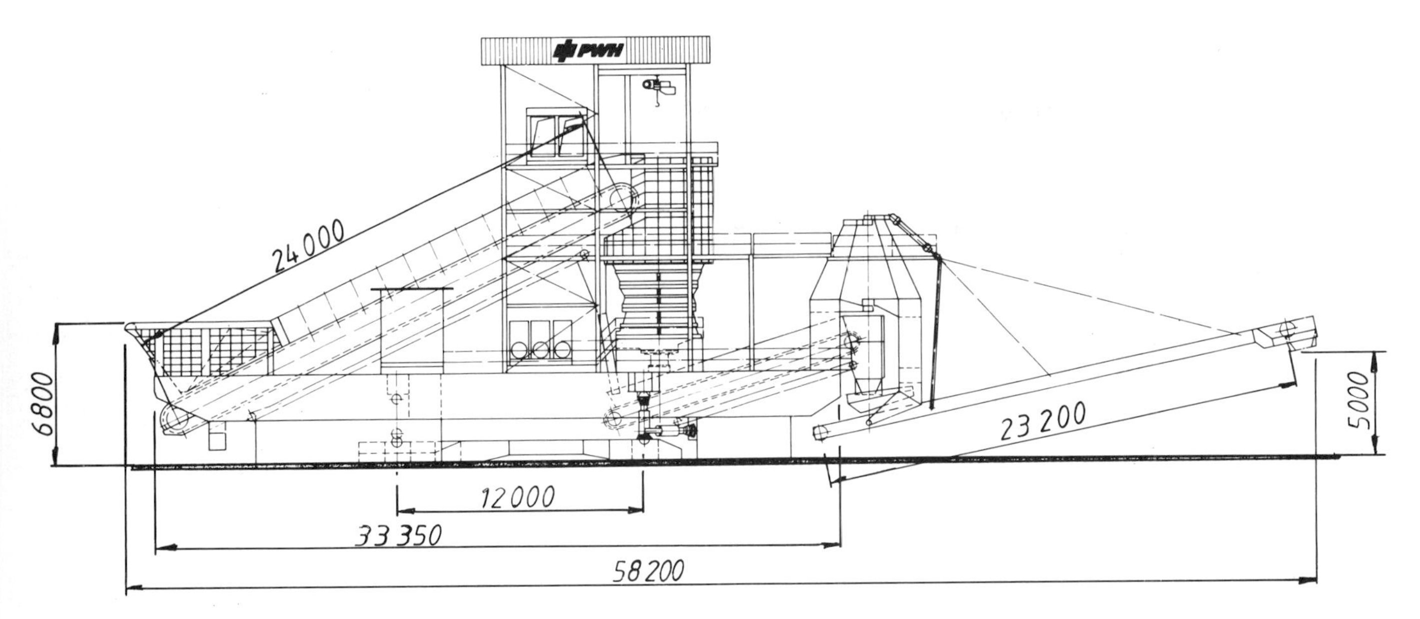

Fig. 11: The mobile crushing unit

Fig. 12: Operation of walking mechanism

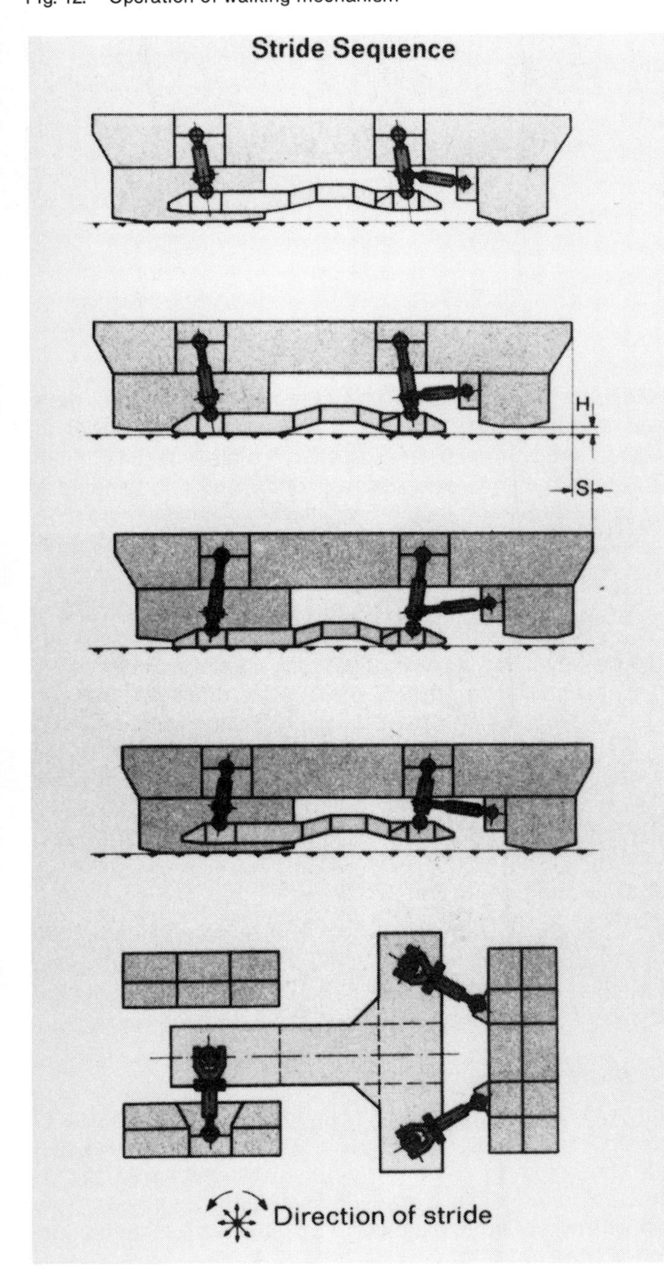

Run-of-mine rock is loaded into a 60 m³ feed hopper directly onto the apron feeder, which has been designed to absorb shock loads in the loading area such that no permanent deformation of apron feeder flights can take place. The feeder is a PHB-Weserhütte design fitted with a Caterpillar type D10 double chain and is driven by two variable speed hydraulic motors to ensure a controlled feed to the crusher. The whole feeder unit can be moved backwards or forwards to enable control of the rock trajectory into the crusher.

The mobile crusher is propelled by means of a coordinated hydraulic walking mechanism, consisting of a T-shaped walking pad, three lifting cylinders and three striding cylinders (Figs. 12 and 13). Through the press of a button any one of eight travel directions can be selected and in addition the installation can be turned on the spot. This high degree of manoeuvrability facilitates accurate repositioning which takes place once per day. During the crushing operation the mobile crusher rests on pontoons so that the walking mechanism is completely relieved.

Invaluable experience has already been gained with this installation whilst certain aspects can only be finally judged over a longer timespan, for example:

— Whether the simpler oil to air crusher lubricating oil cooling system selected (instead of more costly and complicated refrigeration units) is adequate for the worst operating conditions must still be seen.

Fig. 13: Typical part of walking pad with one lift and one stride cylinder

— Oversize boulders are easily loaded into the feed hopper because of the shovel dipper size. A rock breaker on a hydraulic articulated boom, as used in other similar installations, would reduce downtime due to crusher blockages.

— The hydraulic circuit has been designed to operate the three sets of walking cylinders in the walking mode as well as the variable speed motors of the apron feeder in the crushing mode. This design eliminates the need for an additional power source to the apron feeder, but results in a relatively complicated hydraulic circuit which has worked well in practice but still needs to stand the test of time before a final evaluation can be made. The alternative which had been considered was to install a variable speed electric drive to the apron feeder.

— The ground bearing pressure to which the walking pad and the pontoons are designed, needs special thought. If too low a value (conservative value) is selected, the bearing areas are designed unduly large. If the low value does not materialize in practice, undue stresses are induced due to point loading over the larger area. In addition the machine does not bed down as well for proper stability.

6.3 The Bridge Conveyor

The crawler mounted bridge conveyor is a PHB-Weserhütte designed unit with hydraulic travel drives.

Length between pulley centres: 110 m
Belt width: 1,350 mm at 2.62 m/s
Belt drive: 1 x 132 kW

The unit is fitted with an operator's cabin from where the crawlers are controlled. An improvement here would be to install a pendant mounted remote control unit which would allow the operator to relocate the machine from ground level. Accurate positioning is necessary, especially at the discharge end and at least two people are required to carry out this move with the present arrangement.

6.4 Shiftable Face Conveyors

Three shiftable conveyors are used, each having a length of 410 m, 2 x 132 kW drives, 1,350 mm belting at a speed of 2.62 m/s. Careful consideration was given to a single conveyor vs the three conveyor system. Advantages of the three conveyor system are the following:

(i) A straight mining face need not be maintained, which increases the flexibility of the system. During start-up a curved mining face had in fact to be contended with, and future planning necessitates the use of curved mining faces.

(ii) With three conveyors, the shifting operation can commence whilst the system is producing on the conveyor nearest to the extendable conveyor thereby reducing the total conveyor shifting downtime.

(iii) Upstream conveyors not in use can be switched off thereby reducing unnecessary wear and power consumption on the empty belt portions.

The conveyors are shifted by means of a shifting head which is mounted on a crawler dozer (Figs. 14 and 15).

Conveyor old to new positions are normally between 50 and 100 m apart. At present the shifting time is two days per conveyor with no shifting done during the night, mainly because of a shortage of personnel.

As experience is gained, the time required to move a conveyor is expected to decrease. The major problem in shifting

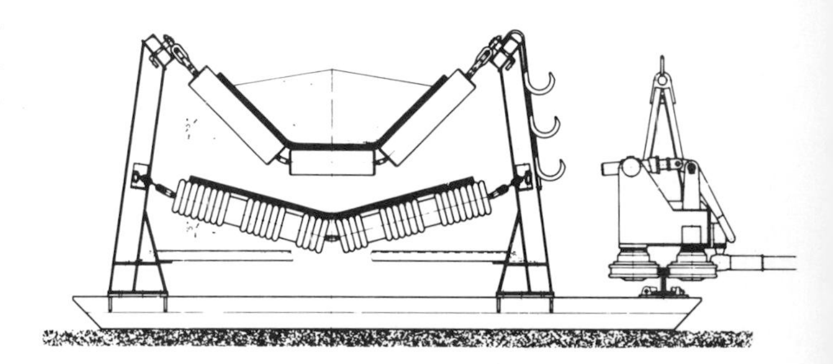

Fig. 14: Typical cross-section through a shiftable conveyor with shifting head

Fig. 15: Typical shiftable conveyor being shifted

the snaking conveyors is to avoid moving the conveyor more than approximately 1 m per traverse. Although larger increments may appear to be quicker, it usually results in subsequent breakages of fishplates, bolts, etc. and bending of the rails connecting the conveyor tables. An even more problematic result is the creep of the conveyor tables along the rails, resulting in uneven spacing of the tables.

6.5 Extendable Conveyor

This conveyor will be extended from its minimum length of 80 m as installed to a maximum of 875 m. The drive station is fitted with 44 m of belt storage, which means that new belt sections will not be spliced in with every extension of the conveyor. One 250 kW drive has been installed initially and two more of 250 kW each will be added as the conveyor length is increased.

6.6 Overland Conveyors

Two overland conveyors deliver the overburden onto the plant waste system where it is mixed with plant waste. Each is fitted with 3 x 250 kW drives and the conveyor lengths are 845 m at 26 m lift and 650 m at 35 m lift respectively.

6.7 Waste and Overburden Stacking

Plant waste and overburden is currently stacked by means of a temporary dual stacking system which had been designed with the object of creating so-called test stockpiles to test the combined stacking of plant waste and overburden. This part of the system is planned to be replaced by the final design in the near future (Fig. 16).

Fig. 16: PHB Weserhütte is building a similar crawler mounted luffing and slewing stacker as part of the final design of the system

7. Operating Practice

7.1 Drilling and Blasting

Blast holes are drilled at 200 mm dia with 6.8 m x 7.3 m burden and spacing, by means of a Bucyrus Erie 40R rotary drilling machine. Penetration rate is approximately 25 m per hour and rotary bit life in the order of 12,000 m.

Sub drilling of 1.5 m is normally applied for adequate floor control and a stemming height of 3.5 m, after filling with Anfex is found to give optimum results.

Flyrock was initially expected to be one of the main problems associated with this continuous mining method. It has been found in practice that the shiftable conveyors could be safely moved to as close as 110 m away from the face to be blasted without undue damage to the equipment.

7.2 Loading

With proper setting up of the shovel/crusher positions instantaneous loading rates of up to 3,000 t/h over short periods have been attained, provided that a short boom swing arc can be maintained. Approximately 65,000 tons can be loaded before a crusher move becomes necessary. Careful planning is therefore essential in order to minimize the number of equipment moves.

7.3 Crushing and Conveying

The crusher has been performing well and it has been found that the very small buffer space between the crusher outlet and the discharge apron conveyor is adequate, due to the controlled feed obtained with the input variable speed apron feeder.

With certain types of overburden the dust created can be a problem. Dust is suppressed by means of water sprays on top of the crusher, but this water addition creates severe problems downstream, due to build-up in chutes and on idlers.

Modifications to chute designs had to be carried out to prevent spillage. The average feed rate is normally set at 75 % of full capacity in order to ensure long runs without blockages or excessive spillage, which results in an average production rate of 1,800—2,000 t/h under present conditions.

8. Economic Evaluation

On the basis of the results of the feasibility study (Fig. 5) the mobile in-pit crusher system was installed at Grootegeluk. After more than a year of production, the question arises — was the right decision made? In order to answer this question, the system as implemented was again evaluated on the basis of the actual capital and working costs as found in practice. Of the systems considered in the original feasibility study, the second best alternative was as described as scheme D, i.e., trucks and semi-mobile crusher (outside the pit). The two systems were again compared on a "with escalation" basis. The waste tonnages used in this study are considerably higher than for the original feasibility study, due to the anticipated increase in the coking coal production from Grootegeluk. On a July 1982 price basis the comparison of the two schemes are as follows:

	Trucks and Crusher Outside Pit	Mobile In-Pit Crusher
Present value of Capital Expenditure (in million Rand)	9.1046	13.702
Total unit cost (after escalation) (Rand/ton)	0.674	0.558

From these results it is apparent that the capital cost of the mobile crusher scheme was higher than anticipated but still shows a D.C.F. rate of return advantage of 22.1 % on the differential capital between the two schemes.

9. Future Applications

An inspection of the final pit shape as shown in Fig. 3 indicates that the haul distance for coal extraction to the primary crusher increases to approximately 6 km and it appears that in-pit crushing offers one alternative method for this part of the operation. In this instance one needs to contend with four mining benches, and an in-pit crusher evaluation would probably resolve around the decision to install a mobile crusher on each bench to eliminate trucks altogether or alternatively, a semi-mobile unit installed at the pit edge to reduce the truck haulage distance. Another alternative method would be to install an electric assist line for the rear dump trucks. A decision to proceed with this latter system was recently taken and it is currently being installed at the mine.

A study was recently carried out for an open pit application with a 60 to 90 m thick overburden layer in the Waterberg coal field. In this study it was found that in-pit crushing of overburden and around-the-pit conveyors and waste stacking compared favourably with a mining system comprising shovels and trucks in a pre-benching operation, together with a dragline in the bottom portion of the overburden. The main advantage of the in-pit crusher in this instance, is that the width of the coal pit could be increased to any given size, whereas the working space becomes very limited at this depth, when using a dragline.

In general it can be stated that, given the correct application, mobile in-pit crushing offers a viable alternative to shovel/truck operations in opencast mining.

10. Conclusions

At the time when the decision was taken to go ahead with the mobile in-pit crushing system for overburden, no other application of similar magnitude and operating conditions existed with which a parallel could be drawn. Now after twelve months of limited operating experience, it can be safely said that most of the objectives have been achieved. The system design capacity has been demonstrated over limited periods and the shifting of components are being carried out within reasonable time with the prospect of further improvement. The question of total system availability over long periods, under full load conditions, is a function of the maintenance and operating stoppages which will only be known once the final design for the stacking system has been installed. There are no indications, however, that the required availability standards could not be achieved and we are of the opinion that this continuous opencast mining system has been proven to be technically and economically viable.

11. Acknowledgement

The authors wish to thank the Management of the South African Iron and Steel Industrial Corporation Limited for permission to publish this paper. Mr. B. C. Alberts, Divisional General Manager (Mining), and other members of Iscor's staff are especially thanked for assistance in completion of the paper.

Volume 2, Number 3, September 1982

In-Pit Crushing and Conveying

W. Guderley, Germany

Summary

The author reviews the factors leading up to the selection of In-Pit Crushing and Conveying technology and discusses the currently available options in equipment selection.

1. Introduction

The enormous rise in oil prices in recent years has led to increasing interest in a technology that has been available for nearly twenty years but which has only slowly been taken up by the mining industry with its rather conservative outlook.

The technology referred to is that of the increased use of rubber belt conveyors to replace heavy trucks for the transport of mined material from hard-rock and mineral open pit mines (see Fig. 1).

Fig. 1: In-pit crushing and conveying system

In this case it is particularly the, often considerable, height difference between the working face and the treatment plant which causes the high transport costs.

At the beginning of the 1960s the accent was clearly most heavily on the reduction in operating costs by a reduction in the number of operating personnel. Today the increased oil prices have driven companies to investigate other possibilities of reducing costs, in particular those concerned with fuel (Fig. 2).

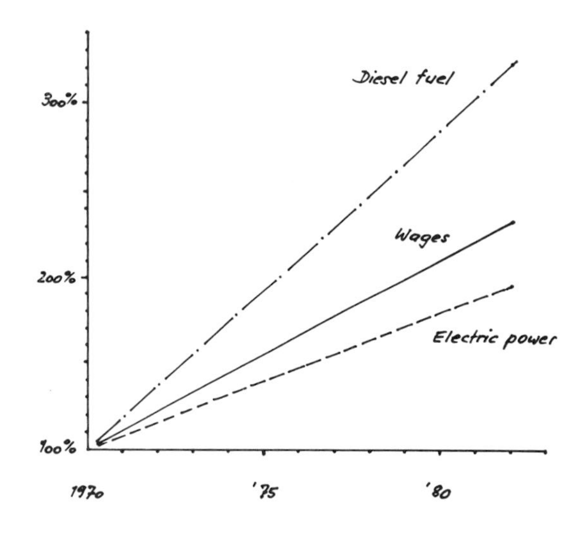

Fig. 2: Comparison of diesel fuel, electric power and wage increases since 1970

One of the major plus points of heavy truck (HT) operation which is constantly emphasised is the increased flexibility of operation and their ability to operate at reduced output (even if half the trucks are out of service the other half are still running). Data from actual operating practice with belt conveyors show an availability factor of 90 %; when coupled with a mobile crusher 85 %; so that these arguments can be seen to be of limited validity.

The cement industry has realised applications for conveyor belt technology in the development of marl and chalk quarries where the ground has shown such a low bearing capacity that the operation with heavy trucks was impossible because of the prohibitive costs involved in laying roads (Fig. 3).

W. Guderley, Chief Engineer, PHB Weserhütte AG,
D-4970 Bad Oeynhausen 1, Federal Republic of Germany

Fig. 3: Mobile crushing unit operating on difficult ground

On the other hand in most cases a large boulder size in the excavated material is a distinct disadvantage for conveyor transport. The material must therefore be crushed to a "belt-ready" size before transport. This crushing should be carried out as near to the work face as possible so as to achieve with the crushing an early blending effect and to regulate the material flow rate.

The smaller the output variations can be held, the better (and hence more economically) the subsequent treatment plants can achieve the required throughput.

For the crushing of such material there are three basic types of crusher available: stationary, semi-mobile and mobile.

2. Stationary Crushers

A first step in introducing conveyor belt technology is often the placing of a fixed crushing plant in the mine. This is usually positioned so as not to inconvenience the extension of the mining operations, i.e., on the edge of the deposit. This often means, however, that in the course of operations the haulage distances become longer and when mining deep-seated deposits steep gradients will have to be negotiated.

The face output and the seam dimensions are the important factors in siting a crushing facility in open pit mines. It is important that, as far as possible, steep trails are avoided, and when unavoidable that they be negotiated with empty trucks.

3. Semi-Mobile Crushers

When operating a number of mining faces concurrently or alternately, a stationary crusher is not to be recommended. In this case a semi-mobile crusher is the better solution. This can be built as a complete unit or in several sub-units and is placed in the main working area. The haulage distances are kept short and the unit can be moved, following the work face, once the haulage distances have become too long. A further advantage of the semi-mobile equipment over fixed equipment is that usually only minimal foundations are required, or even none at all. Thus when the equipment is moved no costs involved in foundation work are lost. Units which are dependent on each other, e.g., feed hopper with feed apron-conveyor or crusher with power unit, can be transported as a single unit thus not requiring extensive set-up and testing procedures (Fig. 4).

Fig. 4: Semi-mobile crushing unit

Once the new site is prepared the move of the semi-mobile plant can be carried out in only a few hours with comparatively minor effort.

For the transport between sites one can use various techniques depending on the plant-layout, the ground conditions and not least the distance involved. The following vehicle types are used:

- Connectable wheeled (tires) vehicle
- Connectable rail vehicle
- Fork-lift car or Piggyback transporter (Fig. 5)
- Crawler-mounted transporter (Fig. 6)

Fig. 5: Connectable wheeled (tires) transport vehicle

Fig. 6: Transport crawler

4. Mobile Crushers

The use of a mobile crusher is indicated where a wandering seam is mined and the loading machinery loads the mined material directly into the crusher, without any intermediate transport. Mobile crusher and loader work as a unit (Fig. 7). Variations in grade which can be controlled by selective mining in the methods above cannot be dealt with here and where necessary a blending pile must be used between workface and processing plant.

Fig. 7: Mobile crusher and shovel working as a unit (see Fig. 1)

This is particularly true when using a dipper shovel for loading operations. The shovel must load that material which is within reach of its shovel. However, due to the permanent availability of the loading point (feed hopper of the mobile crusher) its loading capacity can increase by up to 40 % as compared with heavy truck loading.

A blending effect can be achieved, albeit a comparatively minor one, when the mobile crusher is fed by a wheel loader. This can bridge distances of up to 70 m between feed point and the mobile crusher without any significant loss in throughput. Even with an action radius of 120 m, 60 % of the maximum feed rate is possible (Fig. 8).

Fig. 8: Wheel loader feeding mobile crushing unit

While the type of machinery used (dipper shovel, wheel loader or trucks) decides the layout of the feed assembly, the choice of crusher is governed by the feed boulder size, feed characteristics, required throughput and ground pressures. Mobile crushers are in action with a throughput of up to 3,000 t/h and a service weight of 1,000 t. Projects for throughputs of 6,000 t/h are in progress (Figs. 9, 10).

The soil characteristics and the proposed mining system determine the choice of transporting mechanism. The most useful are:

— Hydraulic walking mechanisms (pads) (Fig. 11)
— Crawler tracked chassis
— Wheeled (rubber tires) chassis (Fig. 12)
— Rail mounted chassis.

Fig. 9: Large mobile crushing unit during erection on site

Fig. 10: Mobile crusher after erection

Fig. 11: Hydraulic walking mechanism

Fig. 12: Wheel mounted mobile crusher

The speeds of travel vary between 2.0 m/min and 30 m/min. Ground pressures of between 60 kPa and 500 kPa are possible.

5. Conclusion

Up to now approximately 80 mobile or semi-mobile crushing units are in operation worldwide. Cost analyses have clearly demonstrated that savings in operating costs of up to 40 % compared to conventional heavy truck operation are possible.

A typical installation of such machinery manufactured by PHB Weserhütte is shown in Figs. 13 and 14. The Phosphate Development Corporation (Foskor) Phalaborwa deposit in the Northeast Transvaal, South Africa, represents one of the most up to date applications of in-pit crushing and conveying technology available.

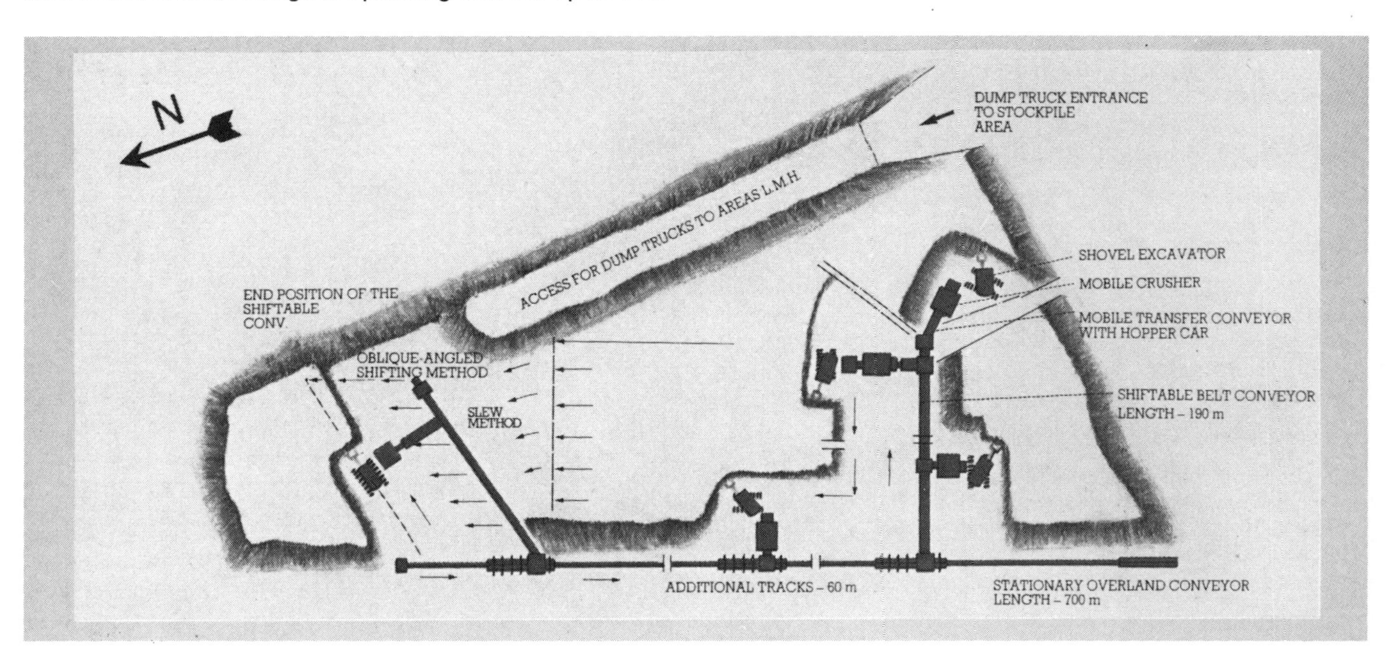

Fig. 13: Foskor Phalaborwa complex, South Africa

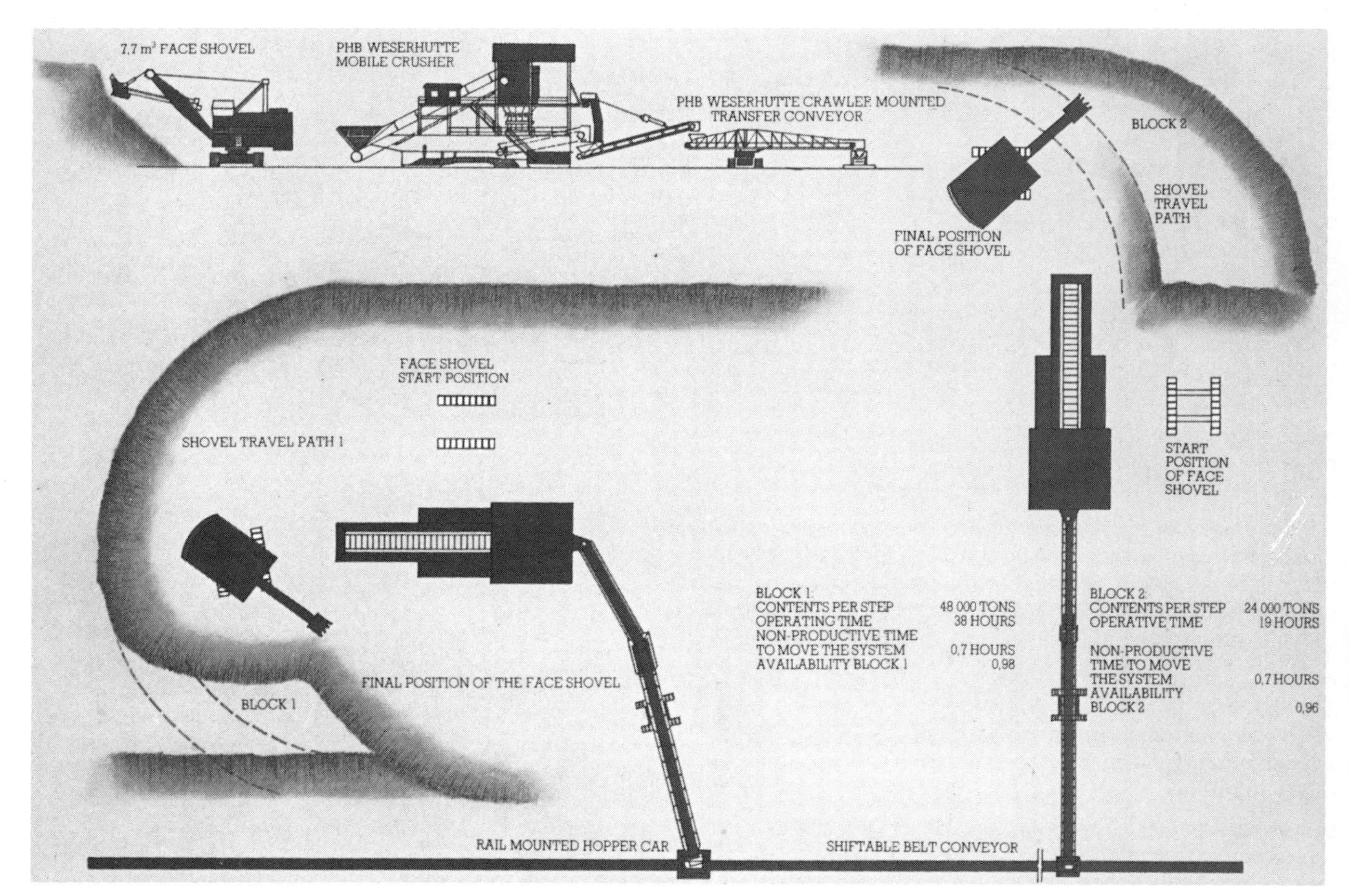

Fig. 14: Foskor Phalaborwa material handling plan

Cost Considerations for In-Pit Crushing/Conveying Systems

Conrad Huss, USA

Summary

This article focuses on continuous materials handling systems used in the hard rock open-pit mining industry. Guidelines for mine planning with continuous systems are outlined. Key parameters affecting capital and operating costs are identified. Commonly encountered trade-offs between operating and capital costs are discussed. A typical range of costs is presented based on average operating conditions.

1. Definitions

Because in-pit/crushing/conveying is relatively new to the industry, the vocabulary used is still in a state of flux and causes considerable confusion at times. The following definitions will be used in this discussion:

Near-Pit Crusher: Crusher located outside the pit but adjacent to the perimeter, e. g. Duval Sierrita's 60 inch gyratory waste crushers, Gibraltar's 54 inch gyratory ore crusher.

In-Pit Crusher: Crusher located within the past or future influence of the ore body, e. g. Anamax Twin Buttes' 54 inch gyratory waste crusher and Cyprus Bagdad's 60 inch gyratory ore crusher.

Mobile Crusher System: Crusher and feeder with integral propelling mechanism such as walking pads or permanent crawler, e. g. Palabora's 54 inch gyratory ore crusher.

Portable Crusher System: Crusher and feeder with independent propelling mechanism such as crawler transporter, e. g. Duval Sierrita's 60 inch gyratory ore crusher.

Movable Crusher System: Crusher/feeder with independent propelling mechanism such as crawler transporter and with relocation costs from 10 % to 15 % of capital costs for site preparation and dismantling, e. g. proposed Utah Mines Island Copper 54 inch gyratory ore crusher and Iscor 60 inch gyratory waste crusher.

Conventional Crusher System: Crusher and feeder housed in reinforced concrete structure which is typically located in large, flat area of pit (e. g. Smoky Valley Round Mountain 42 inch gyratory ore crusher and Chuquicamata 54 inch gyratory ore crusher) or sloping pit wall (e. g. Anamax Twin Buttes 54 inch gyratory ore crusher).

Conveyor Flight: Length of conveyor between transfer.

Relocatable Conveyor: Conveyor that has temporary foundations (such as steel tubes or wooden railroad ties) and which can be moved with some dismantling, e. g. Smoky Valley Round Mountain ore conveyor, 36 inch wide by 2,600 ft long.

Shiftable Conveyor: Conveyor that is designed with sufficient flexibility to be shifted laterally intact with all components, e. g. Majdanpek waste conveyor, 60 inch wide by 4430 ft long.

Portable Conveyor: Conveyor that is designed as a rigid unit to be moved intact with drive and all components in any direction, e. g. Duval Sierrita ore reclaim conveyor, 120 inch wide by approximately 100 ft long.

In-Pit Conveyor: Conveyor which is totally or partially within the pit perimeter, e. g. Duval Sierrita ore conveyors, 60 inch wide by 3,700 ft long.

Around-Pit Conveyor: Conveyor that is routed around the perimeter of the pit and is contoured to the terrain with assistance of bridges and earthwork, e. g. Duval Sierrita ore conveyors, 60 inch wide by 11,200 ft long.

Overland Conveyor: Conveyor that is totally outside of the pit perimeter and is contoured to the undisturbed terrain between pit and concentrator or dumps, e. g. Anamax Twin Buttes' Eisenhower ore conveyor, 42 inch wide by 32,500 ft long; Marcona ore conveyor, 36 inch wide by 50,300 ft long; and New Caldonia's horizontally curved ore conveyor, 30 inch wide by 36,000 ft long.

Belt Conveyor: Conveyors which carry material on flat or troughed reinforced elastomer belts, e. g. the cable-driven systems at Newmont Similkameen and Anamax Twin Buttes, and steel-cable reinforced belting systems at Anamax Twin Buttes, Noranda Lakeshore, and Duval Sierrita at maximum inclines of 25 %, 27 % and 29 % respectively. The maximum incline of these systems is a function of the angle of repose of material, i. e., the angle which the surface of a freely formed pile makes to the horizontal.

High-Angle Conveyor: Conveyors capable of carrying material at any angle between horizontal and vertical by mechanical means, e. g., sandwiching material between two belts, or belts with sidewalls and shovel cleats (vertical coal conveyor at Turris Coal Company).

Permanent Conveyor Transfers: Custom designed area where material is transferred from one belt to another belt.

Movable Conveyor Transfer: Transfer which is designed to be moved to new locations and operate within a range of conditions, e. g., conveyor transfers at Southwestern Illinois Coal Captain Mine.

Rotatable Conveyor Transfer: Transfer which pivots about center pin to accommodate shifting of attendant conveyor, e. g. Majdanpek fan-shaped waste disposal system.

2. Mine Planning Guidelines

2.1 Candidates for In-Pit Systems

In-pit systems are operational today with capacities varying from 550 t/h ore systems to 8,000 t/h waste systems. In general, unit cost savings per ton-mile increase with higher tonnages and with increased distances.

Properly designed conveying systems have a conservatively estimated life of twenty years. In general, the remaining mine life should be at least ten years to warrant a replacement conveying system. New mines or mines with obsolete rail system or truck fleet may justify a conveying system for shorter mine life.

Flat areas with short hauls are still best serviced by trucks. Early mine development and haulage between the working face and in-pit crusher will be predominantly by trucks for the next decade. Mature pits with high lifts favor conveying systems, and unit cost savings can increase exponentially with depth of pit due to escalating truck costs in deep pits. Long haulage systems have a significantly lower unit cost per ton-mile than short haulage systems. Mines faced with high diesel costs or lack of diesel because of government restrictions can justify conveyors under the least advantageous conditions if electrical costs per kilowatt hour are in the range of U. S. $0.05, which is the energy cost assumed for this paper.

2.2 Crusher Location

Because of their prototype nature, the first in-pit crushers were housed in conventional reinforced concrete structures. Such installations are too inflexible for most mines, and often result in locations that become outdated within a few years. Within a given pit, more locations become feasible if the same crushers can handle both ore and waste.

The in-pit crusher should be located as close to the working faces as possible. Required safe distance from blasting for a mobile, portable or movable unit is approximately the same as for a shovel. In general, truck haulage runs should be near horizontal (preferably down one or two benches at the most) or upgrade. Ore crushers should normally be located near the bottom of the pit, and waste crushers should be located near centroids of major overburden removal.

The best locations for crushers should be identified by quantifying the amounts of ore and waste to be removed from each of the bench levels. In general, a greater number of smaller mobile or movable in-pit crushing units are advisable when compared to near-pit or in-pit conventional crusher systems. The reasons for this are twofold:

- First, the greater number of units allows the units to be kept closer to more working faces.
- Second, the larger crushers impact disproportionately on mine planning when installed in mobile or movable systems because of weight and height parameters.

In a conventional system, the crusher can be placed deeper into the ground to handle additional height and the concrete can be increased slightly to accomodate additional loads. Intrinsically, the mobile or movable systems are more sensitive to height and weight because of mine feed requirements and potential structural vibration problems. For example, two movable 60 inch gyratory crusher systems can cost approximately as much as three movable 54 inch gyratory crusher systems, yet the three 54 inch crushers can handle greater tonnage.

Locating the crusher is ideally accomplished without disturbing the mine plan. Crusher location impacts on mine planning differently for the mobile, portable, movable and conventional systems. An improper choice could lead to the additional removal of several hundred thousand yards of waste or the blockage of ready access to a portion of the ore body.

2.3 Conveyor Routing

Few pits are fortunate enough to have a stable, permanent face established early in their development on which a conveyor might be routed out of the pit. Most pits are plagued with combinations of problems including localized instabilities, water seepage, restrictive haul roads, dispersed ore zones and multi-mineral ore bodies which can change configuration with changing market conditions. For these reasons, relocatable conveyors should always be used for in-pit installations.

The following parameters are normally used for routing a conveyor:

1. The conveyor should leave the pit at a 25% to 30% slope to minimize length of ascending conveyor, thereby reducing interferences.
2. Routing should minimize the number of transfers (sometimes at the expense of additional earthwork) to minimize belt wear, reduce energy losses associated with vertical drops through chutework, decrease capital and operating costs and increase reliability.
3. Future mining and crusher relocations should be taken into account. Sometimes a slightly longer conveyor will avoid unnecessary future moves.
4. Uphill and downhill portions of conveyor should be balanced within the same flight when possible to minimize energy usage.
5. Conveyor flights should be designed to ensure standardization of transfer drives, belting and pulley assemblies.
6. In general, a combination of earthwork and vertical curves is more economical than trusses.
7. Conveyor embankments that interrupt drainage areas must have properly sized culverts.

8. Roads should be established alongside all conveyors for maintenance. At conveyor transfers, the turning radii of roads should be checked to ensure crane access. Steep roads will have only downhill traffic.
9. Conveyors are better tunneled under than trussed over mine truck roads for safety reasons, although trussing is more common.

3. Major Cost Items

3.1 Transfer Areas

Conveyor transfers are required to accommodate a change in direction of material flow.

A transfer for 60 inch and 72 inch wide belts requiring a high tension drive can be in the range of $ 1 to $ 2 million for civil, concrete, steel, electrical, mechanical, chutework and indirect costs. The cost of an intermediate drive without transfer or of a transfer without a drive is much less.

For example, in the case of cable-driven conveyors, a drive is not required at a transfer if the drive cable tensions for two successive flights are low enough to be within the range of gearbox and other drive technology. In the case of steel cable reinforced conveyors, long, straight high-tension conveyors do not need in-line transfers if intermediate belt drives are used. The drives relieve high belt tensions through friction of a driving belt against the carrying belt.

3.2 Crusher Systems

A crusher system includes a feed arrangement, a method of control, a crusher and a discharge system. For a 54 inch gyratory crusher, costs can vary from $ 6 to $ 12 million, depending upon whether the system is conventional or completely mobile. In general, the more flexible the system the greater the cost.

3.3 Electrical Distribution System

For a crushing/conveying system requiring 15,000 to 20,000 HP, main substation costs can run in excess of $ 500,000. Any new transmission line costs would be in addition to this. The costs for transfer and crusher area substations are additional and are included in the transfer costs.

3.4 Earthwork Costs

In hilly country, earthwork quantities for conveyors and services roads can easily reach 500,000 yd^3/mile. This figure can include ripped cut, blasted cut, non-structural fill and structural fill. In general, transfer areas should be located on cut or structural fill, whereas conveyors and service road beds are often constructed of select run-of-mine waste. Normally, careful mine planning and conveyor routing can avoid much of the non-structural fill costs.

3.5 Belting, Idlers and Conveyor Frames

The actual conveyor runs can account for less than 25% of total system costs. For example, 60 inch wide conveyor belting and idler costs are in the range of $ 200/ft, and steel support frames with foundations are in the range of $ 100/ft.

4. Economic Trade-Offs

4.1 Impact on Mine Planning

Conveying systems are meant to assist mining and should not become an additional problem for the miners to overcome. In general, the most economical system from a strictly conveyor viewpoint is usually not the best overall system because of adverse affects on mining. For example, small savings in capital costs for a conveying system may be more than offset by unneeded mine operating costs.

In general, any excavation done for the conveyor should be excavation already required by the mine plan. Fill material should not be re-handle material where possible.

To avoid unneeded earthwork, conveyors may be lengthened and crusher systems should be relocatable.

4.2 Crusher Systems

Four crusher system types have been identified: Conventional, movable, portable and mobile (Fig. 1).

Fig. 1: M.S.M.E. movable crusher system

The conventional and movable systems are essentially of the same capital cost, but the movable systems require the additional cost of a transporter which can be handled as a capital cost (buy-out) or operating cost (rental). The portable and mobile systems can cost up to twice as much as a conventional system.

The conventional system is suited for the situation where the crusher location will be valid for a least 10 years and has the usual advantage of an enclosed environment. The movable systems address the market where the crusher needs to be relocated every one and one-half to five years. For relocation costs equivalent to 10% of capital costs, moves can be justified as often as every 18 months. For relocation costs of 15%, relocation can be justified as often as every two years.

The portable system has higher capital costs because of its flexibility and should be moved every three months to two years to justify this flexibility. In general, several systems should be serviced by the same transporter. Relocation costs are minimal. The mobile system should be moved weekly or monthly to justify the capital costs associated with its integral propelling mechanism. Relocation costs are minimal. In summary, flexibility, if purchased, must be used to justify its added cost.

4.3 Method of Accessing Crushers

Conventional crushers that are built into a side slope and movable crushers with retaining walls are accessed from one side from the bench above the crusher level. This requires selective location in a locally wide bench for convenient truck access. With proper design, their location physically impacts on two bench levels. Conventional crushers built into a flat area and movable crushers with bridges are often accessed from two sides above the crusher level. This requires selective location in a locally very wide bench for truck access. Their location physically impacts on at least two bench levels.

Portable and mobile crushers can be fed from below the crusher level with the use of an inclined apron feeder. This system requires a long flat area for equipment placement. Their location physically impacts on only one bench level.

4.4 Transfer Areas

Transfer areas should be eliminated where possible. This can often be accomplished by lengthening conveyor flights slightly and assuming a less contoured shape in reaching the final destination. Increased energy costs due to this longer total run should be compensated for by the lesser number of transfer/lift areas. Decreased capital costs should easily offset any increased earthwork requirements. Maintenance of the resulting system should decrease and availability should be greater.

For narrower belts, horizontally curved conveyors have been used to successfully eliminate transfers.

Transfer areas should be portable where possible because of the likelihood of future relocation, planned or unplanned. Structural/civil costs account for less than 25% of the total costs of a drive area. By modifying the structural steel slightly, transfer areas can be made skid-mounted. For low tension drives, the skids can be placed directly on earth. For high tension drives, the skids must be anchored for overturn; a thick slab on grade being a viable option for achieving this.

5. Representative Operating Costs

Operating costs for in-pit and around-pit conveyors are dependent on the length of the system and will range from $ 0.05 per ton-mile for long systems to $ 0.09 ton-mile for short, high-lift systems, including crushing costs.

When comparing these costs with truck costs, it should be noted that the route transversed by such a conveying system will be shorter than truck hauls because the conveyor can climb approximately three times as steeply as a truck.

Overland conveyor costs can be as low as $ 0.02 per ton-mile for downhill systems to $ 0.05 per ton-mile for relatively flat systems. These costs would be for cable driven systems or steel-cable reinforced belting systems.

Energy costs and labor costs vary widely worldwide. The following breakdown represents average values for operating costs of a "typical" mine.

Energy	40%
Replacement Parts	30%
Labor	30%

6. Concluding Remarks

Mature mining properties with average ore bodies and operating conditions with truck haulage may be forced to switch to in-pit crushing/conveying systems to survive the competition of the 1980s and 1990s. The potential cost savings will force rethinking of mine planning to accommodate a combined truck/conveyor mining system.

By the end of the century, continuous systems (i. e. no trucks) will be operational in the hard rock industry.

Bibliography

[1] Almond, R. M., and Huss, C. E., "Open-Pit Crushing and Conveying Systems", Engineering and Mining Journal, June 1982

[2] Cabrera, V., "Conveyor Belts Make Sense for Long Distance Haulage", World Mining/World Coal, July 1982

[3] Proprietary information, Mountain States Mineral Enterprises, Tucson, Arizona.

Acknowledgements

The author gratefully acknowledges the cooperation of his staff members at Mountain States Mineral Enterprises.

bulk solids handling Volume 4, Number 4, December 1984

In-Pit Crushing and Trolley-Assisted Truck Haulage in Hard Rock Open-Pit Mines

H. Müller-Roden, Germany

1. Introduction

The International Conference on "The Planning and Operation of Open Pit and Strip Mines", organized by the University of Pretoria, the South African Institute of Mining and Metallurgy, and the University's Mining Alumni Society, was held at the University of Pretoria Campus from April 9 to 13, 1984.

The conference consisted of four parts, of which one dealt with new developments in open-pit mining, such as in-pit crushing, trolley-assisted truck haulage, and computer applications.

This article presents a critical discussion on the experience gained from that conference and a visit to several mines in the Republic of South Africa.

2. Trolley-Assisted Truck Haulage

In March 1983, Sishen Iron Ore Mine completed the installation of trolley lines and truck conversion in order to implement their system of trolley-assisted truck haulage. Briefly, "*trolley-assist*" is the direct feed of electrical power into the electric wheel motors of haulage trucks while the Diesel engine generates power only to supply the on-board equipment, thus reducing fuel oil consumption on upgrade haulage roads and increasing truck speed and truck fleet productivity (Fig. 1).

A feasibility study carried out in 1980 for a 10-year period, assuming full operation of some 110 Mt/a total transport capacity, resulted in the following major data:

1. Total savings in fuel consumption approximately 120 million l.
2. 54.1 % internal rate of return out of an investment of R 6.7 million with electrification in priority areas only.

These figures were based on the experiences gained from a 700-m test line, with only a couple of trucks being converted to trolley trucks. The basic data of today's trolley system are outlined in Table 1.

Dipl.-Ing. H. Müller-Roden is Research Assistant at the Institut für Bergbauwissenschaften II, Prof. Dr.-Ing. F.L. Wilke, Technische Universität Berlin, Berlin, Federal Republic of Germany

Fig. 1: Trolley truck travelling on well maintained haulage road

Table 1: Basic data of the Sishen Iron Ore Mine trolley system

Total length of upgrade haulage roads	7.7	km
Grade on haulage roads	8	%
Specific electrification costs/m haulage road	320	R
Number of trucks converted (154 t)	66	
Conversion costs/truck	100,000	R
Fuel consumption:		
— upgrade haulage	360	l/h
— horizontal haulage	160	l/h
— upgrade haulage on trolley	80	l/h
— future development: with blowers run by electrical power, Diesel running at low idle (700 rpm)	40	l/h
Local energy costs:		
— Diesel fuel	0.4	R/l
— electric energy	0.03	R/kWh

The costs/m haulage road include the DC substations as well as the 11 kV 3-phase powerline and the connection to the mine's electrical network. No additional power had to be installed at the mine's generating plant; the maximum 0.5-h demand proved to be controllable and no increase in maximum demand was reported.

2.1 Brief Description of the System

2.1.1 Trolley Lines

The trolley line consists of 161 mm² grooved copper railway conductors, two of which are used in parallel per pole. The output voltage is 1,200 V DC. The charging rate varies between 1,000 and 1,200 A/truck, depending on the total truck mass of normally 250 t. An automatic switch-off is provided in the case of truck overload and, consequently, current overdraw. The lines are not earthed. The substations (rectifiers and transformers) are situated next to the line and the output capacity is equal to the power consumption of a single truck, so that the number of substations equals the number of trucks which are on schedule for that particular line.

The mast construction can almost be regarded as movable. The masts are mounted on approximately 4 m x 2.6 m base plates which are covered with about 12 t ore for stability, anchoring being optional. Straight sections' spacing is about 50 m, with less in curved sections according to the radius. The 11-kV line serves as a counterweight to the trolley hangers and the lines are tensioned by means of weights and pulleys.

2.1.2 Power Collectors

Sishen uses pantographs (Fig. 2), each one being 3 m in width. Pantographs were chosen due to their allowances in both directions, horizontal (maximum deviation 2.5 m) and vertical (deviation ± 50 cm). Their main advantage though is that pantographs enable the trucks to enter (and re-enter!) the line at any section. The pantograph maximum lift is 3.2 m, the spacing between haulage road and line varying between 7.8 and 8.8 m. Sishen started out using conventional railway pantographs which needed reinforcement in terms of links and push-up pressure. A 200-N push-up force ensures current transmission with no interferences on even slightly undulating haulage roads.

Fig. 2: Pantographs mounted on a 154-t truck

2.2 Experiences

2.2.1 Truck Speed and Cycle Times

The increase in truck speed on an 8-% grade averages 42 % (13.3 km/h to 18.95 km/h). This results in a cycle time reduction from 34.8 min to 30.7 min for the 8-km round trip (3 km incline), thus giving a rise in truck productivity just short of 12 %.

Due to the tight situation in the iron ore market, Sishen had to cut down overall production to some 70 Mt material transported. Fewer trucks were needed for operation and therefore no truck obstructions due to higher travel speed on a given road capacity nor truck waiting times at the shovels or crusher stations could be recorded. Earlier investigations on this subject remarked on the necessity for a very well organized truck fleet and a good operating dispatcher system in order to turn the trolley investment to profit.

2.2.2 Diesel Fuel Consumption and Energy Cost

A 64-% reduction in fuel consumption was measured for the above mentioned cycle, while about 770 kWh of electrical energy were used up. This amounts to savings in operating costs of R 8.5/cycle at that particular mine.

However, these improvements decrease rapidly with decreasing haulage road grades. Again, to make maximum use of the system, a certain range of haulage road grades has to be preserved. On the one hand, the optimum grade for a particular type of truck has to be slightly reduced in order to make allowances for truck overloads and road undulations (which often might easily exceed the maximum grade) and which cause thermal damage to the electric wheels. On the other hand, as grades drop to 0 %, the Diesel engine no longer is the limiting factor in terms of travel speed and acceleration and a productivity increase may not be expected. Only in cases where the trucks are operated to their wheel motor limits and the Diesel engines are the limiting factors, will a productivity increase — from the technical point of view — occur.

2.2.3 Life Expectancy and Performance of the Diesel Engine, Wheel Motors, and Tires

At this time, no reliable measurements are available concerning the Diesel engines mechanical life. However, Quebec Cartier Mining experienced an increase of operating hours close to 100 % over a 10-year period.

The wheel motor's mechanical life is likely to be lower on trolley operations compared with conventional haulage, due to the higher acceleration torques at the line entry on an 8-% grade and running at the minimum demanded speed of 7 km/h. No evidence of this has been reported though.

The wheel motor's electrical life, on the contrary, will be considerably increased. Any wheel motor has a definite thermal limit, up to which the heat losses during operation may be generated. Since — under equal circumstances — the electric current drawn by the wheel motors is the same for a trolley as it is for a conventional operation and due to the higher speed — travel time on a trolley is about 30 % less —, the wheel motors will run cooler. This leads to improvements in electrical life.

Secondly, the thermal limit defines the maximum lift for a given truck on a given grade. Assuming 30 % less travel time on a trolley, a deeper pit capability for the same truck is optional.

No definite data exist so far on the life of tires. Higher torques and higher speed will possibly lead to shorter life expectancy if measured in operating hours. However, using the total kilometer figure, no changes in tire life are to be expected.

2.2.4 Availability of the Truck Fleet

The above mentioned problems, dealing with insufficient mechanical panthograph performance, were mainly responsible for a 3-% decrease in system availability. Furthermore, inexperienced drivers found it difficult to follow the line and lift the pantographs at the right time, thus causing damages to both panthograph and line. After several months of practice these problems could be handled and an overall availability drop of about 1.5 % seems to be a stable and acceptable figure.

In fact, the truck accident rate has been reduced since the date of implementation and the system is well accepted by the drivers.

2.2.5 Haulage Road and Entry Sections

Haulage road maintenance is one important factor in off-road haulage. As long as good maintenance is performed (see Fig. 1), no additional efforts are involved with trolley-assist, since short-wave undulations have to be avoided anyway and a ±50-cm tolerance in between road and line takes sufficient care of long-wave undulations.

Any line entry should be on levelled ground in order to maximize speed on entry and minimize acceleration torques. Since the system was implemented into an existing pit design, this postulation could not always be satisfied. Practically, the truck entry into the line on an 8-% grade and even at curved sections proved to be manageable. However, truck entry at curved sections is prohibited and a 50-m levelled entry section should be constructed to avoid unnecessary wear on wheel motors, planetary gears, and tires.

As mentioned earlier, truck overload allowances are provided for up to 7%. Any greater overloads will cause the truck to trip out on entry due to excessive current.

2.3 Conclusion

The trolley-assist system is considered to be capable of potential savings in operational costs and increases in truck fleet productivity. The following features are responsible for the fact that many operators take — and should take — trolley systems into account when they think about reducing their transportation costs:

— comparatively low investment and possibility of fast system implementation
— good adaptability to an existing pit design
— few losses in terms of availability of truck fleet operation
— few losses in terms of flexibility of truck fleet operation
— deeper pit capability
— potential savings in operating costs
— potential productivity increase
— restricted, but manageable movements of auxiliary equipment and shovel moves.

3. In-Pit Crushing

3.1 General Comments on Large-Scale In-Pit Crushers

The installation of the first crushing stage from outside the pit to the inside and belt conveyors taking care of the upgrade haulage to the beneficiation plant is looked upon as a second alternative to conventional truck haulage in hard-rock open-pit mines.

Concerning the classification of mobile or semi-mobile crushing plants, one should talk about semi-stationary crushers rather than calling them semi-mobile. In-pit crushers, suitable for large-scale open-pit mines, face a couple of problems, which considerably reduce their movability (Fig. 3). Basically, these problems are unit size and unit weight and thus time demand for crusher movements which in the case of large units might easily add up to several weeks or months. The most important factors can be broken down as follows:

— crushers have to be dismantled into a few transportable units
— a tremendous amount of civil work is involved, e.g., for stability or truck access ramps
— considerable earth movements are necessary
— belt conveyor shifting or extensions are expensive and time consuming
— in case of ore haulage, large-scale stockpiling has to be done (production buffer)
— maximum truck fleet capacity is in operation just before moving, minimum truck fleet capacity is required right after the crusher is in its new position
— in the case of waste haulage, energy is consumed by dump material which could better be used elsewhere
— the number of dumps then is a matter of purchaseable stackers and it is considerably reduced
— the volume at the dump is increased at the same time and thus the haul distances for waste increase.

Fig. 3: General view of Sishen's large in-pit crusher

Subsequently, a crusher is working at the same location for several years and sometimes operators do not really know whether or not the unit will ever be moved or, in fact, think about purchasing an additional unit to put into operation while the first one is being moved.

It appears to be feasible, indeed, to have two or more crushers employed at the same time, due to the following facts:

— for large-scale operations with capacities of 3,000 to 5,000 t/h hard rock, no movable crushers are available on the market today
— time demand for moving as well as for civil and earth work depends mainly on the weigth of the units; smaller units will require less civil and earth work and relocation will be easier to perform
— irregular requirements of truck fleet capacity may be balanced; expensive stockpiling — necessary to serve as a production buffer during crusher movements — may at least be reduced
— selective mining could be performed and in-pit crushing might be introduced to rather inhomogeneous ore bodies.

3.2 Sishen In-Pit Crusher

3.2.1 Technical Data

Sishen Iron Ore Mine not only implemented trolley-assist but also installed the largest in-pit crusher ever built in 1983. The system consists of the crusher, a 1,000-t transport crawler,

conveyors, and a crawler mounted stacker with a 56-m boom, connected to a 22-m bridge conveyor (for data see Table 2). The system is employed for waste haulage (limestone, dolomite) and scheduled for ore haulage at a later stage of the mine's development.

Table 2: Main data of the Sishen Iron Ore Mine in-pit crusher system

Crusher system		
60 x 109 Allis Chalmers gyratory crusher		
Production rate:		
— average	3,000	t/h
— peak	8,000	t/h
— granted	6,000	t/h
Feed size	1,000 x 1,350 x 2,000	mm
Discharge size	0 to 250	mm
Installed power	2 x 450	kW
Inlet hopper	78	t
Apron feeder:		
— width	2,900	mm
— length	9,100	mm
— power	2 x 55	kW
Discharge apron feeder: same as apron feeder		
Portal overhead crane	140	t
Total weight of system	2,400	t
Conveyor system		
Belt width	1,800	mm
Belt speed	3.5	m/s
Total lift	80	m
Total length	2,000	m
Total investment for system	25,000,000	R

3.2.2 Experiences

The entire system was completed in early 1983 and has been in operation since then. Three months of civil works were required and the total construction time amounted to twelve months. Seven workers are operating the system per shift (crusher 1, conveyor 4, and stacker 1), and a four-man maintenance shift is employed.

The crusher may be transported separately in four parts, using the 1,000-t transport crawler. At its next position, the amount of civil works can be reduced, but will nevertheless remain a major part of the total efforts. Optimistic figures on total time consumption for crusher movements require four months, pessimists say six months. The crusher is planned to stay at its present position for at least five years, future periods may be reduced to three years.

The crusher is fed from two sides (Fig. 4), one of which is equipped with the above mentioned apron feeder. So far, this apron has proved to be superfluous; obviously the gyratory crusher is capable of handling two truck loads, dumped directly at the same time. Due to this, preliminary screening is considered to be unnecessary as well, according to other operators of large in-pit crushers. Apron feeders and screens require additional weight and space and are likely to reduce system availability. As stated earlier, the weight of the unit should be kept to a minimum and any additional equipment advantages are said to compensated for by the problems involved with crusher relocation.

At this time, a cost analysis of the crusher system has not yet been completed. Sishen expects estimated annual savings of some 10 million l Diesel fuel and of about fifteen 154-t trucks. Crushing rather weak material at the time being, the crusher performance has proved to be satisfactory, achieving a peak productivity rate of 8,000 t/h.

Fig. 4: Truck ready for discharge

4. Conclusion

In-pit crushers and trolley-assisted haulage trucks have great potential in terms of fuel oil replacement by less expensive electric energy. Sishen Iron Ore Mine implemented both systems at almost the same time and experienced considerably less fuel oil expenditures.

While trolley systems seem to be very much adaptable to an existing pit and feature relatively low investments, they still do not provide a complete substitute for Diesel fuel. Trolley therefore is a very reasonable improvement on conventional truck operations, but no complete change, and secondly, is bound to steep grades and not profitable at all on horizontal haulage roads. In-pit crushing, on the other hand, features an even greater potential cost reduction, when it comes to mines with a high amount of ton miles, greater pit depths, or large-scale mining in general, but still involves several problems in terms of pit design adaptation, crusher relocation, and rather high investments. Serious attempts should be made to adapt the pit design to this technology in order to make maximum use of the investment.

Volume 3, Number 1, March 1983

Finite Element Structural Analysis of Movable Crusher Supports

Dan Neff and Conrad Huss, USA

Summary

Finite element analysis is applied to two different basic structural arrangements. One configuration utilizes a plate shell structure to carry the loads similar to the construction used for the marine industry. The other configuration utilizes a space frame made up of columns, beams and bracing similar to the construction used for buildings. Both static and dynamic analysis are carried out for the two configurations.

1. Plate Shell Structure

1.1 Geometry

The basic shell structure used to support the crusher and associated equipment is a modified toroid. The modified toroid is a semimonocoque structure characterized by a stiffened plate skin that carries a major part of the loads.

The crusher is supported at the center of the ten foot deep toroid by a series of gusset plates which are welded to the inner ring of the toroid. The toroid is supported on three legs, the bottom approximately twelve feet above the ground. The toroid is fifteen sided on the exterior and approximately 48 ft in diameter. The interior of the toroid is a 21 ft diameter circle. The centers of the legs are on a 58 ft diameter circle.

The upper and lower decks of the toroid consist of floor plate and beams to carry the equipment loads and floor live loads to the toroid walls. The toroid walls act as deep beams to transfer the loads to three shear walls that run radially inside the toroid and continuously down into each support leg.

The legs are box sections. An interior cruciform stiffens the exterior plates against buckling. The three legs rest on rectangular steel footing pads that are placed and leveled prior to the placing of the structure on them. The maximum soil bearing pressure is about 4,000 lbs/ft². A narrow bearing bar at the center of each leg transmits the leg load to the footing pad, assuring concentric loading of the pad.

A work platform at the top of the crusher thirty-five feet above the ground is of standard construction. Floor plate, floor beams, monorails, columns and bracing support a rock hammer, rock backstops and floor live loads. The columns are supported by the upper deck of the toroid. Monorails serve to access equipment hatches in the upper deck.

A control tower, reaching approximately 60 ft above ground, is supported over one of the three legs and houses ventilation equipment.

1.2 Computer Model (Fig. 1)

The basic element used to model the crusher support structure is a quadrilateral isoparametric membrane element which resists inplane forces only. Out of plane bending stiffnesses may be neglected because of the semimonocoque nature of the structure. The maximum element size used is about five feet square. This results in an adequate mesh size for this structural system as the stress and strain distribution in the membrane element is linear.

The computer model was generated in thirds, checking each third with computer plots. The three parts were then combined and duplicate nodes and elements were eliminated at the interfaces. Hatches in the upper deck of the toroid were added. Openings in the exterior panels and openings for doorways were not included in the model. The openings not included have only a small effect on the structure as a whole and their inclusion would only increase the complexity and the computer run time. These openings were analyzed at the time of detailed design.

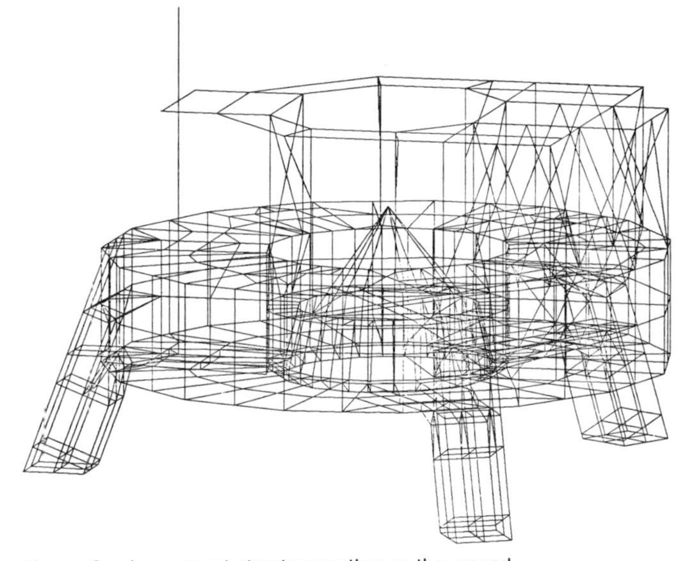

Fig. 1: Crusher support structure resting on the ground — undeformed shape

The upper work platform and control tower completes the computer model. The work platform was modeled using membrane plate elements, beams and bars. Major axis rotations of floor beams were released to simulate a pinned condition at the ends of the beams. The control tower was modeled as a stiff beam with rigid elements connecting it to other elements of the structure.

The crusher was modeled using very stiff beams in a pyramid fashion. The top of the pyramid is located at the crusher center of gravity. The crusher weight is assumed concentrated at the top of the pyramid and is spread out by the stiff beam elements to the series of gusset plates around the inner ring of the toroid.

1.3 Static Analysis

Five static loading conditions were investigated:

1) Dead Load
2) Dead Load plus Live Load
3) Dead Load plus Live Load plus 10 % of the Dead Load in the lateral negative Y direction.
4) Dead Load plus Live Load plus 10 % of the Dead Load in the lateral positive X direction
5) Thermal Load

The Dead Load consisted of the structure weight, automatically calculated and applied by the computer, plus equipment weights. The Live Load consisted of floor live loads. The lateral loading conditions were provided to estimate the effects of possible blasting of ore in the immediate area of the crusher structure. Also, dynamic loads can be equated to a "pseudo-static" loading condition for evaluation of stresses and fatigue. The thermal load is a change in temperature over one third of the structure and immediately around one leg. The thermal loading condition will approximate the magnitude of stresses due to the unequal heating that might be caused by the sun shining on one side of the structure.

Three different sets of boundary conditions were used in the static analysis. The first set was used for vertical loading conditions only, with the structure resting on the ground. All three legs were assumed to be supported vertically at the center of the leg base. Two of the legs were free to move laterally as though on rollers while the third leg was restrained laterally in two directions to create a stable condition.

The purpose of supporting the legs on rollers was to assume a conservative set of boundary conditions for the vertical load case. If all of the legs were restrained against translation, the lateral component of the reaction required to prevent the translation would exceed any friction that could be developed either between the leg and the footing pad or between the footing pad and the soil. The legs, therefore, were allowed to displace laterally. This displacement increases bending stresses in the legs and is therefore conservative.

The second set of boundary conditions supported all of the legs vertically and laterally at the center of the leg base. This set was used only to determine element stresses due to lateral operating loads while the structure is resting on the ground.

The reason these two sets of boundary conditions are justifiable is due to the sequential nature of the loading.

When the structure is in place and under vertical load, the legs will exhibit an initial radial displacement because the frictional restraint is inadequate to overcome the lateral reaction required to prevent the displacement. Once relieved, however, the legs can be expected to develop sufficient frictional resistance to restrain a lateral loading of at least ten percent of the vertical dead load reaction. This is the only means by which the structure can resist lateral loads.

The third set of boundary conditions represented the case when the structure is supported on the transporter. The structure was assumed to be supported vertically and laterally at the bottom of the inner ring of the torus directly below the series of gusset plates that support the crusher. The structure weight is also supported by the gussets. The load path of the structure's weight during transport is just reverse of the load path of the crusher weight when resting on the ground.

Summaries of selected element stresses for the three boundary conditions and various loading conditions are shown in Table 1.

ITEM	ELEMENT	GRAVITY LOADS*		LATERAL LOADS**		
		DEAD LOAD	DEAD LOAD + LIVE LOAD	0.1g IN -Y DIRECTION	0.1g IN +X DIRECTION	THERMAL LOAD
1	Upper Deck (Compression)	9500 psi	12500 psi	320 psi	280 psi	2100 psi
2	Lower Deck (Tension)	6300	6800	390	250	1700
3	Leg Plates (Compression)	4900	6800	200	35C	1300
4	Leg Shear Walls (Shear)	9300	11300	400	860	2600
5	Exterior Panels (Shear)	9000	12900	280	580	2400
6	Exterior Panels (Bending Compression) (Shear)	5300 600	11100 1200	420	250	500
7	Interior Ring (Bending Compression)	6100	7300	360	220	1400
8	Interior Gussets (Shear)	6000	6200	1000	1300	2500
9	Crusher Support Ledge (Compression)	1400	1800	320	690	3000
		VERTICAL		HORIZONTAL		
	Displacement at the Crusher	3/8"	7/16"	1/32"	1/32"	

* Legs are free to move laterally.

** Legs are restrained from moving laterally.

Table 1: Maximum element stress — operating condition

1.4 Elastic Stability

Exterior Panels

Exterior panels close off the interior of the toroid and form the fifteen sides of the basic structure. These panels act as deep beams to transfer the upper and lower deck loads to the legs. A relatively thin plate is sufficient to ensure low stresses and is stiffened to prevent buckling.

A stability analysis of the exterior panels was performed on the computer using plate elements. Vertical channel stiffeners were used at the third points of a typical 123 inch square panel which was modeled with 36 elements. Out of plane displacements were restrained at the edges of the plate. Buckling stresses for four loading cases were obtained.

The load cases were for uniform compression in the horizontal and vertical directions (two cases), pure shear and uniform vertical bending. The results are shown in Table 2.

Loading Case	Critical Stress f'_{cr} psi	Allowable Stress f_{cr} psi
Uniform Compression Perpendicular To Stiffeners	2,218	1,109
Uniform Compression Parallel To Stiffeners	3,995	1,997
Uniform Shear	6,188	3,094
Uniform Bending	25,524	12,762

Table 2: Buckling stress of stiffened exterior panels

The largest shear stresses in the exterior panels occur near the support legs. The maximum shear stress (f_s) was found to be approximately 1,200 psi. The accompanying vertical bending stress (f_b) was approximately 11,000 psi. The uniform compression (f_c) in the principle directions was negligible.

The interaction formula for stresses due to bending and shear is given by:

$$\frac{f_c}{f_{cr}} + \left(\frac{f_b}{f_{crb}}\right)^2 + \left(\frac{f_s}{f_{crs}}\right)^2 \leq 1.0$$

For the stresses given above and allowable stresses given in Table 2:

$$0 + \left(\frac{11{,}000}{12{,}762}\right)^2 + \left(\frac{1{,}200}{3{,}094}\right)^2 = 0.89$$

The result of the interaction formula is less than 1.0, so the exterior panel is within the allowable stresses for this loading condition.

The largest bending stress in the exterior panels occurs at mid-span between support legs. However, this maximum stress is less than half of the allowable bending stress based on buckling (12,762 psi) and is therefore safe.

The stiffening pattern used provides a large increase in strength over the unstiffened plate without adding much weight to the structure.

Leg Shear Walls

The interior leg shear walls are very important structural elements. These shear walls transfer most of the load on the structure down into the support legs. It is necessary for the walls to have doorways in them to allow access to rooms inside the toroid. Prevention of buckling of the walls in shear is a very important aspect of the design.

A stability analysis was performed by the computer for one wall using plate elements. The doorway in the wall was stiffened all around by use of structural tubes that provide a torsional box. The finite element model was made up of 102 plate elements. Out-of-plane displacements were restrained at the edges of the plate as well as along a vertical line supported by interior walls that stiffen the shear wall.

The critical buckling stress in shear was found from the computer analysis to be 65,460 psi. The allowable shear stress considering elastic buckling is then 32,730 psi. The allowable stress considering buckling is larger than the allowable stress in shear considering strength, which is 14,400 psi. Therefore, buckling is not critical for the leg shear walls.

1.5 Dynamic Analysis

The static and buckling analysis established plate thicknesses and stiffening patterns based on static loads due to structure and equipment dead loads, live loads, lateral loads and thermal loads. Element stresses must also be checked for the effects of the dynamic load of the crusher. Vibrational amplitudes caused by the crusher dynamic load are important and must be investigated to ensure that excessive vibration does not occur.

The crusher to be installed in the portable structure is an Allis-Chalmers 60/89 gyratory crusher which weighs approximately 450 tons. Crushing is done by the mantle which swings in a circular motion at an angular velocity of 126 cycles per minute (CPM). The mantle rests in the eccentric at the base of the crusher and crushes the ore against the outer shell. The mantle of the crusher weighs 146,000 pounds.

The crusher dynamic load is the force that is required to keep the mantle rotating eccentrically instead of flying out due to the centripetal acceleration. The dynamic force (P) is given by:

$$P = MW^2r$$

Where M is the mantle mass, W the angular velocity in radians per second and r the eccentricity of the mantle at its center of gravity. The eccentricity, r, for a mantle throw of 1-3/8 inches is 0.58 inch. The dynamic force is then:

$$P = \frac{146{,}000 \text{ lbs.}}{32.2 \times 12}\ 126^2 \text{ CPM} \left(\frac{2\pi}{60}\right)^2 (0.58)$$

$$= 38{,}154 \text{ pounds.}$$

The dynamic load acts at the center of gravity of the mantle and rotates in a horizontal plane. This load was input as two sinusoidal forcing functions ninety degrees out of phase. The load was applied at the center of gravity of the mantle at 126 CPM with the two forcing functions simulating a horizontal rotating load.

The program used a subspace interaction method to calculate five frequencies and mode shapes of the structure. No degrees of freedom were suppressed due to the complexity of the structure. The boundary conditions used for the dynamic analysis were the same as for the static lateral load analysis, that is, all three legs were restrained from motion in three directions. The first five natural frequencies of the structure are listed in Table 3.

The fundamental frequency was found to be about 2.5 times the crusher frequency of 126 CPM. A frequency ratio of at least 2.0 is desirable to prevent occurrence of sympathetic vibrations with the resulting large member stresses and structure displacements.

Frequency Number	Eigenvalue Frequency Extraction from Computer Analysis		Hand Calculated Frequency From Single Degree of Freedom System		Percent Difference
	Cycle/Min.	Period (Sec)	Cycle/Min.	Period (Sec)	
1	321	0.187	307	0.195	4.6
2	331	0.181	324	0.185	2.2
3	388	0.155	400	0.150	3.0
4	455	0.132	-	-	-
5	462	0.130	442	0.136	4.5

Table 3: Natural frequencies

The first two frequencies were primarily a vibration in the Y-coordinate and X-coordinate directions respectively (Fig. 2). The third frequency was associated with the X-direction swaying of the control tower above the toroid. The fourth frequency was a torsional mode about the vertical (Z) axis while the fifth mode was associated with a vertical (Z) vibration. These frequencies were verified approximately by hand by assuming a simplified single degree of freedom system utilizing the stiffness calculated from the static computer results (i.e., $w^2 = k/m$). Close agreement was obtained as shown in Table 3 which gives confidence in the convergence of the computer solution.

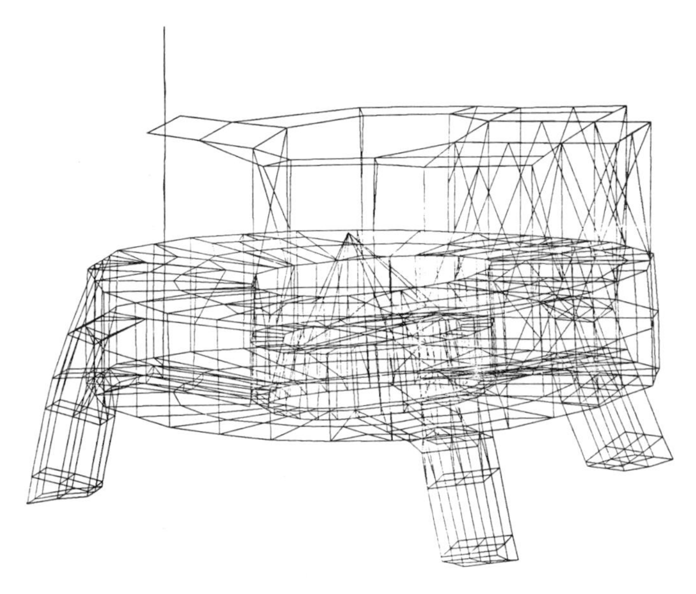

Fig. 2: Crusher support structure resting on the ground — deformed shape of the fundamental mode of vibration

The computer program used the five natural frequencies and the dynamic load to calculate selected member stresses and vibrational amplitudes. The time period considered covered five seconds, or about ten revolutions of the crusher mantle. Structural damping was assumed to be five percent of critical damping. The five second period considered was sufficient to produce steady state conditions.

The result of the forced frequency analysis showed member stresses due to the crusher dynamic load to be quite low. The dynamic stresses were approximately equivalent to a static lateral load (See Table 4) on the structure of three percent of the structure dead load. Comparing this to the static load case of a ten percent lateral structure dead load shows that plate thicknesses established by static load cases are adequate for the dynamic load case.

Element	Stress (psi)
Upper Deck	318
Lower Deck	198
Leg Plates	150
Leg Shear Walls	249
Exterior Panels	187
Interior Ring	217
Interior Gussets	482
Crusher Support Ledge	262

Table 4: Maximum element stresses due to the dynamic load

Vibrational amplitudes were investigated by using the dynamic load and superimposing the five mode shapes. Amplitudes of vibration were obtained for five percent critical damping. The maximum half-amplitude of vibration at the crusher was calculated to be 10 mls.

A second forced response analysis was run to determine what influence the structural damping had on the vibrational amplitudes. A value of 2-1/2 percent of critical damping was used in the second analysis. The resulting vibrational amplitudes did not increase significantly. The amplitudes of vibration are, therefore, not very sensitive to the structural damping assumed in the damping range considered.

Individual structural elements were analyzed on a member by member basis to prevent sympathetic vibrations; beams and exterior panels were designed for natural frequencies greater than 400 CPM.

1.6 Technical Summary and Conclusion

A finite element structural analysis was performed to determine the feasibility of a portable crusher structure. The finite element model consisted of 625 nodes and approximately 1,800 elements. The program solved a system of 3,750 simultaneous equations to obtain solutions.

Static loading cases were evaluated to determine member stresses. Static loading cases included dead load, live load, lateral load and thermal load. Static member stresses were checked against allowable stresses both for strength and for elastic buckling. Plate thicknesses and stiffening patterns were established.

Natural frequencies were calculated to evaluate the effects of the crusher dynamic load. The fundamental frequency of the structure was found to be 321 CPM which is 2.5 times the crusher frequency. Stresses and amplitudes due to the dynamic load were found to be small.

The overall structural system of the portable structure is quite stiff. Member stresses are well within allowable stresses for both strength and buckling. Vibrational amplitudes are small and appear to be tolerable. The weight of the structure is approximately 180 tons.

2. Space Frame Structure

2.1 Geometry

The basic structure used to support the crusher is a four legged space frame. Fig. 3 shows a model of the overall crusher system. This system employs a movable apron feeder which rides over the crusher structure during the transport mode and which also supports the mantle crane during the service mode. Fig. 4 shows the system during the transport mode. The main structure corresponds to the lower platform level.

The four square legs are 20 ft on center by 40 ft on center. Approximately 12 ft of clearance exists between the main support girders and the ground level. The four main girders are 4 ft by 4-1/2 ft. The overall dimensions of the crusher support framing plan including service walkways are 40 ft by 46 ft.

In addition to supporting the crusher, this structure supports the remaining superstructure.

2.2 Computer Model (Fig. 5)

Fig. 5 shows a simplified computer generated finite element model, mainly using quadrilateral, isoparametric membrane elements. The structure was modeled with 420 nodes and 730 elements, that resulted in 2,520 simultaneous equations. Similiar to the plate shell structure, the legs were assumed free to slide for vertical static loading, were assumed pinned against movement for dynamic analysis, and were assumed free when the structure is supported by the transporter. In addition, another boundary condition was considered for this structure; namely, the condition of the structure being solely supported by two diagonally opposite legs. This method of support would correspond to possible differential settlement of one of the legs which could result in the structure rocking on two of its other legs.

2.3 Static Analysis

Five static loading conditions were investigated:

1. Dead Load
2. Dead Load plus Live Load
3. Dead Load plus Live Load plus 10% of Dead Load in lateral X-direction.
4. Dead Load plus Live Load plus 10% Dead Load in lateral Y-direction.
5. Dead Load plus Apron Feeder Load during servicing and transport modes.

These loading conditions were completed with the appropriate boundary condition to generate a definitive structural analysis of the system.

Stress levels were kept low in the main support member to reduce the potential for fatigue failures due to vibration. Maximum stresses for the operating mode (dead load plus live load) were 2,970 psi for the leg columns and 14,300 psi for the longitudinal girders.

Maximum stresses for the service mode (dead load plus live load plus apron feeder) were 6,900 psi for the leg columns and 17,700 psi for the longitudinal girders.

Maximum stresses for the transport mode were considerably lower since the load path for the crusher is almost directly through the transporter to the ground.

Fig. 3: Model of overall crusher system

Fig. 4: Crusher system during transport mode

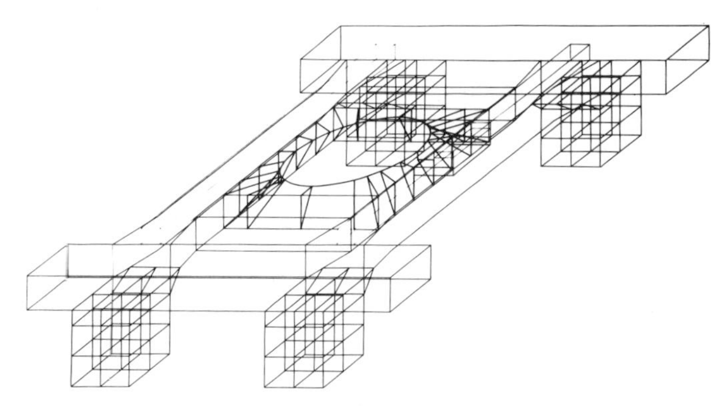

Fig. 5: MSME movable crusher

2.4 Elastic Stability

Because of the more compact nature of this design, detailed elastic stability analysis was not required. The natural configuration is such that stiffeners were not required to achieve allowable code values.

2.5 Dynamic Analysis

Vibrational amplitudes caused by the crusher dynamic load were investigated by a slightly different method than that used for the plate shell structure. This computer analysis utilized the Rayleigh-Ritz technique with consideration given to mass concentrations at each of the 420 nodes. The formula used to calculate the natural frequencies is:

$$w^2 = g \frac{\Sigma m_i d_i}{\Sigma m_i d_i^2}$$

where

m_i = concentrated mass at node i
d_i = deflection at node i corresponding to assumed mode shape
g = acceleration due to gravity
w = frequency

Lowest natural frequencies resulting from this analysis were:

730 cpm in lateral X-direction
580 cpm in lateral Y-direction
640 cpm in vertical Z-direction

Torsional fundamental modes were higher in frequency.

Stress runs were also carried out for the load cases approximating crusher rotation. These lateral loads due to crusher rotation were approximately 4 % of vertical dead loading and the resulting stresses were not critical. The structure weighs approximately 160 tons.

The maximum half-amplitude of vibration at the crusher base was calculated to be six mils.

2.6 Comparison of Structures

The space frame structure as designed houses a 54—74 gyratory crusher; whereas, the plate shell structure houses a 60—89 gyratory crusher. The plate shell structure has the advantage of using only three legs which establishes a definite loading pattern. The space frame is a more rigid (i.e., higher natural frequency) structure, but runs the risk of rocking on two of its opposite legs should significant differential settlement occur. Stress levels and deflections were comparable for both structures.

Both structures certainly deserved a finite element analysis which is fairly routine in this technological era. Such analysis avoids unwarranted conservatism in static analysis and more accurately and safely predicts dynamic response.

 Volume 4, Number 4, December 1984

Giant Open-Cast Coal and Lignite Mines in India

H. Srinivasan, India

1. Introduction

Coal mining in India is nearly 150 years old, but mechanised open-cast mining is of recent origin. A start was made in the early fifties with Sir Lindsay Parkinson, a U.K. firm of contractors being commissioned to operate the first large mechanised open-cast mine in the East Bokaro coal fields in Bihar State.

Since then, nearly 70 mechanised open-cast mines have been opened, with annual production capacities ranging from 0.5 to 10 Mt. But by and large, the capacities have been in the range of 0.5 to 3 Mt only in most of the mines.

Coal production in India, including lignite, is poised to cross the 200-Mt mark by 1990, out of which the share of surface mining production will be more than half, i.e., the existing level of production of 50 Mt from open-cast mines will have to be doubled. Basically, it is proposed to achieve this by opening a number of large-capacity open-cast mines at the Singrauli and Jharia coal fields and by expanding lignite mining activities.

2. Singrauli Coal Field

The Singrauli coal field is situated in the north-eastern part of India in Madhya Pradesh State, an area of 312 km^2, in the form of a hilly plateau, rising over 400 to 500 m above sea level.

There are two main coal seams suitable for open-cast mining, namely, Turra and Purewa. The coal bearing formation is Barakar formation, with a total thickness of 46 to 75 m. The maximum thickness of Purewa's two sections is 21.5 m. The strike of the coal field falls south-west with a gentle dip of 1 to 3° N.

The overburden and inter-bedding between seams consists of a strata of sandstone, sandy clays with a maximum compressive strength of 200 kg/cm^2. Protodjakanov's hardness index ranges up to a maximum of 4.

The mining property is divided into nine blocks, six of which have been selected for open-cast mining for an annual production capacity of 10 to 14 Mt. Out of these, the mines of Jayant, Dudichua, Amlori, and Khadia have already started their operations.

Mr. H. Srinivasan is Technical Adviser, Hindustan Motors Ltd., Madras, India

2.1 Mining Systems

The overburden overlying the top seam at Purewa will be mined by large-size electric shovels, rear dumpers, and large blasthole drills. The capacity of the shovels will be 10 m^3. In the case of Amlori, where the overburden quantity is quite large, the proposed size of shovels is 20 m^3. The size of the dumpers will be 85, 120, and 170 t, respectively. The blasthole drills will be 10.5 to 12 inches in size. A site-blasting slurry-mixing plant will be operated.

The partings between the seams will be handled by tandem walking draglines of 28 to 32 m^3 capacity, with a reach varying from 85 to 95 m. A similar dragline will operate on the overburden dump, rehandling the excess overburden in order to have a clear width for extraction and maintaining quality.

2.2 Coal Extraction and Handling

Coal will be mined by electric shovels with 85-t dumpers and bottom-dump coal haulers. The coal will be brought to the surface by suitable designed haulage roads into a centrally situated receiving pit in each mine.

The central receiving system will have equipment for crushing, screening, and sizing the coal to − 200 mm.

The prepared coal will be stored in ground bunkers from where it will be conveyed to a rapid loading system for transport to thermal power stations constructed near the mine sites.

In the Jayant open-cast mine, "merry-go-round" trains transport the coal to the Singrauli thermal station. The systems for the other projects are still being planned.

At present, there are no proposals for washing the coal to reduce the ash content. But considering the need for maintaining quality consistency, there is a distinct possibility of suitable washing systems being introduced in at least one or two mines at a later stage.

It is estimated that the total production from Singrauli coal fields will be 100 Mt before the turn of the century.

3. Jharia Coal Field

The Jharia coal field is situated in Bihar State in the central eastern part of India and is India's only source of prime coking coal. The coal field is covered by two coal bearing sets of seams, viz., Raniganj and Barakar, containing layers of sandstone and coal seams.

There are 19 main coal seams in this field, which have been partially exploited by a number of underground mines by the board and pillar system. A huge quantity of coal is standing in pillars waiting to be extracted. The extraction of pillars by sand stowing (flushing) has posed serious difficulties in view of the non-availability of sand in sufficient quantities in the nearby rivers.

Thus, open-cast mining has become almost a necessity to increase the prime coking coal production in the area so as to meet the requirements of coking coal for steel mills.

3.1 Open-Cast Mining Layout

Open-cast mining has been found to be totally feasible in the outcrop region of the coal seams. The open-cast mines are designed for a maximum depth of 200 m with a cut-off ratio of 1:10 with a batter at 45°.

Multiple seam mining will be untertaken in each of the blocks, using a shovel dumper transport system for the removal of overburden and coal from upper seams. For mining lower seams, walking draglines of 28 to 32 m^3 capacity will be used and the coal will be transported by large belt conveyors. In some of the projects, conveying the coal entirely by means of very wide steel-cord belt conveyors is contemplated.

Out of the nine blocks, major open-cast mining activity has been planned in blocks IX and VI for the present. The major project in block IX is the Mukunda mine. The average overburden ratio is 1:4.6. The overburden consists of different grades of sandstone with gray shale. The Mukunda mine has been designed to produce 12 Mt/a.

3.2 Mining Plan

The mining area has been divided into three sections, each consisting of the outcrop portion of a limited number of seams out of the total 19 seams. Each of the sections will be worked for the full 3 km strike length. Section I will be operated first, subsequently section II, and so on, so as to take full advantage of the back dumping facility.

It is estimated that the ultimate maximum depth of the mine will be 480 m.

3.3 Mining Equipment

Fully electric or hydraulic shovels of 10 to 12 m^3 capacity with 85 to 120 t dumpers will be used for the removal of overburden and coal from the top seams.

For the lower seams, electric walking draglines of 28 to 32 m^3 capacity will be deployed for the removal of overburden, while coal will also be excavated by electric shovels and dumpers.

However, there are also proposals for avoiding walking draglines and using only electric and hydraulic shovels and dumpers in combination with portable crushers and belt conveyors for both overburden and coal.

3.4 Coal Handling and Beneficiation

As the properties of coal vary in the different seams, it is proposed to construct three separate coal beneficiation plants.

Each beneficiation plant will handle the coal from three to five seams. As such, each of the plants will be provided with blending arrangements for blending coal prior to washing. Subsequently, blending arrangements for blending washed coal outputs from three units will also be provided with a built-in cushion capacity.

3.5 Washing System

Indian coals are difficult to wash; as apart from a high ash percentage, they contain a large percentage of near gravity material. The use of jigs, like Batac jigs, is in the initial stage and therefore heavy media separators and subsequently heavy media cyclones as standard equipment are proposed to be used with necessary auxiliary equipment. Froth flotation systems for handling coal – 0.5 mm will also be standard, with the necessary thickeners, filters, etc.

4. Neyveli Lignite

Lignite deposits occur at Neyveli, approx. 240 km from Madras, the capital of Tamil Nadu State. The total reseves of the Neyveli lignite deposits are about 2 Gt (1 Gt = 10^9 t), spread over an area of 260 km^2. The field is 11 km wide in its east-west axis and 21 km long in its north-west south-east axis. The whole property has been divided into seven blocks. The first open-cast mine was started in an area of 14 km^2 with an estimated reserve of 200 Mt, where the lignite occurs in one single seam without any large partings. The overburden depth varies from 60 to 160 m, consisting of loamy soil, soft laterite sandstone (calcareous and ferroginous), and fire and white clays.

The lignite deposit is just below mean sea level, with an insignificant gradient of 1:100. The lignite is dense and massive, but highly fossiliferrous and friable in nature, with a density of 1.16 kg/m^3. The lignite is of low calorific value (2,400 kcal/kg), low ash content (3%) with traces of sulphur, high moisture (50%), and a high carbon content (22%).

4.1 Ground Water

The Neyveli belt is an aquiferrous artesian basin with a maximum pressure of 8 kg/m^2. Approximately 20 t ground water/t mined lignite have to be pumped out. Casing pipe wells with submersible pumps have been set up on the rise side and on the sides of the mine to keep the mine dry at all times.

The ground water control planning and operation is done by computer for achieving maximum economy.

4.2 Mining System

A production varying from 3.5 to 4 Mt, an overburden ratio of 4.5:1, a maximum overburden transport distance of over 12,000 m, stable slopes, and a mine depth of nearly 100 m are some of the basic parameters which were considered for the adoption of a suitable mining system.

A pilot mine 50 m in depth was excavated using standard small shovels and bottom dumpers. Subsequently, special tailor-made mining equipment, like bucket wheel excavators (BWE), belt conveyors, spreaders, etc., was introduced in the mine.

The overburden benches vary from 8 to 20 m in depth, totalling approximately 65 m to the roof of lignite. Each bench has its own system of bucket wheel excavators, conveyors, and spreader. The cuts are of the terracing type. After one block is completed, the conveyors are shifted with the help of Caterpillar bulldozers, pipe layers, and track shifters.

The belt conveyors are of the steel cord type, with widths varying from 1.8 to 2 m.

4.3 Lignite Production

The lignite is excavated by 350-l BWEs with a set of conveyor/spreaders.

The lignite is transported by a conveyor system to a central bunker for ground storage. A bucket chain excavator retrieves the lignite for dispatch to the thermal station, briquetting plant, fertilizer plant, etc.

4.4 Blasting

About 25 years ago, when Neyveli open-pit started, it made history by introducing bucket wheel excavators in hard lateritic sandstone containing ferroginous nodules. Since then, a fine system of mild blasting has been put into practice to loosen the overburden so that the bucket life is increased.

4.5 Larger Mining Equipment

Experience gained with the initial size of 500-l and 700-l bucket wheel excavators and complementary equipment was good and the mine's equipment fleet has been supplemented by larger sizes like 1,250-l bucket wheel excavators, higher capacity spreaders, and larger conveyors.

This has enabled the mine to re-orient the bench system, stabilise the overburden removal, and provide enough lignite exposure for steady lignite production.

4.6 Increase in Lignite Mining Activities

A second open-cut is well under way for an annual lignite production of 7 to 8 Mt. This open-cut has a reserve of approximately 360 Mt.

The overburden depth has, however, increased approx. to 100 m and the maximum lignite thickness will still be 20 m. The overburden/lignite ratio will be approximately 6:1 (volume of overburden to weight of lignite).

The day is not far off when a third open-cut may be opened.

5. Mining Technology and Equipment

The mining conditions and geological structures of coal in India are different from those in the USA, the U.K., West Germany, or the USSR.

Even so, India has gained immensely by exchange of appropriate technology in open-cast mining with the four countries mentioned above. Many experts from these countries and their expertise have enabled India to make the mines more productive, efficient, and profitable.

Obviously, India has imported open-cast mining equipment from some of the best manufacturers in the world, like Marion, P & H, Ruston Bucyrus, Terex, Wabco, Caterpillar, Komatsu, Krupp, O & K, etc., but indigenous manufacture has advanced considerably over the past fifteen years.

Marion, Demag, Terex, Wabco, and Komatsu were the first to collaborate with Indian manufacturers, like Hindustan Motors Ltd. and Bharat Earthmovers Ltd., in the manufacture of mining equipment. Other manufacturers who followed were Bucyrus Erie, Ingersoll Rand, Chicago Pneumatic, etc.; Caterpillar is soon to follow.

This has led to a steady supply of Indian manufactured equipment for Indian coal mines, but with the increased size of open-cast mines, the scope of collaboration in manufacture/supply of large-size equipment for earthmoving, material handling, and coal preparation will become greater and greater.

 Volume 4, Number 4, December 1984

Material Handling in Mine and Plant at the Malanjkhand Copper Project

S.V.K. Puranik and S.D. Kamra, India

1. Introduction

The Malanjkhand copper mine is located 90 km north-east of Balaghat, a district town of Madhya Pradesh in India. The general relief is flat and the average ground elevation is 575 m, with the highest elevation being 652 m. The deposit area is characterised by seven seperate hillocks, rising 50 to 70 m. The temperature varies from 1 °C in winter to 46 °C in summer, the average being 20 °C; the annual rainfall is 1,500 mm with a daily maximum of 276 mm. The deposit has been explored by 22.7 km of diamond drilling in 91 bore holes before mine construction was commenced.

The deposit area is occupied by a basement complex of granitic rock, belonging to the Archean age, which is folded and traversed by quartz reefs ranging from 60 to 80 m in width and ore overlain by rocks of the Chilpi series. The quartz reef outcrops are prominent on the Malanjkhand hill ranges. The emplacement of the quartz reefs and the subsequent mineralisation result from the formation of fissures in the folded granitic basement rocks due to tectonic movements in various phases. The main quartz reef occupies the axial region of the fold. The bulk of mineralisation is by and large confined to the fractured and sheared zones in the quartz reefs, arcuate in layout, over a north-south strike length of 2.6 km with a dip of 60 to 65 °E. The adjoining granites are also mineralised in minor amounts.

The ore body is fairly continous, exhibiting an arcuate shape. The average thickness is about 70 m. There are three distinct ore zones, viz., the oxidation zone, the primary, and the secondary sulphidization zone. The primary sulphidization zone is represented by chalcopyrite. In the weathering zone, the primary sulphides have been weathered to malchrite and azurite etc. The ore reserve is 58.8 Mt with 1.2% Cu at a cut-off of 0.45%, lean ore 44 Mt with 0.22% Cu, and oxidised ore 5.5 Mt with 0.58 % Cu. The ultimate pit size is 2,200 m in length by 640 m in width by 276 m in depth. The stripping ratio is 4.6 by volume and 4.29 by weight. The bench height is 12 m, conforming to the Indian safety standards.

Mr. S. V. K. Puranik is Deputy General Manager (Mines) and Mr. S. D. Kamra is Mining Engineer, Hindustan Copper Ltd., P.O. Box 16008, Calcutta — 700017, India

The mine is designed for 5 Mm^3/a excavation to produce 2 Mt/a ore. The ore from the open-pit mine is fed to a primary gyratory crusher, where its size is reduced from 1,200 to 150 mm. The ore is then stocked in the coarse-ore stock pile with 18,000 t live capacity.

The coarse ore is reclaimed by three apron feeders and fed to a secondary crusher through a double-deck screen. The −40 mm fraction goes to a surge bin, while the + 40 mm fraction goes to three tertiary cone crushers in closed circuit with three single-deck screens of 12 mm aperture. The screen undersize −12 mm goes to the fine-ore bin, while the oversize is recycled to the tertiary crushers. The final-crushed ore −12 mm is fed to four ball mills in closed circuit with hydrocyclones. The cyclone overflow, containing 30% solids ground to 60% −200 mesh, is fed to four banks of rougher-scavenger flotation cells. As flotation reagents, MIBC (frother), sodium isopropyle xanthate (collector), and machine oil (modifier) are fed through reagent feeders. The rougher tailings are scavenged in scavanger cells and the final tailings are rejected. The rougher concentrates are pumped to the first cleaner cells. The concentrate from the first cleaner cells is taken to the second cleaner cells which yield the final concentrate. This is first pumped to a 25-m diameter thickener and then to two disc filters. The filtered concentrate cakes, with about 13% moisture and 31% Cu, are despatched by trucks to a smelter at the Khetri copper complex in Rajasthan. A suitable tailings disposal system has been incorporated to dispose of the tailings and reclaim the water for recycling in the plant. The slurry is pumped through three rubber-lined 250-mm diameter pipelines to a tailings pond located 4 km away. After the tailings have settled down, the excess water is reclaimed by a marsh pump fitted on a movable trolley and recirculated to the plant.

The total bulk solids handling at Malanjkhand can be subdivided into the three main divisions, viz.:

1. Mine, 2 Mt/a ore and 11.5 Mt/a waste material
2. Concentrator plant, 2 Mt/a ore, 99,200 t/a concentrate with 25% Cu
3. Tailings disposal, 6,330 t/d solids in 24,000 m^3/d water.

Each of these sections is discussed in detail in the ensuing paragraphs.

2. Mine

The Malanjkhand copper mine is basically an open-pit mine working according to the discontinous mining system, i.e., individual operations are done with the following equipment:

The rock is loosened by drilling with a drilling rig, i.e., five Schramm DTH's, one IR DM-20, and five IR CM-20's. Then, after blasting the blasted muck is loaded by rope shovels, i.e., six HEC-EQC's, two P&H 1900 AL's, and one front-end loader FEL Clark, into off-highway trucks — 26 LW-50's and ten LW-85's (Fig. 1).

Fig. 1: Malanjkhand mine — P&H shovel loading into a 50-t dumper (centre bottom); Schramm C 985 H drill (centre left); LW-35 water sprinkler (centre); block ready for blasting (centre right)

The mine is designed to produce 2 Mt/a ore with a total excavation of 13.5 Mt/a rock mass. 11.5 Mt/a waste are transported to various dumps — waste dumps, a lean-ore dump, and an oxidised-ore dump — situated on the hanging wall side of the mine outside the ultimate pit limits.

2.1 Drilling and Blasting

The rock being very hard and abrasive, drilling is done by Schramm down-the-hole percussive drills. There are three Schramm C 985 H's, one T 685 H, one T 985 H, one IR DM-20, and five CM-20's.

The Schramm C 985 H is a crawler mounted rotadrill with two two-stage 425 $ft^3 min^{-1}$/250 psi compressors. These electrical drills are working on 33 kV. All associated rig functions are performed by hydraulic power provided by six pumps.

The compressor of the T 685 H rotadrill, a GMC Diesel-driven truck mounted unit, is also a two-stage model with 850 ft^3 min^{-1}/350 psi. A Detroit Diesel engine of 12 V and 71 t is the power unit for the compressor and the hydraulic pumps; the drilling is done with a double-pass. The maximum hole diameter is 7 $^7/_8$ inches for rotary and 8 inches for down-the-hole drills, while normally it is 6 $^1/_2$ inches.

The IR DM-20, a Diesel-driven crawler mounted unit supplied by Ingersoll Rand, is also a down-the-hole drill. It drills holes of 6 to 6 $^1/_2$ inches diameter.

These six above described units are in use for the primary blasthole drilling. The drilling commences in a staggered pattern. The burden/spacing (S/B) ratio is generally 1.5 to 2.0. The holes are 165 mm in diameter. Usually, vertical holes are drilled, but sometimes inclined hole drilling is also resorted to.

The secondary drilling is done by IR CM-20's and holes of 4-inches to 4 $^1/_2$-inches diameter are drilled. The holes are charged with 125-mm diameter cartridges of slurry explosives and blasted. The bench height is 12 m and sub-grade drilling is 1.2 m (10%). The holes are blasted, using electric delay detonators, detonating relays, and Cordex detonating fuses. The average consumption of explosives is 100 to 120 t/month. Explosives used are water resistant slurry and of a nitroglycerine-based type. For better fragmentation of the rock bulk, with a minimum quantity of oversize boulders, multi-row delay action blasting is done up to a maximum of four rows.

2.2 Loading and Hauling

The Malanjkhand pit is designed to load and haul 13.5 Mt/a total excavation. The fact in mind that the quarry is considerably long (2,400 m) and narrow (600 m wide at the top and 70 to 80 m at the bottom), a trackless mining method with a truck/shovel combination is used. Of the total volume of 106.48 Mm^3 to be mined within the quarry outline, 16.38 Mm^3 are fom the upland and 90.10 Mm^3 from the deep part below-ground level.

The mining bulk haulage system layout has been developed taking into account the scope of stripping operations, the time schedule of the quarry mining, and the location of the concentrator plant, the primary crusher, and the barren rock dumps. On this basis, the system of the quarry motor road is designed in such a way that the barren rock is transported from the quarry towards the east side, while the ore is transported towards the west side.

The equipments in use for loading are two P&H 1900 AL rope shovels of 10 m^3 bucket capacity and six EQC 4.6-m^3 rope shovels. One Clark wheel front-end loader with a 10-m^3 bucket is also used to supplement loading. The equipments for hauling are ten LW-85 T and 26 LW-50 T Haulpak dumpers. Usually, the Haulpak LW-85's in combination with P&H shovels are used for waste rock excavation and the LW-50's in combination with 46-m^3 HEC-EQC's for balancing the waste rock and ore production.

The front-end loader is used specifically for supplementing the production. It is a tyre mounted unit with a bucket size of 10 m^3. Being extremely mobile, it is used at times at the stock pile. Furthermore, it is used for lifting muck from a sinking cut ramp, where sometimes it is not feasible to take the shovel down due to water problems.

The dumpers haul the waste to the respective dumps and the ore to the plant or to the stock pile. Permanent ramps have a 6-% gradient and the in-pit ramp has a 8-% gradient. Komatsu D-355 A-3 and BEML D-120 dozers are employed at the dumps for road maintenance. The envisaged waste hauling is 11.5 Mt/a.

Altogether, there are nine dozers in use, four D-355 A-3 Komatsu dozers, four BEML D-120 dozers, and one Clark wheel dozer. Two graders, one Galion and one BEML, along with road rollers and compactors are used for road maintenance. Three LW-35 water sprinklers are also used for road maintenance and dust suppression.

3. Concentrator Plant

3.1 Primary Crushing

The run-of-mine ore is hauled by dumpers to the primary crushing building (Fig. 2) and discharged into the primary crusher receiving bin designed to accept ore from two sides simultaneously. The primary gyratory crusher, model PGC-1250 x 1900 HEC, with a capacity of 870 t/h at 150-mm open side discharge setting is driven by a 360-kW motor. The 150 mm material is discharged into a surge bin located underneath the crusher. One 1,200 mm x 1,000 mm Stephen Adamsons make manganese steel apron feeder, equipped with a variable speed drive and capable of handling 200 to 1,000 t/h ore, draws the crushed ore from the bin and transfers it to the conveyor. The dual-drive conveyor is 1,400 mm wide and 390 m long. The belt carries the material to the 18,000-t live capacity coarse-ore stock pile.

Fig. 2: Process plant at the time of construction — primary crusher building (left); secondary and tertiary crusher building (centre); concentrator building (right); mine office (foreground)

The primary crusher is served by an overhead crane, which also helps in the removal of oversize rocks from the crusher cavity. Externally fitted wet-type air scrubbers collect the crusher dust from various dust prone points. A ventilation fan caters to the need of fresh air supply to all the floors of the primary crushing building. A tramp iron magnet, a metal detector for tramp iron removal, and a belt scale to keep a weight record of the material going to storage have been installed on the conveyor.

3.2 Secondary and Tertiary Crushing

The primary crushed ore from the stock pile is delivered by three apron feeders of 1,200 mm x 9,000 mm size and a capacity of 200 to 500 t/h over a 1,200 mm wide conveyor belt to a double-deck screen. The conveyor is also provided with a tramp iron magnet, a metal detector, and a belt scale. A secondary crusher, model SEC-2200 HEC, and three tertiary crushers, 7-ft short head Rexnord models, are installed in line in the secondary and tertiary crushing building (Fig. 2). The secondary crusher, directly coupled with a 250-kW motor, has a crushing capacity of 600 t/h at 40-mm closed side discharge setting. The oversize from the double-deck screen is fed to the secondary crusher operating in open circuit, and the crushed material is discharged onto a conveyor.

The three tertiary crushers with a capacity of 450 t/h each at 12 mm closed side discharge setting operate in closed circuit with single-deck vibrating screens. The crushed material is discharged onto a conveyor where it joins the discharge of the secondary crusher and is transported and transferred by three 1,400 mm wide belt conveyors. The −4 mm conveyor distributes the ore into surge bins by means of a tripper. The ore is then taken by three 1,000-mm belt feeders with variable speed to the single-deck vibrating screens.

The −12 mm undersize from the screens is transported and transferred by two 1,200 mm wide conveyors. By means of a discharge tripper on the second conveyor, the ore is distributed into the fine-ore bin of 9 t live capacity. The first conveyor is also provided with a belt scale to keep account of the fine ore produced.

The secondary and tertiary crushers are served by a 30/5-t overhead crane. Wet collectors have been installed to absorb the dust from all the dust prone points in the building. All the equipments of the secondary and tertiary circuit can be regulated from a control room in the same building.

3.3 Grinding Circuit

The grinding circuit consists of four overflow-type ball mills with a diameter of 12 $^1/_2$ ft and a length of 20 ft, operating in closed circuit with eight 26-inches hydrocyclones. The overflow-type ball mills are each driven by a 1,200-kW slip ring type induction motor. The mill speed is 15.10 rpm. The circulating load is designed at 250%. Each mill is fed by its own combination conveyor and a variable-speed belt feeder and can reduce 1,515 t/d −12 mm dry feed to 60% −200 mesh. Each mill is lined with wear resistant rubber lines. The mill discharge overflows into a sump and is pumped by centrifugal slurry pumps to the cyclones. Each mill has one working and one stand-by pump/hydrocyclone combination. The cyclone underflow returns by gravity to the ball mills. The cyclone overflow is delivered to the common distributor, feeding the rougher flotation cells.

3.4 Flotation Circuit

The rougher flotation circuit consists of four rows of Denver 300-ft^3 cells, each row having twelve cells. Four blowers supply air to the cells. From the common distributor, feed can be taken to any of the rows.

The rougher tailings are scavenged and the final tailings are rejected. The rougher concentrate is pumped to the first cleaning stage, whose tailings join the froth product from the scavenger cells and are taken by 100-K Wilfley pumps to the head of the rougher section.

The concentrates from the first cleaning stage are pumped by 100-K pumps to the second cleaning stage, yielding final concentrates which are collected in a single sump and pumped by a 100-K unit to a thickener. The tailings from the second cleaning stage are returned into the head of the first cleaning stage.

The first and second cleaner sections comprise of four rows of 60-ft^3 cells, each row having ten cells. The cyclone overflow, the concentrates from the second cleaning, and the final tailings are sampled by automatic samplers.

3.5 Filtration

The flotation concentrate is thickened in a single thickener of 25 m x 8 ft. The thickened concentrate is drawn with the help of a diaphragm pump and is pumped by a 100-K unit to two vacuum filters with six 8-inches discs of 10 ft diameter each. The overflow from the thickener flows by gravity to a sump and is pumped to an overhead tank.

Three SLM-Maneklal make vacuum pumps (two operating and one stand-by) and two SLM-Maneklal make blowers provide vacuum and compressed air, respectively, to the filters. The filter cake is transferred by two conveyors to the concentrate storage stock pile of 6,000 t capacity.

The final tailings are sampled automatically, before they are taken to the sump of the tailings pump station.

In the main process unit, the ratio of working to stand-by pumps is 1:1. The flotation reagents are fed into the process by means of reagent feeders.

For repair purposes in the main building, there are three cranes, one 30/5-t crane in the mill bay, one 5-t crane in the flotation bay, and one 5-t crane in the filtration bay. The concentrate storage stock pile has been provided with a 10-t grab crane for the loading of the concentrates into lorries.

To meet exigencies, one emergency catch pit for tailings and one for concentrates along with pumping facilities have been provided.

4. Tailings Disposal System

The tailings disposal system provides facilities for the disposal of the copper tailings produced at the Malanjkhand copper complex. The system is designed to handle 1.9 Mt/a tailings, with provisions for expansion for handling 3 Mt/a in the future.

A steel distributor box has been provided to receive slurry from the concentrator for further distribution to the tailings sumps.

Three tailings sumps (two working and one stand-by) of approx. 150 m^3 nominal capacity (effective capacity 100 m^3) have been provided. Each sump is independently connected to a pump series. Agitators of 30 kW rating with rubber-lined impellers are provided for each sump. Instruments have been provided on each sump for monitoring levels. For level control, the dilution is adjusted automatically through a flow control valve linked to the level controller.

The facilities comprise of three identical pump series (two operating and one stand-by) with a capacity to deliver a maximum head of 21 m slurry at 500 m^3/h maximum flow. The pumps are suitable for handling slurry up to a concentration of 45 % by weight.

The motor rating of each pump is 132 kW. The pumps are of the Sala-type HD-507-150 and have a rubber-lined construction to resist erosion. Each series is independently connected to a 250-mm NB 2.5-km rubber-lined main pipeline. The pipelines are connected with two distribution lines, covering the entire periphery of the tailings pond. A number of tappings is provided on the distribution lines for disposal around the pond with hydrocyclones to discharge the tailings at any particular place to build up the level of that section.

To take care of flow and head variations, arrangements have been made to bypass up to three pumps in a series. Balance control is provided through a variable-speed fluid coupling, supplied on the last stage of each series.

Two thirds of the water requirements of the concentrator are to be reclaimed from the tailings pond. For this purpose, four 150-kW Mather and Platt make water pumps (three working and one stand-by) have been provided to deliver water into the reclaim water storage through a 14-inches pipeline. Water from this reclaim storage can be drawn as and when required.

 Volume 4, Number 4, December 1984

Indian Mining
Current State-of-the-Art of Bulk Solids Handling

A.K. Basu and A.K. Ghose, India

1. Introduction

Since Independence, the Indian mineral industry has made giant strides in the expansion of its mineral production capacity. The value of mineral production increased from around Rs. 500 million in 1947 to over Rs. 66,000 million in 1983. The years have also witnessed the emergence of a giant public sector, which has contributed significantly to mineral resource development in the nation and strengthened the technological base of the industry. Currently, the public sector in the mineral industry contributes over 93% of the total value of mineral production. Although the mineral industry today presents a mix of technologies from small-scale and intermediate technology to semi-mechanised and fully-mechanised methods of operation, the trend towards increasingly larger scales of operation, especially in new large investments, is discernible. The giant open-cast mine of 22 Mt capacity of Kudremukh and the planning of a 10 Mt size operation at the Singrauli coal field testify to the trend towards increasingly larger size, which has been supportive of developments in bulk handling of minerals. Major advances have been made in this area in the past decade and there have been major conceptual changes in system design for the bulk handling of minerals at the mine level, and from mine to mill and for port handling facilities with increasingly larger continuous systems to handle larger volumes of material.

The aggregate length of belt conveyors in India increased from around 100,000 m in 1950 to around 200,000 m in 1960 and again to 400,000 m in 1970. Presently, the estimated figure would be in the order of 700,000 m, thus exemplifying the doubling of belt conveyor length every decade. The spectrum of developments in the bulk handling of minerals is impressive, whether we consider the high capacity merry-go-round system at the Singrauli coal complex, the conveying system at the Neyveli lignite mines, or the pipeline transportation system at Kudremukh. The purpose of this paper is to present an overview of major developments in the bulk handling of minerals in different sectors of the Indian mining industry, highlighting some of the technologocial advances.

Mr. A. K. Basu is Assistant Professor at the Department of Engineering and Mining Machinery, and Mr. A. K. Ghose is Professor at the Department of Mining Engineering, both at the Indian School of Mines, Dhanbad, India

2. Review of Bulk Handling Systems

2.1 Neyveli Lignite Mine

The Neyveli lignite mine, the only one of its kind in India, has adopted the fully-mechanised open-cast mining system using bucket wheel excavator technology to handle an excavated overburden volume of 35 Mt/a for a lignite production of 6.5 Mt/a.

The crawler-mounted bucket wheel excavator (BWE) operates in terracing cuts. The travelling transfer-feeder connected to the BWE delivers the material excavated from the face into a transfer hopper, which moves over the fast running receiving belt conveyor line (Fig. 1). The overburden is disposed of at the tail-end conveyor of the series with the aid of a matching mobile tripper to deliver the material at any point over the conveyor length to a spreader.

The spreaders are special-purpose, heavy-duty bulk handling machines of tailor-made design which receive and dispose of the overburden suitably on the dumping ground. The angles of the receiving and discharge boom of the spreader can be varied at will.

Three bucket wheel excavators, of which one is of 700 l and two of 350 l bucket capacity, are deployed for lignite production, while two more of 500 l capacity are deployed in the dumping yard for lignite handling to consumers.

The array of high speed belt conveyors that interlink the total system is characterised as follows:

Belt width	2,000 mm	1,800 mm
Capacity	11,000 t/h	8,000 t/h
Belt speed	4.2 m/s	4.2 m/s
Troughing angle	40°	40°
Idlers	6 sets/frame	6 sets/frame
Belt	steel cord	steel cord
No. of drive heads	3	3

The aggregate line lengths of different categories of conveyors are:

2,000 mm	steel cord	13.0 km
1,800 mm	steel cord	3.2 km
1,500 mm	fabric	6.7 km
1,200 mm	fabric	1.9 km
1,000 mm	fabric	1.8 km

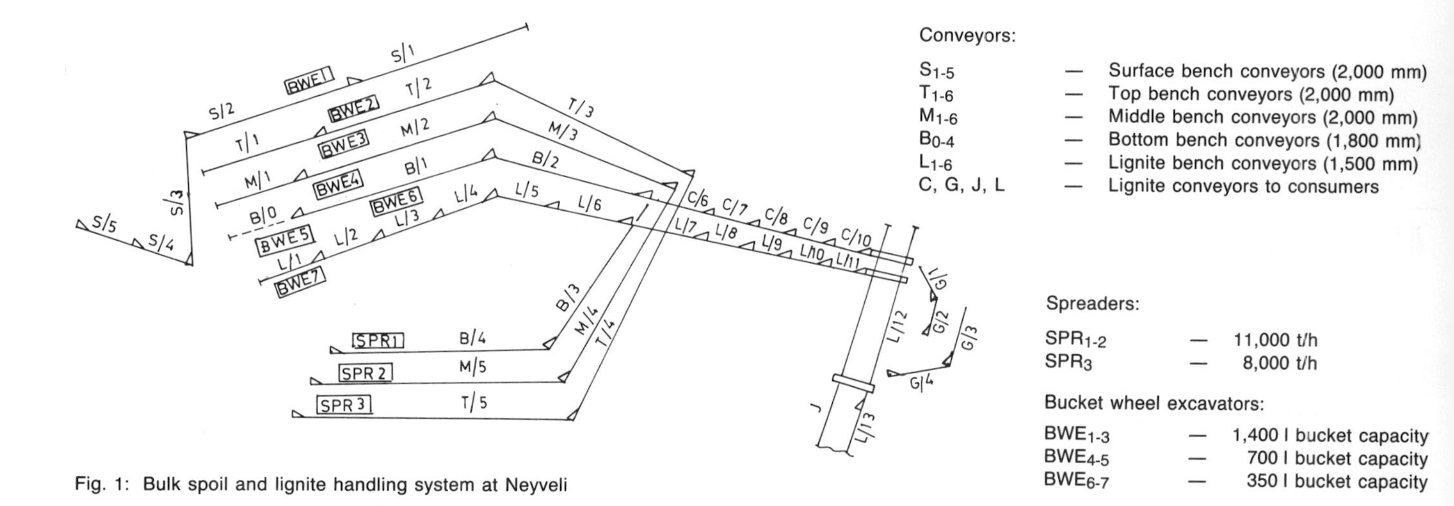

Fig. 1: Bulk spoil and lignite handling system at Neyveli

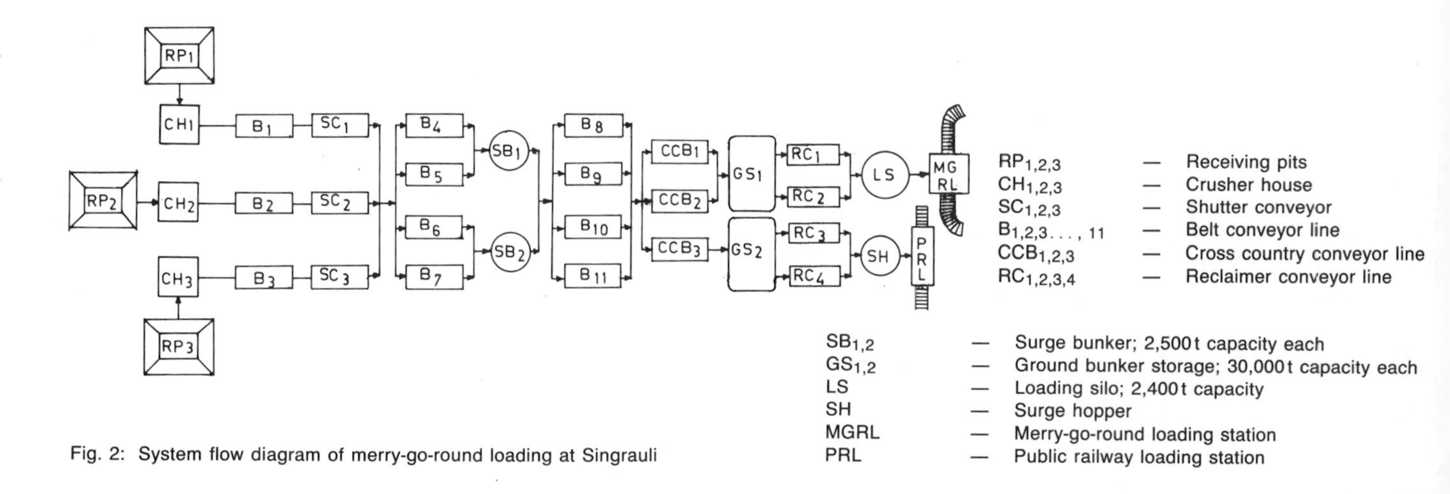

Fig. 2: System flow diagram of merry-go-round loading at Singrauli

Efficient running of equipment in such a capital-intensive continuous series production system warrants strict control on downtime. Interlinking the different systems is a way of trying to avoid high downtime costs and this is a unique feature of the project, which helps in making the diversion of spoil from one line to the other possible, so that both lines can work in unison during the duration of stoppages. To achieve this end, all the link conveyors of the overburden and the lignite system converge to one common platform to effect a centralised distribution using:

— telescopic drive head
— shuttle conveyor
— slewable conveyor.

2.2 Singrauli Coal Handling Complex

Rapid loading of unit trains has been given increasing attention during the last few years. The high capacity coal handling complex at Singrauli coal field is designed to supply coal to the adjacent super-thermal power station from a 2,400 t silo, fed by two conveyors, 1,400 m wide, and each with a capacity of 1,000 t/h (Fig. 2). Cross-country conveyors, 1,200 mm wide, transfer the coal from the feeder breakers of the receiving pits to the tripper conveyor over the storage area. The size and capacity of these cross-country conveyors are compatible with the peak hourly production capacity of the mine. The bunkerage is one of the self-flowing type ground bunkers with 60,000 t capacity.

The complex uses a private merry-go-round (MGR) system, still the only one of its kind in India, with automated rapid-loading chute gates at the rate of 4,000 t/h. The MGR system is comprised of a rake of 30 bottom-discharge type Indian Railway wagons, each of 65 t pay load capacity, which remain permanently coupled to the locomotive. The whole train continually undertakes the circuit between the receiving and supplying ends using its own individual rail loop. The wagons are loaded and emptied on the move at a nominal train speed of 0.8 km/h.

Variable capacity plough feeders at the slit opening of the bunker reclaim coal and feed the MGR reclaim/silo loading conveyor line at the desired rate. The silo is designed to ensure mass flow with a valley angle of 65°. The gates are hydraulic powerpack operated and the design provides for high reliability and a fail-safe arrangement, such that during emergency or power failure the gates would close in a second. The gates are designed for a flow rate of 3 t/s when fully open and can fully open or close in 2 s. Leaving the silo full of material or feeding a preponderance of large lump material into the silo at one time are the factors which contribute to arching. A planned operation demands that the silo is not loaded unless it is known that a train is due; moreover, material size segregation does not take place in order to reduce the chances of ratholing to a minimum. Additionally, there is provision for a public railway loading system with a rated capacity of 1,000 t/h using a conveyor line 1,400 mm wide for supplying coal to thermal power plants in northern and western India.

2.3 Iron Ore Projects: Kiriburu, Donimalai and Kudremukh

India abounds in iron ore of haematite and magnetite deposits. The reserves are so enormous and of reasonably good quality, that they meet the total demand of the Indian steel industry and also offer a vast potential for export.

The Kiriburu iron ore project is the first venture of the National Mineral Development Corporation (NMDC). The ROM ore handling system operates at a rate of 1,500 t/h. Processed lump ore from the 50,000 t ground stockpile is reclaimed by a series of vibratory feeders and a tunnel reclaim conveyor. The reclaim conveyor line makes it possible to load 14 bins, each of 1,400 t capacity, which are hydraulically operated for wagon loading.

The fine ore stockpile in the project has a capacity of 2.5 Mt. Conventional bulk solids stacking and reclaiming methodology is being employed with stacking and reclaiming yard belt conveyors of 1,500 t/h capacity running on each side of the bed (Fig. 3). Both the radial boom stacker and the bucket wheel reclaimer are rail mounted and operate in conjunction with the yard belts. Fine ore loading on wagons is accomplished by a wagon loader of 1,500 t/h capacity fed by the reclaimer (Fig. 4) through a series of belts. An electronic belt-weigher enables automatic recording of the weight of material flow. Further, the travel speed adjustment of the wagon loader is interlinked to the feed rate through the weightometer.

Fig. 3: Single boom stacker in operation at Kiriburu

Fig. 4: Rear view of bucket wheel reclaimer at Kiriburu

The conceptual design of the reclaiming/wagon loading system of the processed lump ore at Donimalai, another NMDC iron ore project, is similar to that of the Kiriburu fines handling circuit. However, the reclaiming capacity at Donimalai has been increased to 1,800 t/h, while the slewable stacker operates at 800 t/h within its travel limits.

Downhill conveyors play a significant role in the Indian iron ore exploitation scenario. While mining operations are carried out at the hilltops, the processing plants are located at the foot hills. The conveyors, usually of long line length and high capacity, use high-tensile strength belting. Moreover, they pass through adverse gradients and rough terrain while operating under hazardous conditions. The conveyor lines are equipped with thrustor brakes, belt speed control devices, pull cord, and side sway limit switches.

Kudremukh is India's largest and most sophisticated iron ore mining project. The crushed ore from the primary crushers is conveyed on two downhill steel cord belts, 1,600 mm wide and with a rated capacity of 4,000 t/h to transfer the load to a common conveyor, 1,800 mm in size, for storing in a covered monsoon stockpile of 0.35 Mt capacity — thus ensuring an uninterrupted running of the concentrator plant.

The concentrates from the thickeners, reclaimed as underflow, are fed to two agitated slurry storage tanks, each of 4,500 m^3 capacity. A bank of five heavy-duty rubber lined slurry pumps feed the slurry of 65% solids content to a 67 km long pipeline. Slime and oxygen scavengers are incorporated as a measure for free oxygen removal and internal corrosion prevention.

The slurry pipeline (Fig. 5) discharges at Mangalore Port into two agitated storage tanks where dewatering in disc filters is carried out. The filter cakes, containing 9% moisture, are conveyed to two covered storage sheds with a total capacity of 0.5 Mt. Two bridge-type bucket wheel reclaimers, each of 3,500 t/h capacity, reclaim the stockpile to convey it to a shiploader for loading ships of 40,000 to 60,000 DWT capacity at an average rate of 6,000 t/h.

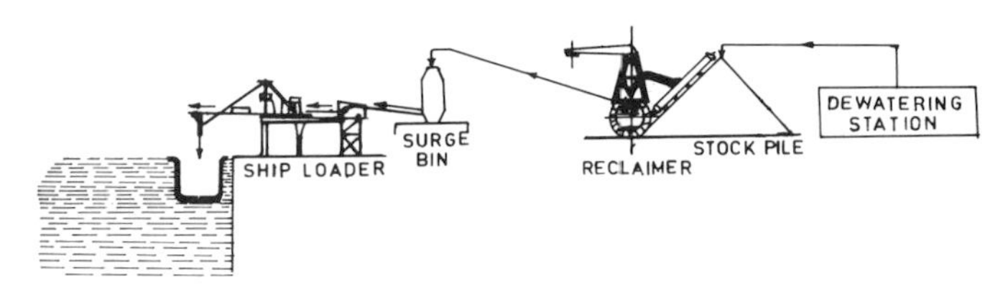

Fig. 5: Concentrate handling at Mangalore Port

3. New Developments in Bulk Transportation Systems

An impressive installation for the bulk handling of bauxite is presently being commissioned by the National Aluminium Company (NALCO) using a Cable Belt system. This is designed to transport crushed bauxite from the mine at Panchpatmali in the Korapur district to the alumina plant located at Damanjodi at the foot of the Panchpatmali hills over a distance of 14.6 km. In negotiating this distance, the system will have a fall of approximately 300 m. The conveying capacity of the system planned is 900 t/h in the first phase which will give an annual output of 2.4 Mt, and 1,800 t/h in the second phase which will yield on annual production of 4.8 Mt. The annual capacity will be doubled in the second phase by doubling the operating speed of the conveyor. For the first phase operations, the cable belt speed will be 2.4 m/s, which will be stepped up to 4.8 m/s in the second phase. The entire system will be driven by a single drive at the mine end consisting of a 2,000 kW motor. The belt width for the system is 1,050 mm and the diameter of the steel cables, which are of the Lang's lay type with a 6 x 19 construction and a fibre core, is 51 mm. About ten curvatures will be negotiated over the total transportation distance of 14.6 km.

4. Concluding Remarks

The past decade has witnessed some major new developments in bulk handling systems in the Indian mining industry. Even if the capacity of systems installed so far has been small by international standards, the increasing scale of operations in the surface mining industry will call for major new developments in bulk handling. With a mining industry set fair to new horizons of development, major breakthroughs in bulk handling systems can be expected.

The Bukit Asam Opencast Mine, Indonesia

G. Meyer and K.-H. Althoff, Germany

1. Description of the Overall Installation

The Bukit Asam coal mine system is a classical example of thorough and systematic opencast mine planning. In the final expansion phase, five bucket wheel excavators will extract and deliver the material via five mobile transfer conveyors, which assist in increasing the block width excavated, onto five shiftable bench conveyors. These conveyors transport the material via various intermediate conveyors, which are also shiftable, to the distribution station. This station comprises five extending conveyor heads, each having three transfer positions. One of these transfer points is at the coal conveyor which links the system with the train loading installation and the upstream buffer stockyard, respectively, whilst the other two transfer points are located at the two stationary conveyors whence the overburden is conveyed on various shiftable intermediate conveyors to the two dump bench conveyors, which are also shiftable. The latter conveyors transfer the overburden via the respective tripper car to an intermediate conveyor, from where it is passed on to the spreader at the spoil tip.

Providing extremely short switchover times, this ingenious system ensures that each bucket wheel excavator can deliver the right material, depending on the deposits, to the coal conveyor line or to one of the two overburden conveyor lines.

2. Technical Data

In the final expansion phase, the installation will mine approximately 3 Mt/a of lignite. This will require excavating, hauling and dumping up to 40 Mt/a of overburden.

To achieve these tonnages, the components in the installation are designed to produce certain outputs, some of which are given, together with further technical data, in the following sections.

2.1 Bucket Wheel Excavator (O&K supply)

Bucket wheel diameter	9 m
Number of buckets	14
Nominal bucket capacity	800 l
Number of bucket discharges/min	65/80
Theoretical output	3,100/3,840 m^3/h (loose material)
Guaranteed output	1,300 m^3(bank)/h
Bucket wheel drive motor	700 kW
Cutting height	15 m
Block width	20 m
Belt width	1,400 mm

Mr. G. Meyer and Mr. Karl-Heinz Althoff, PWH Weserhütte AG, P.O. Box 510 850, D-5000 Köln 51, Federal Republic of Germany

2.2 Conveyor Plant from Excavators to Conveyor Distributor Station (Marobeni supply, PWH design)

Number of conveyor systems	5
Number of conveyors	22
Total length of conveyors	13.72 km
Belt width	1,200 mm
Belt speed	5.5 m/s
Calculated throughput	2,000 m^3(bank)/h ≙ 5,300 t/h
Guaranteed throughput	1,300 m^3(bank)/h
Total installed motor power	47 x 380 kW = 17,860 kW

2.3 Conveyor Plant from Conveyor Distributor Station to Spreaders (Marobeni supply, PWH design)

Number of conveyor systems	2
Number of conveyors	12
Total length of conveyors	13.05 km
Belt width	1,600 mm
Belt speed	5.5 m/s
Calculated throughput	4,000 m^3(bank)/h ≙ 10,600 t/h
Guaranteed throughput	2,600 m^3(bank)/h
Total installed motor power	35 x 600 kW = 21,000 kW

2.4 Spreaders with Tripper Car (PWH supply)

Number	2 each
Outreach of spreader boom	60 m
Conveying length of intermediate conveyor	38 m (max.)
Belt width	1,600 mm
Belt speed of intermediate conveyor	5.5 m/s
Belt speed of spreader conveyor	6.5 m/s
Calculated throughput	4,000 m^3(bank)/h ≙ 10,600 t/h
Guaranteed throughput	2,600 m^3(bank)/h
Total installed motor power per spreader unit	1,500 kW (approx.)
Planned max. dump height	15 m
Planned max. dump depth	15 m
Planned dump block width	20 m

3. The Transport Crawler

The conveyor drive stations in the Bukit Asam opencast mine will be moved with a transport crawler.

The main section of the transport crawler is the lower structure which absorbs the forces resulting from the load. The lower structure is supported on the tractor travel assembly. The twelve travel rollers on each crawler are mounted in bogie assemblies, which distribute the forces over the travel rollers and compensate ground unevenness.

A lifting platform, which is mounted on the lift cylinders and guided by a centre pin, raises the conveyor drive station. The vertical forces exerted by the load are absorbed by lift cylinders, the horizontal forces by the centre pin.

This design principle allows to lift the platform and the drive station on it to be positioned at any desired angle to the longitudinal axis of the vehicle. Thus, the drive station can be shifted at any angle. A hydraulically operated clamping device clamps the drive station tightly to the lifting platform. The nonloaded lifting platform can be rotated at any angle relative to the longitudinal axis of the vehicle by means of a slewing mechanism.

All functions of the transport crawler are powered by a diesel engine. This engine drives the various hydraulic pumps, which in turn powers the travel drive, the lift cylinder, the clamping assembly and the slewing assembly.

The crawler vehicle can be driven at three speeds: 0 to 30 m/min in unloaded condition, 0 to 15 m/min with the station loaded and 0 to 5 m/min for special manoeuvres as well as transportation on steep up- and downhill gradients.

All functions of the transport crawler are controlled and monitored by electrohydraulic means. The load distribution in the supporting area of the lifting platform is displayed electronically in the operator's cab.

The transport crawler can pick up a conveyor drive station without any auxiliary equipment.

To load, the transport crawler is driven into the portal-type space between the supports of the drive station. The lifting platform is turned by the slewing mechanism and manoeuvred at creep speed by the crawler in a to-and-fro pattern into a parallel position to and pressed against the drive station.

The lift cylinders are steered in such a manner that the lifting platform, even when the ground surface is uneven and the transport crawler is standing at an oblique angle to the drive station, can be moored smoothly and evenly under the bottom beams of the drive station without lifting it. Then the drive station and the lifting platform are clamped together by the hydraulically operated clamping device. When the lift cylinder is retracted, the transport crawler is rotated to the desired direction of transportation.

By renewed extension of the lift cylinders, the conveyor drive station is raised to be ready for transportation.

In the case of impermissible relocation of the centre of gravity, i.e., considerably varying load on the cylinders, an optical and accoustic warning is given. The load must then be realigned and the transport crawler repositioned. The pressure indicator of the four lifting cylinders serves as orientation for the load and the centre of gravity.

The incline of the transported load and the corresponding load on the cylinders can, by inclined travel, be reduced to certain limits by extending or retracting one of the cylinders.

 Volume 5, Number 1, February 1985

Surface Mining Systems in Athabasca Oil Sands

A Review of Performance

T.S. Golosinski, Canada

1. Introduction

Commercial production of synthetic crude oil from the Athabasca oil sands in Alberta was initiated by Great Canadian Oil Sands, Ltd. (now Suncor, Inc.) in 1967. At present, there are two plants, producing together close to 30,000 m^3 of crude per day. Both plants mine oil sand by open-pit methods and transport it to the extraction plant, located at the mine-mouth, for extraction of bitumen. The bitumen is then cleaned and refined into synthetic crude oil in the upgrading facilities located nearby.

The layout and technical details of oil sand mining operations as well as some problems faced there can be learned from numerous publications dealing with the topic (e.g., Boehm and Golosinski [2], Fauerbach and Bartsch [5]). The following paper outlines the evolution of mining methods for oil sand mines, which resulted from unexpected problems faced during the initial phase of mine operations. It also discusses the present efficiency and selected performance data of these mines.

2. Peculiarities of Oil Sand Mining

The oil sand surface mining operations applied in Alberta are peculiar in several ways. First, the climate at the mine sites is very severe, yet oil sand mining must be conducted without major seasonal interruptions to keep the price of synthetic oil competitive with that of conventional oil. Second, the oil sand properties differ from those of any other mined commodity. Furthermore, these properties and the behaviour of oil sand depend on its condition as well as on varying ambient temperatures. Third, the mine is only one of the oil sand plant components, with a number of requirements imposed on it by the downstream plant units.

While the first two peculiarities are widely described, the impact of the last one on the performance of an oil sand plant is only seldom taken into consideration. Nevertheless, it has a serious influence on the selection of the mining methods and the way the mines are operated. For example, the mine accounts for 15 % to 25 % of the total capital costs of an oil sand plant; at the same time its operating costs may be as high as 50 % of the production costs of each barrel of synthetic oil. This means that overall savings may be achieved by a substantial overdesign of the mine capacity if, as a result, the operating costs are lowered.

An additional advantage of mine overdesign is the reduction of the idle time of the extraction plant, resulting from shortages of the mine-produced feed. The traditional method of reducing these shortages by providing an ample buffer stockpile capacity between the mine and the extraction plant is not feasible, considering the required production rates and the climatic conditions. Should the production loss result from an extraction plant feed shortage, it cannot be fully compensated by an increased production at a later time due to limited throughputs of both the extraction and upgrading facilities.

Serious consideration must also be given to the control of the extraction plant feed grade, in particular to the rejection of the low grade oil sand, to assure acceptable and profitable levels of bitumen recovery. With no buffer storage, this control can only be implemented in the mine.

Still another peculiarity of oil sand mining is the low overburden-to-ore ratio, presently in the range of 0.25 m^3 (bank) overburden : 1 t ore. This means that much more oil sand has to be mined and handled than overburden. Consequently, to reduce the mining costs much more emphasis must be placed on efficient ore mining than in a typical strip mine operation.

3. Mining of Oil Sand with Bucket Wheel Excavators

The Suncor operation uses bucket wheel excavators (BWEs) to mine oil sand and portions of overburden. The oil sand is then handled by a system of belt conveyors, up to 5 km long and consisting of three to five conveyor flights, delivering the mined oil sand to the extraction plant (Boehm and Golosinski [2]). The main reasons for the selection of this particular mining system are described by Winzer [6]:

- efficiency of conveyors in handling the required large tonnages of mined oil sand
- ability of BWEs to deliver the required tonnages of oil sand at lowest cost, while at the same time producing a conveyor feed of acceptable size

Dr. T.S. Golosinski is Professor at the Department of Mineral Engineering, The University of Alberta, Edmonton, Alberta, Canada

— low and uniform system power demand
— low ground pressures of the used equipment
— insensivity of the BWEs to potential slope failures.

The economic advantages of BWE systems result from specific site factors, such as regularity and uniformity of the deposit, easy mine opening without the need to excavate large opening cuts, lack of ready power supply at the mine site (i.e., the need to build a power plant, thus the need to minimize the power consumption), lack of concern about the stability of the slopes, and a low bearing capacity of oil sand.

Soon after the start-up of the operation, it was learned that, to make it efficient, the mining system and the operating techniques had to be improved considerably. One striking example was that the costs of the excavator cutting teeth were higher than those planned for the whole mining operation. The problems resulted from the properties of the oil sand, which were different from those assumed for planning purpose, and from the underestimation of the influence that climatic conditions would have on the system performance. The major breakthroughs were the introduction of oil sand fragmentation by blasting in the preparation for BWE mining, the development of new BWE cutting organs (teeth and bucket lips) fitted on the buckets, and the development of a belt cleaning technology and of suitable belting.

Of most importance was the oil sand fragmentation by blasting and subsequent surface ripping, because it reduced the frost penetration, the wear and tear of the equipment, and the size of lumps of frozen material created in the process of mining (Fig. 1).

Fig. 1: Blasted and ripped BWE bench, prepared for winter mining

The efficiency of Suncor's BWE mining systems, which have now been operating for over 15 years, is comparable to that of many other BWE operations working under more moderate conditions. Sample performance data of these systems are shown in Table 1.

Table 1: Sample performance indicators for BWE oil sand mining systems

Year	BWE System A			BWE System B		
	Time factor	Load factor	Utili-zation	Time factor	Load factor	Utili-zation
1	0.470	0.415	0.195	0.511	0.420	0.215
2	0.462	0.399	0.184	0.473	0.420	0.199
3	0.448	0.420	0.188	0.468	0.450	0.211

The time factors, load factors, and utilization given in Table 1 are defined as follows:

$$\text{time factor} = \frac{\text{working hours}}{\text{total calendar hours}}$$

$$\text{load factor} = \frac{\text{average hourly productivity}}{\text{theoretical hourly productivity}}$$

$$\text{utilization} = \text{load factor x time factor}$$

A review of Table 1 indicates that the utilization of time by the systems is comparable to that in the Rheinbraun mines in the Federal Republic of Germany, working in favourable climatic conditions and commonly considered the most efficient worldwide. However, utilization of productivity is relatively low. This reflects the system's overcapacity, which is required to assure an efficient utilization of the extraction and upgrading plants, as well as the particular difficulties related to oil sand mining.

4. Mining Oil Sand with Dragline-BWR Systems

The second oil sand plant, belonging to Syncrude Canada, Ltd., was designed and constructed during the mid-1970s, at the time when the BWE systems of Suncor were already in operation for a number of years. A better understanding of the difficulties of oil sand mining and distinct differences in the geology of the orebodies resulted in Syncrude's selection of a different mining system. It includes draglines for direct in-pit dumping of the waste material as well as mining and windrowing of the oil sand, bucket wheel reclaimers (BWRs) for reclamation of the oil sand and loading it onto the conveyors, and a belt conveyor system for delivery of the oil sand to the extraction plant (Boehm and Golosinski [2]). The reasons for the selection of this system were listed by Adam and Regensburg [1]:

— low costs for waste disposal as no waste transportation system was needed
— rapid opening of the mine, made possible by the use of draglines, allowing for a fast production build-up and a reduction of the front-end costs of the operation
— selective mining of oil sand and waste was envisaged to be easier done with draglines than with a BWE
— need for material fragmentation before mining was eliminated
— generation of broken oil sand reserves in the form of dragline-built windrows minimizes the interruptions in the extraction plant feed supply.

The advantages of the system were to outweigh its disadvantages, such as the need for double handling the mined oil sand and continuous monitoring of the dragline slopes to warn of failures, should these tend to occur.

The above points reflect the differences between the two orebodies. Syncrude's was overlain by a thick and fairly uniform layer overburden, creating concerns about the mine opening. It was positioned on top of water bearing sands, creating concerns about the pit bottom stability. The orebody was less uniform with more interbedded rejects, pointing to a possible need for selective mining. Additionally, Syncrude's plant was much bigger than that of Suncor, thus requiring greater reserves of broken ore.

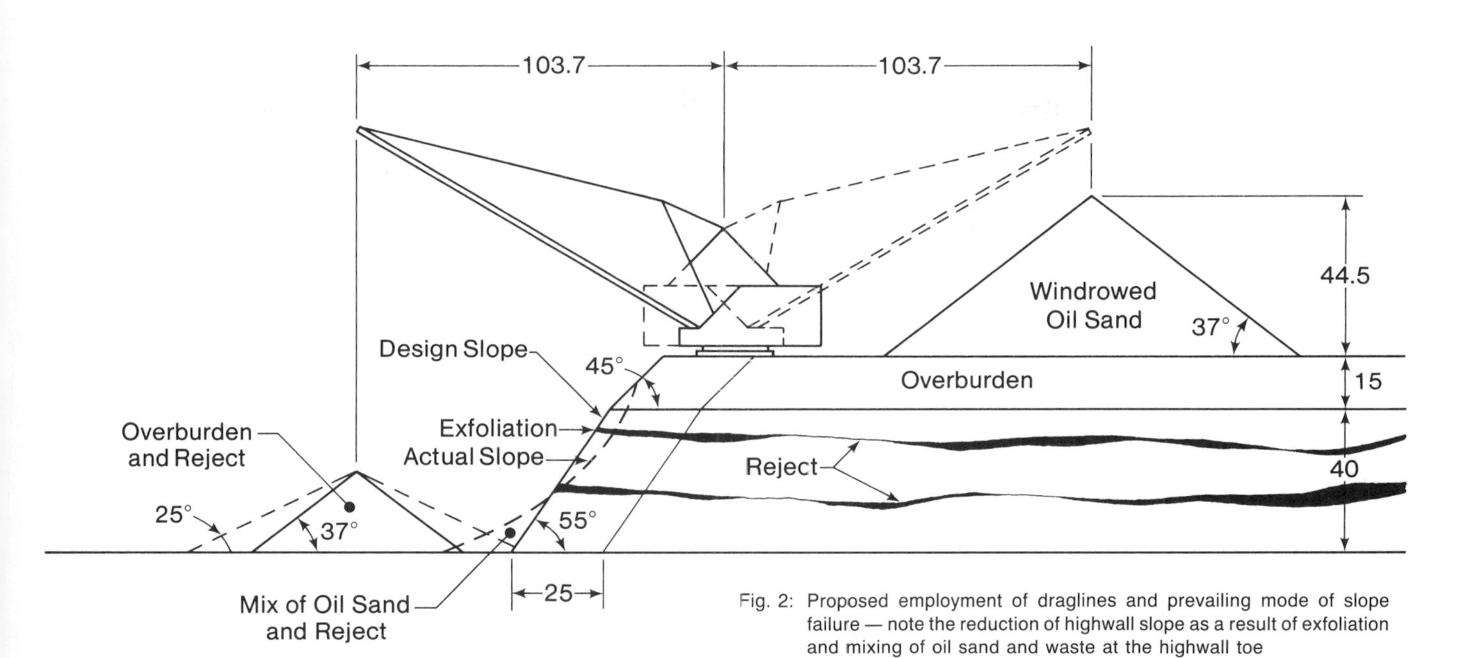

Fig. 2: Proposed employment of draglines and prevailing mode of slope failure — note the reduction of highwall slope as a result of exfoliation and mixing of oil sand and waste at the highwall toe

The key to successful dragline application was the development of correct criteria for the highwall design. Extensive studies were undertaken. These included the excavation of a 55 m deep test pit, which confirmed the feasibility of the dragline highwall design based on 55 ° to 60 ° slopes.

Soon after the start-up of the mining operation, unexpected problems arose. The most important one was the inability to maintain 55 ° highwall slopes. Extensive exfoliation occurred, reducing the effective slopes to 40 ° to 45 ° (Fig. 2). The unique application of draglines, different from that in a typical coal strip mining operation, proved to be very sensitive to this type of slope failure.

It was discovered that the amount of exfoliation depends on the excavation rate, the steepness of the slope, its height, the time of exposure, and the ambient temperature (Brooker [3]). The relationship between the degree of exfoliation and the excavation rate is not fully understood yet. It appears that this behaviour results from high locked-in residual stress, combined with the rapid release of the confining pressure on the gas associated with the oil sand (Brooker [3]). The importance of these phenomena was not understood during the test pit evaluations (Fig. 3).

Other problems faced included a flatter angle of repose for the waste material dumped at the pit bottom, resulting from its high moisture content, and a poor bearing capacity of the overburden, on which both the dragline and BWR were to travel. Problems were also faced with the BWRs; the windrowed oil sand proved to be difficult to reclaim as it contained both the broken rocks and huge lumps of frozen oil sand (65 m³ dragline bucket). Further, the windrow surface was freezing to a substantial depth during winter.

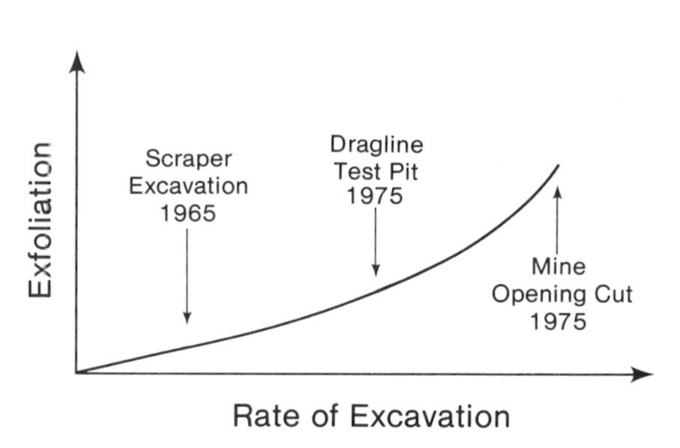

Fig. 3: Qualitative relation between exfoliation and excavation rate

The difficulties were finally solved by modifications of the mining method. These were based on the introduction of an additional mining system for the advanced overburden stripping and positioning of both the dragline and BWR on the oil sand. The modified mining scheme is shown in Fig. 4. The modification eliminates the major advantage of the dragline-BWR system, the direct overburden disposal without transportation. However, the application of this system benefited the operation through rapid mine opening and production build-up.

The present performance of the system matches that of the BWE systems discussed before. In particular, the utilization of BWRs is comparable to that of BWEs. Dragline perfor-

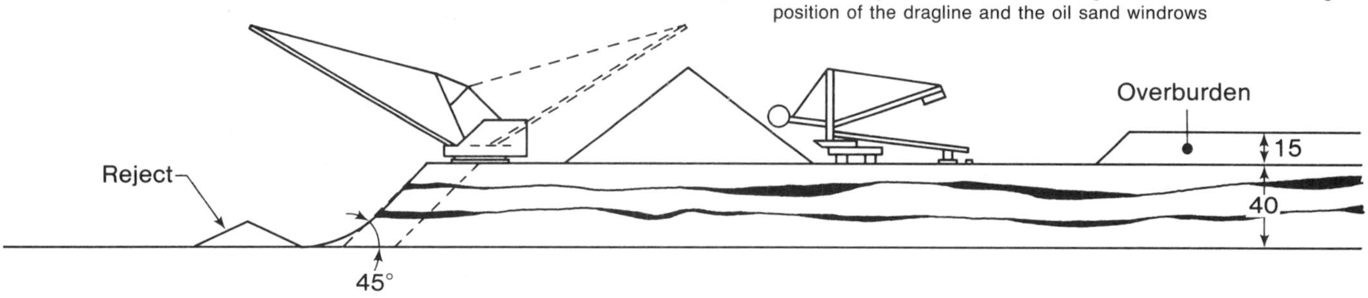

Fig. 4: Present dragline-BWR oil sand mining method — note the changed position of the dragline and the oil sand windrows

mance, however, is somewhat limited when compared to typical coal strip mining operations. This limitation results from the need for selective mining, from large swing angles required by the equipment layout and slope stability considerations, as well as from interference between the dragline and the BWR operations at the block ends. In some situations, time is lost by the dragline when it has to wait for a windrow to be reclaimed before it can move to a new position. The typical performance data of BWRs and draglines, as used presently for planning purposes, is shown in Tables 2 and 3.

Table 2: Sample performance indicators for a BWR reclaiming windrowed oil sand

Year	Time factor	Load factor	Utilization
1	0.39	0.52	0.203
2	0.39	0.53	0.206
3	0.42	0.53	0.222

The definitions of the load factor, time factor, and utilization are the same as above.

Table 3: Sample performance indicators for a dragline mining oil sand

Year	Working hours	Average hourly production m³(bank)/h	Yearly production per m³ bucket capacity m³(bank)/(m³ · a)
1	4,468	2,615	193,000
2	4,129	2,538	173,000
3	4,730	2,581	199,000

5. Conveying of Oil Sand

Both the existing mines use belt conveyors to transport the mined oil sand from the mine to the extraction plant. While this operation is fairly simple and reliable elsewhere, a number of difficulties was faced by the oil sand industry. These included a rapid material build-up on the belts, low resistance of conventional belting to action of the bitumen contained in the oil sand (as well as to kerosene belt spraying), and extensive damage caused by work under low temperatures. Most of these problems are described in detail elsewhere (Fauerbach and Bartsch [5]).

Problems related to belting were dealt with by the development of new belt stocks based on either nitrile or chloroprene polymers. These stocks exhibit excellent low-temperature flexibility and low-temperature impact resistance, combined with a sufficient resistance to the bitumen action. Unfortunately, such belting is some 75% more expensive than comparable belting used in surface coal mines.

Oil sand build-up on belts is prevented by continuously wetting the belt surface in front of the loading chutes, in addition to the use of the whole range of conventional belt cleaning devices. Belt wetting agents, depending on the situation, are either water, antifreeze, or kerosene, which is a byproduct in the bitumen upgrading process.

The uniqueness of the oil sand conveyor transportation is best illustrated by the rolling resistance of idlers and poor performance idlers.

5.1 Conveyor Power Requirements

With dropping temperature, the power needed to flex the belting, turn the idlers, etc., increases. It is a standard North American practice to define this power based on the plot of the temperature factor K_t as a function of the ambient temperature, as provided by CEMA (Conveyor Equipment Manufacturer's Association). The investigations done by one of the mine operators have shown that this plot for oil sand conveying differs substantially when the rolling resistance of the idlers is considered (Costello, [4]). In this case, the value of K_t peaks at —15° to —20°C and drops for lower temperatures (Fig. 5). No explanation for this phenomenon exists. One can only speculate that unfrozen oil sand acts as a glue put between the belt idlers and pulleys, thus increasing the power requirements. On the other hand, when frozen, it becomes brittle, falls off the idlers, pulleys, and belting, and the glueing action ceases.

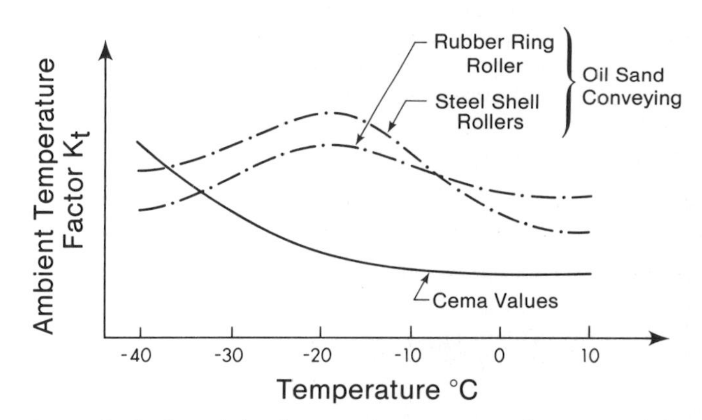

Fig. 5: Qualitative relation between the temperature factor K_t and the ambient temperature related to the idler rolling resistance

Based on past experience, it has become a standard practice to oversize the conveyor drives for oil sand duty. As an example, an 2,280 m long conveyor with a 1,830 mm wide belt is driven by 3,730 kW (5,000 HP), while the standard calculations indicate the power requirements as less than 3,360 kW (4,500 HP).

5.2 Catenary Idlers

Good performance of catenary idlers with steel shells in Rheinbraun and other coal mines prompted the designers of the Syncrude mine to introduce these to their operation. Problems were faced almost immediately, such as excessive vibrations of the idlers and a rapid wear of both the idlers and belting. The sticky oil sand was rapidly building up on parts of the rollers (Fig. 6). This resulted in a change of the diameter along the rollers and the differential speed between the belt and parts of the rollers. As a result, strong idler vibrations took place, rapidly wearing out the links between the rollers and roller shells. Roller wear rates were enormous, and frequent damage to belting from damaged rollers occurred. As a remedy, the steel shell rollers had to be replaced with self-cleaning ones, equipped with rubber rings.

A comparison of different idler types indicates that the best performance is obtained by the use of fixed idlers with self-cleaning rubber ring lagged rollers.

Fig. 6: Oil sand build-up on catenary idlers — note the wear on parts of the roller surface free from build-up

6. Summary

Oil sand experience confirms that the design and start-up of a mine in a new environment, involving mining a material with which the mining engineer is unfamiliar, is always a risky venture. The technical challenges related with such a venture are enormous, despite the availability of elaborate mining equipment and technique. However, the ingenuity of man is able to successfully cope with such challenges, providing a strong commitment is made and the demand for the mined commodity can be sustained at an acceptable price level.

In the case of oil sands, both of the above conditions were met. The mines operate successfully and profitably. Two different systems selected for the two different orebodies are equally efficient in terms of time and productivity utilization. The performance indicators for both systems are somewhat lower than in a typical coal operation, but the differences result from difficult mining conditions and a different mine design philosophy.

References

[1] Adam, D.G., and B.O. Regensburg: Dragline Mining at Syncrude. International Mining Conference and Exhibition IMEC'80, Calgary, Alberta, Canada, August 1980.

[2] Boehm, F.G., and T.S. Golosinski: Ölsandgewinnung in Kanada — eine bergtechnische Herausforderung; Erzmetall Vol. 36 (1983) No. 7/8, pp. 339—347.

[3] Brooker, E.W.: Geotechnical Frontiers in Oil Sand Mining. International Mining Conference and Exhibition IMEC'80, Calgary, Alberta, Canada, August 1980.

[4] Costello, W.R.: Presentation to the CSME Edmonton Branch Meeting, November 1980 (unpublished).

[5] Fauerbach, R., and D. Bartsch: 15 Years of Oilsand Transportation by Belt Conveyor Systems — From a Manufacturer's Viewpoint; bulk solids handling Vol. 2 (1982) No. 4, pp. 733—736.

[6] Winzer, S.: Experience with Bucket Wheel Excavators in the Tar Sands. Symposium on BWE Technology, Calgary/Vancouver, Canada, Oktober 1976.

bulk solids handling Volume 4, Number 3, September 1984

Coal India Expands Opencast Mining and Installs Coal Handling Plants

S.K. Bose, India

1. Coal Production in India

India's coal production reached 137,860,000 tonnes in the financial year April 1983—March 1984. Historically, it means substantial expansion of coal production from a level of 6,000,000 tonnes in 1900 to 32,510,000 tonnes in 1950 and major increases after the nationalisation of coal mines, when the production of 72,950,000 tonnes in 1970—71 increased by about 65,000,000 tonnes in the 13 years up to 1983—84, at a compound growth rate of 5% per annum. In addition, production of lignite from opencast mines recorded 7,340,000 tonnes in 1983—84. India's coal production is about 3% of the world's hard coal production, though its coal resources in coal equivalent terms are about 0.8% of world coal resources.

Coal India Limited shared 88% of all hard coal production in the country, to produce 121,460,000 tonnes in 1983—84. Coal India operates 395 mines in seven states located between longitudes 77°E and 96°E, covering a distance of 1,900 km east-west in the eastern half of the country. Coal India Ltd. manages its mines through four coal producing coal companies: Eastern Coalfields Ltd. (ECL), Bharat Coking Coal Ltd. (BCCL), Central Coalfiels Ltd. (CCL) and Western Coalfields Ltd. (WCL) and one Central Mine Planning and Design Institute (CEMPDIL). During the last ten years there has been a three-fold increase in opencast coal production, when production has risen from 20,770,000 tonnes in 1974—75 to 60,050,000 tonnes in 1983—84.

Opencast production by Coal India is shown in Table 1, with distribution of coal production from subsidiary companies and Cempdil's estimate (R.G. Mahendru — WEC, 1983) of Coal India's opencast production share for 1983—84, 1989—90 and 1999—2000.

Highlights of the Coal India opencast production shown in Table 1 indicate expansion prospects from 60,050,000 tonnes in 1983—84 (49.4% of all coal) to 128,000,000 tonnes in 1989—90 and finally with the aim of 201,000,000 tonnes in 1999—2000, to share 61% of Coal India's expectation of 328 million tonnes by the end of the century. In other words, the aim is to continue the opencast production which has grown at an average rate of 12.5% per annum for the past 13 years aiming for a growth rate of 12.9% per annum during the next six years until 1990; this is indeed a major opencast expansion scenario.

Mr. S. K. Bose is Additional Director (Equipment), Coal India Ltd., Calcutta — 700 001, India

2. Opencast Mining Technology

Opencast mining technology in Coal India encompasses various degrees of mechanisation, from almost manual small old mines to very large mines of 10 million t/a capacity and more. Large machines such as electric shovels (10 m³ bucket capacity) are working in conjunction with 85 ton rear dumpers manufactured indigenously and the first lot of 120 ton capacity imported dumpers have just gone into operation. Large walking draglines with bucket capacities of 15—30 m³ and booms of 90—96 m length are also deployed in several mines (Figs. 1 and 2). So also is complementary equipment: bulldozers, 400 HP; rotary blast hole drills, 250 mm; and motorised scrapers, 11.5 m³ bucket. The pieces of heavy earth moving machinery, about 3,000 in number in Coal India mines, are grouped into different types and sizes of machines with names of manufacturers in Table 2.

Fig. 1: 4.6 m³ bucket shovel made by Heavy Engineering Corporation, based on USSR design

Table 1: Opencast production until the year 2000 (million tonnes)

Coal India Subsidiaries	Coal Production 1974—75	Coal Production 1983—84	Opencast Production 1974—75	Opencast Production 1983—84	Opencast % 1983—84	Expected Opencast 1989—90	Aimed Opencast 1999—2000
ECL	23.16	22.92	2.66	6.39	27.8 %	13	27
BCCL	17.74	21.63	2.10	7.58	35.0 %	14	26
CCL	18.31	36.76	12.34	30.62	83.2 %	64	110
WCL	19.26	39.33	3.53	15.14	38.4 %	36	37
NEC	0.52	0.82	0.14	0.32	39.0 %	1	1
Total/CIL	78.99	121.46	20.77	60.05	49.4 %	128	201
SCCL/TISCO, TISCO, DVC	9.42	16.40					
All India	88.41	137.86					

Table 2: Heavy earth moving machinery in Coal India companies

	Types	Bucket sizes (m³)	Manufacturers
1.	Draglines boom lengths 45—96 m	2 (Diesel) 2, 4, 5, 6, 8, 10, 15, 24, 30 (Elect.)	Marion, Heavy Erie, Ransomes Rapier, USSR, P & H, Page, Tata-RB
2.	Shovels (mech. excavators)	2, 3.2 (Diesel) 1.9, 2.3, 3.2, 4.6, 6.3, 8, 10 (Elect.)	USSR, Heavy Engg.-USSR, Bucyrus Erie., Marion, Ruston Bucyrus, Greaves Cotton, Voltas
3.	Hydraulic excavators	0.9, 3.2, 4.5, 7.5	Poclain — Larsen & Toubro, O & K — HEC
4.	Rear dumpers	25 t, 35 t, 50 t, 85 t, 120 t	BEML. Hind Motor, Unit Rig, Komatsu
5.	Bottom dump coal haulers	32 t, 40 t, 60 t	Athey, BEML, Hind Motor
6.	Bulldozers	180 HP, 275 HP, 320 HP	Komatsu, BEML, HM, Caterpillar
7.	Motor scrapers	11.5 m³	Bharat Earth Movers (BEML)
8.	Blast hole drills, Wagon drills	100 mm, 160 mm, 250 mm	Bucyrus Erie, HEC, Ingersoll Rand, Joy, Scharamm, ReCP, LMP, USSR — HEC
9.	Pay loaders	1.9, 3.4, 5.5, 7.4	Terex — GMMCO, Polish, BEML-Komatsu, Clark Michigan
10.	Cranes	12.5 t, 20 t, 30 t, 40 t	Tractors India — Coles, Escort, Voltas, Marshall, Greaves Cotton, Link-Belt

Fig. 2: USSR dragline with 15 m³ bucket capacity and 90 m boom length, 250 mm rotary blast hole drill, 3.2 m³ hydraulic excavator, payloader and rear dumper

Historically, the first mechanised opencast mine using diesel shovels and dumpers was started during the Second World War in 1945 in the Bokaro coalfield under the organisation Opencast Mechanised Mining (OCMM). When the National Coal Development Corporation was formed in 1956, mechanised opencast mining with the use of shovels and dumpers was encouraged, contributing to a production increase during the second five year plan (1956—61). After nationalisation of the coal industry during 1971—73, there was a further boost to mechanised opencast mining to produce low grade coals from very thick coal seams at shallow depths, to meet the fast expanding coal demand for electricity generation. Today, over 1,900 dumpers, rear and bottom discharging, with capacities of 25, 35, 50, 60, 85 and 120 t, are in operation in Coal India mines.

Draglines came into operation during the third five year plan, when a walking dragline with a 11.5 m³ bucket and 68 m long boom was commissioned in Kurasia Colliery in Madhya Pradesh on April 5, 1961. As many as 20 draglines are in operation in the mines of CCL and WCL now with boom lengths between 45 and 96 m. These draglines handled 22,660,000 m³ overburden during 1983—84. In addition to operating machines, as many as 4 draglines with a 24 m³ bucket and 96 m long boom, 2 draglines with a 20 m³ bucket and 90 m boom and 4 draglines with a 10 m³ bucket and 70 m boom are under erection or on order.

Electric shovels deployed generally have 4.6 m³ to 10 m³ bucket capacity. In future, large electric shovels of 20 m³ bucket capacity will be introduced to the Singrauli coalfield. Even the present population of 361 shovels included mechanised rope shovels which are most popular and since 1976, hydraulic excavators with buckets of 0.9—7.6 m³ capacity have been introduced and their indigenous manufacture organized.

The main burden of maintenance for heavy earthmoving equipment is in the area of dumpers including coal haulers. We are developing close links with indigenous manufacturers for better sales service, supply of spare parts and rehabilitation of equipment. Major organised facilities have been developed in the country for repair and overhauling of diesel engines and transmissions and these are further extended to other sub-assemblies for faster change of parts.

Motorised scrapers with push-by bulldozers are employed in some mines with thick overburden, but scraper operations are being reduced steadily. That shows the trend of the introduction of different types of heavy earth moving machinery. Indigenous manufacturing capabilities with design collaboration from the USSR, USA, U.K., France, Japan and West Germany are meeting most of the demand for draglines, shovels, dumpers, dozers, drills, scrapers, cranes etc. in India.

Central workshops have been established during the past 25 years for servicing opencast equipment in CCL and WCL. We do have problems at times when large machines break down. Early in 1984, a large (30 m³ bucket) dragline's main shaft (7 m long and weighing over 7t), mounting 3 large gears and a clutch, sheared in the middle after 20 years of operation. Fortunately, our engineers and technicians at Bisrampur mine site and central workshop Korba succeeded in dismantling, jointing the broken ends by specially machine tapered pin and welding, and finally recommissioned the dragline that weighed over 1,800 tonnes.

3. Mine Planning and Design

Mine design and equipment technologies for opencast mining are fast changing. The CEMPDIL concept of total planning of coalfields, delineating the opencast blocks of large coal production units of up to $10—12 \cdot 10^6$ t/a, is calling for very large equipment sizes and extensive mine design with mine depths of more than 480 m to exploit thick coal seams. In this process, master plans have been prepared for several coalfields such as Singrauli, Jharia and Raniganj.

The integrated plan for the Singrauli field prepared by CEMPDIL has projected mining of $2.6 \cdot 10^9$ tonnes of opencast coal reserves, in 11 blocks, with individual mine production capacities varying between 1 and 14 million t/a, half of them being of $10 \cdot 10^6$ t/a. The stripping ratios of overburden to coal vary between 1.15 and 3.90 m³ overburden per tonne of coal production. This new coalfield of Singrauli, which produced only $1.17 \cdot 10^6$ tonnes in 1970—71, has expanded production to $9.81 \cdot 10^6$ tonnes in 1983—84 and is aiming for $65 \cdot 10^6$ by the end of the century. Such large production prospects from concentrated coal deposits are becoming possible through application of the concept of integrated coalfield planning and development, large size mining blocks and adoption of large size opencast heavy earth moving machinery.

Jharia coalfield, intended for major reorganisation, is another large coalfield which has been planned on an integrated concept. It is intended to exploit $1.8 \cdot 10^9$ tonnes of coking coal reserves, during which the existing opencast production of $9 \cdot 10^6$ t/a is expected to be expanded to $31 \cdot 10^6$ t/a after completion of the re-construction programme. That calls for very large heavy earth moving machinery of which the first dragline, with a 24 m³ bucket and 96 m long boom, has been ordered.

The Raniganj coalfield integrated plan prepared by CEMPDIL provides for expansion of opencast mining to exploit its $2.16 \cdot 10^9$ tonnes of coal resources. The best quality and greatest quantity of non-coking coals come from this coalfield and are transported to all parts of the country. The integrated coalfield planning concepts are being extended to other coalfields in Orissa and Bihar States.

The trend towards large mine sizes of $10 \cdot 10^6$ t/a and deep opencast mining for better recovery of coal are causing new problems in the operation and maintenance of the mines and equipment. The giant equipment dimensions are calling for new management patterns, better infrastructure for maintenance and intensive appropriate training for operation and maintenance. The very large volume of rock handling for overburden removal and individual large mines handling 40 million m³/a, are calling for extraordinary designs for transporting systems and dumping facilities, keeping in view the environmental impact on the ecology of the surrounding areas. Shales in overburden contain up to 30—40% carbonaceous material and at times fires can break out because of outside heat contact or spontaneous heating, and create enormous problems of smoke, live fire, heat and associated difficulties in the overburden dumping areas. Bench planning and design of overburden dumps are also calling for new experiments in pit slope stability. The depth of the pit adds to the ratio of overburden to coal production. Many of the non-coking coal stacks containing high moisture at 8 to 10% are often susceptible to spontaneous heating. Spontaneous heating or active fires on such coal in seam or in stocks can be a menace and requires research towards solving the problem.

Amongst the problems in the Jharia coalfield, where coal seams have been honeycombed by underground board and pillar mining, the immediate overburden benches and middle partings need special design care so that heavy dumpers can travel with safety. The depth of mines leads to the construction of long haul roads for coal handling and overburden transportation. The problems of multi-seam mining, drainage, mine fires, old workings, removal of surface obstructions and dwellings and re-aligning of surface rail networks and roads are major issues that should be covered in mine design. The application of large in pit crushers and belt conveyor systems may be introduced in new mine designs. Coal conveying through conveyors in drifts in high wall sides and rock crushing and conveying through tunnels underneath overburden spoils are all possibilities for future mine designs. Application of portable surface rock milling machines capable of milling rocks and coal are possibilities for future small and medium sized mines.

Summing up, we can foresee that future large deep opencast mining in India will rapidly add to production but will mean a different opencast mining technology, substantially different from small conventional opencast mines. It involves numerous major problems in design, as also in the organisation of mine operations and in the application of large heavy earth moving machinery. Gigantic training programmes in the operation and maintenance of equipment and new infrastructure capabilities in large workshops, high speed loading of coal, communications etc., will be a part of the diverse tasks which must be implemented to introduce the large mines and machines.

Seen as a whole, mechanised opencast mining has already come of age in Coal India mines, with an average annual increase in production of over 12.5% and aiming at $201 \cdot 10^6$ t/a by the end of the century. By then the country will have a wide spectrum of mine sizes, large draglines, shovels, drills, dozers, conveyors, crushers, cranes etc. in operation. The mine size spectrum will spread from small mines to those with a maximum size of 10—14 million t/a capacity. We will have the use of draglines with 24—30 m^3 buckets, rope and hydraulic shovels from 1 to 20 m^3 buckets and dump trucks with 35 to 170 ton payload capacity and then we will have complex deep opencast mining technology. Commensurate indigenous manufacturing capacities are progressively increasing in each area or large machine design, manufacture and erection. With all this, one can see a major and dynamic opencast development in Coal India mines until the end of the century. No less are the developments of coal handling plants.

4. Construction of Coal Handling Plants

Coal handling plant construction is the other major activity in Coal India that comes within the scope of bulk solids handling. Way back in the early 1960s, the author introduced and built twelve coal handling plants with large ground storages and single point loading of wagons on weighbridges. While this technology is still adopted for the majority of loading points on Indian Railways, high speed loading at 4,000 t/h onto merry-go-round trains owned by super thermal power plants of 2,000 MW capacity and above has been introduced.

Construction of 41 coal handling plants by different subsidiary coal companies of Coal India is shown in Table 3. Two plants will have capacities over 5 million t/a, one with 4.5 million t/a and twelve over 1 million t/a, while 7 + 19 = 26 plants will have capacities of less than 1 million t/a.

Planning and design of many more coal handling plants are under way, as they are meant for mine projects that have already been approved for investment. Table 4 shows the break-up by capacity of the 80 plants to be installed in the four subsidiary coal companies under Coal India.

5. Conclusions

Coal India is planning for a large expansion of opencast mining, indeed a four-fold increase by the end of the century, from the present level of 60 million tonnes in 1983—84. Along with construction of 41 coal handling plants, some with high speed coal loading units, and with a further 80 coal handling plants to be installed, Coal India Ltd. is bringing about a dramatic change in the volume and technology of opencast mine expansion and construction of coal handling plants.

Table 3: Coal handling plants under construction

Subsidiary Company	No. of plants	Size (million t/a)				
		0—0.5	0.5—1.0	1—3	3—5	5—12
ECL	9	2	5	2	—	—
BCCL	7	2	5	—	—	—
CCL	5	—	1	3	—	1
WCL	20	3	8	7	1	1
Total:	41	7	19	12	1	2

Table 4: Coal handling plants to be installed (in planning and design stage)

Subsidiary Company	No. of plants (Approved projects)	Size (million t/a)				
		—0.5	0.5—1.0	1—3	3—5	5—12
ECL	8	2	3	2	1	—
BCCL	50	48	1	1	—	—
CCL	12	1	7	3	1	—
WCL	10	—	9	1	—	—
Total:	80	51	20	7	2	—

Volume 4, Number 3, September 1984

Bulk Handling Systems in Indian Coal Preparation Plants: Status and Outlook

P. Bandopadhyay, India

Summary

With the emergence of coal as the major source of energy for India, massive production of coal is being planned. Coal preparation, an intermediate step between its production and utilisation, will become mandatory for Indian coals. Presently sixteen coal preparation plants are in operation in India with an installed capacity of 32 million t. 18 new plants are scheduled to be installed by the turn of this century with an additional capacity of about 50 million t. A coal preparation plant essentially consists of a bulk handling system and its smooth operation depends on having an appropriate and efficient material handling system in the plant.

The present paper gives a critical evaluation of the status of bulk handling systems in Indian coal preparation plants and suggests modifications to improve the performance of bulk handling systems at present in operation in some major Indian coal preparation plants. The paper also highlights the problems to be encountered in the bulk handling of coal in future plants due to the reconstruction of India's Jharia coalfield.

1. Introduction: Coking Coal Washeries in India

With the advent of the oil-crunch, coal has emerged as the principal source of energy in India. Massive production of coal is being planned to meet the increasing demand for energy by industry. Coal preparation, an intermediate step between coal mining and coal utilisation, has assumed great significance in India, particularly in view of the deteriorating quality of raw coal to be used in coke-making. The limited resources of good quality coking coal have not been able to keep pace with the requirements of the rapidly developing Iron and Steel Industry of this country. This has resulted in the mining of lower seam, inferior grade coking coals and in the establishment of coal washeries during the last thirty years.

P. Bandopadhyay is Assistant Professor in Mineral Engineering at the Indian School of Mines, Dhanbad, India

The first group of coal washeries was set up in India in the early 1950s at West Bokaro, Jamadoba, Lodna und Kargali. Following the recommendation of the Central Coal Washeries Committee (1953—54), Hindustan Steel Ltd., set up five central washeries at appropriate locations at Durgapur, Dugda Patherdih and Bhojudih, respectively. One of the prime considerations in selecting the location of washeries was the ease and convenience of handling and transporting raw coal from mines to washeries and of transporting washed products from washeries to steel plants. Two central washeries (Durgapur and Chasnala) and three pithead washeries were installed by 1970. With the installation of three washeries during the past five years at Sudamdih, Barora and Moonidih, seventeen coal preparation plants are presently in operation in India with an installed capacity of receiving about 32.5 million t of raw coal and disposing of the processed products. Eighteen new plants are scheduled to be installed by the turn of this century, with an additional handling capacity of 50 million t of raw coal. The details of the operating coal preparation plants with respect to annual capacity, material handling systems, washing systems and feed size of the product are given in Table 1.

The total handling operations of a coal washery can be divided into the following three sections:

1. Raw coal handling system
2. Process handling system
3. Product disposal system.

In the present paper, the status of material handling systems in some major central coal washeries in India will be discussed.

2. Raw Coal Handling System

The raw coal handling system consists of dumping, tippling, screening, blending and storing of sized coal for subsequent treatment. Scrutiny of Table 1 reveals the following general features of raw coal handling systems in Indian coal preparation plants:

a) All central coal washeries receive raw coal from different sources through railway wagons and as such are provided with railway sidings/marshalling yards for the receipt of raw coal and the despatch of clean coal wagons.

Table 1: Coking coal washeries in India

Sl. No.	Name of Washery	Annual Capacity t/h	Feed size (mm)	Handling system	Washing system
1.	Durgapur	360	76	Tripper, bunker, bed blending	HM washer, feldspar jig
2.	Dugda I	600	76	Multipoint load, hopper blending	HM washer, baum jig
3.	Dugda II	700	13	Conveyor, rotary flow feeder bunker, single point loading	HM cyclone, hydrocyclone
4.	Bhojudih	500	76	Bed blending, multipoint loading	HM washer, baum jig, HM cyclone
5.	Patherdih	500	76	Rotary flow feeder, bunker blending, multipoint loading	HM washer, baum jig, HM cyclone
6.	Kargali	656	76	Dumper bunker blending	HM washer, baum jig, HM cyclone
7.	Kathara	890	76	Dumper bunker blending	HM washer, HM cyclone, flotation
8.	Swang	750	20	Dumper bunker blending	HM cyclone & hydrocyclone
9.	Gidi	800	150	Dumper bunker blending	HM washer, baum jig, flotation
10.	Durgapur (DPL)	360	13	Tripper, bunker, conveyor	HM cyclone
11.	Jamadoba	350	76	Tripper, bunker	HM washer, HM cyclone, flotation
12.	West Bokaro	160	76	Tripper, bunker	HM washer, HM cyclone, flotation
13.	Lodna	76	13	Tripper, bunker	Feldspar jig, oil agglomeration
14.	Chasnala	550	76	Dumper, ropeway, blending bunker, clean coal mixer	HM washer, HM cyclone
15.	Sudamdih	700	37	Dumper, hopper, silo, feeder	HM cyclone, flotation
16.	Barora	100	13	Dumper, bunker blending, conveyor loading	cyclone, flotation
17.	Moonidih	700	13	Dumper, dump hopper, silo, conveyor loading	HM cyclone, hydrocyclone

b) For raw coal storage in bunkers, the plants are provided with tipplers, vibrofeeders, screens and sets of conveyor belts.
c) The minimum size of raw coal is about 450 mm, which is crushed to a feed size varying from 150 to 13 mm.
d) The central washeries receive feed from twenty to thirty different sources and therefore arrangements for homogenisation of feed have been provided in these plants.

3. Process Handling System

Although each of the four central washeries of Hindustan Steel Ltd. follow different flow-sheets for processing coal, there is no substantial difference in their material handling systems as will be evident from Figs. 1—4.

3.1 Four Central Washeries

The raw coal feed received by Dugda I Washery (Fig. 1) by means of railway wagons is brought to the tippler platform and tippled into the hopper; a screen and picking belt are provided for removal of the larger lumps and shales/stones respectively. Larger lumps (+76 mm size) are passed through a roll crusher to reduce the top size to —76 mm. The —76 mm size coal is stored in a bunker. Blending is carried out in a hopper. The sized and blended coal is sent to the processing section by a conveyor belt for further treatment. The clean coal obtained from the processing section is collected in storage bunkers. Separate bunkers are provided for different size fractions of clean coal. This clean coal is loaded into railway wagons by means of a multi-point loading system. The loaded wagons are despatched to steel plants. The middling products are stored on the ground and sent to nearby Chandrapura Thermal Power Plant by dumpers and a belt conveyor. There is no satisfactory provision for disposal of rejects.

Dugda II Washery (Fig. 2) has a rated capacity of 700 t/h of raw coal and follows the same process as Dugda I for stockpiling the feed. The sized coal is blended by means of a flow feeder.

The blended coal is sent to a secondary crusher (impact roll crusher) after further reduction in size. The secondary crusher product, having a top size of 13 mm, is sent to the processing section. Clean coal obtained is stored in a storage bunker and loaded into wagons by a single point loading system. The middling is dumped outside the plant by means of dumpers.

Patherdih Washery (Fig. 3), with a handling capacity of 500 t/h, treats 75 mm coal for washing. The raw coal handling arrangement is similar to that in Dudga I, while the blending and storage systems are similar to those in Dugda II. The clean coal and middling product are stored in separate storage bunkers and are loaded into wagons by a multi-point loading system.

Bhojudih Washery (Fig. 4) is similar to Patherdih with respect to its capacity, the size of coal to be treated and its raw coal handling system. However, crushed coal, instead of being stockpiled in bunkers, is stored on a long platform on the ground and a special type of blending is done with a harrow

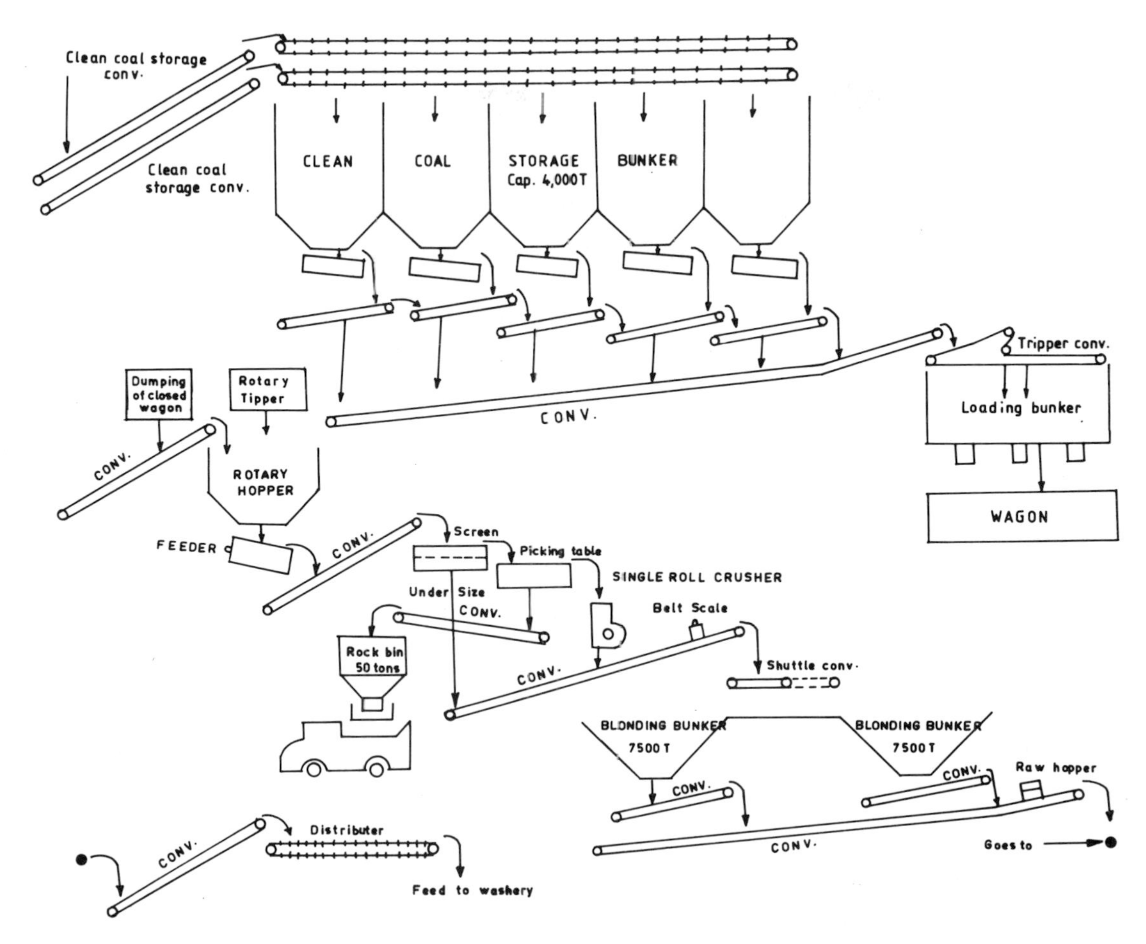

Fig. 1: Coal handling system of Dugda I Washery

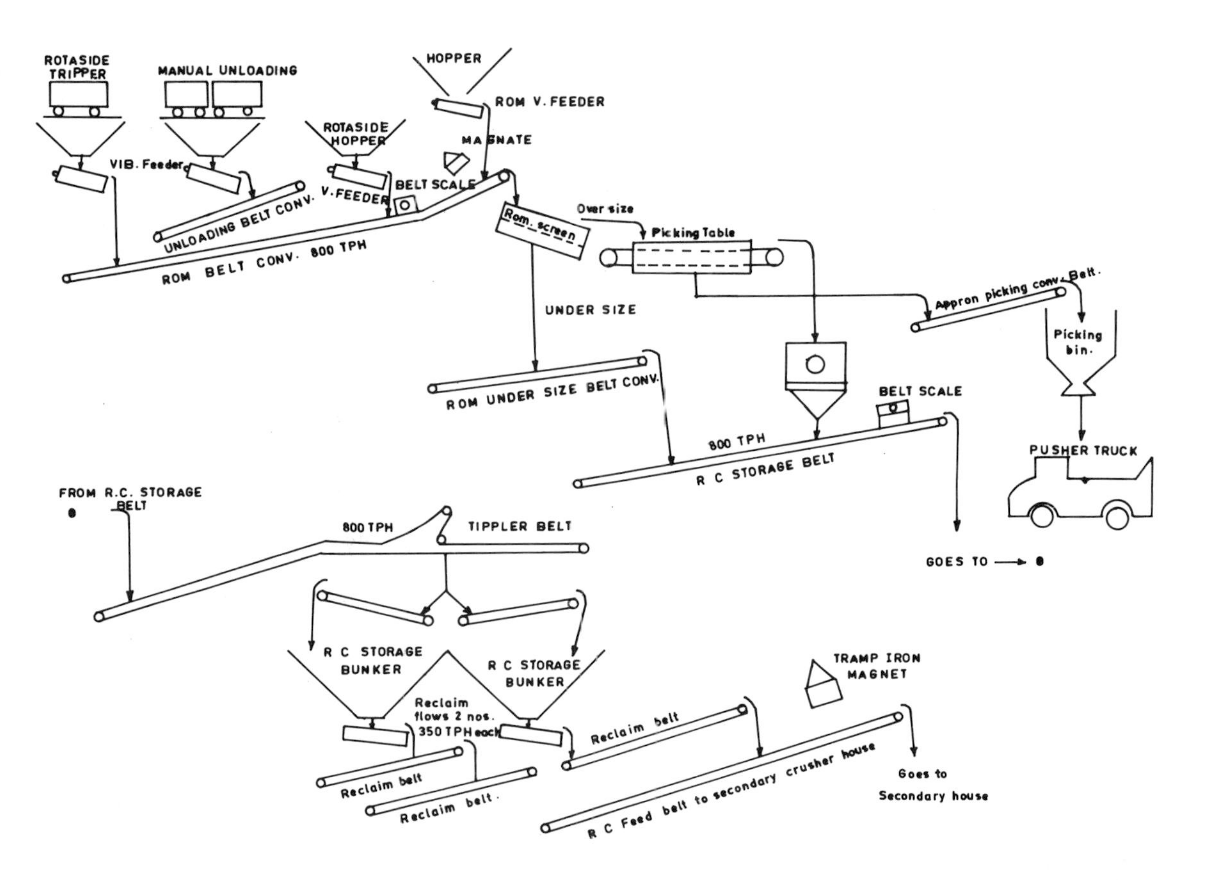

Fig. 2: Coal handling system of Dugda II Washery

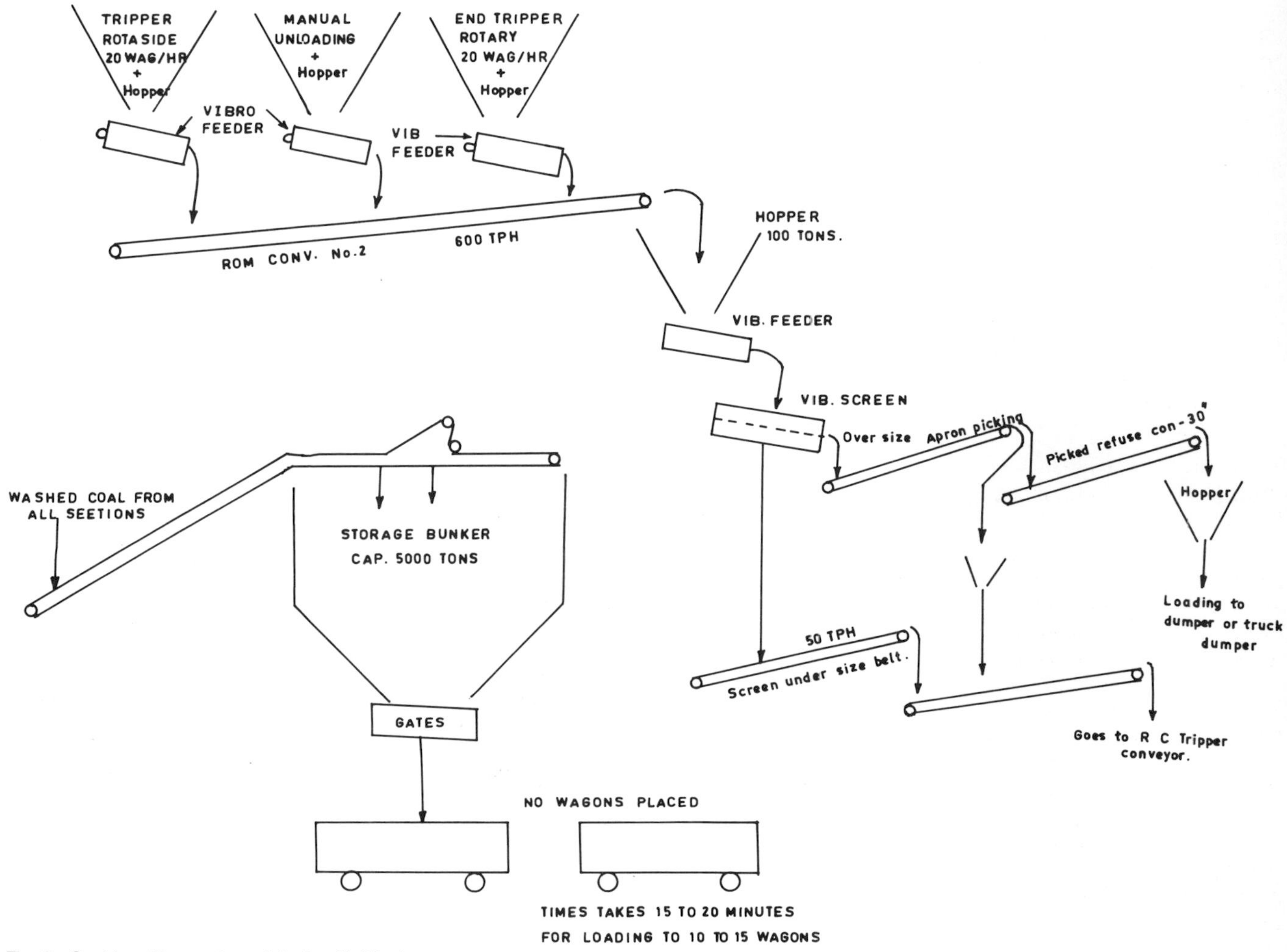

Fig. 3: Coal handling system of Patherdih Washery

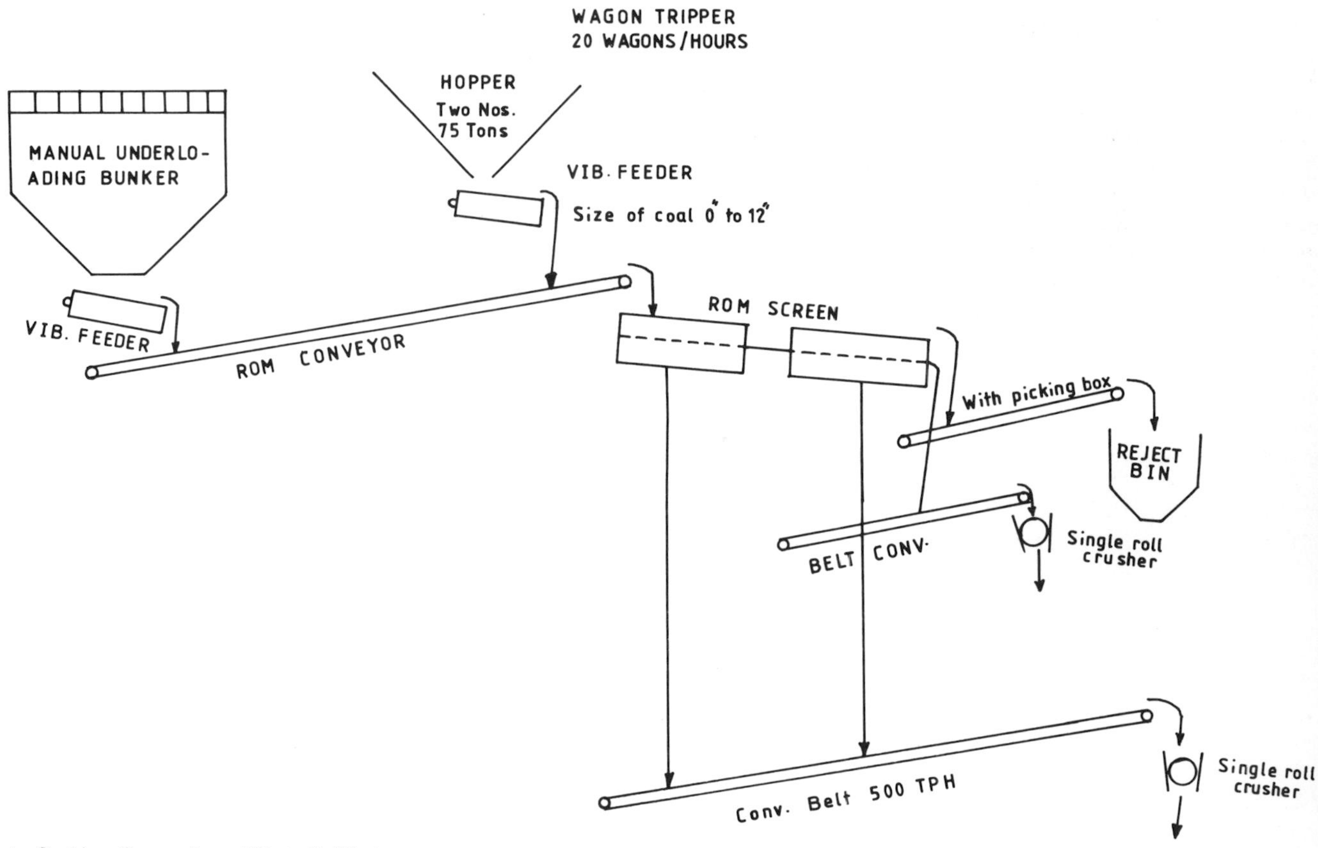

Fig. 4: Coal handling system of Bhojudih Washery

machine, ensuring supply of a consistent size and quality of coal to the processing section. The clean coal and the middling disposal system are similar to those in Patherdih. The rejects are simply dumped outside the washery.

3.2 Blending Systems

One common problem confronted by these coal preparation plants is that of getting a steady supply of raw coal feed of consistent quality at the required rate. As has been mentioned earlier, the central washeries receive coal from 25 to 30 different collieries, each producing coal from two to three different seams. The raw coal ash, the size consistency and the cleaning characteristics of the supplies differ widely from seam to seam, even in the lateral and vertical workings of the same seams.

The quantitative variations can be smoothed out to some extent by providing adequate cushions in the storage bunkers. But in order to ensure a consistent quality of the feed a carefully designed blending system has to be provided in the plant, so that the heterogeneous constituents are combined in a standard manner and the thoroughly mixed product so reclaimed that each portion of the sandwiched feed becomes representative of the whole.

The principal differences between the blending systems adopted by different central coal washeries in India lie in the mode of stockpiling, storage and reclamation, as well as in the time cycle of their operations. Dugda Washery has a storage system with moderate blending. The system consists of bunkers with multi-point discharge through an underground tunnel. The system is economical for handling a large tonnage and when operated simultaneously the multiple discharge gates ensure some reasonable blending. Patherdih and Dugda II have a comparatively improved storage and blending system. The system consists of a trough-bunker provided with a discharge slot running along the full length of the trough. For effective blending they are built in pairs; one of them receives incoming raw coal, while the other discharges in oblique layers along the full face of the pile by means of a rotating plough attached to a mobile discharge car. Bhojudih and Durgapur (DSP) Washeries have a bed-blending system with a harrow-type reclaimer. This is the best blending system available in an Indian washery. The coal is formed into a full-sized pile by forward and backward movement of the tripper in an orderly fashion along the longitudinal axis of the pile. The coal is dug by a rake mechanism along the cross-section of the pile. Thus the harrow frees the coal-layers in the direction of their natural slope and drips the blended material onto a scraper conveyor which delivers the coal to a collective belt. The stockpile formation and reclamation are conducted on two separate tracks. The maximum capacity of a single harrow-type reclaimer does not usually exceed 550 t/h. Bhojudih is the first washery in the world to employ such a bed blending system to homogenise the feed to the coal preparation plant.

4. Product Disposal System

A quick and efficient product disposal system is a prime requirement for the smooth operation of a coal preparation plant. A product disposal system consists of dewatering, storing and loading of clean coal, middling product and the rejects. All the central coal washeries of India are three-product washeries. In central coal washeries, the despatch of the clean coal and middling is done through railway wagons, while the rejects are either stockpiled around the washeries or transported through trucks and dumpers. The loading system adopted by Indian washeries includes either (a) a multipoint loading system or (b) a single point loading system with wagon weighbridges. The multi-point loading system, because of the large number of manual operations involved, is slow, inefficient and costly, and since the weighbridge is located away from the bunker, load adjustment of wagons becomes cumbersome. Usually the loading rate does not exceed 200 t/h. The single point loading system has many advantages over the multi-point loading. The loading rate is much higher (400 t/h), requires fewer operators and weigh adjustment of loaded wagons is eliminated. Dugda II is the only central washery with this system of loading.

5. Reconstruction of the Jharia Coalfield and its Impact on the Bulk Handling System

Bharat Coking Coal has a plan to increase the production of the Jharia coalfield from 23.90 million t (1982—83) to 47.60 million t by 1990. In order to meet this production target there will be considerable changes in the relative contribution of various technologies adopted for coal production. It may be noted from Table 2 that the aim is to increase opencast projects from their current 31 % to 48 % by 1989—90. Introduction of new underground technology will be implemented at a faster rate. Longwall and mechanised board and pillar will replace conventional board and pillar by 50 %. The total underground production from new technology in the form of longwall faces, sub-level caving and mechanised board and pillar will increase production from the present level of 4.71 to 11.52 million t by 1990. Along with the growth in raw coal production, the level of output of washed coal will also be stepped up considerably. Because of this drastic change in mining technology there will be a substantial change in coal characteristics and this will affect the bulk handling systems of future coal preparation plants in the following ways:

Table 2: Contribution of various technologies in coal production at the Jharia coalfield

	1979—80	1989—90
	%	%
Opencast	31	48
Underground	69	52
(a) Longwall	5	33
(b) Mechanised board & pillar	5	18
(c) Conventional board & pillar	90	49

a) Presently, the production capacity of the coal mines is not very high. One central washery, therefore, has to be fed raw coal from 25—30 different coal mines. The large projects planned to be executed in the next few years will have a capacity as high as 3 million t/a as in Pootkee Balihary Project (underground) and 2.5 million t/h in opencast mines as in Block II. Mukunda opencast mine will have an annual production of 12 million t, with integrated coal preparation plants. Thus in future, instead of having one central washery for a number of mines, there will be more than one pit-head washery for one big mine. The problem will be of carrying a larger tonnage over a shorter distance.

b) With the increase in opencast mining, ROM coal coming to washeries will contain an increased percentage of large lumps and boulders. This will create problems for conveyor belts, screens and crushers.
c) The higher moisture content of the raw coal will create problems for coal transport.
d) The generation of an increased quantity of fines may lead to the adoption of pipeline transport for the particulate material.
e) Disposal of washery rejects and black water will become a big challenge for material handling experts.
f) The capacity of future pit-head washeries will have to be higher than that of the present central washeries in many cases.

In view of the changes in mining technology and coal characteristics due to reconstruction of the Jharia coalfield, the following changes have to be incorporated into the designs of future washeries to ensure their smooth operation:

i) The maximum size of ROM coal will be about 1,000 mm. In order to avoid jamming of the conveying system, the belt conveyor, feeder, discharge chutes and primary crusher should be designed to tackle the maximum size of 1,000 mm.
ii) In order to avoid generation of fines and overloading of the system, heavy duty grizzley bars in steps should be provided before the primary crusher for screening out the coal of smaller sizes.
iii) For raw coal storage in washeries, silos in conjunction with ground storage are recommended. The ground storage can be utilized when the washery is not working or when the silos are full. The silos/ground storage should be so designed that while reclaiming, a moderate blending of coal takes place.
iv) For storage of the washed product prior to its despatch by railway wagons, a silo/ground storage system should be adopted in place of a bunker system. This will provide high storage capacity and quick reclamation, which are essential requirements of efficient material handling systems.
v) With the high throughput of future washeries, a loading system with a high loading rate between 2,000 and 3,000 t/h will have to be provided.

6. Conclusion

The availability and efficient transport of coal from mine to washery and the speedy disposal of washery products has been an area of constraint in the past. In the Jharia coalfield, the load on the transport system by the year 2000 will be 392,000 t/d, while the load on the railway will be of the order of 226,000 t/d in contrast to the present level of load of 65,000 t/d. The distribution of load will be 58 % railways, 27 % conveyor and 15 % road. A total re-organisation of transport in and around the Jharia coalfield will therefore be required. It is heartening to note the growing awareness of the Indian Coal Industry of the importance of an effective and efficient transportation and handling system. The development plans include increasing use of continuous coal clearance systems such as belt and chain conveyors from a centralised point to the preparation plant. This will be advantageous because of optimum haul length, easier maintenance and less consumption of liquid fuels. In order to tackle the problem of large lump size, in-pit crushing with the help of a feeder breaker is being introduced.

The technological plan drawn up shows a bright future for the mechanisation and modernisation of bulk handling systems in Indian coal preparation plants.

Volume 5, Number 5, October 1985

Planning Aspects of Coal Handling Plants

L.L. Shrivastava, India

Summary

By the year 2000, coal will have a share of 28% amongst the main sources of global energy. There is a need for most careful planning and investment decisions to ensure an important place for coal. Otherwise, oil may regain its market share, with a resultant oil dependence for the world economy.

Coal handling plants (CPHs) form one of the most important integrated activities contributing to the transformation of the coal scene in India. The design and construction of CHPs now form a very interesting multidiscipline teamwork.

This paper deals with some of the main aspects which are considered when designing a modern coal handling plant. To start with, the coal linkage, the mode of off-take by the customer and the method of loading coal into the customer's mode of transport, which is to be settled between the producer and the customer before making the actual plant design and which virtually governs the entire loading complex, is discussed. The paper also discusses the coal production parameters to the extent they affect the internal structure of the plant. Finally, various subsystems of CHPs contributing to the total effective functioning are discussed step by step. These are: receiving of the coal from the mine and crushing, picking/deshaling of the dirts, coal flow within the CHPs by conveyors, storage methods for the coal and its subsequent reclamation and loading in conjunction with suitable methods for marshalling and weighing. As far as the loading complex is concerned, the rapid loading system has attracted the most coverage in the paper due to its present importance.

1. Introduction

It is now agreed nationally and internationally that, as a future source of energy, coal will play a very dominant role. This opinion has not come about purely because of the oil crisis which started during the 1970s, but is based on a thorough study of the availability of all types of future energy sources. The 12th Congress of the World Energy Conference, held in New Delhi in September 1983, clearly and unambiguously confirmed this.

Mr. L.L. Shrivastava, Head of Division, Central Mine Planning and Design Institute Ltd., Engineering Services Division, Gondwana Place, Ranchi — 834 008, India

Among the deliberations of the various round tables and working groups, those concerning coal deserve particular notice. The round table on coal converged on the theme of the place of coal in the world economy, dwelling on such issues of wide ramification as the status of coal in the present energy market, the future prospects for coal, especially in the developing world, the competitiveness of coal vis-à-vis oil in electric power generation and the environmental effects of coal utilisation.

The message of the round table was loud and clear: an increasing global reliance on abundant and low-cost coal resources, taking cognisance of new energy relationships and energy economics. The stumbling blocks to development for an enlarged coal production and trade must be removed through appropriate policies for an assured future of energy security.

One of the pertinent conclusions of the study carried out by the Conservation Commission of the World Energy Conference was that noncommercial energy consumption will gradually decrease, while commercial energy consumption is going up. Coal will become the leading source of energy. The respective shares of the main sources of global energy in the years 2000 and 2020 may be: coal 28% and 32% (as against 25% in 1978); nuclear power 8% and 13%; hydro power 6% and 7%; new energies 3% and 6%; natural gas may increase in absolute production, but the overall share may remain around 17%; oil 30% and 20% (against 40% in 1978).

It was therefore suggested that the costs of mining and transport of energy resources should be minimised by intensive investments in energy saving. Otherwise, oil may regain its market share with a resultant oil dependence of the world economy.

2. Coal Handling Plants

Coal handling plants form one of the most important integrated activities contributing to the transformation of the coal scene in India. Today, we cannot think of a coal mine and a coal handling plant separately, and it is simply unimaginable that coal produced from a modern mine, especially a big opencast mine, can be directly loaded into any of the recognised devices for coal evacuation. The coal has to be sized, processed, handled and loaded, and all this is done in CHPs. Therefore, planners in the coal industry nowadays give the utmost respect to the necessity of coal handling facilities. The planned capacity of CHPs vis-à-vis the capacity of existing CHPs attracts the attention of the top executives of any company, and there is a constant planning/design/construction activity to narrow the gap between the handling capacity of the existing coal handling plants vis-à-vis the planned capacity, as shown in Fig. 1.

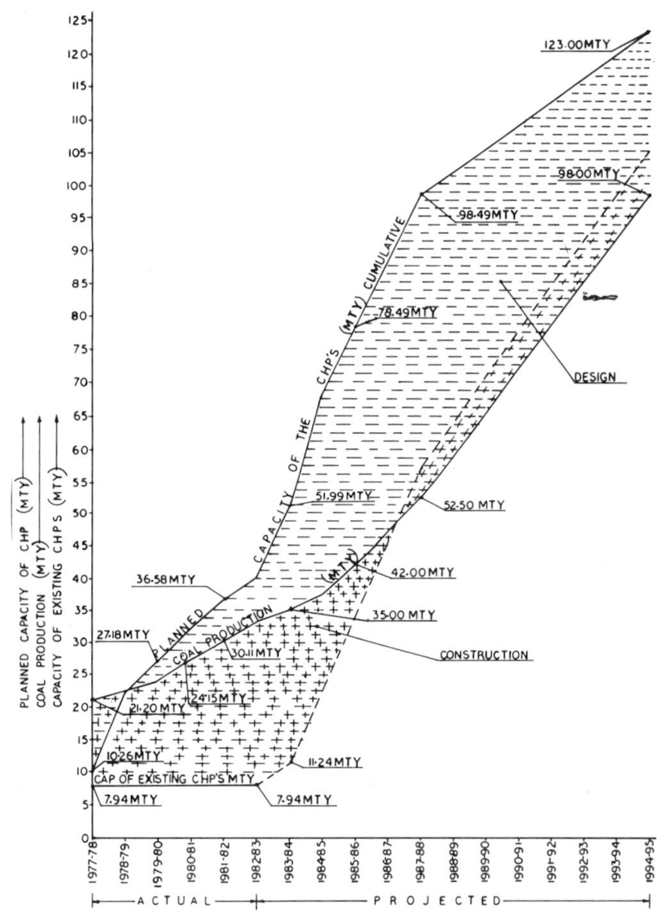

Fig. 1: Planning preparedness

3. Planning of Coal Handling Plants

Gone are the days when an individual sitting alone in his/her office could visualise and finalise the physical and financial dimensions of a coal handling plant. Nowadays, this is very interesting multidiscipline teamwork. There have been cases where the feasibility of opening a mine itself depended entirely on the feasibility of the coal handling and evacuating, and so a couple of months were spent in in-depth analyses of all types of variants to find out the optimal alternative for coal handling and loading.

For the planning of CHPs, normally the following factors are invariably considered:

a) linkage of the coal
b) mode of off-take by the customer
c) method of loading the coal into the mode of transport
d) quality parameters envisaged in the sale agreement of the coal
e) method of mining
f) characteristics of the coal, i.e., hardness, presence of shale bands
g) internal structure of the CHPs governed by the above mentioned factors a) to f), which, among other things, primarily relates to:

— receiving and crushing of the coal
— storage of the crushed coal
— loading of the coal into the customer's mode of transport
— coal flow within the CHP by conveyors
— picking/deshaling of the dirts.

4. Details of the Planning Approach

It will not be possible to cover the entire gamut of the planning of CHPs in this paper. Only some of the important aspects will be covered here:

1. Effect of the coal evacuation on the CHP design
2. System capacity of a CHP
3. Internal structure of a CHP:

a) receiving pit and crusher
b) deshaling
c) storage
d) loading
e) considerations for CHP construction
f) technoeconomics.

The main factors mentioned above, as well as other subfactors which form part of them, must carefully be weighed before giving a final shape to the coal handling plant. This has become all the more necessary in the case of big opencast mines where during emergency demand a shovel and a few dumpers could be introduced for producing extra coal output. A CHP, therefore, should also have some inherent cushion for such future extra load from the mine.

5. Coal Linkage, Mode of Evacuation and CHP Design

A coal handling plant receives the coal from the mine and loads into the mode of transport to send the coal to the customer. A very clear picture of the customer's requirements in terms of quantity and quality of the coal to be despatched and also the mode of despatch is required for designing a CHP. Some of the factors are:

1. Size of the coal
2. Picking/deshaling required to maintain quality consistency
3. Mode of despatch:

a) belt conveyor
b) aerial ropeway
c) road transport
d) combination of any two of the above
e) railroad
f) customer's own merry-go-round (MGR) system
g) combination of e) and f)
h) combination of any three or four of the above.

While the parameters regarding size and quality of the coal will govern the receiving pit, including the crushing complex and the deshaling methods to be adopted, the mode of evacuation of the coal will also dictate the design of storage, reclamation, loading and, to a very great extent, the schedule of construction of the plant.

The effect of coal evacuation on coal handling plant design can to a great extent be explained by Fig. 2. In this figure, the customer's decision to evacuate the coal by an MGR system

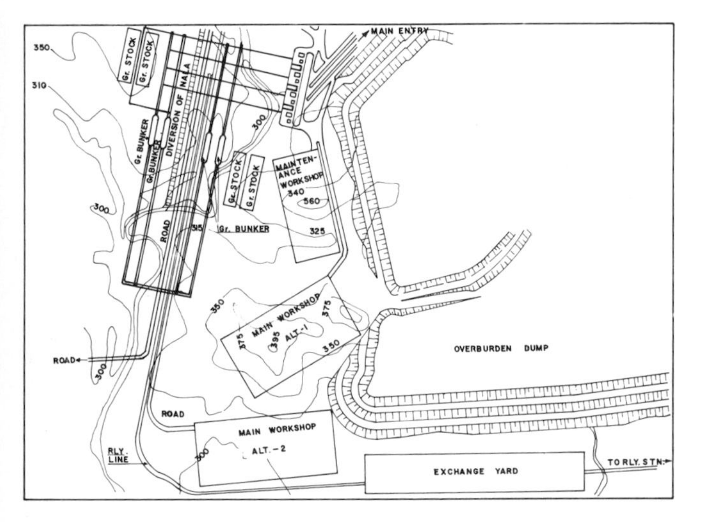

Fig. 2: Layout of a coal handling plant

in an MGR bulb has left no alternative but to locate the entire CHP within the MGR bulb itself. In the present case, however, the railway's suggestion for evacuation of the coal by an MGR system could not be accepted simply because of the dictates of the unfavourable terrain and the location of the mine entry, and as such, a rapid-loading system with shuttle arrangement has to be planned. Fig. 2 also indicates that huge costs had to be incurred for developing an exchange yard as the coal loaded from the coal handling plant by the rapid-loading system has to be transported to long-distance customers. The exchange yard became a must in this case to supply rakes at 90-min intervals to meet the daily despatch schedule of fourteen to fifteen rakes.

Another possibility can be explained by the case depicted in Fig. 3, where both conventional loading arrangements and a rapid-loading system on a MGR had to be planned in view of linkage of the mine to more than one customer and according to the customers' respective despatch methods.

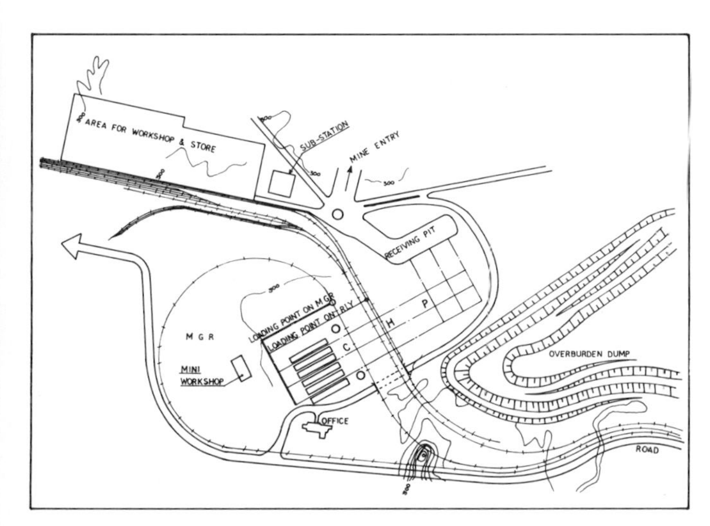

Fig. 3: Layout of a coal handling plant

6. System Capacity of a CHP

The "system capacity" denotes the average capacity of a CHP which can be maintained theoretically over days, weeks, months and years. In practice, however, this average is simply a "philosopher's average", and the actual capacity is decided based on considerations dictated by the realities. As such, a coal handling plant is designed not only for an average capacity, but with an inherent flexibility to absorb peak loads.

The ratio between the peak load and the average has always been a matter of dispute. It reflects the approach of the management to the total mining activities including the CHP, the social habits and behaviour and the circumstances under which the working personnel is engaged in the total mining complex including the CHP, the agreed cushion to be kept in the CHP design for future marginal expansion of the mine during emergencies and the operating conditions of the plant, whether it is a two-shift or three-shift operation, whether it serves a single mine or a group of mines etc. Before discussing the individual fluctuation factors to be povided in the internal structure of a CHP, it will be worth discussing Fig. 4, which shows how an Indian mine is affected by different factors.

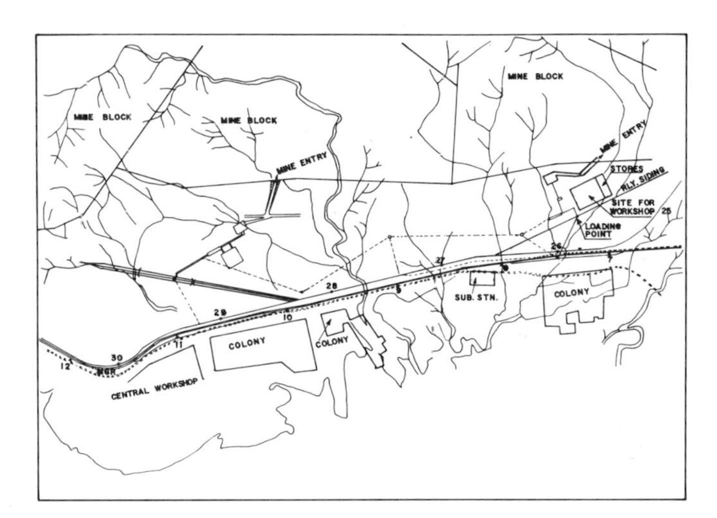

Fig. 4: Layout of a coal handling plant

This figure was prepared while planning a CHP for an underground mine. The actual monthly production of the mine was converted into the equivalent hours per day (h/d), based on the designed tons per day (t/d) of the winder taking the coal to the surface. It can be seen that, against the planned assumption of the mine working for an average of 15 h/d, it worked in actual practice from around 10 to virtually 23 h/d (Fig. 5).

These fluctuations were not special to this mine only, but apply to the entire mining industry. In brief, the mine production fluctuations are due to various factors:

1. Hourly fluctuations due to the operator's efficiency
2. Shift fluctuations, accounted for by fluctuations in the coal transport system from hour to hour, from mine to CHP
3. Shift time availability factors due to time spent on routine maintenance, operators having a tea break etc.
4. Daily fluctuation factors accounted for by variations in coal output between various operating shifts, variation in

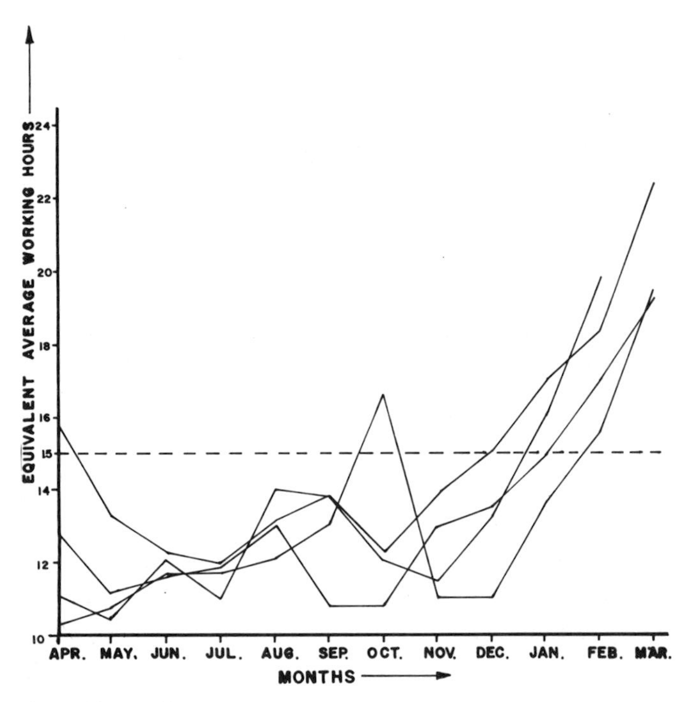

Fig. 5: Reserve factor

operator's efficiency, occurrence of unforeseen breakdowns and repairs etc.

5. Monthly fluctuation factors accounted for by day to day fluctuations during a month of an average of 25 working days, shifting of equipment to match with the shifting of coal faces, seasonal effects such as heavy downpour, excess temperature or cold wave
6. Yearly fluctuation factors accounted for by such seasonal items as absence of workers due to harvesting or festivals; this may also include work stoppages during monsoons, especially in opencast mines, major breakdowns and overhauling of the equipment.

In short, while the mine planners will take care of these factors to provide coal winning and transport equipment with an adequate cushion, the planners of the coal handling plant cannot loose sight of those factors which will apply to the coal handling plant as well as to the mine itself. While designing the internal structuring of the CHP, the main places needing careful consideration for peak factors are:

a) crushing complex — approx. 1.5
b) conveying complex — approx. 1.3
c) screening complex — 1.3 to 1.5
d) loading complex — 1.3 to 2

All the cushion provided in the system in the form of peak factors needs further backup when the size of the mine becomes too big, say, bigger than 4 to 5 Mt/a. In such cases, the normal practice is to provide an additional cushion in the plant in the form of standby circuits, having a capacity of 25 to 30 % as the case may be. Table 1 indicates the effects of such peaks and reserve factors while selecting the conveyor capacities and bunker sizes for the coal handling plant. The effect of fluctuations in the mine production and other factors will be further elucidated when discussing the internal structuring of coal handling plants.

7. Receiving Pit

7.1 Factors Affecting Receiving Pit Design in Opencast Mines

1. Type and size of coal carrier
2. Time cycle of coal carrier
3. Reliability of coal carrier
4. Type and size of face equipment
5. Size distribution of run-of-mine (ROM) coal
6. Flow characteristics and stickiness of ROM coal
7. Method of handling coal lumps bigger than the size acceptable at the receiving pit
8. Topography of the site
9. Economics of installation.

7.2 Location

In general, the location of the receiving pit should be as near to the quarry entry as possible. This will result in a higher availability of the coal carriers, whose operational costs are

Table 1: Calculation of belt and bunker capacities for two- and three-shift operation of the CHP

Annual Mine Production	Rated Daily Mine Production	Max. Daily Mine Production	Max. Mine Production per Shift	Capacity of Belt to ROM Bunker	Capacity of ROM Bunker	Total Capacity of Belt after ROM Bunker	Min. Capacity of Storage Bunker (with Suitable Ground Stock)		Capacity of Wagon Loading Conveyor
							Steam Coal	Slack Coal	
Mt/a	t/d	t/d	t/shift	t/h	t	t/h	t	t	t/h
Two-Shift Operation of CHP									
0.5	1,670	2,505	850	170	925	167	3,500	3,000	800
1.0	3,340	5,010	1,700	350	2,000	334	4,500	3,000	800
1.5	5,000	7,500	2,500	500	2,450	500	5,500	4,500	800
2.0	6,700	10,050	3,350	670	3,350	670	6,500	4,500	1,200
2.5	8,400	12,600	4,200	840	4,250	840	7,500	6,000	1,200
3.0	10,000	15,000	5,000	1,000	5,150	1,000	8,500	6,000	1,200
Three-Shift Operation of CHP									
2.5	8,400	12,600	4,200			840	7,500	6,000	1,200
3.0	10,000	15,000	5,000			1,000	8,500	6,000	1,200
5.0	16,700	25,000	8,400			1,750	20,000	20,000	6,000
10.0	33,400	50,000	17,000			3,500	40,000	20,000	6,000

Mine operation: 300 d/a, 3 shifts/d, 5 h/shift

high. Sometimes, a compromise has to be made for the following reasons:

1. The receiving pit should be at a safe distance from the quarry edge so that the installation is not affected by blasting. A minimum distance of 100 to 150 m is used for the purpose.
2. A normal receiving pit has a total depth of 16 to 18 m. Out of the various possible designs of construction of the receiving pit, it has been found that the most economical receiving pit is the one which is partly aboveground and partly underground. Usually, 10 to 12 m aboveground and 6 to 8 m underground are chosen. In order to meet the gradient of the haulroads and also to provide some level distance for movement just on the surface after leaving the quarry, a distance of 200 m is generally required, which also governs the overall height of the receiving pit.

7.3 Topography of the Site

The topography governs the selection of the site to a great extent. The site should be selected so that there is minimum total transportation in moving the coal from one point to another. As far as possible, the help of gravity flow is taken to reduce the overall transport distance. However, the site should be cleared of flood levels and at a safe distance from eventually adjoining nallahs, rivers, dumping grounds etc.

7.4 Capacity

The minimum capacity of the receiving pit is the capacity which is just sufficient to accommodate the difference between the intake from the coal carrier and the offtake by the feeders. Theoretically, therefore, the minimum capacity of the receiving pit is the capacity of the biggest coal carrier employed.

But in practice it will be seen that the frequency of the coal carrier in reaching the receiving pit fluctuates quite widely, whereas the offtake from the feeders etc. remain more or less constant, depending upon the system capacity and its design. The wider the fluctuation, the bigger will be the requirements of the receiving pit. In practice, the capacity of the receiving pit is assumed to be three to five times the capacity of the biggest coal carrier employed. This works out at 150 to 250 t in the case of 50-t coal dumpers.

If necessary, the capacity of the receiving pit could be further reduced on economic considerations by providing a surge capacity immediately after the receiving pit, from where the remaining circuit of the coal handling plant starts. In this case, the receiving pit is not required as a surge, and, therefore, smaller sizes can be adopted. The surge hopper has the added advantage of providing continuous flow to important equipment in the system (e.g., crushers, screens etc.).

7.5 Grizzly over the Receiving Pit

Theoretically, only coal lumps of a predetermined maximum size should reach the receiving pit, but in actual practice, despite all efforts in the quarry, oversize pieces get unloaded into the receiving pit. This normally occurs with rear discharge dumpers, and, therefore, a grizzly is installed on top of the receiving pit to eliminate the possibility of abnormally big lumps being conveyed to crushers and other delicate equipment.

The designer has to strike a balance by anticipating how far the sizes can be controlled by blasting in the quarry and how far the system should absorb unusually big pieces on being received at the pit. It is worth remembering that face equipment such as shovels is very costly and thus should be allowed to handle coal to its fullest capacity. The shovel should not be used for breaking lumps of coal. A blasting pattern should be selected so as not to be very expensive and at the same time not to produce very big coal lumps.

7.6 Feeder

Feeders are required to deliver a uniform stream of ore from a reservoir of any kind through a gate. The feeder ensures that the subsequent equipment in the handling system receives a uniform flow of coal, thereby ensuring optimum output and the most economic production.

The requirements of a good feeder are:

— capability to deliver at a predetermined rate, irrespective of the total amount of coal lying before it
— suitability for quick adjustment to vary its delivery rate
— ability to start under load and stop without spillage
— adaptability to the size of the material to be handled
— ability to stand the impact of the material falling over it.

The various types of feeders being used in coal handling plants are:

— apron feeders
— belt feeders
— rotary feeders
— reciprocating feeders
— screw feeders
— revolving disc feeders
— travelling endless chains.

In opencast mines, the use of apron feeders is preferred for the following reasons:

— very robust design
— ability to sustain part of the load left behind in the receiving pit
— suitability for handling big-size lumps
— ability to stand up to the impact load of falling boulders from the dumper into the receiving pit
— broad parameters for selection of given requirements
— easier feed controls through a varying-speed drive mechanism
— positive feed and unaffectedness by sticky and wet coal.

8. Crushers

8.1 Lump Size

Coal from opencast mines varies to a great extent in its size as it comes from the mine to the receiving pit. The biggest coal lump in ROM coal depends on:

— seam formation and quality of coal
— blasting pattern and type of explosive used
— type and bucket capacity of coal face equipment
— type and size of coal carrier.

Therefore, the selection of a crusher becomes very important so that it is not unnecessarily selected for the occasional arrival of the biggest boulders, but at the same time not purely for low sizes, thereby leaving quite a huge amount of big boulders at the receiving pit.

The factors affecting the crusher selection are:

— characteristics of the material
— feed size available
— product size required
— production capacity desired.

8.2 Material Characteristics

These include:

— crushing strength of the ROM coal
— hardness of the coal
— moisture content of the coal
— flow characteristics of the coal.

8.3 Feed Size

As already discussed, the feed size depends on many factors. Having fixed all the parameters earlier, it would become very difficult to select the proper type of crusher. So we are left to assume the available feed size as guideline for selecting the crusher.

It is obvious that crushers cannot be selected for occasional large pieces coming to the receiving pit. Therefore, when we talk of the maximum feed size, we assume that other factors affecting the size of the coal will be controlled and monitored suitably at the mine face itself, so that, when occasionally bigger sizes occur, they will not reach the coal handling plant. As regards an occasional big piece, the grizzly at the receiving pit will take care of it and manual/mechanised means can be employed for breaking it up.

In practice, the feed opening of a crusher is 1.3 to 1.4 times the maximum recommended size of the available lump in the ROM coal. For example, for a recommended feed size of 1,200 mm, the crusher opening size should be 1,500 to 1,600 mm.

It is obvious that the larger the percentage of bigger sizes of coal in the ROM, the bigger the size of the opening becomes, thus leading to prohibitive costs and also to underutilisation of the crusher.

Similarly, fines also play an important role in crusher selection. The high percentage of coal sizes smaller than the recommended feed size will affect the crusher capacity. In this case, hardly any crushing takes place in the upper zone of the crusher, though the crusher mouth appears full of coal. Therefore, a mixed feed always gives the best results. In this case, voids are reduced, better crushing takes place all over the crushing zone and a higher output is available. A mixed feed means pieces of varying sizes, but not bigger than the recommended feed size and not smaller than the crusher setting. For selection of a crusher and its feed opening, screen analysis of the ROM coal is a must. In the case of a working mine, it may be possible to have a screen analysis done and study the results. But more often the planner is required to make the selection without knowing the results of an actual screen analysis. In such cases, approximate empirical thumb rules and/or recommendations of manufacturers are used for adopting the best possible parameters.

8.4 Product Size

The product size depends entirely on the customer's requirements. Washeries normally require a size of —75 mm, railways and industries +50 mm to —300 mm and power houses —50 mm coal.

From the above, it can be seen that customers' requirements cover a wide range. To meet the specific requirements of a customer, the crusher has to be carefully selected and necessary crushing may have to be done in two or three stages.

8.5 Production Capacity

Normally, the production capacity of a crusher is expressed in m^3/h. Depending upon the specific gravity of the material, the output in t/h is obtained. Usually, the manufacturers give recommended capacities for specified feed sizes and product sizes. These recommendations are for an assumed mix feed. Therefore, in actual practice, a different output is received from the one recommended by the manufacturers.

One of the important factors affecting the crusher capacity is the gap setting. The higher the gap setting, the larger is the output. It is therefore obvious that any output of the crusher should be read along with the feed size and the crusher gap size.

When we talk of gap setting this means that everything that passes through the gap is equal in size to or smaller than the setting size. From practical experience, it is seen that in a roll crusher with a setting of —200 mm at least 5 to 10% of the product are bigger than 200 mm. If we really want to obtain 100% coal of —200 mm, we may have to set the crusher at —100 mm, or at least —150 mm, depending upon the quality of the coal.

However, this is quite a complicated affair to be judged involving the material to be crushed, the crusher being used and the operation conditions demanded from the crusher.

The next important factor affecting the crusher selection is the undersize coal passing through the grizzly which does not reach the crusher. This grizzly eliminates unnecessary choking of the crusher and also forms a cushion on the belt conveyor onto which the crushed material falls. This adds to the efficiency of the crusher as well as the belt conveyor transporting the coal from the crusher to further handling points in the coal handling plant. But this is not the end of the story. The undersize does not necessarily pass through the grizzly, but more than often, especially in the rainy season, it passes through the crusher, too. Therefore, while selecting the crusher capacity, these aspects must be carefully considered. Some manufacturers recommend that for the purpose of crusher capacity the entire coal, including the undersize, should be considered. Perhaps this comes to the rescue of the crusher when some abnormally big lumps are required to be crushed by the crusher despite all our precautions. It may be of interest that in a crusher of 300 to 400 t/h capacity an occasional big lump may demand a temporary duty of 800 to 1,000 t/h.

Apart from the above main considerations, the crusher capacity varies according to the hardness of the coal and the pattern in which it gets fractured. Designers have to be very careful about selecting the feed arrangements to the crusher, which also affect the capacity of the crusher.

8.6 Selection of Crushers for Opencast Mines

To date, the following types of crushers are in use in the coal industry as primary crushers in opencast mines:

— jaw crushers
— gyratory crushers
— roll crushers
— impact crushers.

Of the above, the roll crushers find a place in 95% of India's coal handling plants at opencast mines. However, these crushers have limitations of capacity and availability in the country. Consequently, of late, other crushers are also being seriously considered for selection for Indian opencast mines. A brief description of the various types of crushers is given in the following sections to give an idea of how and why they are selected.

8.7 Jaw Crushers

A jaw crusher is a reciprocating pressure breaker. It finds a wide range of use as a primary crusher, especially when big lumps are to be handled. It can take big sizes of coal lumps continuously without getting clogged at the mouth. The angle between the jaws rarely exceeds 24°, whereas the nip angle for ordinary coal is around 33°. Therefore, this does not become an important factor unless the material is very slippery. In this case, the nip angle should be considered as 18°.

For a better grip, the thickness of the largest particle should not exceed 80 to 90% of the gap.

Jaw crushers give quite a good reduction ratio, and, for an approximate estimation, 1:7 is normally taken as an achievable figure for this type of crusher.

Quite large sizes of jaw crushers are available, such as 60 inch x 84 inch, and even bigger than this can be made to order. These crushers can handle lump sizes of up to 1,500 mm in any one direction. The capacity of the above-mentioned jaw crushers with a setting of —200 mm is around 500 to 600 m³/h for coal. Of course, all the pieces will not be —200 mm, as discussed earlier, and if the customer's requirement is fully known, some percentage will need rehandling through the secondary crusher, because the jaw crusher has a tendency to produce flaky coal sizes.

Being costly equipment, normally we do not consider its duty below 600 t/h nor where the reduction ratio does not exceed 1:4.

8.8 Gyratory Crushers

Like jaw crushers, gyratory crushers are also reciprocating-pressure-type crushers, but while jaw crushers have two flat jaws moving to and fro, in the gyratory crushers a cone is moving gyratorily around a vertical axis in a conical chamber. In this type of crusher, the annular opening (outside diameter of the gap) specifies the size of the crusher. Naturally, the circumference of the cone provides a much larger crushing surface in gyratory crushers, resulting in a much higher output. Due to this, gyratory crushers are chosen as primary crushers, especially for hard materials, where they can reach reduction ratios of up to 1:12. Such high reduction ratios coupled with high outputs and capacities make gyratory crushers much more suitable as primary crushers than other crushers. However, the initial costs of installing this type of crusher are quite prohibitive, and, therefore, it has not been used in coal processing unless the high-duty conditions necessitated its application. By rule of thumb, gyratory crushers should be considered for feed rates of around 1,000 t/h or more at a reduction ratio of approximately 1:12.

8.9 Roll Crushers

Usually, roll crushers are employed as secondary or tertiary crushers. Their reduction ratios generally do not exceed 1:4. They are selected in quite a good number of cases because of their high output at low initial costs, low power requirements, low weight and low head room. Their maintenance is easy, and they are ideal crushers for materials such as coal where a minimum amount of fines is desirable.

Roll crushers are available in various types such as single-roll crushers, double-roll crushers etc. Some manufacturers have tried to combine two crushing stages into one, thereby achieving an overall reduction ratio of as high as 1:6.

8.9.1 Feed Size

The maximum size of the feed to the crusher is governed by the nip angle, the roll diameter, the gap setting, the hardness of the feed material and the type of the teeth/spikes. If a considerable percentage of the feed is larger than the recommended maximum feed size, spillage and choking of material occurs, which will result in loss of production and excessive wear of the crusher. The diameter of the roll and its width are selected according to the feed size, which is normally given by the respective manufacturers. To help in selecting the proper crusher type, some empirical formulae are also available. The setting of the rolls should be about 65 to 85% of the output size.

8.9.2 Advantages of Roll Crushers

1. They produce uniformly crushed material sizes.
2. They feature a wide range of possible adjustments, making it possible to meet difficult specifications of the crushed material required by the customer.
3. As the material is crushed only once, the percentage of fines is minimal.
4. By careful selection of the feed size and the setting, recirculation load can almost be avoided.
5. As the whole circumference of the rolls is used for crushing, complete utilisation of the wearing surface is possible.
6. Maintenance is simplest as compared to other crushers. Within the limitations of its capacity and reduction ratio, the roll crusher is a most economical crusher type.

8.10 Impact Crushers

With this type of crusher, crushing is effected by impact and not by pressure, as is the case in the crushers described above. Crushing by impact features several advantages. Pressure is applied instantly during crushing, and, therefore, breaking by impact almost eliminates the development of hidden cracks in the product.

Impact crushers are normally available in the form of impact breakers or hammer mills. These crushers are used both for primary as well as secondary crushing in most of the cement plants. In the coal processing industry, they have not found much application because they produce too much fines.

Impact crushers can give reduction ratios of up to 1:6 with a matching output of 1,600 t/h.

8.11 Feeder Breakers

A new concept of coal breaking has come into use in the recent past in the United States. It consists primarily of a robust chain conveyor on the top of which a rotating toothed drum revolves. The coal travels on the chain conveyor and is crushed while passing below the feeder breaker roll. Coal

is crushed primarily by fracture. The main advantages of feeder breakers are:

- low capital investment
- low power requirements
- small quantity of fines in the product
- low profile and, therefore, low installation costs.

9. Storage Systems

The main requirements of bunkers are that they should be suitable for storing the required quantity of coal and that the coal should be able to flow freely at the required rate so that the full capacity is utilised without the need of any flow inducing devices. The problems of coal flow are numerous. Some of the factors that can effect the coal flow are:

- rank of the coal, i.e., type of the coal according to the age (anthracite, bituminous coal, subbituminous coal, lignite or brown coal)
- sizing of the coal and percentage of fines
- ash content and amount of clay in the ash
- distribution of clay in the coal (surface, fine partings, heavy partings and pockets of clay)
- moisture content of the coal (equilibrium and surface moisture), time the surface moisture has been on the coal and leaching of sulphur
- coal freshly mined or stored and method of stockpiling
- temperature of the coal
- amount of tramp iron in the coal.

9.1 Common Bunker Types

The most common types of bunkers adopted in colliery coal handling plants are:

- ground bunkers
- overhead bunkers and silos
- ground stacking and reclaiming arrangement.

Since ground bunkers and overhead bunkers/silos are comparatively cheaper in capital costs as well as in operating costs up to a certain storage capacity, which, in most cases covers the normal requirements of a colliery, the usual choice of the designer and the investor falls for either of these two types. Ground stacking and reclaiming arrangement, due to its excessive capital costs, is usually adopted where the storage capacity is over 50,000 t, as, e.g., in power houses.

9.2 Capacity Determination for Bunkers

The determination of the optimum storage capacity of a bunker catering for eventualities like unduly long failures of the despatch and production systems is not always possible due to many indeterminates. So, the only instrument handy for the designer is the experience gained with similar installations in the vicinity of the proposed plant after ignoring infrequent peaks, whose frequency is usually small and which may be handled by other means. It may, however, be mentioned that, though the costs of handling additional coal which cannot be accommodated in the bunkers selected in the manner described above is high, yet, considering the total coal handled during a year, the impact on the handling costs may be insignificant (Fig. 6).

From our experience we have found that the storage capacity equivalent to one day of production plus the carrying capacity of one rake of box wagons for rail despatch usually strikes the balance. This is in view of the fact that the railways allow a limited time of five hours for loading a rake of 30 box x 44 box wagons, each of 56 to 58 t capacity, and dislocation of rail transport for more than one day is rare. In the case of road despatch, the bunkerage capacity is decreased suitably as the continuous availability of trucks throughout the day is better ensured.

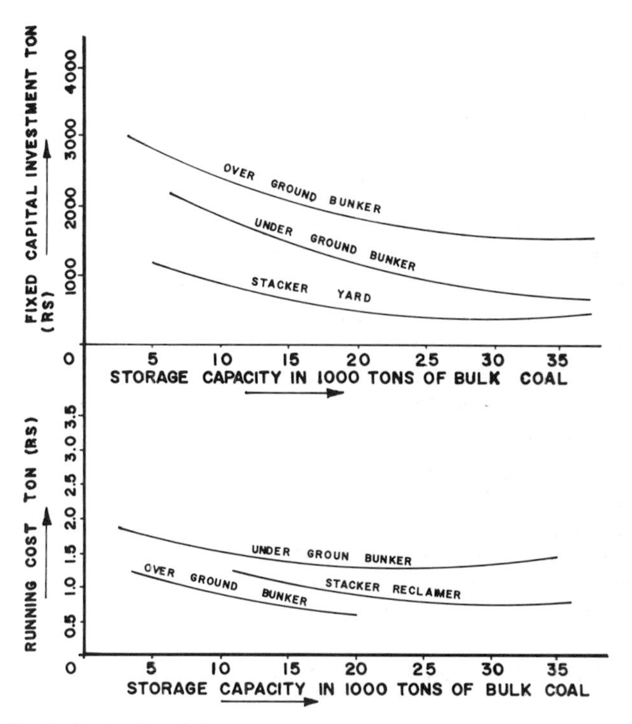

Fig. 6: Costs of coal storage

Cost analyses of bunkers reveal that for capacities less than 1,200 t or so overhead bunkers are suitable, while for capacities between 1,200 and 50,000 t ground bunkers are favoured, and for capacities higher than 50,000 t stacker/reclaimer systems are ideal.

10. Deshaling

The problem of deshaling has a rather peculiar history. As long as mines were not very big, the obvious choice was hand picking. Then came the age of mechanisation, especially in big opencast mines. Since hand picking was not a suitable solution for big mines, a heavy demand for washeries arose. Washeries, being heavily capital-oriented projects, needed the concurrence of the customers for taking up the extra costs involved. Many customers insisted that the coal supply were the responsibility of the producer and, as such, the washery costs should be borne by them. There is no ready-made solution to that problem even now, and decisions are taken depending upon the merit of individual cases.

Even if a washery is not to be considered because of heavy capital costs and the boilers of the power houses are to be designed to suit ROM coal, some degree of deshaling becomes a must to maintain consistency of the coal quality.

10.1 Hand Picking

Hand picking is the simplest and the traditional type of deshaling. In this process, ROM coal/crushed ROM coal is screened normally at —50 mm to get rid of the fines, and picking is carried out for the +50 mm size class on a slow-speed conveyor. Based on experience, it has been found out that for hand picking belt conveyor speeds from 0.3 to 0.5 m/s are effective. The success of this system depends upon the coal composition making the dirts clearly distinct from the clean coal. A good layout of the picking plant with proper lighting and working conditions adds to the efficiency of the manual pickers. On an average, one manual picker can be expected to pick about 1 t of material per shift. Normally, belt widths of 1,200 and 1,400 mm are used, while the lengths are kept between 15 and 20 m.

10.2 Rotary Breakers

When the quantum of dirt increases beyond the application range of manual picking, rotary breakers/Bradford crushers can be used. They consist of a rotating drum of predetermined diameter and length having some inclination to the horizontal. From the intake side, the ROM coal moves spirally until it reaches the end of the drum. During this spiral movement, it drops down many times the height equivalent to the diameter of the drum, and the coal pieces are broken to smaller sizes and find their way out of the holes provided all around the drum, whereas the stones/shales go out of the rotary breaker separately.

The system is very efficient and attractive, but very much dependent on the composition of the coal, which should be such that the coal is effectively reduced in size in drop and shatter tests.

10.3 Wet Deshaling

The wet deshaling process has been developed in the Federal Republic of Germany, and it may also be used in India in the near future.

The wet deshaling system consists of a vibrating plate in a liquid media which receives the ROM coal through the inlet. After the contact of the ROM coal with the vibrating plate, this is made to travel upwards against the gravity of the liquid. In this process, the coal and the dirts attain different particular heights, so that the coal and the dirts can be collected separately, thus obtaining deshaling of the coal.

11. Coal Loading Complex

The coal loading complex is one item which has gone through tremendous changes during the recent past. The subject is so vast that it commands the respect of being presented in a separate paper. However, in this article, the subject has been dealt with briefly.

The loading complex must be discussed along with:

1. Reclamation from the coal storage system
2. Marshalling device for moving the wagons (in the case of loading into railway wagons/MGR unit train system)
3. Weighment arrangement
4. Loading system.

11.1 Reclamation

Reclamation from the storage bunkers/yards is predominantly carried out by means of belt conveyors. The parameters of the reclamation system such as speed, width and t/h are practically in all cases the same as those of the loading conveyor. The following devices are normally used to reclaim the coal:

— rake and pinion chute gates
— vibro feeders/reciprocating feeders
— plough feeders
— stacker/reclaimers (Fig. 7).

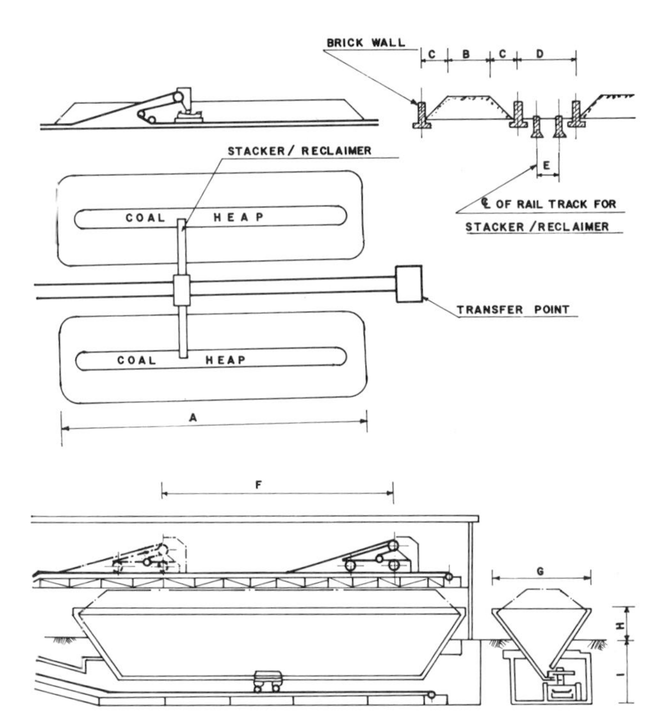

Fig. 7: Stacker/reclaimer

11.2 Marshalling

Hand pushing used to be the marshalling device during the past when there was no particular limitation posed by the railways for duration of coal loading etc. Gradually, wagon haulers became the most important equipment of marshalling the rake for loading the coal into Indian Railways wagons at a rate of about 600 to 800 t/h, the speed of the wagon hauler rope being in the neighbourhood 8 to 10 ft/min to synchronise with the loading rate.

With the advent of fast-loading systems, the use of locomotives with a creep control speed has become very popular. Currently, a creep control speed of 0.8 km/h is used for loading rates of around 6,000 t/h. It is expected that locomotives will have to be manufactured with faster creep control speeds to match with higher rates of loading of 10,000 to 15,000 t/h or even more.

11.3 Weighment

In the case of conventional loading with a wagon hauler as marshalling device, 100 t capacity mechanical weighbridges have been the most popular equipment. With the introduction of rapid-loading systems, weighment has been overtaken by in-motion weighment arrangements with load cells etc. In these systems, electronic gadgets are used to ensure automatic weighment recording and displaying of the coal loaded into the container, which again is interlocked with the loading and marshalling arrangements.

11.4 Loading Complex

11.4.1 Conventional Loading

With "conventional loading" we mean the existing practice of loading coal into railway wagons. The loading conveyors are normally designed for capacities of 600 to 800 t/h in order to complete the loading and allow the rake to clear the siding within the time of five hours prescribed by the Indian Railways. Wagon haulers and mechanical weighbridges form the normal compliments to effect marshalling and weighment.

Recently, the Kumar Mangalam Committee has suggested that the minimum rate of loading should be designed at about 1,200 t/h for all mines with a capacity of 2 Mt/a or less. For mine capacities bigger than 2 Mt/a, a loading rate of more than 1,200 t/h has been suggested by the committee to be achieved by suitably designed rapid-loading systems.

11.4.2 Rapid-Loading Systems

Rapid-loading systems are the need of the hour in the national interest. They help the producer the despatch maximum amounts of coal per day. They help the railways to transport maximum amounts of coal per day with minimum moving stock. In the case of super thermal power stations, there is no choice except to supply a rapid-loading system in such a way as to synchronise with the loading arrangements at the end of the power station.

Rapid-loading systems and unit trains have become synonymous. So far, they have been designed and used for super thermal power stations. However, planning is also being done to evacuate the coal by Indian Railways by rapid-loading systems. Fig. 2, mentioned earlier, is an example of such planning.

11.4.3 Requirements of High-Speed Loadout Systems

Here, high-speed loadout systems of 6,000 t/h capacity in conjunction with unit trains with self-discharging hopper wagons shall be discussed. The Rail India Technical and Economic Service has suggested the following requirements for high-speed loadout systems:

1. The length of the hopper body at the top is smaller than the overall length of the wagon (between coupling faces of the couplers). The loading mechanism must be designed for opening only while the hopper wagon is passing under the loading chute. The discharge doors must be quick acting to ensure full loading without spillage between the wagon hoppers.
2. The train will travel through the loading station at a speed of 0.8 km/h, i.e., at a rate of approximately 1 min/wagon. As the wagon hopper body is shorter than the overall length of the wagon, the door of the loadout may remain open between 70 to 80% of the time the wagon passes under the loadout. Thus, the coal will have to flow through the loadout opening at a rate of approximately 5,500 to 6,000 t/h to discharge 60 t of coal in approximately 40 s.
3. The rate of coal flow into the loadout, in the case of a surge bin, should match the rate of coal flow from the surge bin to the wagons. In the case of a silo-type loadout, the rate of coal flow into the silo should permit uninterrupted loading of trains from the silo at the closest expected spacings of trains without detention to empty incoming trains.
4. The loading system must ensure an even distribution of the payload throughout the length of the wagon hopper and provide for automatic trimming of the top of the coal loaded in the wagon. The height of the trimmed surface above the top of the wagon shall be adjustable in a range of 25 to 250 mm.
5. The loadout mechanism shall have spillage preventive arrangements, both in longitudinal and transverse directions.
6. Should there be a variation in the train speed or a total stoppage of the train during the loading process, the loading system must have an inbuilt feature to stop coal flow into the wagons and prevent spillage.
7. A permissible tolerance of +2 t for the payload of a wagon is stipulated, and, as such, the loading mechanism should be sensitive enough to meet these requirements.
8. To facilitate unhindered movement of the wagons and locomotive under the loadout, the fixed structures and all connected members should be outside the minimum clearance dimensions.
9. The overall height of the wagons is smaller than the minimum clearance dimensions, and, as such, the loading chute, for achieving a levelled heap, would infringe the permissible maximum dimensions. For ensuring safety against fouling of the loco/wagon with the loading mechanism, the chute, while not in use, must be locked in the full retracted position or rolled out of the position as applicable to the design.
10. An interlock must be provided to ensure that the train does not come under the loadout till the chute is out of the minimum clearance dimensions.
11. Permission to enter the loadout must be conveyed through indicators "red/green" provided at a suitable distance in advance of the loadout. While the indicator is "red", the driver shall bring the train to a halt and proceed only after the indicator has been put to "green". "Green" should go to "red" automatically as soon as the loading operation has started.
12. Adequate means of communication have to be provided to ensure that the driver of the train receives audio signals over and above visual signals to start or restart the train in the event of any emergency at the loading station.
13. The control room shall be elevated and so positioned that the loading station operator has a clear continuous view of the interior of the wagon being loaded. Location of the transparent front wall of the control room would be just outside the minimum clearance.

11.4.4 Elements of Rapid Train Loading Systems

The main elements of rapid-loading systems are:

1. Minimum rate of loading
2. Silo
3. Load gates
4. Loading and weighment systems.

11.4.4.1 Minimum Rate of Loading

The present Indian installations have minimum rates of loading of 5,500 t/h, which is achieved with silo outlets of 1.2 m, times of 2 s for loading gate opening and closing, respectively, creep control speeds of the loco of 0.8 km/h, 62 t capacity and wagon hoppers of 10.8 m length.

Table 2: Comparison of rapid-loading systems at different CIL projects

Project	Gevra (Ordered)	Jayant (Ordered)	Rajmahal (Tender Specifications)
Quantity of coal to be loaded (in Mt/a)	10	7	10.5
Unit train size (in wag./train; t/wag.)	30—35; 60	30; 60	28—57; 60
Unit train payloadt (in t)	1,800—2,100	1,800	1,680—3,420
Unit train speed during loading (in km/h)	0.5 (up to 1.2)	0.8	0.8
Loading time per wagon: — at 0.8 km/h (in min) — at 1.2 km/h (in s)	 1 20	 1	 1
Loading rate (approx.) (in t/h)	5,500	5,500	5,500
Loading system	mass loading	volumetric loading	mass loading
No. of silos	2	1	2
Capacity per silo (in t)	2,400	2,400	4,000
Inner dia. of silo (in mm)	15,000	15,000	18,000
No. of openings per silo	1	2	2
No. of loading gates per silo opening	1 double clamshell gate for loading wagons (silo no. 1 has an add. opening for loading PR wagons)	1 single guillotine gate for loading wagons	4 single guillotine gates for loading preweighing hopper
No. of loading gates on preweighing hopper	not applicable	not applicable	1 single guillotine gate for loading wagons
Operation of loading gates	automatic, employing track logic	automatic	automatic
Maintenance gate	1	not provided	not provided
Retraction system of loading for passengers of locomotive	telescopic, automatic, with track switch	telescopic and horizontal, operation from control room	telescopic and horizontal, operation from control room
Lining of silo interior	8-mm stainless steel, up to 8,000 mm dia., of conical portion, balance up to max. coal level, 10-mm Tiscral	20-mm ferrosite/ ironite coating up to max. coal level	8-mm stainless steel of conical portion, balance up to max. coal level, 25-mm ferrosite.
Silo hopper slope (in °)	70	60	over 60
Silo feed conveyors: — number — width (in mm) — capacity (in t/h)	 2 1,400 1,800	 2 1,200 1,200	 2 1,400 1,800
Weighing system	electronic, axle weight integration, with the following facilities for printout of: — identification of individual wagon — tare weight per wagon — gross weight per wagon — net weight in each wagon — total payload per unit train — total despatch from 0.00 h to 24.00 h — time of entry of each train — time of departure of each train	electronic, axle weight integration, with the following facilities for printout of: — gross weight per wagon — total gross weight of unit train — date and time	electronic, preweighing hopper, with the following facilities for printout of: — net weight loaded — date and time
Level sensors in silos	ultrasonic, with the following facilities: — stock of coal in silo — high level interlocked with feed-in conveyors — low level interlocked with loading gates	electrical (Bindicator), with the following facilities: — high level interlocked with feed-in conveyors — low level interlocked with loading gates	ultrasonic, with the following facilities: — stock of coal in silo — high level interlocked with feed-in conveyors — low level interlocked with loading gates
Arch detection and interlocking system	provided by means of level sensors and a weighing system, loading gates close automatically at arching	not provided	not provided
Arch breakers	pneumatic gun	not provided	pneumatic gun
Sampler system	automatic, final sample 6 kg/1,800 t, 80 % —8 mesh, automatic sample collector	automatic	automatic, final sample
Temperature sensors	not provided	not provided	provided

11.4.4.2 Silo

In view of the high rates at which the wagons are required to be loaded, it becomes necessary to have an overhead silo of sufficient capacity equipped with loading conveyors, flood gates, sampling devices etc. The capacity of the silo should be more than the payload of the unit train so that during the loading period the unit train is independent of the backup arrangement which feeds coal into the silo.

11.4.4.3 Loading Gates

There are many different types of silo outlet gates, varying from horizontal sliding gates (guillotine-type) through retractable pivoted chutes to clamshell-type gates. It may, however, be remembered that, though the gate is a relatively inexpensive part of the loading system, yet it is one of the most critical parts because it can create problems out of all proportions to its cost. It can:

- restrict the material flow from the silo, thus leading to a decrease in the loading speed
- fail to close or fully close, leading to discharge of huge quantities of material upon the rail track, thereby shutting down the whole system for a long duration.

The size of the gate should be adequate to allow for unrestricted and uninterrupted passage of material and the minimum rate of flow, compatible with the quantity of material to be loaded into wagons moving at a specified speed. Usual flow rates for rapid-loading systems are 3 t/s.

The operation of the gate should be fully synchronised with the movement of the wagons. It should as well be "fail to safety", so that it closes under any electrical or mechanical failure. The operation can be manual in the case of low despatches, but it should be automatic when the despatches are high, with provision for manual operation.

One further critical consideration is the size of the coal to be handled by the gate. For small crushed coal, single gates can serve the purpose. However, where crushed ROM opencast mine coal with hard impurities is to be handled, it is advisable to use double gates so that, in the case that one gate fails to close because of a hard lump getting into it, the other closes and spillage is avoided.

11.4.4.4 Wagon Loading and Weighment

In respect of rapid-loading systems, two main principles are used for wagon loading:

1. Flood loading, i.e., loading by volume
2. Mass loading, i.e., loading by weight.

These principles are according to two distinct and fundamentally different philosophies of wagon loading, as is self-evident. Flood loading is suitable where the bulk density of the material is fairly uniform so that the wagons are loaded within specified limits to avoid overloading or underloading the train. In the case, however, that the bulk density is expected to vary widely, mass loading is the only solution.

Mass loading can be effected in two ways, viz.:

a) Loading and weighing operations are done simultaneously, and the operation of the silo opening gate is governed by the weighing system.
b) A preweighing hopper is employed between the silo and the wagon. The operation of the silo opening gate is governed by the weighing system, and the preweighed quantity of material is then loaded into the wagons.

11.4.4.5 Other Facilities in Rapid Loading

Other facilities in rapid-loading systems are:

- sampling systems for primary samples at rates of 2 t for every 1,800 t (crushing to 8 mesh) and secondary samples at rates of 6 kg for every 2 t of primary samples
- bucket elevators for feeding the surplus material from the sampling system back to the silo
- dust suppression at the feed-in and loading points at the silos
- annunciator units for indication of fault conditions in the system by audio and visual alarms
- additional outlets at one of the silos for loading wagons on the public railway system
- battery backup power supply for retention of the memory in the microprocessor
- pneumatic arch breakers
- 1,000-kg lifts for passenger and freight movement
- maintenance monorail hoists
- air-conditioned control rooms
- pressurised motor control centre rooms.

In Table 2, rapid-loading systems at the Gevra, Jayant and Rajmahal projects of Coal India Ltd. (CIL) are compared.

12. Economics

The capital costs of a coal handling plant are of the order of 3 to 5 crores (Rs 30 to 50 million) for a project of 1 to 2 Mt/a, 15 to 20 crores (Rs 150 to 200 million) for a project of 2.5 to 4.5 Mt/a and 40 to 50 crores (Rs 400 to 500 million) for a project of about 10 Mt/a capacity. Table 3 shows the approximate present-day costs of some of the mines and coal handling plants recently designed. The operating costs are normally of the order of Rs 6 to 7 per tonne for small- and medium-size projects and Rs 7 to 12 per tonne for bigger projects.

Table 3: Capital costs of mine and CHP

Mine Production Mt/a	Total Costs of Mine Project Rs 10 million	Costs of CHP Rs 10 million	CHP Costs as Percentage of the Project Costs %
1.0	38.7	6.3	17.45
1.5	61.8	3.8	6.15
1.5	38.0	4.2	11.28
2.0	23.1	2.8	12.2
2.25	30.5	3.9	12.8
2.5	52.9	3.6	6.81
4.0	400.0	27.9	6.99
4.5	136.0	18.9	13.97
10.0	710.0	37.7	5.32
10.0	400.0	40.2	10.05

Acknowledgements

The author is grateful to the management of the Central Mine Planning and Design Institute Ltd. (CEMPDIL), Ranchi, for giving him permission to present this paper and for the assistance received from his colleagues.

The author would like to mention that the views expressed in this paper are his own and should not be attributed to CEMPDIL/CIL.

bulk solids handling Volume 1, Number 1, February 1981

Truck Dispatching by Computer Simulation

Young C. Kim and Miguel A. Ibarra, USA

LKW-Einsatzsteuerung durch Computer Simulation
Contrôle de l'expédition des camions par simulation sur ordinateur
Simulación del despacho de camiones mediante un ordenador
コンピュータシミュレーションを利用した台車運搬
卡车运货的计算机模拟
تحميل الشاحنات باستخدام الاستقصاء بالحاسب الالكتروني

LKW-Einsatzsteuerung durch Computer Simulation

Der Beitrag beschreibt die Ergebnisse einer Computer-Simulationsstudie, in der der Einfluß einer vorgeschlagenen LKW-Einsatzsteuerung auf die Gesamtproduktivität eines bestehenden Tagebaues untersucht wurde.

In dieser Studie wurde der Tagebaubetrieb zuerst über einen Monat lang simuliert, auf der Basis der tatsächlichen und der berechneten LKW-Zeitpläne. Anschließend wurde der Betrieb mit LKW-Einsatzsteuerung nochmals gerechnet und mit den tatsächlichen Werten verglichen. Es ergab sich eine 10%ige Produktivitätserhöhung für den Fall der LKW-Einsatzsteuerung.

Contrôle de l'expédition des camions par simulation sur ordinateur

Cet exposé décrit les résultats d'une simulation sur ordinateur dans laquelle on a étudié l'impact de l'expédition des camions sur la productivité d'ensemble d'une mine pour une mine à ciel ouvert en cours d'exploitation. Le système existant a été comparé avec l'utilisation de l'expédition des camions. Les résultats montrent une amélioration substantielle de l'ensemble de la productivité avec contrôle de l'expédition; un gain d'environ 10% pour l'exploitation.

Simulación del despacho de camiones mediante un ordenador

Este artículo describe los resultados de un estudio realizado con un ordenador con el fin de investigar el impacto del despacho organizado de camiones en la productividad global de una mina a cielo abierto en explotación actual. Se simuló el sistema existente durante un periodo de un mes y se hicieron comparaciones con la explotación utilizando el nuevo sistema de despacho de camiones. Los resultados arrojan una mejora importante de una productividad total con el nuevo sistema de despacho organizado, con una mejora del 10% aproximadamente.

Summary

This paper describes the results of a digital computer simulation study in which the impact of the proposed truck dispatching on the overall mine productivity was investigated for an operating open pit mine.

In the study, the existing operation was first simulated for a one-month period, using both the actual time study data and the computed cycle times. By adjusting certain input parameters to the simulation program, the actual one month production was duplicated through simulation. Afterwards the operation was again simulated in the dispatch mode.

Comparison of the simulation results under the dispatch mode, with the initial base case simulation results, showed a definite improvement in overall productivity with dispatching; that is, approximately 10% gain for the operation. As expected, the results also showed that the extent of possible improvement did vary with the particular pit configuration being investigated.

Prof. Dr. Y. C. Kim, Department of Mining and Geological Engineering, University of Arizona, Tucson, AZ 85721, USA; M. A. Ibarra, Mining Engineer, Pincock, Allen and Holt, Tucson, AZ 85714, USA

1. Introduction

Truck haulage is the most widely used means of transportation in an open pit mining operation, but is often the single most expensive process in a truck-shovel mining system. According to Michaelson (1974), truck-fleet productivity in open pit copper mines has the lowest improvement rate among the three major unit operations: drilling, loading and hauling. In addition, trucks require much labor, high maintenance and relatively frequent replacement making them sensitive to inflation. Most operating shovels experience either some insufficient or excessive truck capacity or a combination of both in truck-shovel mining systems. To meet required production with increasing depth of pit or changing ore-waste stripping ratios, additional equipment is required each year. As a result, management is faced with the problem of buying additional trucks or shovels if there is an improper balance of equipment in the mining operation. This problem usually results from inadequate use of haulage resources. Recent increases of the energy cost together with projected future increases will further increase truck-fleet capital and operating costs in the future. Therefore, it seems appropriate to test any strategy for optimizing truck-fleet performance.

The state-of-the-art in computing technology has advanced to a point where there are several truck dispatching systems which offer the potential of improved truck productivity and subsequent savings. Truck dispatching systems in the mining industry can solve the problem of inefficient use of resources by reducing waiting times in the haulage operation. Several open pit mines around the world have successfully implemented truck dispatching systems in their haulage Baron [1], Beaudoin, [2]; Crosson, Tonking, and Moffat, [3], Hobday, [4] Ibarra, [5]; Naplatanov, et.al., [9]; Naplatanov, Sgurev, and Petrov, [10]; Mueller, [8]; Schlosser, [12]. The ultimate question, of course, is whether the savings are sufficient to warrant the cost of such systems.

The effects of implementing a truck dispatching system in any mining operation can always be observed by physical implementation of the proposed system. This could be very costly. Computer simulation is probably the best tool that can be used to predict changes in system behavior for any contemplated changes to the existing system. This study investigated the impact on productivity by employing a truck dispatching system in an open pit mining operation.

2. General Steps in the Simulation Study

To obtain the end results of any simulation study, the following three basic tasks, as presented by Pritsker [11], should be performed:

1. Determine that the problem requires simulation.
2. Build and program a model to solve the problem.
3. Use the computer simulation program as an experimental device to solve the problem.

In Task One, the analyst is concerned with a possible mathematical solution to the problem under consideration. A system analysis of a truck-shovel open pit mining operation is usually far too complex to be conducted analytically. Therefore, it is safe to say that the problem requires simulation.

Task Two, consisting of building and programming a truck-shovel open pit mine operations model, was not performed since a Mine Operations Analysis Model developed by Kim and Dixon (1977), was available. The first two segments of this model are Haul-Cycle Simulation Program and GASP IV Open Pit Simulation Program.

Consequently, the main emphasis of this investigation was given to Task Three in which real data from an open pit mine were used to simulate stochastically the haulage under two different systems of truck control: one with a non-dispatching system and one with a dispatching system.

The real haulage system operates in a non-dispatching mode in which a certain number of trucks are assigned to each operating shovel throughout the entire shift unless a significant event occurs such as a shovel or truck breakdown. The proposed change to the real system is to operate in a dispatching mode. In a dispatching mode, the trucks will be allowed to serve different operating shovels throughout the entire shift. Each time a truck becomes empty, it is assigned from a dispatch point to the next available operating shovel or to the shovel that has been idle the longest.

In this study, the real system plays an important role as an integral part of the analysis because the parameters input to the simulator are first adjusted and validated by comparing the generated results with the real results of the system in a non-dispatch mode. When the non-dispatch simulation results appear representative of reality, the simulator is switched to operate in a dispatch mode. The results between non-dispatching and dispatching are compared. Consequently, the analyst relies heavily on comparative results between non-dispatching and dispatching rather than in advanced statistical techniques.

3. Data Collection, Analysis and Generation

Production and operating data needed for the study were obtained from a large mining operation which currently operates under a non-dispatching system of truck control. These data included the following:

1. Haulage road profiles and characteristics, i.e., distances, grades, rolling resistances, efficiencies, speed limits, right-of-way rules.
2. Equipment characteristics and availabilities, i.e., speed-rimpull curves, motor-current curves, mechanical availabilities, empty weights.
3. Field observations of shovel's loading time, truck's dumping time and load weights.
4. Pit configuration, equipment configuration and associated production during the time simulated.

The amount of data collected was considered sufficient to simulate the operation stochastically. Furthermore, access to a number of reports proprietary to the mining company made possible a more accurate adjustment and validation of input data later during the study.

Next, the obtained data were analyzed for their respective distributions.

3.1 Shovel Loading Time

The loading time observation starts when an empty truck begins backing up to the shovel and ends when the same truck starts on its way to an unloading point (or material destination). Because the operation currently utilizes a single truck type and a single shovel size, data from only one loading combination were collected. A total of 527 loading observations are grouped into the histogram shown in Fig. 1a. The shape of the histogram suggests that the loading time may be approximated by a lognormal distribution. Fig. 1b illustrates the histogram of transformed loading time data that confirmed the lognormality assumption.

3.2 Truck Dumping Time

Similarly, the dumping time observation starts when a loaded truck begins backing up to the dumping point and ends when the empty truck starts the return trip to a shovel. Three types of materials are mined in this operation: ore, waste and leach. Densities of ore and waste are 2.57 t/m^3 and 2.54 t/m^3, respectively. All unloading points present about the same dumping conditions. Therefore, it was considered valid to group all dumping observations in one having a single truck-shovel-material combination. The histogram of untransformed data (Fig. 1c) suggests that the 679 dumping time observations may also be approximated by a lognormal distribution. Fig. 1d shows the histogram of transformed dumping time data that also confirmed the lognormality assumption previously made.

3.3 Load Weights

Measurements of load weights were obtained from a field application weighing study performed by WABCO field applications engineering. Again, these data include a single combination of truck-shovel-material type in the mining operation. A total of 77 payloads was first analyzed in the study. Fig. 1e illustrates that grouped data in a histogram form. As may be noticed, this histogram presents suspicious underloading and overloading situations for the trucks (see left and right tails of histogram). Further investigation revealed a possible bias in the observed data. At the time of the

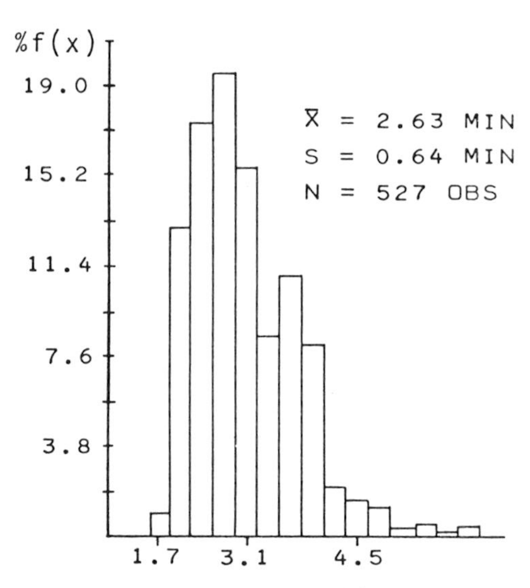

Fig. 1a: Loading Data

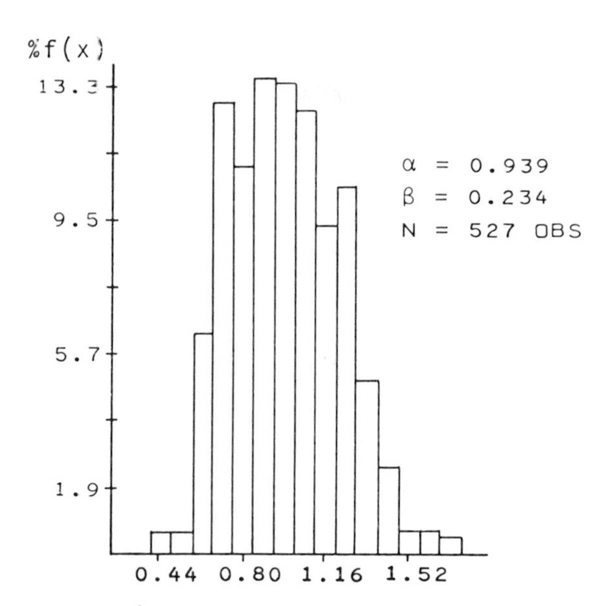

Fig. 1b: Transformed Loading Data

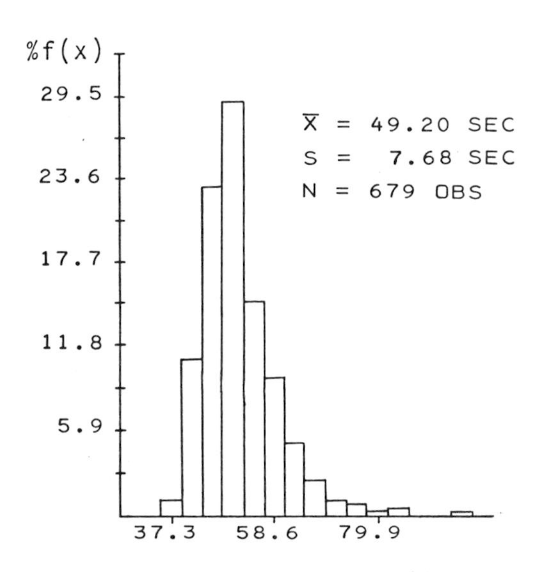

Fig. 1c: Dumping Data

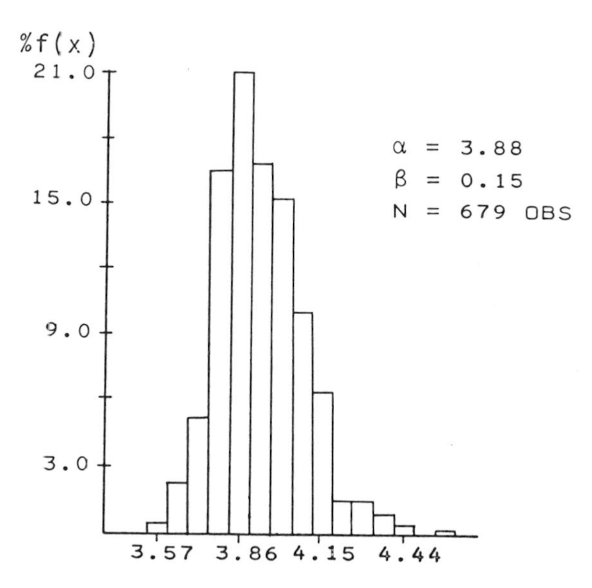

Fig. 1d: Transformed Dumping Data

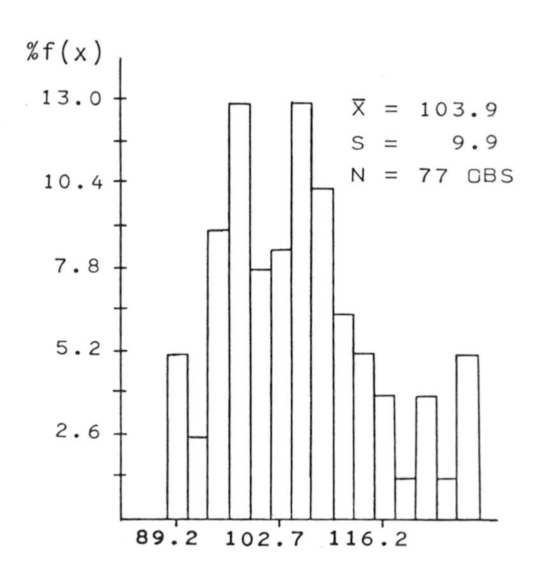

Fig 1e: Load Weight Data

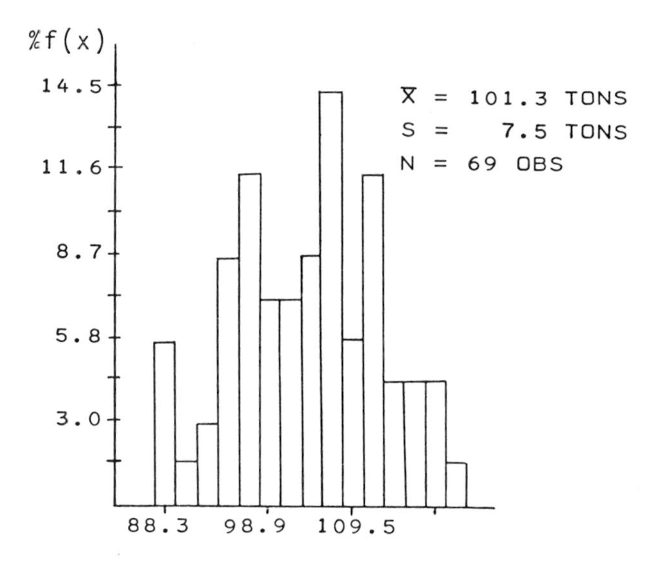

Fig. 1f: Reduced Load Weight Data

Fig. 1: Histograms of input data before and after transformation

weighing study, the operation had just returned to normal after a 3-month labor strike. As a consequence of that strike, most of the experienced shovel operators had been laid off. Consequently, new operators were on the shovels. This fact probably accounts for the obvious observed underloading and overloading of trucks. To correct for the bias in the load-weight data, a total of 11 high payloads were arbitrarily not included in the second load-weight analysis. Fig. 1f shows these grouped data, which suggests that the load weights are distributed normally.

To determine how well the assumed distributions fit the actual data, the Kolmogorov-Smirnov (K-S) test was performed on each set of data for goodness of fit. For each set, the 0.05 critical value of the K-S test was always greater than the K-S computed value, thus confirming the validity of each assumed distribution.

The analysis of the obtained data for their respective theoretical distributions was required because these distribution parameters are part of the input data to the Open Pit Mine Simulation Program which is the second segment of the Mine Operations Analysis Model.

3.4 Travel Time Generation

The travel times needed in the simulation study had to be calculated because the mining operation did not possess them. Knowing the exact amount and type of material to be mined at each shovel location and the respective material destination, it was easy to determine what travel times would be needed during the study. The use of a Haul Cycle Simulation Program (MCYCLE), which is the first segment of the mine operations analysis model, made possible the generation of travel times. This program uses the manufacturer's equipment performance characteristic curve and is designed to perform cycle time calculations in the discrete event simulation technique Pritsker [11]. The program considers the existence of switch-backs, interim stop points, and speed limits in the haul roads. This capability is possible by defining a set of velocity limits that overrides the truck's speed capabilities under appropriate situations. Other capabilities are the handling of multiple runs, deterministic or stochastic simulation, and generation of haul road profiles internally.

Next, the haulage roads were defined by breaking them into segments having equivalent characteristics as required by MCYCLE. A general layout of the segmented haulage roads in the mine being simulated is given in Fig. 2. By counting Node B3 twice because it appears in two haulage patterns, there are seven shovel locations (Nodes: D4, E2, C3, B3, N6 and N7) and four material destinations (Nodes: CRUSHER, LEACH, DUMP and RAMP).

A careful observation of the present pit and haul road configurations show that there are two well defined haulage traffic patterns in the pit (Fig. 2). The first pattern is located in what is called the north sector of the pit where three shovel locations exist (Nodes: N6, N7 and B3) with just one material destination (Node: RAMP). The second pattern covers the east, central and west sectors of the pit where there exists four shovel locations (Nodes: D4, E2, C3 and B3) and three material destinations (Nodes: CRUSHER, LEACH and DUMP). As mentioned earlier, travel times between each shovel location and its possible material destinations were calculated using the haul cycle simulation program.

The calculated results were always validated by comparing the *computed speeds* under various grades against the *actual speeds* of the trucks on the haulage roads of the mining operation. At first, it was noted that by allowing the trucks to go as fast as they could upgrade, the simulation results always underestimated the travel time. This result was probably due to such factors as the empirical truck manufacturer's speed-rimpull curve overestimating the performance of the trucks or unknown mechanical-electrical

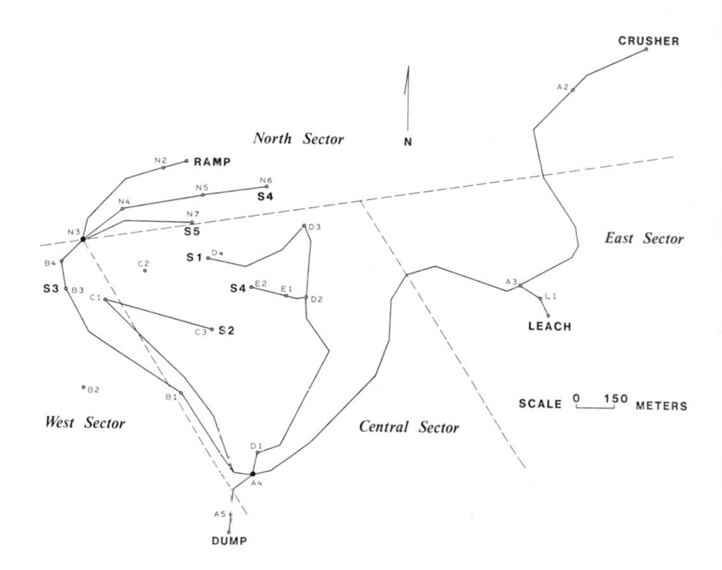

Fig. 2: Haul road layout

inefficiencies of the haulage units. To approximate the performance of the simulation model with the real system, the upgrade speed of the trucks was limited so that the *computed speeds* always matched the *real speeds* of the trucks on the mining operation. As a result of this adjustment, the *computed speeds* matched the *real speeds* under all conditions (level and incline) and the computed travel times were accepted as representing reality.

Because travel time distributions were required in the open pit simulation program, the multiple run capability of the haul cycle simulation program was utilized to obtain a distribution which approximated a normal distribution.

4. Simulation of Pit Configuration

To determine the impact on productivity of a dispatching system in the mine, the haulage operation was simulated first in a non-dispatch mode using the Open Pit Mine Simulation Program. The simulated results were next validated by comparing them with the results from the real system. If the two results did not match, certain adjustable parameters in the input data were varied and the haulage simulation using a non-dispatch mode was repeated. Only when the non-dispatch results were considered representative of reality was the haulage operation simulated in a dispatch mode and the differences between dispatching and non-dispatching observed.

At this point, it is important to note the presence of only one intersection in each haulage pattern mentioned earlier (Nodes: N3 for the north sector and A4 for the other sectors). Consequently, all trucks coming from any material destination have to pass one of these two points to get to any shovel location. These node intersections were, therefore, used as dispatch points in this study. In dispatching, trucks were assigned to the next available shovel.

During that month of operation which was being simulated in the study, some shovel either changed the location or handled the material from the same location but to different destinations. For example, shovel 4 began the month mining waste from Node N6 (North Sector) which was hauled to build a ramp on the north haulage road. By the middle of the month, the shovel was relocated at Node E2 (Central Sector) where ore was mined and was hauled to the crusher. Finally, by the 20th day of the month, this shovel started mining waste from the same location, which was hauled to the waste dump. As a result, six different cases involving various combinations between trucks and shovels had to be investigated separately during the operating month simulated.

As an example, one of the cases investigated is explained below. Pit configuration of this case is shown in Fig. 3a. There existed three shovels located at Nodes D4, C3 and B3 and two material destinations at CRUSHER and LEACH. Shovels 1 and 2 handled ore which was hauled to CRUSHER and shovel 3 handled waste which went to LEACH.

First, the non-dispatch mode was simulated. The output of the simulator included the total and daily ore and waste production and production statistics from each individual machine and from each location to which the material was hauled. Other outputs were statistics on truck and shovel down times and wait times.

The non-dispatch results were then compared with actual production of the real system, as shown below:

Production (metric tons)	Shovel 1	Shovel 2	Shovel 3
Actual	140,500	140,500	149,900
Simulated Non-dispatch	145,800	140,100	154,300

With less than a 4% difference between actual and simulated production, the system simulation was considered representative of reality.

Next, the dispatch mode simulation was performed with the trucks allowed to serve any of the three shovels. The resulting production figures of each shovel are shown below:

Production (metric tons)	Shovel 1	Shovel 2	Shovel 3
Simulated Dispatch	148,500	159,600	171,300

Using the non-dispatch and dispatch output statistics as the basis for comparison of the two systems, dispatching improved productivity up to 14% and reduced shovel and truck waiting times more than 30%. Tables 1 and 2 show computer printouts for this illustrative case.

The same simulation procedures were used to determine the effects of dispatching in each one of the other five cases given in Fig. 3.

5. Analysis of Simulation Results

As discussed previously, the analysis relied heavily on comparative results between non-dispatching and dispatching. Furthermore, a decision was made to observe more closely the behavior of equipment waiting times with the introduction of dispatching in the haulage system. Any reduction of equipment waiting time means a possible production gain for the operation.

The results for each of the cases investigated showed a definite improvement in productivity with the dispatching system. Fig. 4, which summarizes the results, clearly shows this improvement. However, the extent of improvement varied with the particular pit configuration of each case. For instance, if the pit configuration contained a combination of long and short hauls (Cases a, b and c in Fig. 3) dispatching was definitely better than non-dispatching. When the pit configuration contained only long hauls or only short hauls (Cases d, e and f in Fig. 3), dispatching still showed a significant improvement over non-dispatching; however, the extent was not as great as in the combination of long and short hauls. Fig. 5 illustrates the equipment waiting time of the experimental results. A straight line connection between points on the plots does not mean fractional trucks from point to point. This connection merely helps to distinguish between the non-dispatching and dispatching results of the simulation. Furthermore, the position of the operating shovels was arranged in such a way that the shovel nearest the dispatch point starts from the left on these figures.

As expected, dispatching considerably reduces equipment waiting time. In addition to the overall reduction in truck-fleet waiting time, dispatching also reduces variability by evenly distributing the waiting time between the trucks.

From Fig. 5a and 5b, it is apparent that shovel 3 is overtrucked and the other two shovels are undertrucked during non-dispatching. In contrast, truck waiting time was evenly distributed among the trucks with the use of dispatching, and there was a reduction in shovel waiting time. However, the extent of reduction varied greatly depending on the length of travel from the dispatch point to the shovel locations. The more distant shovels seem to accumulate shovel waiting time more than the nearer ones. All the figures show a substantial reduction in equipment waiting time with dispatching. In Fig. 5c and 5d, however, non-dispatching seems to do as good a job as dispatching in terms of reducing the variability in waiting time for the equipment. Again, in these two pit configurations the most distant shovels accumulated a little more idle time than the shovels nearer the dispatch point.

The effects of dispatching on equipment waiting time for Cases e and f (Fig. 5e and 5f) are similar to the one obtained in Cases a and b (Fig. 5a and 5b) in that one of the two shovels is overtrucked. Notice that dispatching evenly distributed the waiting time among the trucks. In both pit configurations, shovel 5 accumulated substantial idle time with non-dispatching, but dispatching substantially reduced the amount of idle time for this shovel even though the other shovel maintained the same amount of idle time in both non-dispatch and dispatch simulations. Again, the more distant shovel accumulated more idle time than the one nearer the dispatch point.

Reduction of equipment idle times was achieved with the introduction of dispatching in all cases studied. However, the

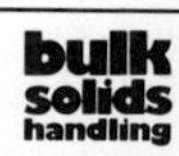

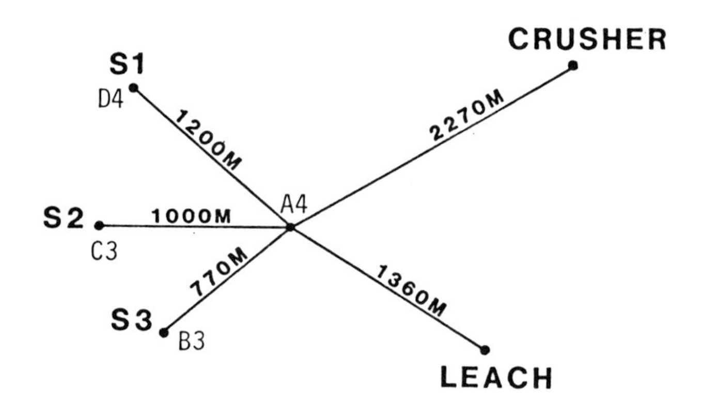

CRUSHER
S1
D4
1200M
2270M
S2
1000M
A4
C3
770M
1360M
S3
B3
LEACH

Figure 3a.

CRUSHER
S1
1200M
2270M
S2
1000M
A4
190M
770M
DUMP
S3
B3

Figure 3b.

CRUSHER
S1
D4
1200M
2270M
A4
S2
1000M
C3
190M
820M
S4
DUMP
E2

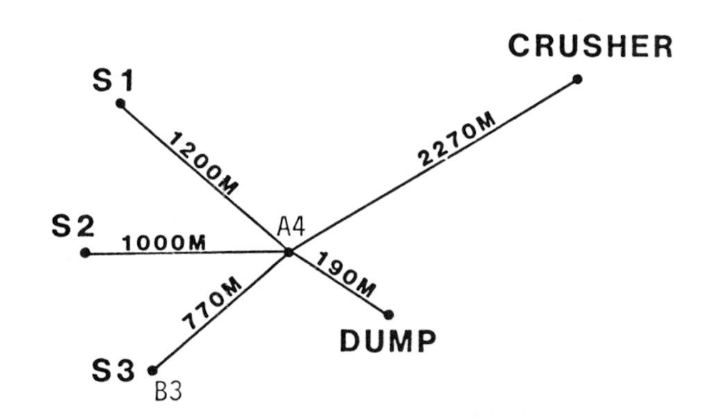

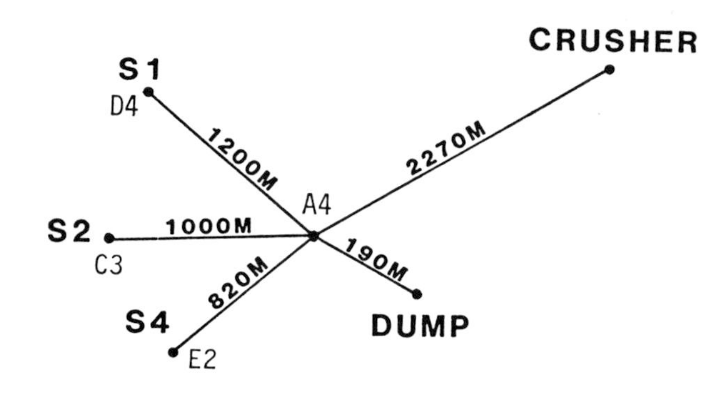

Figure 3c.

S1
D4
1200M
CRUSHER
S2
1000M
A4
2270M
C3
820M
S4
E2

Figure 3d.

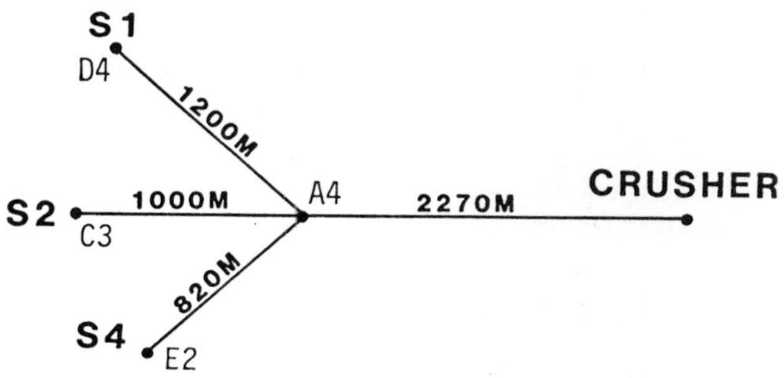

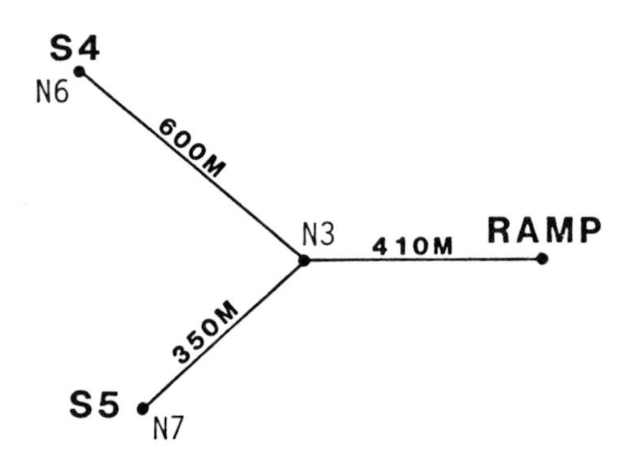

S4
N6
600M
N3
410M
RAMP
350M
S5
N7

Figure 3e.

S3
B3
180M
RAMP
N3
410M
350M
S5
N7

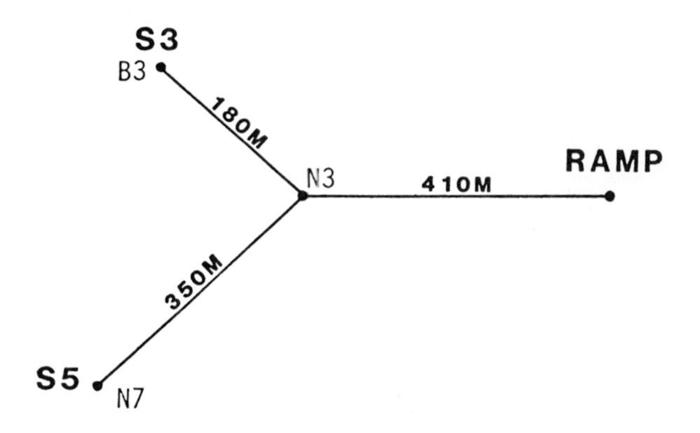

Figure 3f.

Fig. 3: Pit configuration of cases studied

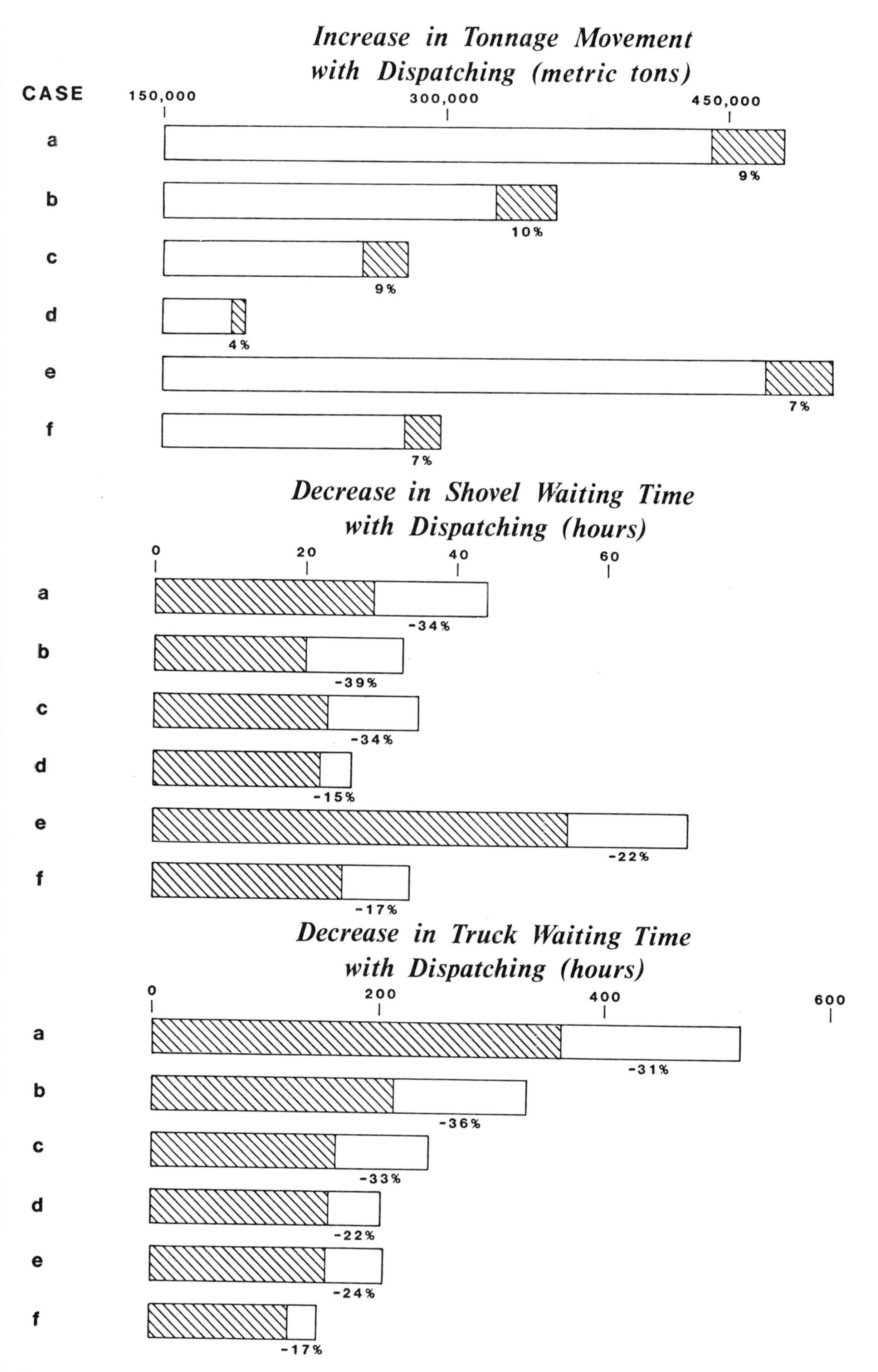

Fig. 4: Summary of simulation results for the 6 cases investigated

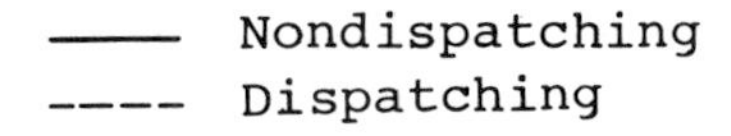

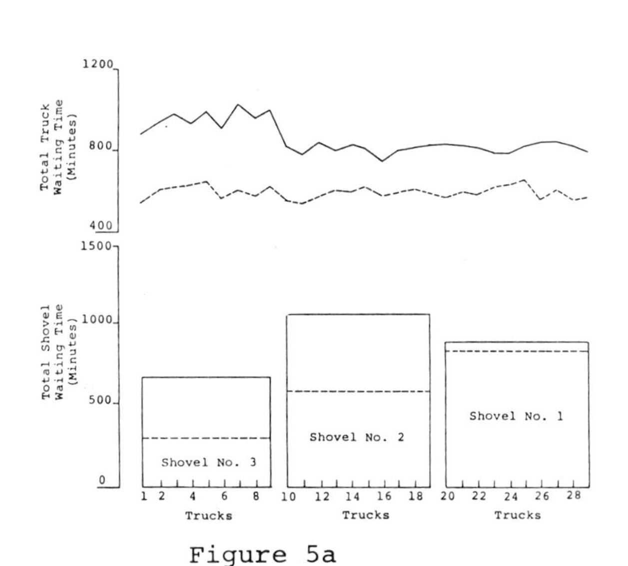

Figure 5a
(For Pit Configuration of Figure 3a)

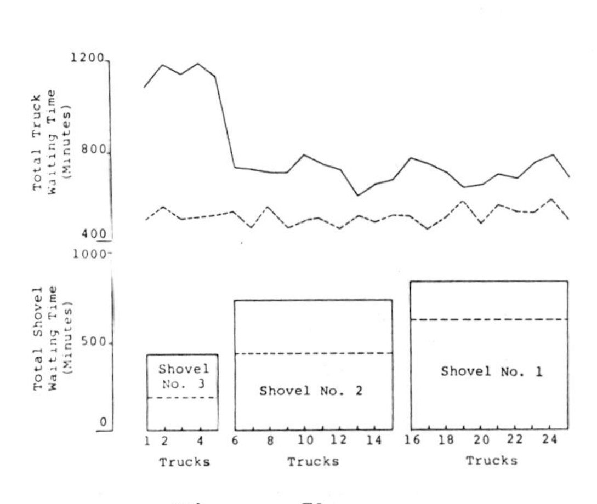

Figure 5b
(For Pit Configuration of Figure 3b)

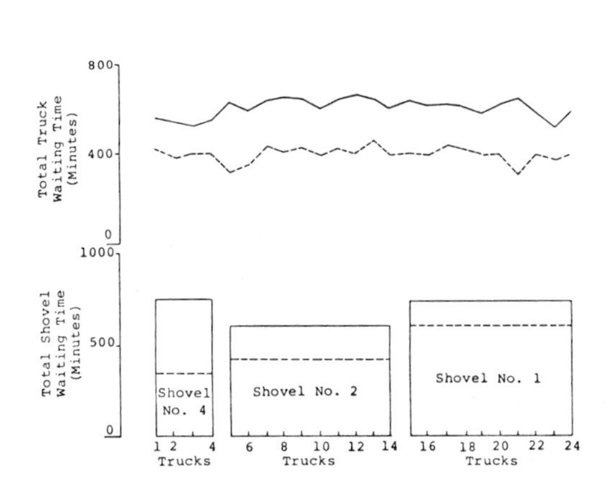

Figure 5c
(For Pit Configuration of Figure 3c)

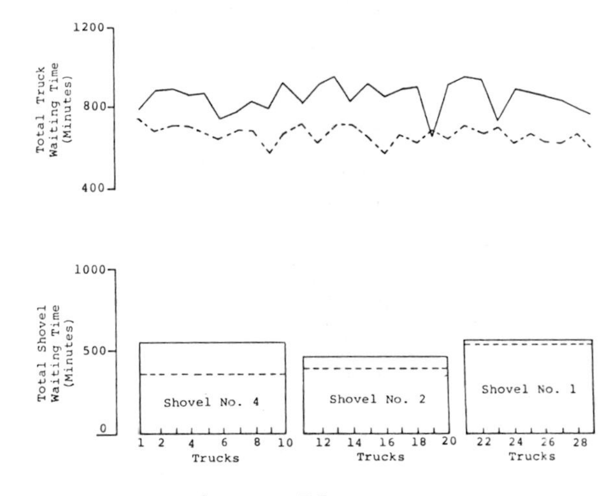

Figure 5d
(For Pit Configuration of Figure 3d)

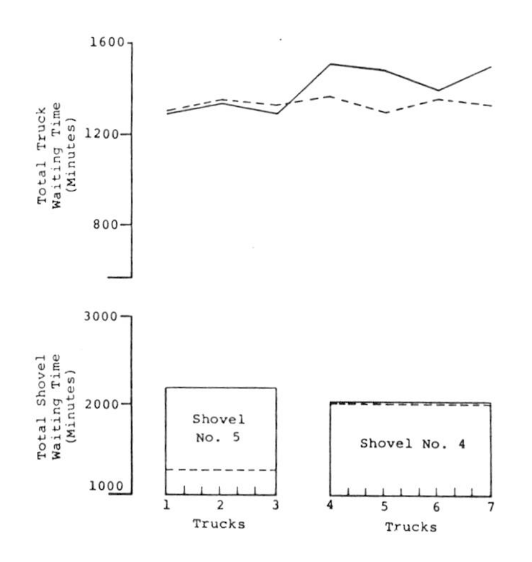

Figure 5e
(For Pit Configuration of Figure 3e)

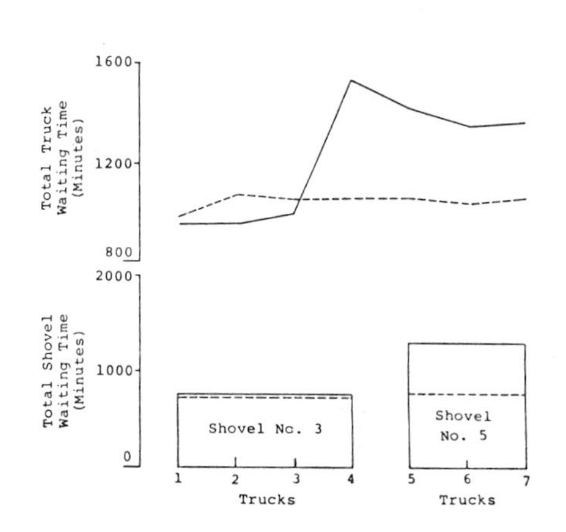

Figure 5f
(For Pit Configuration of Figure 3f)

Fig. 5: Equipment waiting times under nondispatching and dispatching

RESULTS OF SIMULATION RUN 1

```
ORE PRODUCTION DURING SIMULATION (TONS)           =   285950.
WASTE PRODUCTION DURING SIMULATION (TONS)         =   154343.
TOTAL SHOVEL WAITING TIME (MIN.)                  =     2635.
TOTAL TRUCK WAITING TIME (MIN.)                   =    31214.
TOTAL TRUCK QUEUING TIME ON HAUL ROAD             =        0.
TOTAL NO. OF TRUCKS LOADED                        =     4351.
```

INDIVIDUAL SHOVEL STATISTICS:

SHOVEL	TOTAL ORE (TONS)	DAILY ORE (TONS)	TOTAL WASTE (TONS)	DAILY WASTE (TONS)	TOTAL ON TM (MIN)	DAILY ON TM (MIN)	TOTAL WT TM (MIN)	DAILY WT TM (MIN)
1	145820.	24303.	0.	0.	926.3	154.	911.	152.
2	140130.	23355.	0.	0.	930.4	155.	1059.	177.
3	0.	0.	154343.	25724.	931.8	155.	664.	111.

INDIVIDUAL TRUCK STATISTICS:

TRUCK	TOTAL ORE (TONS)	DAILY ORE (TONS)	TOTAL WASTE (TONS)	DAILY WASTE (TONS)	TOTAL ON TM (MIN)	DAILY ON TM (MIN)	TOTAL WT TM (MIN)	DAILY WT TM (MIN)
301	13364.	2227.	0.	0.	1340.3	223.	983.	164.
302	13274.	2212.	0.	0.	1331.9	222.	1023.	170.
303	13126.	2188.	0.	0.	1319.8	220.	1047.	175.
304	13142.	2190.	0.	0.	1300.7	217.	1071.	178.
305	13516.	2253.	0.	0.	1281.4	214.	1041.	174.
306	13420.	2237.	0.	0.	1299.0	217.	989.	165.
307	13311.	2218.	0.	0.	1361.2	227.	994.	166.
308	13327.	2221.	0.	0.	1333.2	222.	1019.	170.
309	13218.	2203.	0.	0.	1343.9	224.	1041.	173.
310	13047.	2174.	0.	0.	1355.9	226.	1042.	174.
311	13982.	2330.	0.	0.	1313.8	219.	1043.	174.
312	13912.	2319.	0.	0.	1331.8	222.	1017.	169.
313	13914.	2319.	0.	0.	1377.5	230.	1003.	167.
314	14152.	2359.	0.	0.	1376.9	229.	938.	156.
315	13997.	2333.	0.	0.	1337.8	223.	1010.	168.
316	14023.	2337.	0.	0.	1332.8	222.	1045.	174.
317	13991.	2332.	0.	0.	1347.0	225.	1006.	168.
318	13972.	2329.	0.	0.	1308.7	218.	1056.	176.
319	14259.	2377.	0.	0.	1320.5	220.	985.	164.
320	13927.	2321.	0.	0.	1330.8	222.	1039.	173.
321	0.	0.	19080.	3180.	1344.1	224.	1268.	211.
322	0.	0.	19525.	3254.	1335.4	223.	1193.	199.
323	0.	0.	19085.	3181.	1331.0	222.	1277.	213.
324	0.	0.	19477.	3246.	1369.7	228.	1154.	192.
325	0.	0.	19347.	3224.	1304.3	217.	1249.	208.
326	0.	0.	19712.	3285.	1331.0	222.	1177.	196.
327	0.	0.	18939.	3156.	1380.2	230.	1230.	205.
328	0.	0.	19179.	3196.	1371.5	229.	1175.	196.
329	13075.	2179.	0.	0.	1332.7	222.	1099.	183.

Table 1: Non dispatching cases computer printout

RESULTS OF SIMULATION RUN 1

ORE PRODUCTION DURING SIMULATION (TONS)	=	308144.
WASTE PRODUCTION DURING SIMULATION (TONS)	=	171307.
TOTAL SHOVEL WAITING TIME (MIN.)	=	1724.
TOTAL TRUCK WAITING TIME (MIN.)	=	21662.
TOTAL TRUCK QUEUING TIME ON HAUL ROAD	=	0.
TOTAL NO. OF TRUCKS LOADED	=	4732.

INDIVIDUAL SHOVEL STATISTICS:

SHOVEL	TOTAL ORE (TONS)	DAILY ORE (TONS)	TOTAL WASTE (TONS)	DAILY WASTE (TONS)	TOTAL DN TM (MIN)	DAILY DN TM (MIN)	TOTAL WT TM (MIN)	DAILY WT TM (MIN)
1	148533.	24755.	0.	0.	929.0	155.	860.	143.
2	159611.	26602.	0.	0.	882.2	147.	604.	101.
3	0.	0.	171307.	28551.	934.9	156.	260.	43.

INDIVIDUAL TRUCK STATISTICS:

TRUCK	TOTAL ORE (TONS)	DAILY ORE (TONS)	TOTAL WASTE (TONS)	DAILY WASTE (TONS)	TOTAL DN TM (MIN)	DAILY DN TM (MIN)	TOTAL WT TM (MIN)	DAILY WT TM (MIN)
301	11568.	1928.	4926.	821.	1345.9	224.	720.	120.
302	10867.	1811.	5876.	979.	1354.8	226.	703.	117.
303	9867.	1644.	6737.	1123.	1352.3	225.	761.	127.
304	10402.	1734.	6013.	1002.	1360.0	227.	706.	118.
305	10714.	1786.	5318.	886.	1325.4	221.	834.	139.
306	10512.	1752.	5947.	991.	1343.3	224.	807.	135.
307	10161.	1693.	6497.	1083.	1336.7	223.	777.	130.
308	11295.	1882.	5104.	851.	1326.6	221.	740.	123.
309	11100.	1850.	5167.	861.	1356.4	226.	746.	124.
310	11285.	1881.	5001.	833.	1363.8	227.	721.	120.
311	10830.	1805.	5725.	954.	1328.8	221.	752.	125.
312	10039.	1673.	6886.	1148.	1347.1	225.	768.	128.
313	9863.	1644.	6954.	1159.	1345.5	224.	751.	125.
314	10716.	1786.	6063.	1011.	1387.8	231.	676.	113.
315	10615.	1769.	5721.	954.	1331.9	222.	782.	130.
316	10514.	1752.	5704.	951.	1389.7	232.	759.	127.
317	9758.	1626.	6985.	1164.	1330.6	222.	767.	128.
318	10651.	1775.	5816.	969.	1348.7	225.	730.	122.
319	11110.	1852.	5788.	965.	1340.3	223.	691.	115.
320	10243.	1707.	7068.	1178.	1342.0	224.	691.	115.
321	9741.	1623.	6937.	1156.	1364.6	227.	777.	129.
322	10123.	1687.	6630.	1105.	1341.5	224.	723.	121.
323	10783.	1797.	5867.	978.	1323.9	221.	759.	126.
324	11337.	1890.	4848.	808.	1365.8	228.	722.	120.
325	10083.	1681.	6437.	1073.	1351.9	225.	821.	137.
326	10521.	1754.	6195.	1032.	1317.2	220.	775.	129.
327	11130.	1855.	4746.	791.	1342.9	224.	781.	130.
328	10606.	1768.	5486.	914.	1370.9	228.	750.	125.
329	11712.	1952.	4865.	811.	1343.8	224.	673.	112.

Table 2: Dispatching case computer printout

extent of reduction also varied with the particular pit and equipment configuration. If the shovels presented undertrucking or overtrucking conditions, dispatching balanced the truck-to-shovel ratio by evenly distributing idle time between the pieces of equipment. In addition, dispatching controlled variability of truck waiting times in some of the cases studied. The shovels farthest from the dispatch point accumulated more idle time than the nearer ones. This result can be explained by the differences in travel times. If a truck is dispatched to a nearby shovel and another to a more distant shovel at the same time, the nearer shovel will accumulate less idle time than the far one.

6. Conclusions

This study has shown that a truck dispatching system gives greater productivity than a non-dispatching system for this particular mining operation. The simulation results clearly show about a 10% improvement in truck-shovel productivity with dispatching. A 10% gain in productivity in this mining operation is equivalent to half the production of eight 109t trucks and one 16yd³ shovel under the present non-dispatching system or about 200,000t/month.

Analysis of the experiment results showed that combinations of long and short hauls are the most favorable to improvement with dispatching. Most open pit mining operations have combinations of long and short hauls, so they should find it advantageous to adopt a truck dispatching system for their haulage systems. On the other hand, the results revealed that if the hauls are of equal length, dispatching still produces substantial improvement in productivity over non-dispatching. Therefore, operations with these conditions, also, may find it advantageous to implement such system. In addition, the results show that use of dispatching reduces waiting time variability for the mining system under study. Most mining operations recognize that variability in waiting time reduces production. Variability is a natural consequence of complex systems and/or a consequence of inadequate planning, and can be reduced by dispatching.

In any open pit mine, the haulage operation will probably become more and more complicated year after year because of longer hauls and acquisition of more equipment. Therefore, it is likely that dispatching systems would have a greater impact on productivity later in the life of a mine.

Acknowledgments

The authors would like to thank the management of Mexicana de Cobre, La Caridad Mine, for permission to publish this paper and the Department of Mining and Geological Engineering, The University of Arizona, in Tucson, Arizona, for the availability of its Mine Operations Analysis Model.

References

[1] Baron, J., 1977, "Automatic Truck Dispatching. Lake Jeannine Operations", CIM First Open Pit Operators Conference Paper No. 1, 9pp.

[2] Beaudoin, R., 1977, "Automatic Truck Dispatching. Mount Wright Operations", CIM First Open Pit Operators Conference Paper No. 2, 12pp.

[3] Crosson, C.C., Tonking, M.J.H., and Moffat, W.G., 1977, "Palabora System of Truck Control", Mining Magazine, Vo. 1, No. 2, Feb., pp 74—82.

[4] Hobday, P.J., 1978, "The Computer in Their Hands", a paper presented at the Australian Computer Society Conference, Aug.

[5] Ibarra, M.A., 1977, "Despacho de Camiones", Asociacion de Ingenieros de Minas, Metalurgistas y Geologos de Mexico. Mem. Tec. XII Conv., Oct., 24pp.

[6] Kim, Y.C. and Dixon, W.C., 1977, "Mine Operations and Financial Analysis Models for Surface Mining", proprietary report prepared for the Kemmerer Coal Company, Frontier, Wyoming, Department of Mining and Geological Engineering, The University of Arizona, Tucson, Arizona.

[7] Michaelson, S.D., 1974, "Wanted: New System for Surface Mining", Engineering and Mining Journal, Vol. 175, No. 10, Oct., pp. 63—69.

[8] Mueller, E.R., 1977, "Simplified Dispatching Board Boosts Truck Productivity at Cyprus Pima", Mining Enginnering, August, pp. 40—43.

[9] Naplatanov, N.D., Sgurov, V.S., Petrov, P.A., Nicolov, Z.A., Ganchev, A.I., Trendafilov, S.P., 1976, "Method of and System for Rationalizing the Operations of Open-Pit Mines", U.S. Patent, 3,979,731, Sept.

[10] Naplatanov, N.D., Sgurev, V.S., and Petrov, P.A., 1977, "Truck Control at Medet", Mining Magazine, Vol. 2, No. 8, July, pp. 12—18.

[11] Pritsker, A.A.B., 1974, "The GASP IV Simulation Language", John Wiley and Sons, ed., New York, 451 pp.

[12] Schlosser, R.B., 1975, "Use and Vision of Future Uses of Haulage Truck Location System" 10th Annual Symposium Instrument Society of America, Duluth, Minnesota.

Volume 1, Number 1, February 1981

A Computer Simulation Model for a Surface Coal Mine

S. Bandopadhyay,
R. V. Ramani and C. B. Manula, USA

Ein Computer-Simulationsmodell für einen Steinkohlen-Tagebau
Modèle de simulation sur ordinateur pour une mine de charbon à ciel ouvert
Modelo de simulación para una mina de carbón a cielo abierto
露天掘炭鉱用コンピュータシミュレーションモデル
一个露天煤矿的计算机模拟模型
نماذج الاستقصاء بالحاسب الالكتروني لمناجم الفحم السطحية

Ein Computer-Simulationsmodell für einen Steinkohlen-Tagebau

Die Einsatzmöglichkeiten eines Simulators für ein Tagebau-Fördersystem zwecks Analyse von Abbau- und Terminplanung durch Erstellen und Auswerten von Alternativlösungen wird beschrieben. Im Tandem-Betrieb werden Abraum durch einen Schaufelradbagger und Kohle durch einen Löffelbagger gewonnen. Anhand des Modells ließen sich wertvolle Aussagen über mögliche Produktivitätssteigerungen durch den Einsatz weiterer Geräte und Verbesserung bereits im Einsatz befindlicher Maschinen machen.

Modèle de simulation sur ordinateur pour une mine de charbon à ciel ouvert

Application d'un simulateur de manutention de matériaux dans une mine à ciel ouvert pour l'analyse des problèmes de programmation et de planification à travers son développement et évaluation des autres possibilités. La surcharge est supprimée en utilisant ensemble un excavateur à roue à godets et une pelle excavatrice.

Modelo de simulación para una mina de carbón a cielo abierto

Se explica la aplicación de un simulador des sistemas de manutención para una mina de carbón a cielo abierto destinado a analizar los problemas de planificación y programación de la mina mediante la generación y la evaluación de diversas soluciones posibles. El descombrado se hizo con una excavadora de rueda de cangilones y una pala mecánica trabajando en tándem.

Summary

This paper describes the application of the open pit materials handling simulator (OPMHS) to a surface coal mine in Illinois. The mining method practiced is classified under the general heading *Area Mining*. The overburden is removed by a bucket wheel excavator (BWE) and a stripping shovel operating in tandem. The specific objective of the study was to demonstrate the application of the simulator to analyse mine planning and scheduling problems through the generation and evaluation of alternatives on the basis of simulated results.

On the basis of the simulation of the existing system, it is concluded that on the average, the system can perform to within 85% of the designed capacity. The performance of the stripping shovel was identified as the bottleneck in the system. However, the production can be increased to over 95% of the designed capacity by increasing the availability of the stripping shovel from 71% to 80%. Also, a number of plans to increase the prodution capacity of the mine by the introduction of new equipment was analysed.

S. Bandopadhyay, Graduate Student in Mining Engineering; Prof. Dr. R.V. Ramani; Prof. Dr. C.B. Manula; Department of Mineral Engineering, 104 Mineral Sciences Building, The Pennsylvania State University, University Park, PA 16802, USA.

1. Introduction

Coal production from surface mining has been on the increase ever since 1960; during the last 10 years coal production [1] from surface mining has grown enormously (Table 1). In fact, since 1974 the production from surface mining has exceeded that from underground mines. The US Federal Energy Administration [2] has predicted that surface mined coal would play an even more important role in meeting the projected production increase for the coal segment of the energy market (Table 2). Large production requirements from surface mines necessitate deployment of many large pieces of equipment. This equipment must be utilized efficiently by maintaining production at or near designed capacities. Basically, the problem is to select, size and schedule equipment to maximize production and minimize adverse environmental impacts [3].

The time, manpower, and cost limitations required to analyze a wide variety of situations and alternatives in planning designing, and scheduling equipment and methods for surface mines have been overcome with the application of simulation methods using a digital computer. Since the digital computer can be programmed to simulate the situations which are to be analyzed, many alternatives can be evaluated in a relatively short time. Input variables to the computer model (or simulator) are easily changed and the resulting changes in the system performance can be evaluated. When standard methods of analysis are applied, weeks and many man-hours are involved in analyzing a single design for the mining operations. Even then, many simplifying assumptions must be made since these operations are dynamic and transient. Hence, to make sound engineering decisions and to take proper and timely corrective actions, simulation methods are the only recourse.

This paper is concerned with the application of the Open Pit Materials Handling Simulator (OPMHS) developed by the Mining Engineering Section, The Pennsylvania State University.

2. The Open Pit Materials Handling Simulator (OPMHS)

A generalized flow diagram of the OPMHS model is shown in Fig. 1. The simulator consists of a number of interrelated sub-assemblies which represent various unit operations of a complex mining system. These include a BWE sub-assem-

Table 1:
Bituminous and lignite coal production in the USA (National Coal Association, 1979)

Year	U.S. Total Coal Production (tonne)	Coal Production by Surface Mining Method (tonne)	Coal Production by Underground Method (tonne)	Percentage Distribution Surface	Percentage Distribution Underground
1968	494,638,000	182,521,390	318,811,550	36.9	63.1
1969	508,482,000	193,731,470	314,750,080	38.1	61.9
1970	546,971,000	239,573,150	307,397,520	43.8	56.2
1971	500,940,000	250,470,070	250,470,070	50.0	50.0
1972	540,125,000	263,213,960	276,003,910	48.9	51.1
1973	527,744,000	265,723,750	274,401,320	49.5	50.5
1974	547,401,000	297,785,970	249,614,710	54.4	45.6
1975	588,253,000	322,362,660	265,890,370	54.8	45.2
1976	603,278,000	337,835,580	265,442,240	56.0	44.0
1977	627,177,000	386,400,810	241,265,770	61.6	38.4
1978	603,393,000	383,693,770	219,699,270	63.6	36.4

Table 2:
New mine requirements (1975—1990) [1]
(after Federal Energy Administration, 1974)

	Business as Usual[3]	Accelerated Development[2]
Underground Mines:		
1 million tons	153	445
3 million tons	74	190
Surface Mines:		
1 million tons	110	195
3 million tons	25	90
5 million tons[2]	98	219
Total	460	1139

1 Including new mines to replace depleted productive capacity and new mines to increase existing productive capacity.

2 Although there are new 10-million ton surface mines in the West, and others are on the drawing board, for the purposes of this report nothing larger than a 5-million ton mine was considered. Checks with western surface mine operators indicate that the economy of scale is such that the cost of producing coal at a 10-million ton mine was considered the equivalent of two 5-million ton mines for the purpose of determining minimum selling prices, man-power requirements, equipment and supply requirements, etc.

3 Production targets for two cases.

1 short ton = 0.907 tonne

bly, shovel and dragline sub assembly, truck haulage sub-assembly and conveyor sub-assembly and the train sub-assembly. OPMHS simulates the total material handling of a mine operation and furnishes production and performance data for each system sub-assembly. For example, the basic aspects of the system that has been modeled may be brought into focus by referring to Fig. 2, a schematic diagram of a typical surface mining operation. A multi-stage materials handling scheme is employed to mine and move materials from multiple origins (Pits 1 and 2) to several destinations (bins, waste disposal sites, etc). The system is a complex, dynamic network with several interconnected networks. Consequently, in OPMHS all decision points are defined and coordinated such that the system status (data, information, frequency of operations, etc.) is updated for each time interval Δt. The length of the interval, Δt, and the total simulation time, T, are defined by the user at the beginning of simulation. The system is interrogated and updated from 0 to T in interval of Δt. For example, during any time period (t) to ($t + \Delta t$) the system status at (t) is updated by the information generated during the interval Δt to provide the system status at ($t + \Delta t$). Thus, the information generated at a time interval Δt affects all subsequent decisions. Each of the sub-assemblies may or may not be active in any particular simulation since not every surface mine would have all of these methods for materials handling. Only those sub-assemblies which are required can be used. Besides, the sub-assemblies are used in the particular sequence the user defines. One major advantage is that once the basic mining plan and sequences have been selected, rapid evaluation of various projected mining configurations can be made. This information can then be used for equipment selection, operational procedures, and monitoring and control of the selected mining practice. Thus responsive management information is provided as well as sound engineering decisions made as illustrated in the following application [5].

2.1 Application to an Illinois Mine

As is typical throughout the central coal basin of the USA, the topography of the area where the selected mine is located is generally flat. The strata overlying coal seam (Illinois No. 6) consists of clay, sand and gravel, medium hard shales, sandstone and some limestone. A stratigraphic column of the overburden and coal seam at the selected mine is shown in Fig. 3. The mining method praticed at this location can be classified as *area mining.* The operation consists of a single pit extending east to west with highwall to the north. The pit which is approximately 1920.24 m long, varies in width from 39.62 m to 45.72 m, with coal loading restricted to a width of 18.29 to 24.38 m. The mine has a current annual production of 1.09 million tonne of coal. The major stripping equipment includes a German BWE and a 61.16 m³ stripping shovel. The BWE and the stripping shovel operate from the top of coal. Fig. 4 shows a section view of the cut. Exposed coal is loaded into 90.71 tonne bottom dump coal hauler by a 5.35 m³ loading shovel. For overburden removal the BWE and the stripping shovel are scheduled for 24 hours a day, seven days a week and 364 days a year. The coal loading shovel, however, is scheduled for two shifts a day, five days a week, and 240 days a year. At the mine, the BWE and the stripping shovel take varying heights of overburden. The height of the BWE bench can vary from 4.87 m to 6.09 m. The shovel bench can vary from 12.19 m to 21.33 m. In effect, the required ratio between the shovel and BWE production can vary from 2.0:1 to 4.4:1. In 1976, for example, the overall stripping ratio

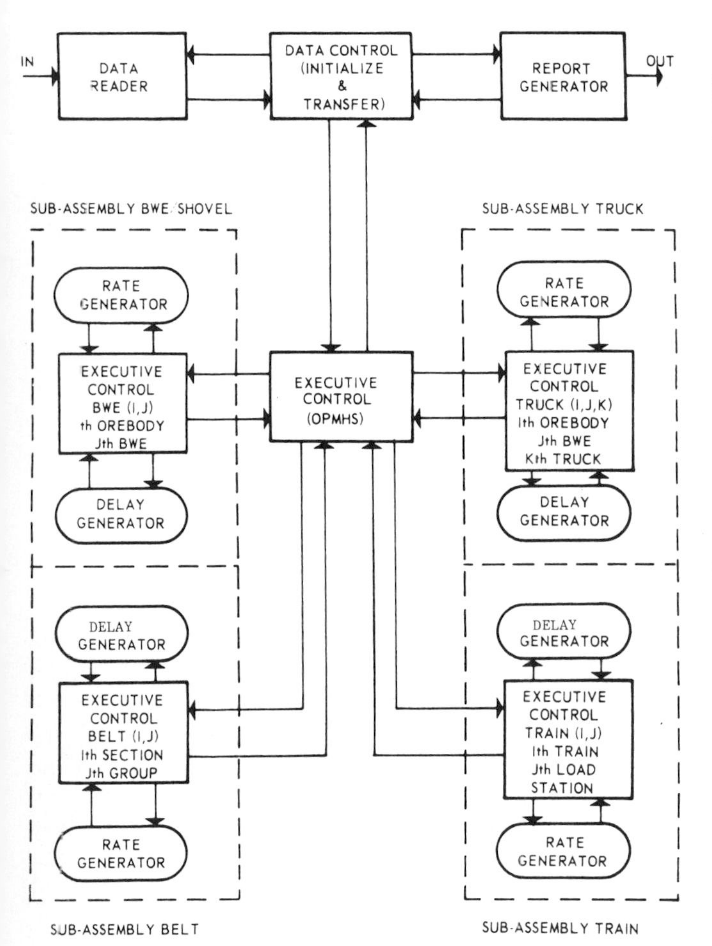

Fig. 1: Generalized flow model of OPMHS

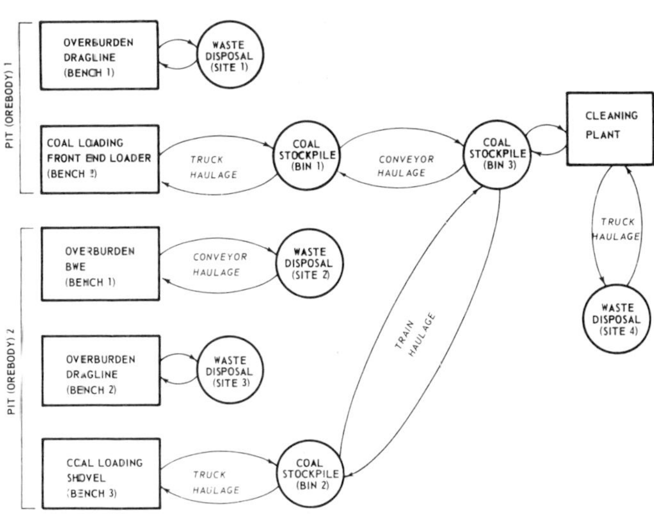

Fig. 2: Conceptual operating system

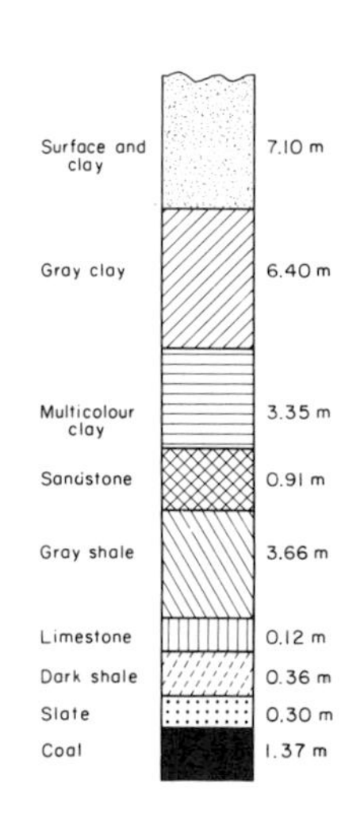

Fig. 3: Stratigraphic column

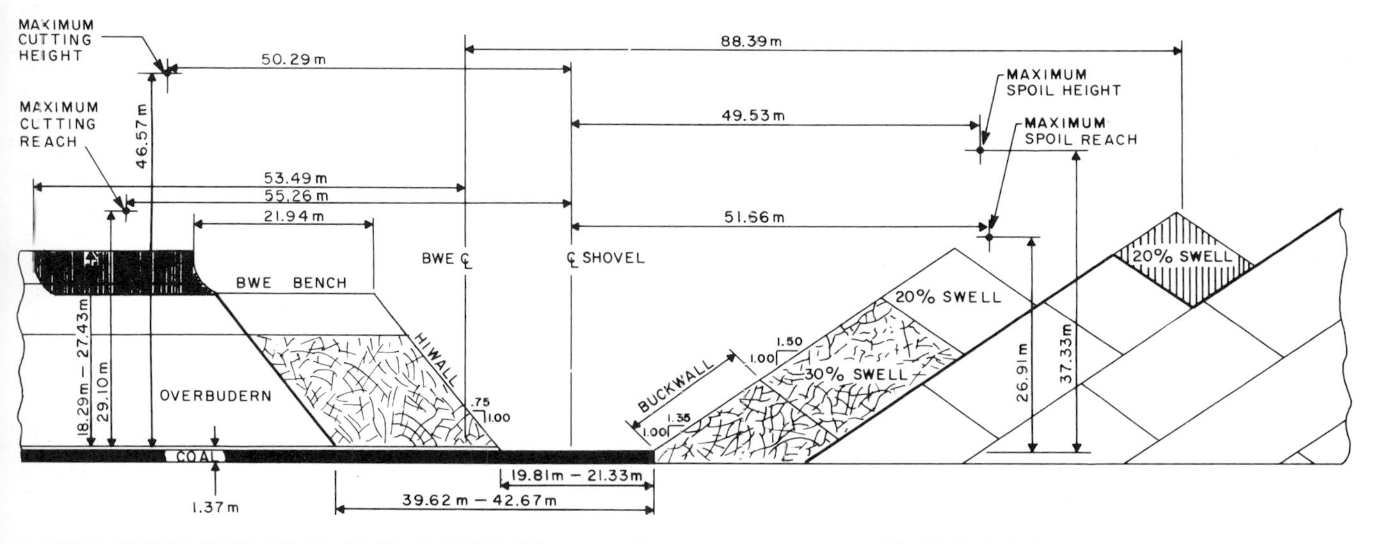

Fig. 4: A section view of the cut

(m³/tonne) was 17.22, and the match ratio was 4.4:1. The stripping shovel, however, can take overburden up to 21.33 m and helps out the BWE where the total overburden is less. In view of the great variability in the match ratio and in the operations, the analysis chosen for the mine is based on the following average conditions:

— Average overburden thickness	21.33 m
— Average coal thickness	1.37 m
— Height of BWE bench	6.09 m
— Match ratio	2.5:1
— Annual coal production	1.09 million tonne

The stripping ratio, computed directly from drill hole maps for an overburden thickness of 21.33 m is 12.14 m³/tonne. Since the coal has a reject of 20%, the stripping ratio for clean coal, assuming 100% recovery is 15.17. The average stripping ratios for raw and clean coal at the time of the mine visit were 12.47 and 15.59 respectively.

On the basis of the above planned annual clean coal production, the average raw coal production per shift is:

$$\frac{(1{,}090{,}000)}{(0.8)\ (240)\ (2)} = 2{,}835 \text{ tonne/shift}$$

On the basis of 21 shifts of overburden removal per week, 10 shifts of coal removal per week and an average stripping ratio of 12.14 for the required raw coal production of 2835 tonne/shift the overburden volume to be removed each shift is:

$$(2835)\ (12.14)\ \frac{(10)}{(21)} = 16{,}389 \text{ m}^3$$

On the basis of the match ratio of 2.5:1, the stripping shovel should remove each shift 11,706 m³ and the BWE, 4,683 m³.

The mine haul roads are generally flat and appeared to be in good condition. Two areas having significant grade are at the pit incline (6%) and at the ramp to the dump (5%). The average one way haul distance is 6,678.78 m. Coal preparation plant refuse is trucked to previous cuts or haulage inclines in the mined area by 58.96 tonne end dump trucks. The one-way haul distance is approximately 2,414.01 m. Fig. 5 is a flow diagram of the material handling system.

3. Computer Simulation

The following sub-assemblies of the OPMHS were used in this application:

1. BWE sub-assembly
2. Shovel sub-assembly
   a) Overburden removal
   b) Coal removal

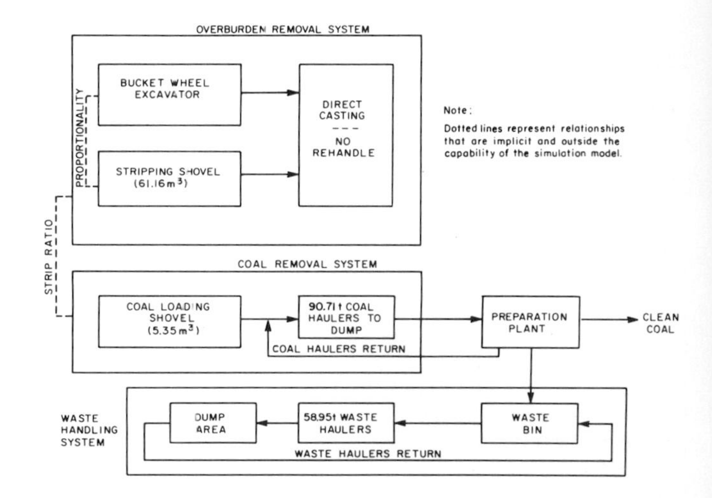

Fig. 5: Flow diagram of the materials handling system

3. Truck sub-assembly
   a) Coal transport
   b) Waste transport

Data for the application was obtained through time studies, discussions with mine personnel, from mining engineering records, and from equipment manufacturers' catalogs. The input requirements for operation of the BWE sub-assembly are shown in Tables 3 and 4. Table 5 shows the input requirements for the shovel sub-assembly. The input requirements for the truck sub-assembly (for coal and waste transport) are given in Table 6. Additional input information required for the truck sub-assembly is the profile of haulroads and truck speed-rimpull information which were obtained from the mine maps and manufacturer's catalog, respectively.

An initial simulation of the existing system (base case) was made to validate the simulation model. In each computer run the operation was simulated for ten shifts but for the purpose of the analysis, the results were averaged per shift. Table 7 shows a production summary of the output from this run. The total overburden removed is 14,332 m³/shift which is less than the required production of 16,389 m³ by 2,057 m³. Of this amount, the tripping shovel removes 9,873 m³, and the BWE removes 4,459 m³. The simulated results indicate that in the overburden removal system, the stripping shovel and BWE productions are fairly matched. However, the match ratio 2.21 indicates that in the longrun, the BWE will wait on the shovel.

The required overburden removal for a coal production of 2,868 tonne/shift at a strip ratio of 12.14 is 16,580 m³. However, the actual amount of overburden removed is only 14,332 m³. The effect of this imbalance between the overburden and coal removal systems can be seen in the decreased strip ratios. In the long run, the coal loading

Table 3: Soil characteristics of top 6.09 m of overburden

Soil Type	Specific Cutting Resistance kg/cm	Boulder Occurrence Frequency	Wheel Speed Fraction of Maximum Speed	Bucket Fill Factor Fraction of Theoretical Capacity	Swell Factor (Bank Volume/ Loose Volume)
Topsoil and Clay	Range 28—42 Mean 35	0	0.40	0.95	0.80

Table 4:
Input data for BWE subassembly

Description	
Number of soil types in the system	1
Soil Type	1
Bucket filling capacity of the soil	0.95
Cutting resistance of the soil, kg/cm	35
Ratio of allowable cutting speed to maximum cutting speed	0.40
Swell factor of the soil	0.80
Number of observation towers for trucks	0
Number of BWEs, shovels or draglines in the orebody	3
Wheel diameter of the BWE(m)	9.2
Crawler speed of BWE(m/s)	0.073
Length of the wheel boom for BWE(m)	48.31
Capacity of the bucket of BWE (m^3)	0.99
Number of buckets on wheel of BWE	8
Maximum advance of BWE before next cut(m)	18.28
Weight of the superstructure of BWE(kg)	249,480
Total weight of the BWE(kg)	2,169,115
Crowding BWE (Non-crowding BWE = 0)	1
Frictional resistance at the ball race for BWE(%)	3
Radius of the ball race for BWE(m)	6.0
Average stop time for boulder hit for BWE(s)	0
Mean maneuvering time for BWE(s)	600
Maximum slewing angle to left of line of advance (radians)	0.52
Maximum slewing angle to right of line of advance (radians)	0.32
Slope of bench for BWE	0
Mechanical availability	0.46
Continuous run time for BWE(s)	600
Initial material being cut (1 = ore, 2 = waste)	2
Surge bin capacity of BWE(tonne)	272,155 ore 272,155 waste
Number of benches to be cut by BWE	1
Height of bench in orebody for BWE(m)	6.09 for left side 6.09 for right side
Probability of not hitting a boulder in ore body	1
Material density (kg/m^3)	2.40
Ore to waste ratio	0
Floor rolling resistance(%)	2
Floor slope encountered by BWE(%)	0
Tail bin number	1

system will wait on the overburden removal system. Since the coal shovel works for over 93 % of the time, the system is also reaching a limiting capacity with regards to coal loading.

At the time of the mine visit, coal was loaded only three days a week, two shifts/day. Unless the demand for coal increases to the extent that the mine starts loading coal 10 shifts a week and at a rate 2,835 tonne/shift, the imbalance between the BWE and shovel productions, and the overburden and coal removal systems should not pose major problems. However, on the basis of the above analysis, assuming that demand for coal will increase to the planned capacity, the recommended short term plan called for the increase in the production from the shovel so that the shovel and BWE productions are better matched, and the BWE does not wait on shovel.

Table 5:
Input data for shovel subassembly (for stripping and coal)

	Stripping Shovel	Coal Shovel
Diameter of the wheel(m)	0.0	0.0
Mean cycle time(s)	60.0	30.0
Deviation from cycle time(s)	10.0	2.0
Maximum cycle time(s)	72.0	32.0
Minimum cycle time(s)	53.0	28.0
Bucket capacity(tonne)	93.44	4.71
Deviation of bucket capacity(tonne)	4.53	0.0
Maximum bucket capacity(tonne)	97.97	4.71
Minimum bucket capacity(tonne)	88.90	4.71
Ore to waste ratio	0	1
Mechanical availability	0.71	0.95
Continuous run time(s)	3000.0	3000
Capacity of surge bin (ore),(tonne)	272,155	0.0
Capacity of surge bin (waste),(tonne)	272,155	0.0
Tail bin number for face excavators:	2	3

1 short ton = 0,907 tonne

Table 6:
Truck data

Truck Type	Type 1[1]	Type 2[2]
Mean payload(tonne)	58.96	90.71
Empty weight tonne)	36.87	54.43
Mean dump time(s)	10.0	10.0
Mean maneuvering time(s)	15.0	10.0
Maximum acceleration rate(m/s^2)	0.15	0.15
Mechanical availability	0.9	0.9
Continuous run time(s)	3000	3000

1 WABCO 65A
2 Athey Model PH660

In simulation run 2 all input data used in the base case were held constant except the stripping shovel's availability which was increased to 80 %. The results are abstracted in Table 8.

The only way to increase the availability of a machine is to make a detailed analysis of the machine application to identify areas for potential improvements. This entails collection of records for the machine activities over extended periods of time. The operating times and the delay times should be broken down into specific independent categories and data collected on each category. These categorized records can then be evaluated for possible improvements. Even small changes in the mode of operation can increase the availability and performance of the machine. For example, deadheading is usually a major source of operational delays. The deadheading operation is greatly dependent on the pit layout and mine dimensions. Advance planning and preparation can reduce the total overall maintenance delays and therefore increase the machine availability.

From Table 8, it can be seen that in the overburden removal system, the performances of shovel and BWE are fairly matched. If the coal shovel loads 2,836 tonne, it will catch up with the overburden removal system since more coal is loaded than is being uncovered as indicated by the 12.14 stripping ratio. Therefore, the overburden removal capacity for the system must be increased. Since the stripping shovel is operating at its maximum availability, an increase in the shovel production is not possible. In the longrun, increases in BWE production will not increase total overburden

Table 7: Production summary for simulation run 1

	Availability			Average Overburden Depth (m)	Average Raw Coal Production (tonne)	Strip Ratio[2] (m³/tonne)		Overburden Removal			Match Ratio[1]
	BWE	Stripping Shovel	Coal Shovel			Raw Coal	Clean Coal	BWE (m^3)	Stripping Shovel (m^3)	Total (m^3)	
Actual data averaged for a shift	0.46	0.71	0.95	21.33	2835	12.14	15.17	4679	11705	16384	2.5
Simulation run 1 (base case)	0.46	0.71	0.95	21.33	2868	10.50	13.12	4460	9873	14333	2.21

$$1.\ \text{Match Ratio} = \frac{\text{Shovel production, m}^3}{\text{BWE production, m}^3}$$

$$2.\ \text{Raw Coal Strip Ratio} = \frac{\text{Total overburden removed, m}^3}{\text{Average raw coal production, tonne}}$$

Raw coal strip ratio = clean coal strip ratio x 0.8

Calculations are based on 10 shifts of coal production and 21 shifts of overburden removal per week.

Table 8: Production summary for simulation run 2

	Availability			Average Overburden Depth (m)	Average Raw Coal Production (tonne)	Strip Ratio[2] (m³/tonne)		Overburden Removal			Match Ratio[1]
	BWE	Stripping Shovel	Coal Shovel			Raw Coal	Clean Coal	BWE (m^3)	Stripping Shovel (m^3)	Total (m^3)	
Actual data averaged for a shift	0.46	0.71	0.95	21.33	2835	12.14	15.17	4679	11705	16384	2.5
Simulation run 1 (base case)	0.46	0.8	0.95	21.33	2836	11.87	14.38	4430	11598	16028	2.61

$$1.\ \text{Match Ratio} = \frac{\text{Shovel production, m}^3}{\text{BWE production, m}^3}$$

$$2.\ \text{Raw Coal Strip Ratio} = \frac{\text{Total overburden removed, m}^3}{\text{Average raw coal production, tonne}}$$

Raw coal strip ratio = clean coal strip ratio x 0.8

Calculations are based on 10 shifts of coal production and 21 shifts of overburden removal per week.

Table 9: Production summary: long term plan

	Availability			Average Overburden Depth (m)	Average Raw Coal Production (tonne)	Strip Ratio (m³/tonne)		Overburden Removal			Match Ratio
	BWE	Stripping Shovel	Coal Shovel			Raw Coal	Clean Coal	BWE (m^3)	Stripping Shovel (m^3)	Total (m^3)	
Simulation run 3 Plan 1	0.50	0.70	0.95	21.33	3551	10.71	13.38	4811	13309	18120	2.76
Simulation run 4 Plan 2	0.55	0.75	0.95	21.33	3576	11.56	14.44	5282	14397	19679	2.72
Simulation run 5 Plan 3	0.65	0.80	0.95	21.33	3507	12.88	16.10	6240	15269	21509	2.44
Simulation run 6 Plan 4	0.65	0.80	0.95	21.33	4010	11.33	14.16	6240	15398	21638	2.46

1 short tonne = 0.907 tonne

removed per shift since the performance match in the overburden removal system will be unfavorable to the shovel and the BWE has to wait on the shovel.

These simulation runs confirm that the maximum productive capacity of the present system is approximately 2,722 tonne/shift, assuming availabilities of 95% for the coal shovel, 46% for the BWE and 80% for the stripping shovel. The total overburden removal capacity under these assumptions is 16,028 m³. Since the designed capacity of the mine is only 1.09 million tonne/year of clean coal (2,835 tonne/shift of raw coal), the present system can perform to within 97% of its designed capacity, if the stripping shovel availability is increased to 80% from the current 71%.

4. Longterm Plan

Any plan for production improvement over 2,722 tonne/shift will require new equipment for overburden removal. In an earlier study at this mine, performed by a management consultant [6] under contract to US Bureau of Mines, it was recommended to replace the existing 61.16 m³ shovel with a 85.63 m³ shovel, and the existing 5.35 m³ coal shovel by a 7.64 m³ shovel. No changes were proposed with regard to the BWE, the coal haulers and the waste haulers. These recommendations were evaluated on the simultator. Several plans (Plan 1, Plan 2, Plan 3, and Plan 4) were designed and the system performance was studied. The results of the simulations are summarized in Table 9. Plan 1 and Plan 2 are not satisfactory for following reasons:

1. From the strip ratios, it can be seen that more coal is being loaded than is being exposed. There is not adequate overburden removal capacity in the system to sustain this production.
2. More importantly, from the match ratios, it can be seen that the overburden removed by shovel is higher than that removed by the BWE. The performance match in the overburden system is unfavorable to shovel. In the longrun the shovel will wait on the BWE.

The following conclusions can be made based on the alternatives designed for longterm improvement:

1. The maximum production capacity that can be achieved is approximately 3,511 tonne/shift, assuming a 95% availability for the coal loader, 65% availability for the BWE and 80% for the stripping shovel.
2. Although the coal loading shovel is waiting on coal hauler any increase in the number of coal haulers will not increase production. In practice, the coal loading system will wait on the overburden removal system.

5. Final Comment

The complexity of surface mining operations is increasing as attention is directed towards mining deeper coal seams with larger equipment than heretofore. Whereas selection of equipment and initial pit design can be done through standard engineering procedures, the interactions in a complex system due to changes in operating procedure and equipment can be accurately evaluated only through the application of simulators such as OPMHS.

Acknowledgements

Financial assistance from the US Bureau of Mines under Grant No. G0254030 to develop OPMHS applications to coal stripping is greatfully acknowledged. The permission and cooperation of the Norris Mine of Consolidated Coal Company to use their mine for this application is much appreciated and acknowledged.

References

[1] National Coal Association, Coal Data, 1979. Washington. D.C.

[2] "Project Independence Final Task Force Report — Coal" Nov., 1974, Federal Energy Administration, Washington, D.C.

[3] Surface Mining Control and Reclamation Act of 1977, Public Law 95—87, 95th Congress.

[4] Manula, C. B., Albert, E. K. and Ramani, R. V., "Application of a Total System Surface Mine Simulator to Coal Stripping. Volume II: User's Manual", Final Report on Project No. G0254030, 1977, Bureau of Mines, U.S. Department of the Interior, Washington, D.C.

[5] Ramani, R. V., Bandopadhyay, S. and Manula, C. B., "Application of a Total System Surface Mine Simulator to Coal Stripping. Volume V: Application to an Illinois Mine." Final Report on Project No. G0254030, 1977, Bureau of Mines, U.S. Department of the Interior, Washington, D.C.

[6] Theodore Barry and Associates, "Operations Study of Selected Surface Coal Mining in the United States", Final Report on Project No. S0241048, February, 1975, Bureau of Mines, U.S. Department of the Interior, Washington, D.C.

Dredging of Heavy Minerals

U. Hahlbrock, Germany

Summary

After a short discussion of the mining prodedure used for dredging heavy minerals, the current dredging techniques are reviewed. Successful full-scale tests proved that an underwater bucket wheel can be applied for dredging purposes up to depths of 100 m and that this bucket wheel shows the same favourable results as bucket wheels used on land-based excavators.

1. Introduction

Heavy mineral deposits, called placers, contain a large number of different metals such as gold, tin, titanium, and iron. Dredging of these minerals has a long history, the most spectacular example being the gold-rush of the last century.

Mining in the early days was fairly easy because in some areas the deposits forming sand dunes or river banks were easily accessible. Today, from a general point of view, the placer deposits include a wide range of different soil types with different geotechnical properties. Despite the same origin, during the weathering of a primary deposit a variety of transport, settlement and rebuilding processes may occur, forming the actual deposit.

There are various classification methods possible for placer deposits, e.g., by gravity or by the transportation and settlement process. With the gravity classification, heavy mineral placers — such as gold and cassiterite — and light mineral placers — such as magnetite, rutile and ilmenite — can be separated.

For the transportation process it is important whether the barren material or only the minerals of value have been removed from the location of the primary deposit.

2. Problems Encountered in Dredge Mining

Decisive for the mining procedure is the question, whether the minerals are continuously distributed over a large area or concentrated in pockets in the bedrock. In marine estuaries the process of transport and settlement is dominated by the drag forces of waves and river currents. Due to variations of the sea level during the ice age periods, these deposits can be found to a depth of approximately 150 m.

The diverse soil properties in placer deposits create problems for the choice of a suitable dredging tool. The difficulty is that despite a large number of standardized soil mechanical test procedures, only very few criteria have been developed which give information as to by which method a special type of soil should be dredged.

This data will have to be determined separately for different kinds of soil as well as for different dredging processes; a task hitherto only partly completed. At present one manages by obtaining standard soil mechanics data and trying to interpret these for the special dredging task. (see Table 1).

Empirical data for the classification of soils for dredging purposes

Type of soil	Grain size (mm)	Shear strength (N/cm²)	Specific cutting resistance (N/cm)
Sand and gravel low compactness	0.2-20	-	200 - 500
Sand Medium compactness	0.06-2	-	300 - 800
Sand and silt High compactness	0.006-0.2	-	400 -1000
Clay, very soft	-	less than 1.7	100 - 200
Clay, soft	-	1.7 - 4.5	200 - 500
Clay, firm	-	4.5 - 9.0	500 -1000
Clay, stiff	-	9.0 -13.4	1000 -1500
Clay, hard	-	more than 13.4	more than 1500
Consolidated material e.g. limestone, sandstone	-	-	more than 2000

Table 1: Empirical data

An example of this is the correlation of the specific cutting force per unit cutting blade length of cutter heads to measured mechanical properties of soils. It is, for coherent soils, sensible to determine the cohesion, the unconfined

Dipl.-Ing. Udo Hahlbrock, Chief Engineer of Offshore Technology Division, O&K Orenstein & Koppel AG, Einsiedelstr. 6, D-2400, Lübeck, Federal Republic of Germany

compression strength and the angle of internal friction. For incoherent soils, the dredging process is best investigated with static and dynamic penetration tests.

Besides the soil characteristics, the action of waves and currents are major influencing factors for dredge design. As in most cases the dredging tool is suspended from a floating pontoon, the motion must be kept within special limits in order to get a constant maximum output and to keep the floating treatment plant operating.

The first step in this direction is to carefully examine the sea conditions at the location of the deposit. If these data are known then adequate measures to overcome the wave problem can be taken. Some of these are:

- design of special hulls or semi-submersibles;
- design of flexible connections between dredging tool and hull;
- use of passive and active motion compensators;
- use of seasonal change of the main wind direction for dredging in sheltered areas.

3. Review of Current Dredging Techniques

Selecting a mining method for a known placer deposit should of course start with a review of existing dredgers that have proved their operating performance in previous mining ventures.

The basic types will be discussed using as examples some civil type dredgers recently delivered by O&K. Although the operating conditions of civil type dredgers are to some extent different from mining applications, the dredging procedure is the same.

3.1 Grab Dredges

The grab dredge (Fig. 1) is one of the oldest dredging tools. The main characteristics are: simple construction, comparatively great water depth, moderate production. To increase the production the dredge is equipped with a number of grabs. The grabs positioned on the port and starboard sides are staggered and operated in a cyclic programme. However, the control of a large number of grabs operating in great water depths creates problems. Especially in severe water conditions a steady production seems to be difficult.

Fig. 1: Multi grab dredger "GAZA"

For a period of 10 years a large grab dredge was operating offshore of Phuket at a maximum water depth of 36 m. In 1967 the dredge was replaced by a bucket chain dredge.

3.2 Backhoe Dredge

If very hard soils are encountered, a backhoe dredge (Fig. 2) is the most efficient tool. Mounted on a pontoon this dredge is able to precisely mine pockets and traps and even remove the top level of the bedrock. The disadvantage, however, is the discontinuous process and the moderate production. Backhoe dredges are often used in gold dredging projects.

Fig. 2: Pontoon-mounted hydraulic backhoe as bucket dredger

3.3 Cutter Suction Dredge

Very high production and very high digging forces are the main characteristics of the cutter suction dredge (Fig. 3). The dredge shown is designed for a mixture discharge of 8,600 m³/h achieving an output of up to 2,500 m³/h in sandy bottoms.

Fig. 3: Cutter suction dredger

The main disadvantage of this system in mining operations is that due to the suction process the heavier mineral grains may escape being transported. Attempts to overcome this problem were made using different shapes of cutterheads but the principle "once moved — no escape" could not be completely fulfilled.

The most spectacular example of a cutterhead dredge in mining operations is the "TEMCO II" operating offshore of Phuket.

3.4 Hopper Suction Dredge

Trailing hopper suction dredgers (Fig. 4) do not play a major role in mining heavy minerals. They are, however, used with considerable success in mining marine aggregates such as sand and gravel.

Fig. 4: Hopper suction dredger

The great advantages of a hopper dredge are the great flexibility, the good performance in rough sea, and the possibility to mine large areas of a thin deposit. Difficulties emerge with the attempt to combine the dredging procedure with an onboard treatment plant. The gravel dredgers discharge the barren material during the dredging process, spoiling the deposit for the next working cycle.

Consequently the only hopper dredger used in tin mining, which was operated by Southern Kinta Consolidated to mine the Takuapa deposit, was systematically guided across the deposit by side mines discharging tailings into an area that had already been completely mined.

3.5 Bucket Chain Dredge

Up to now the bucket chain dredger (Fig. 5) is the most common mining equipment for placer deposits. Over many decades bucket chain dredgers have seen a continuous development resulting in a remarkably high state of technology. Today these dredgers provide a maximum operational reliability, a long service lifetime, a high digging capacity and high production rates. The latest designs even comprise heave compensators and automatic operation control.

Fig. 5: Bucket chain dredger for tin placer mining

The operational limits of the bucket chain dredgers determine the targets which are to be matched by new developments.

For tin mining, most important actual targets are for mining at greater depths (down to 100 m), for better performance in rough seas, and for avoiding any spillage of minerals.

To achieve this, the development of a new mining system with separate elements for loosening, picking up and transporting the soil is essential. An underwater bucket wheel in connection with a hydraulic transport system seemed to be the best solution.

The excellent experience made with onshore bucket wheels is one reason for the attempt to introduce this earth moving tool for placer mining operations.

It is of great value for the realization of this idea that O&K has extensive know-how in both fields of technology — onshore excavators and floating dredgers. A great number of questions emerging with the design of an underwater bucket wheel can be answered from the experience obtained from onshore excavator operation.

4. Development of a Bucket Wheel Dredge

The O&K development of the underwater bucket wheel started with comparative model tests to find out the optimum shape and location of the suction chamber. It was realized that the best results could be achieved with closed buckets discharging into a suction chamber situated at the upper part of the wheel body. By using closed buckets the principle "once moved — no escape" is completely realized. The transition of soil from the buckets into the suction chamber is effected by gravity and by suction forces.

The resulting design is quite similar to existing excavator wheels with discharge elements in the wheel's upper part and drive elements in the centre. The prototype underwater bucket wheel (Fig. 6) was therefore built according to a very successful excavator series and mounted on an existing suction dredger (Fig. 7) with an underwater dredge pump.

The first impression gained from operation of the underwater bucket wheel was the very high production achieved with astonishingly quiet running of the dredger.

In sandy bottoms it was, for example, possible to reach a constant output of 2,000 m^3/h (soil) with a wheel of 5 m diameter. This output was achieved with a very steady transport concentration of 45 % which corresponds to a mixture density of 1.4 t/m^3; even in very hard boulder marl an average production of 1,250 m^3/h was possible.

The outstanding performance of the underwater bucket wheel can be summarized as follows:

- high effective output due to the continuous operation;
- low losses when picking up the soil;
- exact soil cutting conditions resulting in:
  high digging forces
  low lateral forces
  digging capacity not affected by the depth
  equal digging capacity in both slewing directions;

Fig. 6: Underwater bucket wheel — prototype

— greater achievable depth due to lower weight of the ladder;

— less vibration as compared to bucket chain dredgers.

This summary shows that there is no economical alternative to an underwater bucket wheel with hydraulic transport system when dredging hard soils at great dredging depths. As its operation has been examined and its design tested under the hardest conditions, its profitable application in dredging and marine mining can be predicted.

5. Practical Example

One example for such an adaptation is the development of a dredger for mining of cassiterite placers, equipped with an underwater bucket wheel and reaching down to depths of 100 m (Fig. 8).

The type of wheel used has a diameter of 6.3 m and achieves a maximum production of 3,050 m³/h with 600 kW power installed. The design of the wheel is such that special ancillary features for application in sticky clay mentioned before can be incorporated as needed.

In the case on hand, a rigid ladder was chosen for the operational depth of 100 m as it was designed for operation on a dredge pond. For offshore operation, however, a flexible ladder is planned, compensating any movements of the hull with respect to the seabed.

A dredging depth of 100 m seems to be the limit for any system using a bucket wheel attached to a floating pontoon. For greater depths, seabottom vehicles can be developed which serve as supporting structures for underwater bucket wheels.

The successful full-scale test on an underwater bucket wheel has proved that it is possible to use the favourable properties of on-land bucket wheels for marine applications.

Fig. 7: Suction dredger with underwater bucket wheel

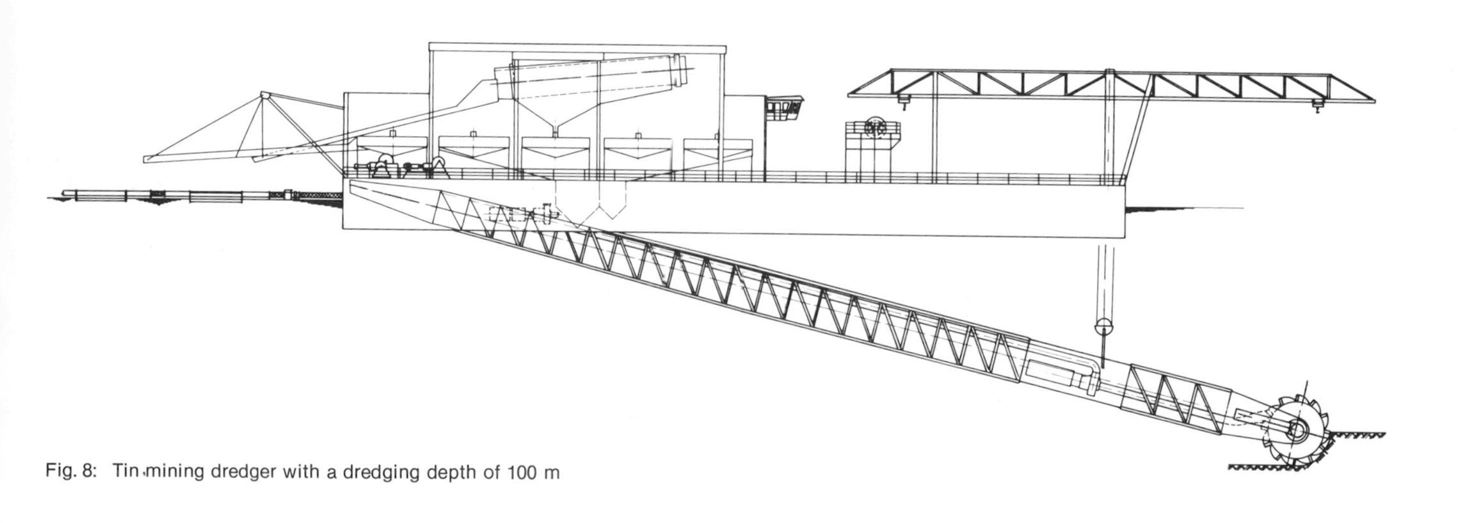

Fig. 8: Tin mining dredger with a dredging depth of 100 m

Professional services

Bulk Handling in Open Pit Mines & Quarries

Other Volumes Published within the Series

The Best of bulk solids handling 1981—1985

A/86

Silos, Hoppers, Bins & Bunkers for Storing Bulk Materials

B/86

Stacking, Blending & Reclaiming of Bulk Materials

C/86

Continuous Ship Unloading & Self-Unloading Ships & Vessels

D/86

Pneumatic Conveying of Bulk & Powder

E/86

Hydraulic Conveying & Slurry Pipeline Technology

G/86

Mechanical Conveying, Transporting & Feeding

H/86

Bulk Port Development, Design & Operation

I/86

Conveyor Belt Technology

K/86

Sampling, Weighing & Proportioning of Bulk Materials